STERLING
Test Prep

SAT
BIOLOGY E/M
REVIEW

Complete
Content Review

3rd edition

www.Sterling-Prep.com

Copyright © 2019 Sterling Test Prep

3 2 1

ISBN-13: 978-1-9475560-3-4

Sterling Test Prep products are available at special quantity discounts for sales, promotions, academic counseling offices, and other educational purposes.

For more information, contact our Sales Department at:

Sterling Test Prep
6 Liberty Square #11
Boston, MA 02109

info@sterling-prep.com

Congratulations on choosing this book as part of your SAT Biology preparation!

Scoring high on SAT subject tests is important for admission to college. To achieve a high score on SAT Biology, you need to develop skills to properly apply the scientific knowledge you have to solve each question. Understanding key concepts, having the ability to extract information from the provided data and distinguishing between similar answer choices is more valuable than simply memorizing terms.

This book provides a detailed and thorough review of topics tested on the SAT Biology E/M Subject Test. The content covers foundational principles and theories necessary to answer related questions on the test. The information is presented clearly and organized systematically to provide students with targeted SAT Biology review tool. You can focus on one knowledge area at a time to learn and fully comprehend important concepts and theories or to simply refresh your memory. By reading these review chapters thoroughly, you will learn important biology concepts and the relationships between them. This will prepare you for the SAT Biology E/M, and you will significantly increase your score.

All the material in this book is prepared by our science editors who possess extensive credentials and experience. This team of science experts analyzed the content of the test, released by the College Board, and designed essential review that will help you build and solidify the knowledge necessary for your success on the test. The content was reviewed for quality and effectiveness by our science editors to ensure strict adherence to the topics and skills outlined by the College Board for the current SAT Biology E/M Subject Test.

We wish you great success in your future academic achievements and look forward to being an important part of your successful preparation for the SAT Biology!

Sterling Test Prep Team

181024gdx

Our Commitment to the Environment

Sterling Test Prep is committed to protecting our planet's resources by supporting environmental organizations with proven track records of conservation, ecological research and education and preservation of vital natural resources. A portion of our profits is donated to help these organizations so they can continue their critical missions. These organizations include:

 Ocean Conservancy

For over 40 years, Ocean Conservancy has been advocating for a healthy ocean by supporting sustainable solutions based on science and cleanup efforts. Among many environmental achievements, Ocean Conservancy laid the groundwork for an international moratorium on commercial whaling, played an instrumental role in protecting fur seals from overhunting and banning the international trade of sea turtles. The organization created national marine sanctuaries and served as the lead non-governmental organization in the designation of 10 of the 13 marine sanctuaries.

For 25 years, Rainforest Trust has been saving critical lands for conservation through land purchases and protected area designations. Rainforest Trust has played a central role in the creation of 73 new protected areas in 17 countries, including the Falkland Islands, Costa Rica and Peru. Nearly 8 million acres have been saved thanks to Rainforest Trust's support of in-country partners across Latin America, with over 500,000 acres of critical lands purchased outright for reserves.

Since 1980, Pacific Whale Foundation has been saving whales from extinction and protecting our oceans through science and advocacy. As an international organization, with ongoing research projects in Hawaii, Australia, and Ecuador, PWF is an active participant in global efforts to address threats to whales and other marine life. A pioneer in non-invasive whale research, PWF was an early leader in educating the public, from a scientific perspective, about whales and the need for ocean conservation.

With your purchase, you support environmental causes around the world.

This book should be supplemented by

"SAT Biology E/M Practice Questions" book

or online practice questions

at www.Sterling-Prep.com

To access these and other SAT questions online at a special pricing for book owners, see page 611

Table of Contents

Table of Contents (*cont.*)

Table of Contents (*cont.*)

Table of Contents (*cont.*)

We want to hear from you

Your feedback is important to us because we strive to provide the highest quality prep materials. Email us if you have any questions, comments or suggestions, so we can incorporate your feedback into future editions.

Customer Satisfaction Guarantee

If you have any concerns about this book, including printing issues, contact us and we will resolve any issues to your satisfaction.

info@sterling-prep.com

We reply to all emails – please check your spam folder

Thank you for choosing our products to achieve your educational goals!

Please, leave your Customer Review on Amazon

Chapter 1

Eukaryotic Cell

- **Cell Theory**

- **Nucleus and Other Defining Characteristics**

- **Membrane-Bound Organelles**

- **Plasma Membrane**

- **Cytoskeleton**

- **Cell Cycle and Mitosis**

- **Control of Cell Cycle**

- **Tissues Formed from Eukaryotic Cells**

Cell Theory

Cell theory, the scientific theory which describes the morphological and biochemical properties of cells, is a fundamental doctrine of biology.

Classical cell theory includes three key tenets derived from the research of early biologists:

1. *All living organisms are composed of one or more cells.*

    Multicellular organisms are composed of many cells, while unicellular organisms, such as bacteria, are composed of only one cell.

2. *Cells are the smallest, most basic units of life.*

    Cells are the smallest units of life because they are the smallest structures capable of carrying out the fundamental metabolic process (e.g., reproduce and divide, extract energy from their environment).

3. *Cells may only arise from pre-existing cells and cannot be created from non-living material.*

    The process of creating new cells is cellular division and is involved in both sexual and asexual reproduction.

As the modern understanding of biology evolved, so too did the tenets of cell theory. Modern cell theory adds the following concepts to classical cell theory:

4. *Cells pass on the genetic material during replication in the form of DNA.*

5. *The cells of all organisms are chemically similar.*

6. *Cells are responsible for energy flow and metabolism.*

The tenets of cell theory are somewhat dynamic and, as such, some scientists may omit some of these tenets or include others not mentioned here.

History and development

Robert Hooke first observed the small structural units, which he called "cells," under the light microscope in 1665. However, it was not until nearly 200 years later that significant progress was made in the understanding of these cells. In the 1830s, botanist Matthias Schleiden discovered cells in the tissues of plants and declared that cells are the building blocks of all plants. Simultaneously, Theodor Schwann published his work on cell theory, generalizing it to both plants and animals. Shortly after that,

Rudolf Virchow overturned the predominating belief that cells were spontaneously generated from non-living matter. He proclaimed that all cells must arise from other living cells.

From there, cell biology was consistently advanced by new findings in membrane physiology, mitosis and other cellular and molecular processes.

In the 1950s, James Watson and Francis Crick published their discovery of the molecular structure of DNA, which revolutionized the entire field of biology. Other researchers, including Rosalind Franklin, went uncredited for their seminal work on this monumental discovery.

Impact on biology

Cell theory is an important unifying concept which provides biologists with a common understanding from which they can make further discoveries. The relatively slow progress of biology before the mid-1800s demonstrates the difficulties of advancing in a field where the framework is not understood. Once the foundation of cell theory was laid, along with advances in laboratory techniques and instrumentation, scientific progress in the field of biology rapidly accelerated.

By understanding the basics of a single cell, researchers were able to add context to all the life sciences. This is because cells represent the building blocks of all organisms. Multicellular organisms exhibit *emergent properties*, meaning the whole is greater than the sum of its parts. Cells form tissues; tissues form organs, organs form organ systems and organ systems form multicellular organisms. For example, the individual cells in the lungs are not of much use by themselves, but when combined as a working unit, they create a highly sophisticated set of lungs essential for the organism's survival.

Defining characteristics
(membrane-bound nucleus, presence of organelles and mitotic division)

Cells can are into two taxa: *Eukarya* and *Prokarya*. Prokaryotes may be further divided into the domains Bacteria and Archaea.

The most salient difference between prokaryotes and eukaryotes is that prokaryotes lack a true nucleus with a nuclear membrane or any membrane-bound organelles. The main intracellular components of a prokaryotic cell are its single double-stranded circular DNA molecule, its ribosomes, and its cytoplasm. Prokaryotic cells include a plasma membrane and peptidoglycan cell wall. They are usually much smaller than eukaryotic cells and contain smaller ribosomes (the 30S and 50 S subunits; 70S as assembled).

Eukaryotic cells have linear DNA enclosed in a membrane-bound nucleus, along with many other membrane-bound organelles. Eukaryotic cells replicate via mitosis or meiosis, while prokaryotes replicate via *binary fission*, a form of asexual reproduction.

Many similarities exist between the two cell types. Both prokaryotes and eukaryotes contain cytoplasm, ribosomes, and some form of DNA, and they can be either unicellular or multicellular, although multicellular prokaryotes are rare. Some eukaryotes are capable of asexual reproduction, albeit in a different way than prokaryotes. Furthermore, plants and fungi (eukaryotes) both have cell walls like prokaryotes. Cell walls in fungi are a glucosamine polymer of chitin, plants have cellulose cell walls, while bacteria have peptidoglycan cell walls.

Comparison of prokaryotes to eukaryotes

Prokaryotes	Eukaryotes
Domains: Bacteria and Archaea	Domain: Eukarya
Cell wall present in all prokaryotes	Cell wall is in fungi, plants and some protists
No nucleus, circular strand of dsDNA	Membrane-bound nucleus housing dsDNA
Ribosomes (subunits = 30S and 50S; 70S)	Ribosomes (subunits = 40S and 60S; 80S)
No membrane-bound organelles	Membrane-bound organelles

The cell's metabolic activity describes the many biochemical reactions occurring within the cell. Substances need to be taken into the cell to fuel these reactions, while the waste products of the reactions need to be removed. When the cell increases in size, so do its metabolic activity. The surface area of the cell is vital because it affects the rate at which particles can enter and exit, with a larger surface area

[handwritten: ⚡ surface area: affects amt going in & out]
[handwritten: ⚡ volume: affects rate at which things are made (chemical activity/unit)]

resulting in a higher rate of uptake and excretion. However, the volume affects the rate at which materials are made or used within the cell, known as the chemical activity per unit of time.

[handwritten: ⚡ volume increases, surface area increases]

As the volume (i.e., chemical activity) of the cell increases, so does the surface area, but not to the same extent. When the cell gets bigger, its surface area-to-volume ratio gets smaller. If the surface area-to-volume ratio gets too small, substances are not able to enter the cell fast enough to fuel the reactions. Waste products are produced faster than they can be excreted, and they accumulate inside the cell. Also, cells are not able to lose heat fast enough and may overheat. It is clear that the surface area-to-volume ratio is critical for a cell. It is the physical limitation of the area-to-volume ratio that limits the size of cells.

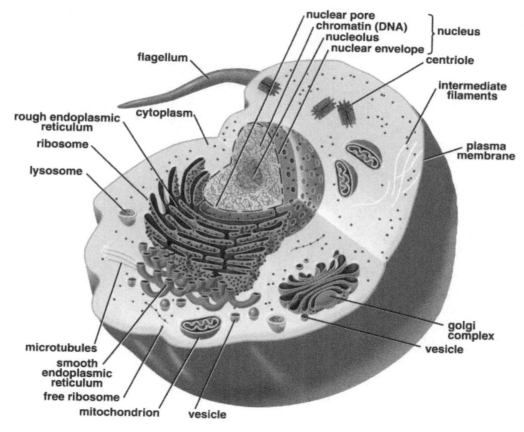

A eukaryotic animal cell with no cell wall or chloroplasts as in eukaryotic plant cells

The nucleus for compartmentalization of genetic information

The *nucleus* is the largest membrane-bound organelle in the center of most eukaryotic cells. It contains the cell's genetic code—its DNA. The function of the nucleus is to direct the cell by storing and transmitting genetic information. Cells can contain multiple nuclei (e.g., skeletal muscle cells), one nucleus or rarely, none at all (e.g., red blood cells).

Inside the nucleus is the *nuclear lamina*, a dense network of filamentous and membranous proteins that associate with the nuclear envelope and its pores. The lamina provides mechanical support and is involved in crucial cell functions, including DNA replication, cell division, and chromatin organization.

The *nucleoplasm* is the semifluid medium of the nucleus, analogous to the cytoplasm of the cell proper. In the nucleoplasm, DNA and proteins interact to form *chromatin*.

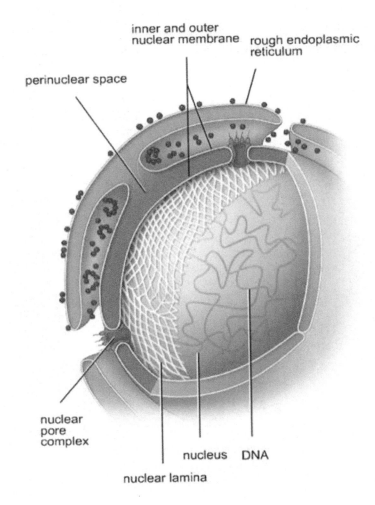

A double membrane surrounds the nucleus of the cell with
nuclear pores for select transport of a substance in and out of the nucleus

Nucleolus (location, function)

The *nucleolus* is a region inside the nucleus where ribosomal RNA (rRNA) is transcribed, and ribosomal subunits are assembled. Here, rRNA joins together to form the subunits of a complete ribosome. These subunits are then exported to the cytoplasm for final assembly into the complete ribosome used for translation of mRNA into proteins.

Nuclear envelope, nuclear pores

The nuclear envelope (nuclear membrane) is a double membrane system composed of an outer and inner membrane. The space between these membranes is the *perinuclear space.* Nuclear pores are selective and allow the passage of certain particles through the nuclear envelope, so that essential cell processes and communications can occur. The number of nuclear pores is not static but rather is subject to change based on the needs of the cell. Through the pores, signal molecules, nucleoplasm proteins, nuclear membrane proteins, lipids, and transcription factors can enter the nucleus, while mRNA, rRNA and ribosomal proteins exit into the cytoplasm.

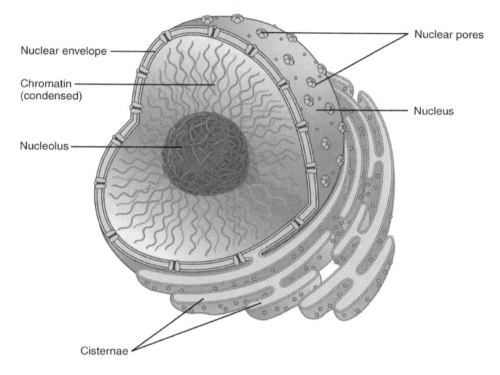

The nucleolus is located within the nucleus and assembles ribosomal subunits in eukaryotic cells

Membrane-Bound Organelles

The cytoplasm is the cellular material outside the nucleus and within the cell's plasma membrane. It includes the *cytosol*, which is the fluid medium of the cell, and the *organelles*, which are small, usually membrane-bound subunits with specialized functions (ribosomes are not membrane-bound). Among other functions, organelles structurally support the cell, facilitate cell movement, store and transfer energy and exchange products in transport vesicles. Mitochondria in animal cells and chloroplasts in plant cells are organelles that contain their genetic material and replicate independently of the nucleus.

The *endomembrane system* is a series of intracellular membranes that compartmentalize the cell. Vesicles bud off from the endomembrane system to transport molecules within the cell. Products synthesized in the cell pass through at least some portion of the endomembrane system.

A typical pathway through the endomembrane system is:

1. Proteins produced in rough ER (endoplasmic reticulum) and lipids from smooth ER are carried in vesicles to the Golgi apparatus.

2. The Golgi apparatus modifies these products and then sorts and packages them into vesicles that are transported to various cell destinations (e.g., organelles or exported from the cell).

3. Secretory vesicles transport products to organelles or to the membrane, where they are secreted via exocytosis.

Aside from the Golgi apparatus, smooth and rough ER, and secretory vesicles, the endomembrane system includes the membranes of lysosomes, peroxisomes, and all other organelles within the cell.

While most cells have the same organelles, their distribution may differ depending on the cell's function. For example, cells that require much energy for locomotion (e.g., sperm cells) have many mitochondria; cells involved in secretion (e.g., pancreatic islet cells) have many Golgi apparatuses; and cells that primarily serve a transport function (e.g., red blood cells) may have no organelles.

Mitochondria

Mitochondria (sing. mitochondrion) are responsible for aerobic respiration, which is the conversion of chemical energy into ATP (adenosine triphosphate) using oxygen. ATP is used as the major energy source within cells. Mitochondria vary in shape; they may be long and thin or short and broad. Mitochondria can be fixed in one location or form long, moving chains. They have a double membrane, with the outer membrane separating the mitochondria from the cytoplasm. The inner membrane has folds as *cristae*, which project into the inner fluid, the *matrix*. Between the outer and inner membrane is the intermembrane space. This region is high in protons, creating a proton gradient, which drives the synthesis of ATP.

The cristae are dotted with *ATP synthase* protein complexes, which are powered by the proton gradient and convert ADP into ATP. This process is essential to producing the energy that all organisms require for metabolic functions. Thus, cells with higher energy needs require more mitochondria.

Mitochondria are unique in that they have their genome, distinct from the genome within the nucleus. They have their own circular DNA, inherited exclusively from the mother, which contains the genes for the synthesis of some mitochondrial proteins. Mitochondria can replicate their DNA independently from the nucleus. They have ribosomes, different from the host cell's ribosomes in both sequence and structure.

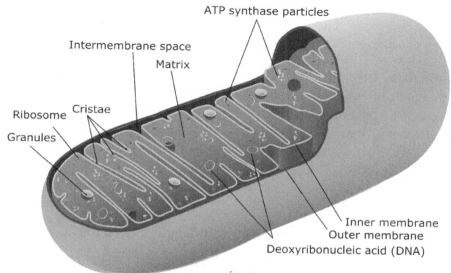

Mitochondria have a double membrane enclosure with ATPase embedded in the inner membrane

The unique characteristics of mitochondria support the *endosymbiosis theory* for the origin of eukaryotic cells. The endosymbiosis theory states that mitochondria were once free-living aerobic prokaryotes that were consumed by a cell about 1.5 billion years ago. Within the cell, the prokaryote (probably a proteobacterium) became an endosymbiont, providing the anaerobic host cell with ATP via aerobic respiration. In return, the host cell provided the endosymbiont with a stable environment and nutrients. Over time, the endosymbiont transferred most of its genes to the host nucleus, to the point that it became obligate (i.e., could no longer survive outside the host cell) and evolved into a mitochondrion.

Biologists largely accept the endosymbiosis theory. One of the most compelling pieces of evidence is the fact that mitochondrial DNA does not encode for its proteins. Rather, many of its genes are in the nuclear DNA; therefore, many proteins must be imported into the mitochondria. Furthermore, mitochondrial DNA, ribosomes, and enzymes are similar to bacterial forms, and mitochondria even replicate by a process similar to binary fission. Additionally, some of the proteins within the plasma membrane of the mitochondria are similar to prokaryotes, which are different from proteins in the eukaryotic plasma membrane.

Chloroplasts, the organelles that conduct photosynthesis, exhibit strong evidence of an endosymbiotic origin, although they are hypothesized to have descended from cyanobacteria rather than proteobacteria. Chloroplasts and other plastids can undergo secondary and even tertiary endosymbiosis, causing the development of extra membranes.

Lysosomes (vesicles containing hydrolytic enzymes)

Lysosomes, only in animal cells, are membrane-bound vesicles produced by the Golgi apparatus. These small organelles contain hydrolytic enzymes (low pH) for the digestion of macromolecules: proteins, nucleic acids, carbohydrates, and lipids. These macromolecules may originate from food, from the waste products of cells or foreign agents, such as viruses and bacteria. After these particles enter a cell in vesicles, lysosomes fuse with vesicles and digest their contents by hydrolyzing the macromolecules into their monomers.

Lysosomes are especially important in specialized immune cells. For example, white blood cells that engulf foreign agents use lysosomes to digest the invaders. *Autodigestion* is the process by which lysosomes digest parts of the body's cells, either due to disease or trauma or for immune purposes (e.g., programmed cell death). Mutations in the genes that encode for lysosomal enzymes cause *lysosomal storage disorders*. When a mutation renders certain lysosomal enzymes inefficient (or completely inoperable), waste products accumulate in the cells and cause debilitating, often incurable complications.

Rough and smooth endoplasmic reticulum

The *endoplasmic reticulum* (ER) is a system of membrane channels, *cisternae*, that is continuous with the outer membrane of the nuclear envelope. The space enclosed within the cisternae, the *lumen*, is thus continuous with the perinuclear space.

The rough ER, so-called because of its rough appearance, is studded with ribosomes on the cytoplasmic side. Here, proteins are synthesized and enter the ER interior for processing and modification. Modifications may include folding the protein or combining multiple polypeptide chains to form proteins with several subunits.

The smooth ER is usually interconnected with the rough ER but lacks ribosomes, hence its smooth appearance. It is the site of various synthesis, detoxification and storage processes, such as the synthesis of lipids and steroids, and the metabolism of carbohydrates and other molecules.

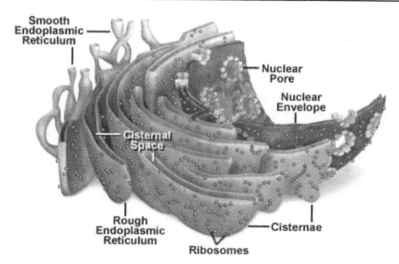

The ER forms transport vesicles for trafficking particles to the Golgi apparatus

RER - site of ribosomes

Ribosomes are organelles composed of proteins and ribosomal RNA (rRNA). They can be either floating free in the cytoplasm, attached to the surface of the rough ER, or within mitochondria and chloroplasts. They translate messenger RNA (mRNA) to coordinate the assembly of amino acids into polypeptide chains, which then fold into functional proteins.

Ribosomes of eukaryotic cells are 20 to 30 nm in diameter, while those in prokaryotic cells are slightly smaller. They are composed of one large and one small subunit; each with its mix of proteins and rRNA. *Polyribosomes* are several ribosomes simultaneously synthesizing the same protein; they may be attached to the ER or floating freely in the cytosol.

Membrane biosynthesis: SER for lipids and RER for transmembrane proteins

The smooth ER and rough ER synthesize key membrane components. The smooth ER synthesizes the major lipids of a membrane: phospholipids, glycolipids, and steroids. Some lipid products are already in the correct form for incorporation into a membrane once secreted by the smooth ER, while others require additional modification by the Golgi apparatus. Either way, lipids synthesized in the smooth ER must pass through the Golgi apparatus before heading to their destination at the plasma membrane or a membrane-bound organelle.

The rough ER synthesizes the protein components of cell membranes. This includes not just the plasma membrane, but the membranes of the ER, Golgi apparatus, lysosomes, and other organelles. Membrane proteins can be divided into several classes (discussed later), but some of their functions include membrane transport, cell-to-cell adhesion, cell signaling, and catalysis. Similarly to lipids synthesized on the smooth ER, proteins synthesized on the rough ER follow a set pathway through the Golgi apparatus towards their final destinations.

RER's role in the biosynthesis of transmembrane and secreted proteins that are cotranslationally targeted to RER by a signal sequence

Proteins destined for the plasma membrane, Golgi apparatus membrane, ER membrane or lysosomal membranes are inserted into the ER membrane immediately after synthesis on the cytosolic side of the rough ER membrane. These proteins are transported as membrane components rather than soluble proteins. ER membrane proteins end their journey here, but the others proceed to the Golgi apparatus. Upon post-translational processing, Golgi membrane proteins remain in the Golgi apparatus, while the remaining proteins (secretory pathway) travel to either lysosome, the plasma membrane or undergo exocytosis to leave the cell.

Proteins for secretion and proteins destined for the *lumen* of the ER or Golgi apparatus are released into the ER lumen following ER synthesis. ER lumen proteins remain in the ER lumen, Golgi lumen proteins travel to the Golgi lumen, and secretory proteins travel to the Golgi and then to the plasma membrane or are secreted out of the cell.

Not all protein synthesis takes place on the rough ER. Free-floating ribosomes synthesize proteins designated for use in the cytosol and some organelles (e.g., nucleus, mitochondria, chloroplasts, and peroxisomes) in the cytosol. After synthesis, cytosolic proteins are released directly into the cytosol, while nuclear, mitochondrial, chloroplastic, and peroxisomal proteins are escorted to their destinations by receptor molecules.

Protein synthesis always begins on free ribosomes. Therefore, proteins that need to be synthesized in the ER must be translocated there. *Posttranslational translocation* to the ER occurs after a free-floating ribosome has fully synthesized a polypeptide. *Cotranslational translocation* is more common in mammalian cells and occurs as the polypeptide is being synthesized.

Cotranslational translocation is facilitated by a *signal sequence* on the growing polypeptide chain. This sequence is a short chain of amino acids, mostly hydrophobic. As soon as the signal sequence emerges on the polypeptide from the ribosome, a protein-RNA complex as a *signal recognition particle* (SRP) recognizes and binds to the signal sequence and ribosome, halting translation. The SRP then targets the ribosome and polypeptide chain to the rough ER membrane, where the SRP binds to an SRP receptor. Binding to the SRP receptor releases the SRP from the ribosome and polypeptide, allowing the ribosome itself to bind to a protein translocation complex next to the SRP receptor called *Sec61*. After binding, the signal sequence is inserted into the Sec61 membrane channel (part of the translocation complex), and polypeptide synthesis resumes. As it grows, the polypeptide chain is translocated through the membrane channel. A *signal peptidase* enzyme cleaves the signal sequence from the rest of the polypeptide, allowing the finished polypeptide to be released into the lumen, where it undergoes folding and modification.

If the polypeptide is a membrane protein that must enter the ER membrane and not the lumen, it is inserted into the ER membrane in a variety of ways. For example, transmembrane proteins may contain a *stop-transfer sequence*, which anchors the polypeptide in the ER membrane partway through synthesis so that the polypeptide is anchored to the ER membrane rather than located in the ER lumen.

Double membrane structure

While most organelles of the eukaryotic cell are composed of a single bilayer membrane, three key organelles have a double membrane: mitochondria, chloroplasts and the nucleus.

Mitochondria have a double membrane structure due to their proposed evolution from an endosymbiotic prokaryote. This double membrane is crucial for creating the proton gradient that drives ATP synthesis. The intermembrane space is high in proton concentration, while the matrix (like cytosol) within the mitochondria is relatively low in proton concentration. This proton gradient powers ATP synthases, with cytochrome proteins dotting the cristae of the inner membrane that combine ADP with Pi and O_2 to form ATP by oxidative phosphorylation.

Chloroplasts, the sites of photosynthesis and ATP synthesis in plant cells, have a double membrane of endosymbiotic origin, the *chloroplast envelope*. The two membranes regulate the passage of particles into and out of the chloroplast. The outer membrane is permeable to ions and metabolites, while the inner membrane is highly specific to transport proteins. Unlike mitochondria, all chloroplasts contain *thylakoids* as additional membrane-bound structures. The thylakoid membranes are analogous to mitochondrial inner membranes, and the spaces within the thylakoids (lumen) are analogous to the mitochondrial intermembrane space. The high proton concentration in the lumen creates the proton gradient that drives the synthesis of ATP on the thylakoid membranes.

A sophisticated, double-membrane nuclear envelope surrounds the nucleus. The highly selective nuclear protein pores dotting the envelope regulate gene expression by controlling the passage of transcription factors, biomolecules, and mRNA into and out of the nucleus. Since the outer membrane is continuous with the rough ER, no vesicle transport is required to transport ER proteins into the nucleus. As a result, the energy requirements of the cell are lower than they would be if transport were required.

Golgi apparatus (general structure; role in packaging, secretion, and modification of glycoprotein carbohydrates)

The *Golgi apparatus*, named for the scientist Camillo Golgi, consists of a stack of many flattened sacs. The Golgi acts as an intermediary in the secretion of biomolecules. The Golgi apparatus receives transport vesicles from the ER and then may further modify their contents before packaging the protein or lipid to be sent in vesicles to its final destination.

Glycosylation is the process by which the Golgi apparatus modifies a protein by adding carbohydrates. Glycosylation affects a protein's structure and function and protects it from degradation. The Golgi apparatus can glycosylate proteins as well as modify existing glycosylations. The finished product of glycosylation is a *glycoprotein* and is a protein with attached saccharides.

Peroxisomes: organelles that collect peroxides

Peroxisomes are membrane-bound vesicles that contain enzymes for a variety of metabolic reactions. They are involved in the catabolism and anabolism of many different macromolecules, including fatty acids, proteins, and carbohydrates. When peroxisomes were first discovered, they were defined as organelles that produce hydrogen peroxide through oxidation reactions. It was believed that peroxisomes use the enzyme catalase to further break down the produced hydrogen peroxide into water and oxygen, or use the hydrogen peroxide to oxidize another compound. Today, this process is merely one of their many functions.

Peroxisomes are most abundant in the liver, where they are notable for producing bile salts from cholesterol and metabolizing alcohol. They are believed to be involved in lipid biosynthesis. In germinating seeds, peroxisomes convert oils into sugars to be used as nutrients by growing plants. In leaves, peroxisomes give off CO_2 that can be used in photosynthesis. The functions of peroxisomes are vast and varied, making them a vital part of eukaryotic cells.

Plasma Membrane

General function in cell containment

The *semi-permeable plasma membrane* separates cell contents from the extracellular environment and regulates the passage of materials into and out of the cell. In plant cells, the outer boundary of the plasma membrane is surrounded by *cellulose* (polysaccharide) cell wall. Fungi and prokaryotes have *cell walls* of alternative polysaccharides. A cell wall gives the cell strength and rigidity but does not interfere with the function of the plasma membrane.

This plasma membrane surrounds the cell, providing support, protection and a boundary from the outside environment. It is primarily composed of lipids and proteins, forming a dynamic bilayer of lipids with membrane proteins. The function and composition of the two layers of a plasma membrane differ; therefore, the membrane is asymmetric.

Lipid components: phospholipids and steroids

Lipids, a large group of naturally occurring hydrophobic molecules, are vital components of the plasma membrane. Lipids in the plasma membrane include phospholipids, steroids, and glycolipids. The foundation of the plasma membrane is the *phospholipid bilayer*. Phospholipids are molecules with a phosphate head and two long hydrocarbon tails. The head is hydrophilic and attracts water and other polar molecules, while the hydrocarbon tails are hydrophobic and repel water molecules.

The special dual nature of phospholipids, *amphipathic*, allows them to align into a bilayer when placed in water spontaneously. In the bilayer, the hydrophilic phosphate heads point out towards the aqueous solution, while the hydrophobic tails point inward towards one another. Thus, the extracellular and intracellular surfaces of a plasma membrane are hydrophilic, while the interior of the membrane is hydrophobic. Hydrophobic interactions in the interior of the membrane hold the entire structure together.

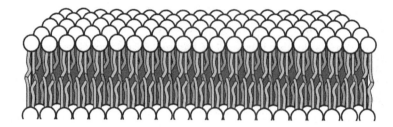

Phospholipid bilayer (absent in the schematic are embedded proteins)

The bilayer not only provides the plasma membrane with stability but with extraordinary flexibility. Lipids exhibit free lateral diffusion about the bilayer, resulting in varying lipid compositions across different sections of the membrane. Generally, cell membranes have a consistency similar to that of olive oil at room temperature. Increasing the concentration of lipids with unsaturated hydrocarbon tails increases membrane fluidity; the addition of saturated hydrocarbon tails makes the membrane more rigid. Cells may regulate membrane fluidity by lengthening phospholipid tails, altering the cytoskeleton, changing their protein composition and adding steroids (e.g., cholesterol).

Steroids are a class of lipids that regulate membrane fluidity by hindering phospholipid movement. Cholesterol, a steroid in animal plasma membranes, plays a crucial role in maintaining membrane fluidity despite temperature fluctuation. At high temperatures, membranes become dangerously fluid and permeable unless cholesterol interferes with extreme phospholipid movement. At low temperatures, membranes may freeze unless cholesterol prevents phospholipids from becoming stationary due to strong hydrophobic interactions. Cholesterol molecules facilitate cell signaling and vesicle formation.

Glycolipids are lipids modified with a carbohydrate. In the plasma membrane, they assist in various functions and anchor the plasma membrane to the *glycocalyx*, a layer of polysaccharides that are linked to the membrane lipids and proteins. Essentially, the glycocalyx is a carbohydrate coat present on the extracellular surface of the plasma membrane and the extracellular surface of the cell walls of some bacteria. The carbohydrate chains of the glycocalyx face outwards, providing markers for cell recognition and adhesive capabilities to the cell.

Protein components and the fluid mosaic model

In the early 1900s, researchers noted that lipid-soluble molecules entered cells more readily than water-soluble molecules, suggesting that lipids are a component of the plasma membrane. Chemical analysis later revealed that the membrane indeed contained phospholipids. The amount of phospholipid extracted from a red blood cell was just enough to form one bilayer. The analysis suggested that the nonpolar tails were directed inward and polar heads outward. To account for the permeability of the membrane to non-lipid substances, researchers initially proposed a *sandwich model*, describing a phospholipid bilayer in between two layers of protein.

After investigation with an electron microscope, the *unit membrane model* was proposed, which was based on the "trilaminar" appearance of two dark outer lines and a light inner region visible under the electron microscope. The dark outer lines were believed to be protein monolayers, while the inner region was thought to be a phospholipid bilayer. The unit membrane model was essentially in agreement with the sandwich model. However, both of these models failed to explain permeability satisfactorily.

In 1972, Garth L. Nicholson and Seymour J. Singer published the current *fluid-mosaic model*, which describes a plasma membrane as a phospholipid bilayer embedded with proteins. Electron micrographs of the freeze-fractured membrane (and other evidence) supported the fluid-mosaic model. In this model, it is the lipid portion of the plasma membrane which gives it its "fluid" characteristic. Thus, fluidity describes the lipids that diffuse freely throughout the membrane and regulate consistency.

The membrane's protein components contribute to the "mosaic." Protein composition in the plasma membrane is dependent upon the function of the particular cell. Some proteins are held in place by cytoskeletal filaments, but most migrate within the fluid bilayer.

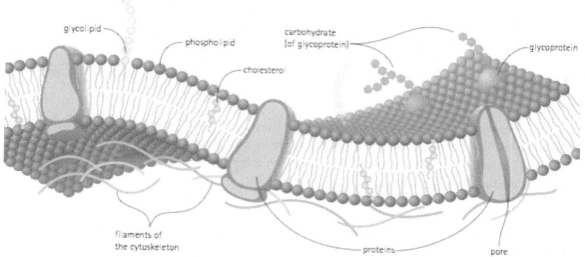

The membrane is a phospholipid bilayer with cholesterol between the hydrophobic lipid tails and embedded proteins – some with modifications (e.g., glycolipid or glycoprotein)

The two classes of proteins embedded in a membrane by location.

Peripheral membrane proteins are on the membrane surface, mainly the intracellular side, that interact with cytoskeletal elements to influence cell shape and motility. These proteins are not amphipathic; they interact only with the hydrophilic heads of the membrane. Peripheral membrane proteins can be removed from the membrane with relative ease using high salt or high pH, and therefore are not permanently attached to the membrane.

Integral membrane proteins are permanently attached to the membrane and thus cannot be removed without disrupting the lipid bilayer. They possess hydrophobic domains which are anchored to hydrophobic lipids. Most integral membrane proteins are *transmembrane* proteins, spanning the entire membrane.

Membrane proteins can participate in cell signaling, cell-to-cell adhesion, transport through the membrane, enzymatic activity, and some other activities.

When divided by function, several key classes of membrane proteins emerge.

Receptor proteins provide a binding site for hormones, neurotransmitters, and other signaling molecules. Receptor proteins are usually specific in that they bind to only a single molecule or certain class of molecule. Binding of the signal molecule to the receptor triggers a cellular response that corresponds to a specific biochemical pathway.

Adhesion proteins attach cells to neighboring cells for cell-to-cell communication and tissue structure. These proteins generally attach to the cytoskeleton of one cell and extend through the plasma membrane to the extracellular environment, where they bind and interact with the adhesion proteins of another cell.

Transport proteins move materials into and out of the cell. These include *channel proteins* and *carrier proteins*. Channel proteins provide a passageway large, polar or charged molecules that cannot pass through the lipid bilayer without assistance. These proteins facilitate only the passive transport of molecules, that is, they do not require ATP to operate. Carrier proteins, however, may facilitate both passive and active (energy-requiring) transport of molecules. They bind to specific molecules on one side of the cell membrane and then change conformation to release the molecule on the other side of the membrane.

Enzymatic membrane proteins carry out metabolic reactions at the cell membrane. For example, many enzymatic membrane proteins help digest membrane components for recycling. The mitochondrial membrane contains enzymatic proteins (e.g., the protein complexes that are part of the electron transport chain of aerobic cellular respiration).

Recognition proteins are glycoproteins which identify a cell to the body's immune system. They allow immune cells to recognize a substance as either belonging to the body or as an invasive foreign agent to be destroyed. Recognition proteins of antigens are the basis for A, B, and O blood groups in humans. Immune cells recognize the sugars attached to these proteins and attack any red blood cells with foreign sugars, which is why patients of some blood groups cannot donate blood or receive blood from people with other blood groups.

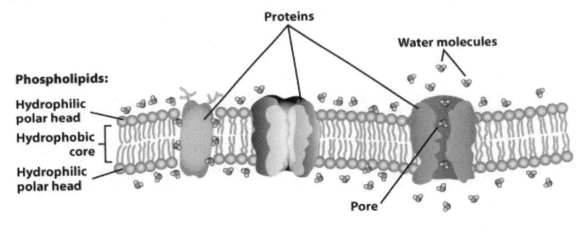

Osmosis: colligative properties and osmotic pressure

All fluids of the body are *solutions*: they contain dissolved substances (solutes) and a fluid (water) in which the substances dissolve (solvent). *Diffusion* is the movement of a solute from an area where it is in higher concentration to an area where it is in lower concentration. *Osmosis* is the diffusion of water from an area of low solute concentration to an area of high solute concentration. Note that solutes diffuse to an area of lower solute concentration, while water (a solvent) diffuses to an area of higher solute concentration. Both solute and solvent work together to offset unequal solute concentration and restore the solution to equilibrium.

The natural inclination of solutes is to diffuse until they are evenly distributed. However, different areas of the body require different concentrations of specific solutes. Separation of these areas is accomplished via the complex system of membrane compartmentalization in organs, tissues, and cells. The body must prevent all body compartments from reaching equilibrium with each other but must allow the passage of certain atoms, ions, and molecules through membranes to the areas where they are needed. To accomplish this, membranes are highly selective and tightly regulated.

Colligative properties of solutions depend on the concentration of a solute in the solution. They do not depend on the chemical nature of the solutes. The most important colligative property for the regulation of body fluids is *osmotic pressure*, the pressure that must be applied to prevent the net flow of water into a solution separated by a membrane (i.e., a measure of water's tendency to flow from one solution to another). This property is related to several other key terms in osmoregulation (the regulation of osmotic pressure). *Osmolarity* is the total solute concentration of a solution and is measured in units of *osmoles*. Osmolarity takes both penetrating and non-penetrating solutes into account and can be used to describe a single solution in addition to comparing different solutions. A *hyperosmotic* solution has a higher osmolarity, while a *hypoosmotic* solution has a relatively low osmolarity. *Isosmotic* solutions have the same osmotic pressure.

Tonicity describes the relative concentration of two solutions separated by a selectively permeable membrane, and explains how diffusion occurs between them (e.g., between intracellular and extracellular fluid). Unlike osmolarity, tonicity refers only to non-penetrating solutes (solutes which cannot cross a membrane), and always describes how one solution compares to another. A *hypertonic* solution has a relatively higher concentration of solute. Conversely, a *hypotonic* solution has a relatively lower concentration of solute. Therefore, a cell with a lower concentration of solute than the extracellular fluid is hypotonic to the fluid, while the extracellular fluid is hypertonic to the cell. The reverse is true for a cell with a higher concentration of solute than the extracellular fluid.

A cell placed in a hypertonic solution shrinks through a process of *crenation* (i.e., shrinking of the cell due to the decrease of cytoplasm), as water diffuses out of the cell to offset the high concentration of the outside solution. Conversely, a cell in a hypotonic solution swells as water rushes into the cell for plasmolysis. Plant cells rely on hypotonic extracellular fluids to keep their cells *turgid* (swollen), maintaining pressure against the cell wall and keeping the entire organism upright. Too much water can enter the cell, resulting in *lysis* (breakage of the cell). An *isotonic* solution has an equal solute concentration to the solution it is being compared to. In this situation, there is no net water movement.

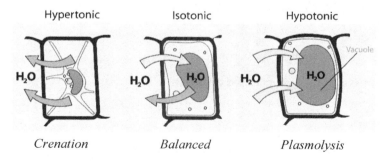

Passive transport

The plasma membrane is selectively permeable, meaning only certain molecules can pass through. A molecule's ability to diffuse through the plasma membrane depends on the molecule's size, charge, and polarity. The greater the lipid solubility of the diffusing particle, the more easily it passes through the membrane. Generally, smaller particles diffuse more rapidly than larger ones, and hydrophobic particles diffuse more rapidly than hydrophilic particles.

Many particles cannot diffuse through the plasma membrane without assistance. Small, non-charged or non-polar molecules pass through the membrane freely. Large, charged and/or polar molecules usually require assistance to pass through the membrane.

Passive transport enables the movement of molecules across a membrane without the expenditure of energy by the cell. The methods include simple diffusion, osmosis, and facilitated diffusion. Passive transport utilizes a *concentration gradient*, whereby particles diffuse from an area of higher to an area of lower solute concentration.

Simple diffusion is the process by which smaller, lipid-soluble molecules freely diffuse through the phospholipid bilayer unassisted. For example, oxygen and carbon dioxide pass through the membrane via simple diffusion.

While water is a polar molecule, it is small enough that it can diffuse freely across a plasma membrane. *Osmosis* is the passive diffusion of water molecules. Osmosis occurs when water moves from a region of lower solute concentration to a region of higher solute concentration and is facilitated by *aquaporins* as channel proteins. Osmosis is often classified as simple diffusion, despite requiring the transport proteins characteristic of facilitated diffusion.

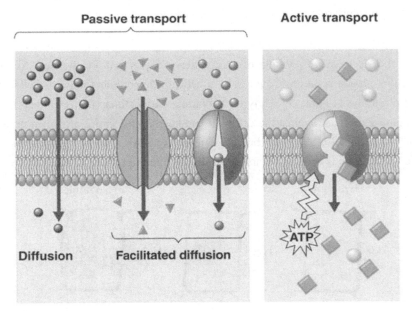

*Passive transport includes diffusion and facilitated diffusion with
the expenditure of energy as molecules move down the concentration gradient*

Facilitated diffusion allows larger, lipid-insoluble molecules which cannot freely pass through the phospholipid bilayer (e.g., sugars, ions, and amino acids). During facilitated diffusion, a molecule is transported down its concentration gradient with the assistance of either a carrier protein or channel protein, often as a *uniporter*.

Active transport: sodium - potassium pump

Active transport requires cellular energy to move solutes against their concentration gradient. Unlike passive transport, which exploits the natural inclination of molecules to move down their concentration gradient, active transport requires the expenditure of energy to resist this opposing force. Carrier proteins are the transmembrane proteins that mediate movement of molecules that are too polar or too large to move across a membrane by diffusion, thereby governing active transport. During this process, a solute (molecule to be transported) binds to a specific site on a transporter on one surface of the membrane. The transporter then changes shape to expose the bound solute to the opposite side of the membrane. The solute then dissociates from the transporter and is then on the opposite side from which it started.

Depending on the membrane and the needs of the cellular environment, many types of transporters may be present with specific binding sites for particular types of substances. *Solute flux magnitude* (i.e., the rate at which the solute flows) through a mediated transport system is positively correlated with the number of transporters, the rate of conformational change in the transporter protein and the overall saturation of transporter binding sites, which depends on the solute concentration and affinity of the transporter.

There are two types of active transport: primary and secondary. *Primary active transport* utilizes energy generated directly from ATP. The carrier proteins for primary active transport are pumps. An example is a sodium-potassium pump, which works by moving 3 Na^+ ions out and 2 K^+ ions into a cell, resulting in a net transfer of positive charge outside the membrane.

For a cell at rest, intracellular K^+ concentration is high, and Na^+ concentration is low, while extracellular K^+ concentration is low and Na^+ concentration is high. These concentration gradients facilitate transport across the plasma membrane and help the cell manage its *membrane potential*, the difference in electrical charge between the outside and inside of the cell.

Cellular sodium and potassium concentrations are maintained via active transport by the sodium-potassium pump (Na^+ / K^+ ATPase). When the intracellular Na^+ concentration is too high, and K^+ concentration is too low, the sodium-potassium pump must pump Na^+ out of the cell and K^+ into the cell to restore the appropriate concentration gradients.

The energy for *secondary active transport* comes from an electrochemical gradient established by the action of primary active transport. Secondary active transporters work via a mechanism of *cotransport*. Cotransport occurs when one molecule moves with (down) its concentration gradient, while another molecule moves against (up) its concentration gradient. The energetically favorable movement of the molecule moving with its concentration gradient powers the movement of the other molecule against its concentration gradient. *Antiporters* (e.g., sodium-calcium exchanger) move molecules in opposite directions, i.e., one is transported out while the other is transported into the cell. *Symporters* move both molecules in the same direction.

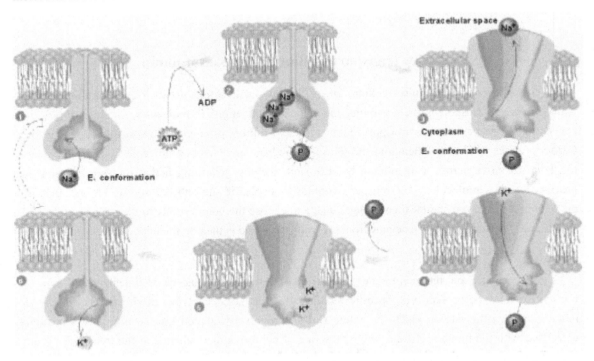

The K^+ / Na^+ pump maintains a cell's membrane potential. Na^+ bind to induce a conformation change in the transmembrane protein and pump Na^+ out of cell up its concentration gradient using ATP

One example of a symporter is the sodium-glucose linked transporter (SGLT), which transports sodium *with* its concentration gradient from the exoplasmic space to the cytoplasmic space, and transports glucose *against* its concentration gradient from the exoplasmic space to the cytoplasmic space. These movements are energetically coupled. Note that while both molecules are moving in the same *physical* direction in a symporter, the molecules are moving in opposite directions but are energetically favorable with one process driving the other.

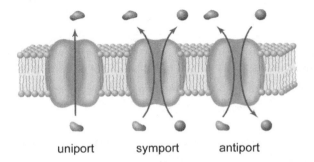

uniport   symport   antiport

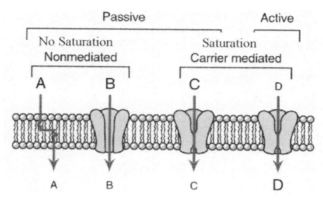

Comparison of passive and active transport: A is diffusion with movement unaided by a protein,
B is passive transport down the concentration gradient through a channel (no saturation),
C is passive transport through a carrier protein (saturation),
D is active transport up a concentration gradient (saturation possible and energy needed)

Membrane channels

Membrane channels are transmembrane proteins that allow ions to diffuse across the membrane via passive transport. Different cells have different permeabilities, depending on their membrane channels. The diameter of the channel and the polar groups on the protein subunits forming channel walls determine the permeability of the channels for various ions and molecules.

Porins are a type of channel proteins that are less chemically-specific than many other channel proteins; generally, if a molecule can fit through the porin, it can pass through it. *Ion channels* allow for the passage of ions. Channel gating is the opening and closing of ion channels to the molecules they transport. Changes in membrane potential modulate *voltage-gated* channels. Ligand-gated channels are modulated by the allosteric or covalent binding of ligands to the channel protein. *Ligands* are small molecules that bind to a protein or receptor, usually to trigger a signal. Mechanically-gated channels are modulated by mechanical stimuli such as stretching, pressure or temperature. Several factors may influence a single channel, and the same ion may pass through several different channels.

Membrane potential

Membrane potential is the electrical potential difference between the intracellular and extracellular environment. Membrane potential is mediated by channels and pumps, which alter electrochemical gradients as needed by each cell. Whenever there is a net separation of electric charges across a cell membrane, a membrane potential exists. The concentration gradient influences all molecules, but ions, because of their charged nature, are influenced by differences in membrane potential. The *electrochemical gradient* is the combined forces of membrane potential and concentration gradient. These forces may oppose one another, work independently from one another or work in conjunction.

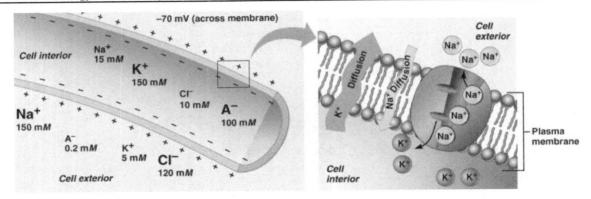

The differences in ion concentrations maintain membrane potential (i.e., voltage)

Nearly all eukaryotic cells maintain a non-zero membrane potential. In animal cells, this value ranges from −40 mV to −80 mV; thus, with respect to the extracellular environment, the inside of the cell has a negative voltage. Membrane potential is especially important for neurons, which have a *resting membrane potential* of about −70 mV. Changes in this resting potential allow for the electrical communications of neurons.

Membrane receptors, cell signaling pathways and second messengers

Cells must communicate with their neighbors as well as their environment. *Cell signaling* is the system by which cells receive, integrate and send signals to communicate information. These signals are sent through the *extracellular matrix*, a collection of polysaccharides and proteins secreted by cells in multicellular organisms. This matrix fills the space between cells, providing structure, facilitating cell signaling and allowing the cells to move and change their shape.

The composition of the extracellular matrix is highly variable depending on tissue type, as different tissues have different functions. For example, the extracellular matrix of bone is highly calcified, while the extracellular matrix of blood is fluid, containing dissolved proteins and other molecules. The most extensive extracellular matrices are of connective tissues, which are throughout the body and include bone and blood. Connective tissues are largely composed of their extracellular matrix and are only sparsely populated with cells.

A typical connective tissue extracellular matrix includes fibrous proteins and *proteoglycans*, glycoproteins that form a packing gel around the fibrous proteins. The primary fibrous proteins are *collagen* and *elastin*, which provide structure and flexibility, along with *fibronectin* and *laminin*, which assist in adhesion and cell migration. Transmembrane proteins such as *integrins* bind to specific proteins in the extracellular matrix and membrane proteins on adjacent cells. Integrins help organize cells into tissues. They are responsible for transmitting signals from the extracellular matrix to the cell interior.

Membrane receptors may be on the plasma membrane or intracellular membranes. Not all signal molecules bind to membrane receptors at the plasma membrane. For example, steroid hormones and gases, diffuse across the plasma membrane and bind to intracellular receptors. Other signal molecules enter the cell via endocytosis. In addition to hormones and gases, signal molecules include neurotransmitters, proteins, and lipids.

Membrane receptors are crucial in the cell-signaling pathways. When a signal molecule binds to a membrane receptor, it may initiate a metabolic response, change the membrane potential or alter gene expression. *Signal transduction* is when one signaling molecule triggers a multi-step chain reaction, which indirectly transmits the initial signal to its destination. *Second messengers* are the molecules that relay the signal.

At steps in the signal transduction pathway, second messengers can greatly amplify the strength of the original signal by increasing the number of molecules they activate. For example, one signal molecule at the membrane receptor may produce a second messenger, and these second messengers may each produce another second messenger, and so on, amplifying the signal significantly.

Cell signaling is a complex process but is into four general categories. *Endocrine signaling* is when a cell secretes a signal that travels to a distant target cell. *Paracrine signaling* is when the target cell is nearby, but not in direct contact. *Juxtacrine signaling* is the signaling of a target cell in direct contact with the secreting cell. *Autocrine signaling* targets the same cell that secreted the signal.

Exocytosis and endocytosis

The fluidity of the plasma membrane allows it to change shape, pinch off and reform. This fluidity enables substances to exit the cell via *exocytosis* and enter the cell via *endocytosis*. Both processes require cellular energy and are therefore considered forms of active transport.

During exocytosis, membrane-bound vesicles in the cytoplasm fuse with the plasma membrane and their contents are released outside the cell. The vesicle then assimilates into the plasma membrane, replenishing portions of the membrane that would otherwise be lost. The process of exocytosis is triggered by stimuli, leading to an increase in cytosolic calcium concentration, which activates proteins required for the vesicle membrane to fuse with the plasma membrane. Exocytosis provides a route for the release of the proteins it produces for extracellular secretion, as well as the proteins and lipids destined for the plasma membrane.

Endocytosis is essentially the opposite process of exocytosis. During endocytosis, extracellular molecules destined for the plasma membrane or the cytoplasm are imported into the cell. In preparation for endocytosis, a region of the outer side of the plasma membrane indents and encloses the material for import. This indentation then folds and pinches off into a membrane-bound vesicle inside the cell. *Pinocytosis* ("cell-drinking" or fluid endocytosis), is a form of endocytosis which cells perform regularly; it is the import of small amounts of extracellular fluid that contain molecules able to be readily absorbed by the cell. The process of pinocytosis may be non-specific or mediated by receptors on the plasma membrane.

Phagocytosis is performed only by a few types of specialized cells. Phagocytosis ("cell-eating") involves the import of much larger particulate matter than that imported during pinocytosis. The particulate matter must be broken down before it can be absorbed by the cell. As such, phagocytes digest bacteria, viruses and cell debris as a function of our immune system. Some unicellular eukaryotes, such as amoeba, rely on phagocytosis for nutrient intake.

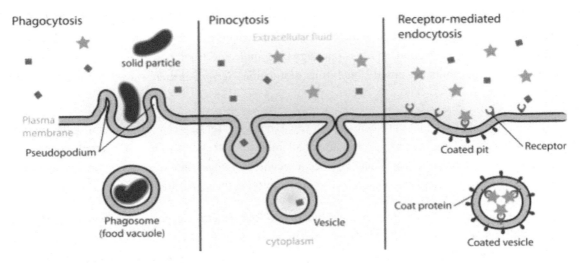

Types of endocytosis with substrates taken into the cell

Intercellular junctions: gap junctions, tight junctions, and desmosomes

Cell junctions are points of contact that physically link neighboring cells. Animal cells have three types of intercellular junctions: gap junctions, tight junctions, and anchoring junctions.

Gap junctions are protein channels that directly link the cytoplasms of adjacent cells. Gap junctions are communicating junctions because they allow for rapid cell-to-cell communication. They are formed by the joining of two membrane channels on adjacent cells, allowing the movement of small molecules and ions between cells, while still preventing their cytoplasms from mixing. Gap junctions are important in tissues such as cardiac muscle, where electrical impulses must be transmitted through cells extremely rapidly so that the muscle fibers contract as a single unit.

The *tight junction* has plasma membrane proteins attach in zipper-like fastenings, holding cells together so tightly that the tissues become barriers to molecules. Tight junctions are formed by the physical joining of the extracellular surfaces of two adjacent plasma membranes, producing a seal that prevents the passage of materials between cells. Materials must enter the cells by passive or active transport to pass through the tissue. Tight junctions are important in areas where more control over tissue processes is needed (e.g., the epithelial cells in the intestine involved in nutrient absorption).

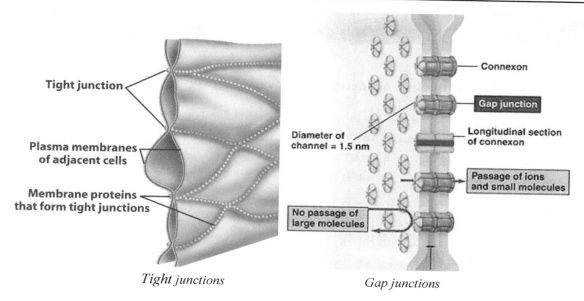

Tight junctions   *Gap junctions*

Anchoring junctions use proteins extended through the plasma membrane of one cell and attached to another cell. Anchoring junctions are firm but still allow for spaces between adjacent cells.

Desmosomes, a type of anchoring junction, are created by dense patches of protein on the plasma membranes of two cells. Internally, proteins anchor to the cytoplasm of each cell, while externally the proteins adhere to one another. The purpose and function of desmosomes are to hold adjacent cells firmly in place in tissue areas that are subject to stretching, (e.g., bladder, skin and stomach).

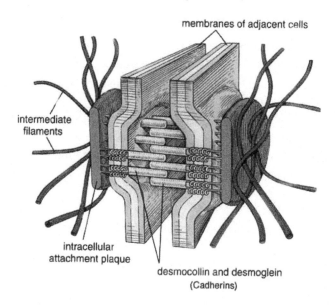

Desmosome

Cytoskeleton

The general function of the cytoskeleton in cell support and movement

The *cytoskeleton* is a scaffold of flexible, tubular protein fibers extending between the nucleus to the plasma membrane in eukaryotes. This vast network of fibers maintains the shape of the cell, provides support and facilitates the transport of vesicles. The cytoskeleton is the cellular analogy to the bones and muscles of an animal. It anchors organelles and enzymes to specific regions of the cell to keep them organized in the cytosol.

The cytoskeleton can change shape to facilitate contractility and movement, allowing the cell to divide, migrate or undergo endocytosis and exocytosis. During cell division, the cytoskeletal elements can rapidly assemble and disassemble to form spindles for the organization of chromosomes and to cleave the cell into two daughter cells. The long fibers of the cytoskeleton serve for intracellular transport, upon which vesicles and organelles move via motor proteins (e.g., dynein and kinesin).

Microfilaments: composition and role in cleavage and contractility

Microfilaments (actin filaments) are the thinnest and most abundant of the cytoskeleton proteins. They are composed of the contractile protein *actin*. These long, thin fibers, about 7 nm in diameter, may be in bundles or meshlike networks. Each microfilament consists of two chains of globular actin subunits twisted to form a helix.

Microfilaments can be assembled and disassembled quickly according to the needs of the cell. They are involved in cell motility functions, such as the contraction of muscle cells, the formation of amoeba pseudopodia and cleavage of the cell during cytokinesis. While flexible, microfilaments are strong and prevent deformation of the cell by their tensile strength.

Microfilaments provide tracks for the movement of myosin. Myosin attaches to vesicles (or organelles) and pulls them to their destination along the microfilament track. Additionally, the interaction between microfilaments and myosin is crucial to the function of the cell.

Intermediate filaments: role in support

Intermediate filaments are thicker than microfilaments but thinner than microtubules. Typically, they are 8 to 11 nm in diameter. These rope-like assemblies of fibrous polypeptides are most extensive in regions of cells that are subjected to stress. Most intermediate filaments are in the cytoplasm, supporting the plasma membrane and forming cell-to-cell junctions. However, *lamins*, a class of intermediate filaments, are responsible for structural support within the nucleus. Unlike microfilaments and microtubules, intermediate filaments are not capable of rapid disassembly once assembled.

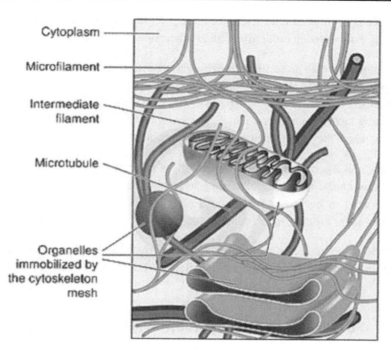

Cytoskeleton as a network of tubular proteins to provide shape and facilitate transport of vesicles

Microtubules: composition and role in support and transport

Microtubules (tubulin) are hollow protein cylinders about 25 nm in diameter and 0.2–25 μm in length. They are the thickest and most rigid of filaments. Microtubules are composed of the globular protein, *tubulin,* which occurs as α tubulin and β tubulin. The microtubule assembly brings these two together as dimers, and the dimers arrange themselves in rows.

The strength and rigidity of microtubules make them ideal for resisting compression of the cell. However, these fibers serve functions similar to microfilaments. Microtubules act as tracks for intracellular transport, but they interact primarily with *kinesin* and *dynein* motor proteins, rather than myosin. The transport function of microtubules is especially important for trafficking neurotransmitters throughout nerve cells. Like microfilaments, microtubules can be rapidly assembled or disassembled. Regulation of microtubule assembly is under control of a *microtubule-organizing center* (MTOC). Microtubules radiate from the MTOC and extend throughout the cytoplasm. During cell division, the centrosome generates the microtubule spindle fibers necessary for chromosome separation.

Composition and function of eukaryotic cilia and flagella

Cilia and *flagella* are two types of microtubule complexes that protrude from the cell body and serve cell motility and sensory functions. In eukaryotes, they are membrane-bounded cylinders that enclose a matrix of nine pairs of microtubules encircling two single microtubules, a *9 + 2 pattern.* Movement occurs when these microtubules slide past one another. At the plasma membrane, a *basal body* anchors the cilium (or flagellum) to the cell body. The basal body is derived from a centriole, which is formed by nine pairs of microtubules without central microtubules, the *9 + 0 pattern*. Eukaryotic cilia and flagella grow by polymerizing (adding) tubulin to their tips.

Cilia are short, hair-like projections. Nearly every human cell has at least one cilium, which is *non-motile* and functions as a sensory antenna important in cell signaling pathways. Non-motile cilia lack the 2 central microtubules and thus are have a 9 + 0 pattern. Many cells are covered with some *motile* cilia, which move in an undulating fashion to transport particles across the cell surface. For example, motile cilia are in the respiratory tract, where they push mucus and irritants out of the lungs, trachea, and nose.

Essentially, eukaryotic flagella are structurally similar to eukaryotic cilia, so distinctions are often not drawn between the two. Like cilia, they have both sensory and motility functions. However, eukaryotic flagella tend to be longer and move with more of a whip-like motion. An example of a eukaryotic flagellum is the tail of a sperm cell.

Prokaryotic flagella are notably different from their prokaryotic cilia counterparts. These flagella are noted for their long, whip shape and functional differences from cilia.

Centrioles and microtubule organizing centers

The *centrosome* is the main microtubule organizing center at the poles during mitosis and meiosis of the cell. Each centrosome contains a pair of barrel-shaped organelles of *centrioles*. The centrosome has a key role in mitosis when it divides into two centrosomes, which then interact with chromosomes via microtubules to form the mitotic spindle.

Centrioles are components of the centrosome. Microtubules radiate from these barrel-shaped structures, which are made of microtubules themselves. They are short cylinders with a ring pattern (9 + 0) of microtubule triplets. In animal cells and most protists, a centrosome contains two centrioles oriented at right angles to each other. Plant and fungal cells contain the equivalent of a centrosome, but do not contain centrioles. Centrioles serve as basal bodies for cilia and flagella.

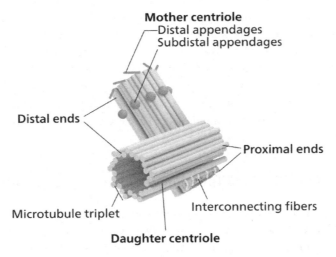

Centrosome contains two centrioles oriented at right angles to each other

In mitosis, the terms centrioles and centrosomes are often used interchangeably because centrioles form the most important parts of a centrosome.

Cell Cycle and Mitosis

The cell cycle describes the lifetime of a cell, detailing the life stages from the creation of a new cell to its division into two daughter cells. Most of an organism's cells divide throughout their lifetime (i.e., an exception is nerve cells), as cell division is the process by which organisms grow, repair tissues and reproduce.

Mitosis and meiosis are two types of cell division. *Mitosis* is cell division when new somatic (body) cells are added to multicellular organisms as they grow, and when tissues are repaired or replaced. When a cell is preparing for mitosis, it grows larger, the number of organelles doubles and the DNA replicates. *Meiosis* is the production of gametes (reproductive cells of egg and sperm) by organisms that reproduce sexually. Meiosis is discussed later in further detail.

Mitosis does not produce genetic variations. A daughter cell is identical in chromosome number and genetic makeup to the parent cell. The purpose of mitosis is to distribute identical genetic material to two daughter cells. What is remarkable is the fidelity with which the DNA is passed along, without dilution or error, from one generation to the next. All eukaryotes divide by mitosis, but prokaryotes undergo a different mechanism of *binary fission*, a form of asexual reproduction. Binary fission is a simpler process wherein the parent cell replicates its DNA and then divides. Since prokaryotes have a single and circular DNA without a nucleus, there is no complex steps of chromosome formation and separation by mitosis with microtubules between the centromere and centrioles as in eukaryotic cells.

Mitotic process: prophase, metaphase, anaphase, telophase, and interphase

Interphase is the stage before mitosis and represents the majority of a cell's life. This is not a static state but rather a progression towards mitosis. However, in some cases, a cell halts its progress through interphase, either temporarily or permanently. This may be because it is a non-dividing cell (e.g., nerve cell), or because the cell is not healthy enough to perform the growing and replicating functions of interphase.

During interphase ($G_1 \rightarrow S \rightarrow G_2$), the cell prepares to divide by growing, replicating its DNA and many of its organelles, and synthesizing mRNA and proteins. When it has completed all the functions of interphase, the cell exits interphase and enters mitosis.

There are four phases of mitosis: prophase, metaphase, anaphase, and telophase (PMAT).

1. Prophase = *Prepare*: cell *prepares* for mitosis

2. Metaphase = *Middle*: chromosomes align in the *middle* of the cell

3. Anaphase = *Apart*: centromere splits and sister chromatids are pulled *apart* by microtubules to the opposite poles of the cell

4. Telophase = *Two*: two daughter nuclei are re-formed with separate nuclei

Prophase, the first phase of mitosis involves chromatin condensation, nucleolus dissolution, nuclear membrane fragmentation, and centrosome movement. *Chromatin condensation* is the process by which loose euchromatin condenses into chromosomes. At this time, the chromosomes have no particular orientation in the cell. Upon chromatin condensation, the nucleolus dissolves, and the nuclear membrane begins to fragment, exposing the chromosomes to the cytoplasm of the cell. Simultaneously, centrosomes begin to migrate to opposite sides of the cell. Microtubules start to extend from the centrosomes, forming the *spindle apparatus.*

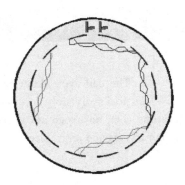

Prophase

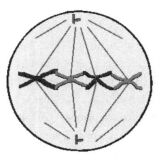

Metaphase

After prophase, the cell enters *metaphase*, when nuclear fragmentation completes and the centrosomes station themselves at opposite poles of the cell. Microtubules emerge from the centrosomes and attach to the chromosomes, aligning them along an imaginary line in the center of the cell, the *metaphase plate*, or *equatorial plate*. This completes the formation of the spindle apparatus. All chromosomes must be attached and lined up along the metaphase plate before the cell can proceed to the next phase.

During *anaphase*, sister chromatids (chromatin strands replicated during S phase of interphase but attached at the centromere) are pulled apart to opposite poles of the cell. The sister chromatids become detached from one another at their centromere and travel towards opposite centrosomes by action of the spindle microtubules. By the end of anaphase, equal numbers of sister chromatids are stationed by both centrosomes.

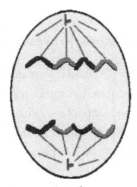

Anaphase

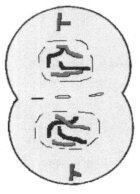

Telophase

Telophase is when the cell reverses the actions of prophase and prepares to divide. The spindle apparatus disassembles, and two daughter nuclei reform within the cell. As spindle microtubules disassemble, two regions of identical chromatids are present in the cell. A nuclear envelope develops around each of these regions, forming two daughter nuclei. Within the nuclei, chromosomal DNA uncoils into chromatin, and nucleoli form. The single cell now contains two identical daughter nuclei, and cell division proceeds.

Cytokinesis is the division of the cytoplasm to create two daughter cells. This usually coincides with the end of telophase but is *not* a phase of mitosis. Rather, it is a separate event which does not always occur. When mitosis occurs but cytokinesis does not, it results in a multinucleated cell; (often in plants, but also in skeletal muscle cells of animals).

In animal cells, cytokinesis occurs by process of *cleavage.* First, a *cleavage furrow*, a shallow groove between the two daughter nuclei, appears. The cleavage furrow deepens as a band of microfilaments, *contractile ring*, constricts between the two daughter cells. A narrow bridge exists between daughter cells during telophase until constriction completely separates the cytoplasm. The result is two daughter cells enclosed in their plasma membrane and with their own, identical nuclei. Recall that the parent cell replicated its organelles before mitosis; thus, each daughter cell contains a full set of organelles, smaller than the cell size of the parent.

Cytokinesis in plant cells is different because plant cells have rigid cellulose cell walls which do not permit cytokinesis by furrowing. In plants, vesicles containing cellulose move to the middle of the cell. Additional vesicles arrive and coalesce, building a *cell plate* of cellulose (cell wall). When the cell plate is complete, the parent has divided into two separate daughter cells.

Nuclear membrane breakdown and reorganization

Before mitosis begins, the cell's DNA is contained inside the nucleus, inaccessible to the mitotic spindle. During prophase and metaphase, the nucleolus disintegrates, chromatin condenses, and the nuclear membrane breaks down. This process exposes chromosomal DNA to the cell's cytoplasm and allows mitosis to proceed. The nuclear membrane does not reform until telophase when the two daughter nuclei are each enclosed within their nuclear membranes from fragments of the parental cell's nuclear membrane. At this point, chromosomes uncoil into chromatin and the nucleoli reform, making nuclear reorganization complete.

Centrioles, asters, and spindles

As previously discussed, centrioles are the units of the centrosome, the microtubule-organizing centers of the cell. Centrioles are replicated during interphase in preparation for mitosis. During prophase, *polar microtubules* emerge from both pairs of centrioles and centrosomes and push against each other to move the centrosomes to opposite sides of the cell. *Astral microtubules* extend from the centrioles to assist in orienting the mitotic spindle apparatus. At the end of prophase, kinetochore microtubules originate from the centrioles and attach to kinetochores of the chromosomes.

The entire spindle apparatus consists of the two centrosomes (composed of two centrioles each), polar microtubules, astral microtubules (asters), and *kinetochore microtubules* (k-fibers), which attach to the kinetochores on chromosomes.

Like animal cells, plant cells have a spindle apparatus, but many do not have centrioles or astral microtubules. Centrioles are not strictly necessary for mitosis even in animal cells.

Chromatids, centromeres, and kinetochores

As the cell enters metaphase, kinetochore microtubules extend from the centrosomes and attach to *kinetochores* on the chromosomes. Kinetochores are assembled on the *centromere*, the region of a chromosome that links two sister chromatids. Technically, there are two sections of the kinetochore: the inner kinetochore, which associates with the DNA of the centromere, and the outer kinetochore, which interacts with the kinetochore microtubules. During metaphase, these microtubules pull on the chromosomes with equal tension, eventually aligning them at the metaphase plate. Upon successful attachment of all chromosomes to the spindle via the kinetochore and microtubules, proteins are released from the kinetochore, which signals the end of metaphase and the beginning of anaphase.

Mechanisms of chromosome movement

In anaphase, the centromere holding the sister chromatids (S phase of interphase when the chromosome replicated to form two sister chromatids) dissolves, and the sister chromatids are released from their attachment point. The chromosomes are pulled to opposite sides of the cell by shortening of the kinetochore microtubules. Shortening occurs when the motor protein attached to a kinetochore "walks" along the kinetochore-microtubule, dissembling the microtubule into tubulin subunits as it passes. Polar microtubules assist the separation of chromosomes by lengthening the spindle. Wherever the ends of two polar microtubules from opposite poles overlap, motor proteins interact between the fibers and push them in opposite directions, thus pushing the entire spindle apart.

Phases of the cell cycle (G_0, G_1, S, G_2, M)

Interphase is divided into three phases: G_1, S, and G_2. G_0 is another, resting phase when the cell is not dividing nor preparing to divide. G_0 is the static state in which some cells remain permanently (e.g., nerve cells) and others only temporarily. A cell can exit G_0 and reenter the active cell cycle upon receipt of suitable signals from *growth factor* proteins. Under a microscope, a cell can be recognized as being in interphase by the lack of visible chromosomes, because the DNA is uncoiled as loose chromatin.

During the *G_1 phase* (*Gap phase 1*), the cell continues normal function as it grows larger and replicates its organelles, including ribosomes, mitochondria, and chloroplasts (if a plant cell). The mitochondria generate sufficient storage of energy for the functions of mitosis. In G_1, the cell synthesizes mRNA and proteins in preparation for the *S phase*. In the S phase (*Synthesis*), the cell replicates its DNA to produce two sister chromatids attached at the centromere. During the S phase, any DNA nucleotide damage must be detected and fixed before the cell may proceed.

G_2 phase (*Gap phase 2*) is where the cell continues to grow and synthesize proteins needed for mitosis. Completion of G_2 marks the end of interphase and the beginning of mitosis, often abbreviated as M. In summary, the sequential phases of the cell cycle are:

G_0 = no DNA replication or cell division

G_1 = making of organelles, increase in cell size (growth)

S = DNA replication, complete duplication of chromosomes (sister chromatids)

G_2 = making of organelles, increase in cell size (growth), cell committed to mitosis

M = mitosis (PMAT); prophase, metaphase, anaphase, and telophase

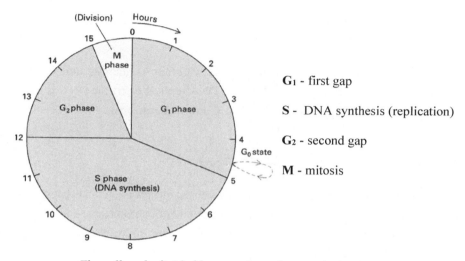

G_1 - first gap

S - DNA synthesis (replication)

G_2 - second gap

M - mitosis

The cell cycle divided between interphase and mitosis

Growth arrest

Cell growth refers to *cell proliferation* (populations) or the growth of an individual cell, whereby biomolecules are synthesized.

Cell proliferation is the ultimate goal of unicellular organisms. These organisms are mostly limited by nutrient availability. However, in multicellular organisms, cell proliferation must be carefully monitored to prevent tumor formation and invasion into nearby tissues. *Contact inhibition*, the tendency for cells to cease dividing when they come into physical contact with their neighbors, regulates this. Therefore, lack of free space signals growth arrest.

The growth of the individual cells regulates cell populations. For individual cell development, growth arrest is *cellular quiescence* or *cell cycle arrest.*

During the cell cycle, the cell encounters checkpoints at which the cell cycle is halted before proceeding. There are three major checkpoints: G_1, G_2, and M:

1. The G_1 Checkpoint – Restriction Point

    Partway through G_1, the cell reaches the restriction point. Here, the cell checks for cell size, nutrients and growth factors. If the cell is not sufficiently healthy and prepared for mitosis, the cycle halts and the cell returns to G_0. Additionally, if DNA damage is detected, this triggers *apoptosis* (cell death) if the DNA is not repaired. If everything is as it should be, the cell clears the checkpoint and proceeds toward DNA replication (S phase). G_1 is the only checkpoint mediated by extracellular signals. After this point, only intracellular signals direct the cell cycle to proceed or halt progress.

2.   The G_2 Checkpoint – DNA Damage Checkpoint

At the end of G_2, before the cell proceeds with mitosis, there is another checkpoint. The cell checks for cell size and proper DNA replication. If the DNA is not finished replicating or if DNA has been damaged and requires repair, the cell remains in G_2 until these issues are resolved.

3.   The M Checkpoint –Mitotic Spindle Checkpoint

After the cell has entered mitosis and has reached metaphase, a final checkpoint occurs. The M checkpoint ensures that the proper number of chromosomes are aligned at the mitotic plate and secured to the mitotic spindle. Errors during chromosome segregation can cause defects resulting in genetic conditions (e.g., Down syndrome). The M checkpoint reduces the occurrence of defects by arresting the cell in metaphase until all chromosomes are properly attached to the spindle apparatus and aligned for anaphase.

Control of Cell Cycle

The cell cycle is controlled by intracellular and extracellular signals that either stimulate or inhibit metabolic events. Extracellular stimulatory signal molecules are growth factors; these are proteins or hormones that promote cell growth and differentiation. Extracellular inhibitory signal molecules are growth suppressors, or more often *tumor suppressors* because they prevent the rampant growth of cancer cells. Tumor suppressors inhibit growth by halting the cell cycle or directing the cell to destroy itself via apoptosis.

Intracellular signaling directs the cell cycle and involves the activation and inactivation of proteins; *cyclin-dependent kinases* (CDKs).

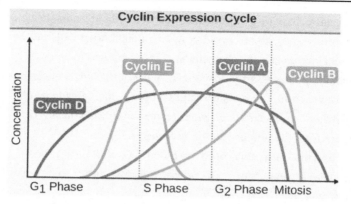

Relative concentrations of cyclin proteins during phases of the cell cycle

Apoptosis (programmed cell death)

Apoptosis is programmed cell death initiated by the organism. Apoptosis destroys cells which pose a threat to the organism, such as infected cells, cells with DNA damage, cancerous cells, and immune system cells no longer needed and are unnecessarily attacking other body cells. While it may seem paradoxical, apoptosis is important for growth. For example, cells must die to create spaces in the webbed hands of a fetus for the formation of separate digits.

Some internal or external pathways and signals causes cell death, but the morphological changes during apoptosis are consistent whatever the cause. These changes include shrinkage and *blebbing* (bulging) of the plasma membrane, of the nuclear envelope and DNA fragmentation. Engulfment by nearby phagocytic cells ultimately occurs as a result. Apoptotic cells release signals which attract phagocytic cells. The engulfment of fragments of the dying cell prevents viruses or other dangerous cell contents from spilling out of the damaged cell.

Loss of cell cycle controls in cancer cells

Cancer cells are abnormal cells with a variety of dangerous properties, making them a serious threat to the body. They invade and destroy normal tissue, causing serious illness and death. Cancerous cells do not normally respond to the body's control mechanism. They no longer respond to inhibitory growth factors and do not require as many stimulatory growth factors. They may produce the required external growth factor (or override factors) themselves or possess abnormal signal transduction sequences which falsely convey growth signals, thereby bypassing normal growth checks. Due to their irregular growth cycles, if the growth of cancer cells does occur, it does so at random points of the cell cycle.

Cancer can kill the organism because these cells can divide indefinitely (i.e., immortalized cell) if given a continual supply of nutrients. Normally, DNA segments as *telomeres* form the ends of chromosomes and shorten with each replication, eventually signaling the cell to stop dividing. However, cancer cells produce the enzyme telomerase, which keeps telomeres long and allows the cells to continue dividing. The cell is "*immortal*."

Unlike normal cells which differentiate, cancer cells are non-specialized. Cancer cells do not exhibit contact inhibition; they do not avoid crowding neighboring cells but rather pile up and grow on one another. This behavior creates the characteristic tissue mass as a tumor.

Not all tumors are necessarily dangerous. A *benign tumor* is encapsulated and does not invade adjacent tissue. However, benign tumors can still compress and damage nearby tissue, and some benign tumors have the potential to become *malignant* (cancerous). Malignancy occurs when new tumors are spread to areas distant from the primary tumor by *metastasis.*

Angiogenesis, the formation of new blood vessels, is a process required for metastasis. Angiogenesis is triggered when cancer cells release a growth factor that causes nearby blood vessels to grow and transport nutrients and oxygen to the tumor. Because of angiogenesis in metastasis, angiogenesis inhibitors are an important class of cancer drugs.

Cancer cells have abnormal nuclei that may be enlarged and have an abnormal number of chromosomes, as some chromosomes are mutated, duplicated or deleted.

When the DNA repair system fails to correct mutations during DNA replication, damage to crucial genes may occur. *Oncogenes* encode growth factor proteins such as Ras, which stimulate the cell cycle, while tumor-suppressor genes encode proteins such as p53 that inhibit the cell cycle. Mutations of oncogenes or tumor suppressor can cause cancer.

Mutation of proto-oncogenes may convert them into *oncogenes*, which are cancer-causing genes. An oncogene can cause cancer by coding for a faulty receptor in the stimulatory pathway, for an abnormal protein product or for abnormally high levels of a normal product that stimulates the cell cycle. About 100 oncogenes have been identified; the *ras* gene family includes variants associated with lung, colon, and pancreatic cancers, as well as leukemia and thyroid cancers.

Mutation of tumor-suppressor genes results in unregulated cell growth. For example, the *p53* tumor-suppressor gene is more frequently mutated in human cancers than any other known gene; it normally functions to trigger cell cycle inhibitors and stimulate apoptosis. However, if it malfunctions due to mutation, cell growth is not suppressed, and cancer may result.

Biosignaling

Rhythmic fluctuations in the abundance and activity of cell-cycle control molecules pace the events of the cell cycle. One example of these control molecules are *kinases*, proteins which activate or deactivate other proteins by phosphorylation. They are responsible for a procession through the G_1 and G_2 checkpoints. To give the signal, the kinases themselves must be activated by a *cyclin* protein. Because of this requirement, these kinases are cyclin-dependent kinases (CDKs). Cyclins, named for their cycling concentration in the cell, accumulate during the G_1, S, and G_2 phases of the cell cycle. By the G_2 checkpoint, enough cyclin is available to form a complex of cyclin and CDK are the *maturation-promoting factor* (MPF), or the M-Phase-promoting factor. MPF initiates progression from the G_2 to the M phase by phosphorylating key proteins involved in mitosis. Later in mitosis, MPF switches itself off by initiating a process which leads to the destruction of cyclin. Cdk, the non-cyclin component of MPF, persists in the cell in an inactive form until it associates with the new cyclin molecules synthesized during interphase of the next round of the cell cycle.

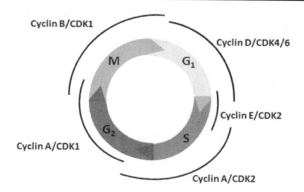

Cyclin-dependent kinases (CDK) and cyclin relationship to regulate the cell cycle

Platelet-derived growth factor (PDGF) is another protein that regulates cell growth and division. PDGF is required for the division of *fibroblasts*, connective tissue cells essential in wound healing. When an injury occurs, platelet blood cells release PDGF, which then binds to fibroblast receptors and activates a signal-transduction pathway that leads to a proliferation of fibroblasts and healing of the wound.

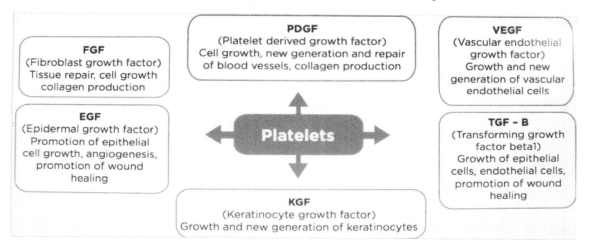

Growth factors released by platelets in wound healing

The extracellular environment has a direct effect on cell division. Cells grown in culture rapidly divide until a single layer of cells is spread over the area of the petri dish. However, if cells are removed, those bordering the open space begin dividing again and continue to do so until the gap is filled. This propensity to avoid division when in contact with neighboring cells is *density-dependent inhibition* of growth. When a cell population reaches a critical density, the amount of required growth factors and nutrients available to each cell becomes insufficient to allow continued cell growth.

Anchorage is another extracellular factor that controls cell division. For most animal cells to divide, they must be anchored to a substratum, such as the extracellular matrix of a tissue or the inside of a culture plate. Anchorage is signaled via pathways involving membrane proteins and the cytoskeleton.

Tissues Formed from Eukaryotic Cells

Epithelial cells: simple epithelium and stratified epithelium

Epithelial cells make up epithelial tissues, which are surfaces that line structures throughout the body, particularly organs and blood vessels. Several types of epithelial cells perform a variety of functions. *Squamous epithelial cells* appear flat, *cuboidal epithelial cells* are cube-shaped and *columnar epithelial cells* are column-shaped.

Simple epithelium is a single layer of epithelial cells connected by tight junctions. The function of the simple epithelial layer is highly dependent on the types of epithelial cells involved. Simple squamous layers are often involved in passive diffusion, lining surfaces such as the alveoli during oxygen exchange. Simple cuboidal layers are involved in secretion and absorption (e.g., gland ducts and kidney tubules). Simple columnar epithelial layers form a protective layer in the stomach and gut.

Types of epithelium

Epithelial cells can form layers as *stratified epithelium*. Layered epithelial layers allow for more protection and complex function. For example, stratified columnar epithelium lines the vas deferens, protecting the glands and assisting in secretion. Stratified columnar and cuboidal epithelium is rarely seen in human anatomy; stratified squamous covers the entire body as the skin.

Endothelial cells

The *endothelium* is a layer of simple squamous cells that forms the interior lining of lymphatic vessels and blood vessels. It acts as a semi-selective barrier that controls the passage of materials. *Lymphatic endothelial cells* are in direct contact with lymph. *Vascular endothelial cells* are in direct contact with blood, and line every part of the circulatory system, from the tiniest capillaries, to larger arteries, veins and to the heart itself. These endothelial cells have many functions, including blood clotting, the formation of new blood vessels, blood pressure control and inflammation control. Endothelial cells have a strong capacity for cell division and movement, and they proliferate quickly.

Connective tissue cells: major tissues and cell types, loose fiber vs. dense fiber and extracellular matrix

Connective tissue holds structures of the body together. It consists of specialized cells, ground substance, and fibers. The cells in connective tissue secrete the extracellular matrix, which is held together by ground substance. The fibers, made mainly of collagen, give the matrix its strength. Several types of connective tissue cells exist, making up bone, fat, tendons, ligaments, cartilage and blood. For example, chondroblasts make cartilage, fibroblasts make collagen, and hematopoietic stem cells make blood.

The nomenclature of the numerous types of cells in connective tissue helps to differentiate their function. Cells that contain the suffix *blast* describe a stem cell that actively produces a matrix, while the suffix *cyte* describes a mature cell. For example, while osteoblasts are specialized connective tissue cells that build the matrix in bone, osteocytes are mature, immobile osteoblasts involved in bone maintenance.

Various types of fiber make up connective tissue. The most common protein fiber is collagen or *collagenous fibers*. These coiled fibers give collagen its rigidity. There are many collagenous fibers, including *elastic fibers*, which give connective tissue its flexibility, and *reticular fibers*, which mainly join one connective tissue to an adjacent organ or blood vessel. Connective tissue is either "loose" or "dense." *The loose connective tissue* has a higher concentration of ground substance and cells and fewer fibers. It provides protective padding around the internal organs, as well as fat.

The *dense connective tissue* has a higher concentration of collagenous fibers than loose connective tissue and is needed in anatomical structures that require great strength (e.g., ligaments and tendons). *Cartilage* is a connective tissue that is produced and maintained by chondrocytes. Cartilage can absorb shock and is seen on the ends of bones and in the spinal disks. Since it is more flexible than bone, cartilage is advantageous for structures that do not require as much protection, such as the nose or ears. The *extracellular matrix* (ECM) exists outside of cells. Cells secrete molecules that make up the matrix, which include proteins and polysaccharides. In connective tissue, the ECM gives surrounding cells a support system both physically and chemically.

Notes

Please, leave your Customer Review on Amazon

Chapter 2

Cell Metabolism

- Glycolysis, Gluconeogenesis and the Pentose Phosphate Pathway

- Krebs Cycle

- Oxidative Phosphorylation

- Principles of Metabolic Regulation

- Metabolism of Fatty Acids and Proteins

- Enzyme Structure and Function

Glycolysis, Gluconeogenesis and the Pentose Phosphate Pathway

Glycolysis (aerobic), substrates and products;
feeder pathways: glycogen and starch metabolism

In *cellular respiration*, cells release the energy in chemical bonds of food molecules and transfer this energy to ATP molecules, which allows for efficient use of an organism's energy. The ATP generated during cellular respiration is then used for life's essential processes.

Cellular respiration can be *aerobic* (with oxygen) or *anaerobic* (without oxygen). *Glycolysis* (glycol for sugar, lysis for breaking) is the first step in cellular respiration and is the same in both aerobic and anaerobic cells because glycolysis does not require oxygen. Glycolysis occurs in the cytosol of the cell and catabolizes a six-carbon glucose molecule into two three-carbon pyruvate molecules. During glycolysis, energy is transferred through phosphate groups undergoing hydrolysis reactions (breaking) and condensation (joining).

The overall reaction of glycolysis is:

$+ 2 \text{ [NAD]}^- + 2 \text{ [ADP]} + 2 \text{ [P]}_i$

2

$+ 2 \text{ [NADH]} + 2 \text{ H-} + 2 \text{ [ATP]} + 2 \text{ H}_2\text{O}$

D-Glucose   *Pyruvate*

From the reaction above, glycolysis yields two ATP molecules and two NADH molecules per molecule of glucose. The ten steps of glycolysis are below. However, the summary of the steps is:

1. The glucose molecule goes through several enzymatically regulated reactions and becomes a double-phosphorylated fructose molecule.

2. The 6-carbon fructose molecule is cleaved into two glyceraldehyde 3-phosphate molecules (3-carbon molecules with a phosphate group), abbreviated GA3P (or PGAL). This transformation requires 2 ATP.

3. Hydrogen and water are removed from the two PGAL, leaving two pyruvate molecules. This creates 2 NADH and 4 ATP (net ATP production is 2 ATP).

Glucose ATP ⟍ ⟩ Hexokinase ADP ↙	*Reaction 1, Phosphorylation — First ATP invested:* Glucose is converted to glucose-6-phosphate when ATP is hydrolyzed to ADP. The reaction is catalyzed by hexokinase.

P — OCH₂

OH

HO OH

OH

Glucose-6-phosphate

Phosphoglucoisomerase

$$P = \text{phosphate group} = -\!\!\!\{-O-\overset{\displaystyle O}{\underset{\displaystyle O^-}{\overset{\displaystyle \|}{P}}}-O^-$$

Reaction 2, Isomerization:

The enzyme phosphoglucoisomerase converts glucose-6-phosphate (an aldose) to its isomer fructose-6-phosphate (a ketose).

P — OCH₂ CH₂OH

HO

OH

OH

Fructose-6-phosphate

ATP

ADP Phosphofructokinase

P — OCH₂ CH₂O — P

HO

OH

OH

Fructose-1,6-bisphosphate

Reaction 3, Phosphorylation — the Second ATP invested:

A second ATP is hydrolyzed to ADP, and the phosphate is transferred to fructose-6-phosphate forming fructose-1,6-bisphosphate. "Bisphosphate" indicates two phosphates. This reaction is catalyzed by phosphofructokinase.

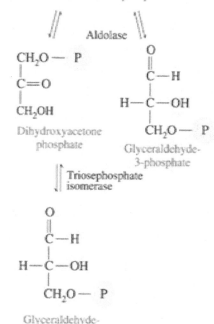

Aldolase

CH₂O — P

C=O

CH₂OH

Dihydroxyacetone phosphate

O
‖
C—H

H—C—OH

CH₂O — P

Glyceraldehyde-3-phosphate

Triosephosphate isomerase

O
‖
C—H

H—C—OH

CH₂O — P

Glyceraldehyde-3-phosphate

Reaction 4, Cleavage — Two trioses are formed:

Fructose-1,6-bisphosphate is cleaved into two triose phosphates (dihydroxyacetone phosphate and glyceraldehyde-3-phosphate), catalyzed by aldolase.

Reaction 5, Isomerization of a triose:

Triosephosphate isomerase converts one of the triose products (hydroxyacetone phosphate) to the other (glyceraldehyde-3-phosphate). Now, all 6 carbon atoms from glucose are in two identical 3-carbon triose phosphates.

O ‖ C—H | H—C—OH | CH₂O—P Glyceraldehyde-3-phosphate Pᵢ + NAD⁺ ⟩ Glyceraldehyde- NADH + H⁺ ⟩ 3-phosphate dehydrogenase O ‖ C—O—P | H—C—OH | CH₂O—P 1,3-Bisphosphoglycerate	*Reaction 6, First energy production yields NADH:* The aldehyde group of glyceraldehyde-3-phosphate is oxidized and phosphorylated by the enzyme glyceraldehyde-3-phosphate dehydrogenase. The coenzyme NAD^+ is reduced to the high-energy compound NADH (and H^+) in the process.
ADP ⟩ Phosphoglycerate ATP ⟩ kinase O ‖ C—O⁻ | H—C—OH | CH₂O—P 3-Phosphoglycerate	*Reaction 7, Next energy production yields ATP:* The energy-rich 1,3-bisphosphoglycerate drives the formation of ATP when phosphoglycerate kinase transfers one phosphate from 1,3-bisphosphoglycerate to ADP.
⟩ Phosphoglycerate ⟩ mutase O ‖ C—O⁻ | H—C—O—P | CH₂OH 2-Phosphoglycerate	*Reaction 8, Formation of 2-phosphoglycerate:* A phosphoglycerate mutase transfers the phosphate group from carbon 3 to carbon 2 to yield 2-phosphoglycerate.
⟩ Enolase H₂O ⟨ O ‖ C—O⁻ | C—O—P ‖ CH₂ Phosphoenolpyruvate	*Reaction 9, Removal of water makes a high-energy enol:* Enolase catalyzes the removal of water to yield phosphoenolpyruvate, a high-energy compound that transfers its phosphate in the next step.
ADP ⟩ Pyruvate ATP ⟩ kinase O ‖ C—O⁻ | C=O | CH₃ Pyruvate	*Reaction 10, Third energy production yields a second ATP:* ATP is generated in this final reaction when the phosphate from phosphoenolpyruvate is transferred. The pyruvate kinase catalyzes the reaction.

Glucose is not always immediately available, as it is often stored in skeletal muscles and the liver as the polysaccharide glycogen. Hormones control whether glucose enters the anabolic pathway to form glycogen or the catabolic pathway to undergo glycolysis to form pyruvate. If the signals for the catabolic pathway are initiated, cellular respiration begins (glycolysis → Krebs → electron transport chain).

With the help of glycogen phosphorylase, glycogen can be converted into glucose 6-phosphate, and enter the second step in the ten-step pathway of glycolysis. Fructose can undergo glycolysis. In the liver, fructose catabolism is unregulated and can produce excess products that become stored in the body as fat. In the muscles, fructose is converted to fructose-6-phosphate, entering glycolysis at step 3. In the liver, it is converted to the trioses used in step 5 of glycolysis.

Starch (amylose and amylopectin) begins to be digested in the mouth by enzymes in the saliva. The enzyme α-amylase hydrolyzes some of the α-glycosidic bonds in the starch molecules, producing glucose, disaccharide maltose, and oligosaccharides. Only monosaccharides are small enough to be transported from the gastrointestinal system into the bloodstream. To complete the digestion of starch, enzymes in the small intestine hydrolyze starch and disaccharides into monosaccharides.

The most important regulatory step in glycolysis is step 3. The enzyme phosphofructokinase, which catalyzes the phosphorylation of fructose-6-phosphate to fructose-1,6-bisphosphate, is tightly regulated by the cells because this step is irreversible and commits the pathway to glycolysis. ATP acts as an inhibitor of phosphofructokinase, so if cells have sufficient ATP levels, glycolysis slows down. If there is not much ATP, then glycolysis continues. The step after glycolysis depends on the type of respiration (i.e., aerobic or anaerobic).

Fermentation (anaerobic glycolysis)

Glycolysis produces two pyruvate molecules, two NADH molecules, and two ATP molecules. At this point, it is possible to either begin the aerobic part of the cellular respiration (Krebs Cycle) or to continue with anaerobic respiration. *Fermentation* is the next step in anaerobic (i.e., without oxygen) cellular respiration, which takes place in the cytoplasm. It predominantly occurs in yeast and bacteria but occurs in oxygen-starved muscle cells of vertebrates.

The goal of fermentation reactions is to oxidize NADH produced in glycolysis back to NAD^+ by reducing the pyruvate. Then, the NAD^+ molecules are used in another round of glycolysis to produce two more ATP and two more NADHs. It is clear that fermentation is a slow and inefficient method of making ATP, since only two ATP are produced in each cycle (compared to ~36 ATP that are produced in each complete cycle of aerobic respiration).

There are two types of fermentation. *Lactic acid fermentation* produces two molecules of lactate (or lactic acid as the acidic form of the molecule) and occurs in bacteria and some types of fungi. It is used in the production of some foods (e.g., yogurt). *Lactobacillus* is one genus of bacteria known for lactic acid fermentation. Lactic acid fermentation occurs in the muscle cells of humans and other mammals during demanding physical activities (e.g., sprinting), when the rate of demand for energy is high; lactic acid fermentation provides a quick burst of energy via ATP synthesis needed for the muscular activity.

Lactic acid is toxic to mammals; this is the "burn" felt when undergoing strenuous activity. When blood cannot remove all of the lactate from muscles, this decreases the pH, causing muscle fatigue. *Oxygen debt* is the oxygen that the body needed, but that was not delivered to the cell. At this point, oxygen is needed to restore ATP levels and rid the body of lactate (thus "repaying" the oxygen debt), which is one reason why a person might breathe harder after exercise. Recovery occurs after lactate is sent to the liver, where it is converted into pyruvate; some pyruvate is then respired or converted into glucose.

During lactic acid fermentation, the middle carbonyl (C=O) in pyruvate is reduced (i.e., hydrogen is added) to an OH group, and lactate is formed, as catalyzed by lactate dehydrogenase. The hydrogen (and energy) required for this reaction is supplied by NADH, producing NAD^+. NADH passes its electrons to pyruvate, and NAD^+ then returns to the glycolysis pathway to receive more electrons. In lactic acid fermentation, pyruvate is the final electron acceptor.

Alcoholic fermentation is the production of two ethanol molecules and two carbon dioxide molecules. Yeasts and some types of bacteria perform alcoholic fermentation. Some yeasts perform aerobic cellular respiration when oxygen is available, and only utilize alcoholic fermentation in anaerobic environments. However, many yeasts prefer fermentation, even if oxygen is available. These yeasts are utilized in the production of bread and alcoholic beverages. *Saccharomyces cerevisiae,* the yeast used in baking, consumes sugars in the dough of bread (converting glucose to pyruvate during glycolysis) and then reduces the pyruvate, creating carbon dioxide and ethanol as waste products. The carbon dioxide causes the dough to rise, and the ethanol evaporates when the bread is baked. *S. cerevisiae* is used in the production of beer. This yeast consumes the grain starches during glycolysis, and the carbon dioxide that is produced along with the ethanol during alcoholic fermentation is carbonation in the beer. Just as lactic acid is toxic to mammals, ethanol is toxic to the microorganisms that produce it.

Unlike lactic acid fermentation, where the pyruvate directly accepts electrons from NADH, there is an intermediate compound in alcoholic fermentation. The two pyruvate molecules are first converted into two molecules of acetaldehyde, catalyzed by pyruvate decarboxylase. The byproduct is carbon dioxide. The acetaldehyde is the final electron acceptor (from NADH). Then, the two acetaldehyde molecules are converted into two molecules of ethanol, catalyzed by the enzyme alcohol dehydrogenase.

$$CH_3-\overset{\overset{O}{\|}}{C}-\overset{\overset{O}{\|}}{C}-O^- \ + \ NADH \ + \ 2H^+ \ \longrightarrow \ CH_3-\overset{\overset{HO}{|}}{\underset{\underset{H}{|}}{C}}-H \ + \ CO_2 \ + \ NAD^+ \ + e^-$$

  Pyruvate   *Ethanol*

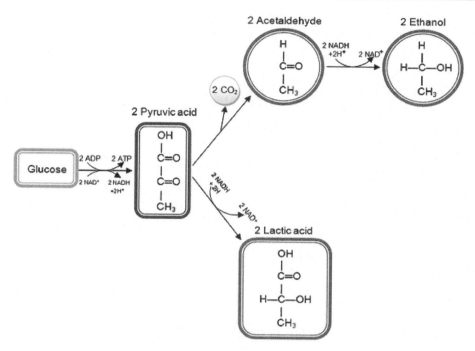

Alcoholic fermentation for ethanol production and lactic acid fermentation pathways to regenerate NAD⁺

Gluconeogenesis

In addition to catabolizing glucose, many organisms can produce it from non-carbohydrate substances, as *gluconeogenesis*. Gluconeogenesis occurs in the mitochondria and cytoplasm of many organisms, including animals, plants, fungi, and bacteria. In vertebrates, gluconeogenesis occurs primarily in the liver and, to a limited extent, in the kidneys. This process helps maintain the glucose concentration in blood.

A variety of non-carbohydrate carbon substrates, including pyruvate, glycerol, lactate, and certain amino acids, can be used as the starting molecule in gluconeogenesis. If the starting molecule is not pyruvate, the first step in gluconeogenesis is to convert the precursor (e.g., lactate) to pyruvate. It is possible for an amino acid precursor to enter the gluconeogenesis metabolic pathway at oxaloacetate or later in the pathway (for glycerol).

The steps of gluconeogenesis are displayed below, along with enzymes that catalyze each step. The entire metabolic pathway of gluconeogenesis uses two ATP, two GTPs and one NADH. As seen from the similarity of the two pathways, gluconeogenesis can be thought of as "reverse glycolysis." However, three of the enzymes used in the two pathways are different, so the energy cost of gluconeogenesis is not too great to be energetically favorable for the organism. Instead of using hexokinase, phosphofructokinase and pyruvate kinase (used in glycolysis), gluconeogenesis uses glucose-6-phosphatase, fructose-1,6-bisphosphatase and PEP carboxykinase/pyruvate carboxylase. This replaces three highly endergonic reactions in glycolysis with reactions that are exergonic and therefore favorable.

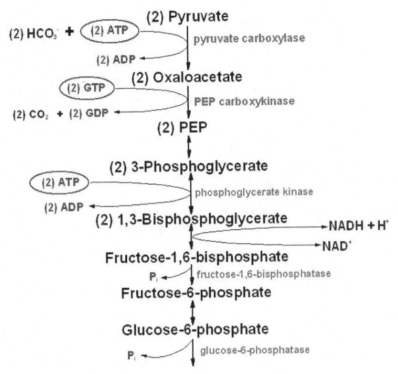

Glucose is produced via gluconeogenesis

In addition to building up glucose from carbon substrates (i.e., gluconeogenesis), organisms derive glucose by breaking down stored energy sources, such as the polysaccharides of starch and glycogen, with hormones controlling these processes. The hormone glucagon promotes glycogen degradation and inhibits glycolysis in the liver, which causes the glycolytic intermediates to be used in gluconeogenesis. Glucagon does this by inhibiting the phosphofructokinase (PFK) enzyme, since PFK is integral to glycolysis, to increase the concentration of glucose in the blood. Conversely, insulin inhibits gluconeogenesis by activating the PFK enzyme, so that the concentration of glucose in the blood is reduced.

Pentose phosphate pathway

The *pentose phosphate pathway* is a metabolic pathway that produces nicotinamide adenine dinucleotide phosphate (NADPH) and five-carbon sugars. In most organisms, it takes place in the cytosol, the exception being plants where it occurs in plastids. This pathway can be thought of as "parallel" to glycolysis, but rather than being catabolic, the pentose phosphate pathway is anabolic, since it synthesizes five-carbon pentose sugars.

The first phase of the pentose phosphate pathway is the oxidative phase, where two $NADP^+$ molecules are reduced to two NADPH molecules, and glucose-6-phosphate is oxidized to ribulose-5-phosphate. The production of NADPH in the pentose phosphate pathway is vital because it is the primary source of NADPH in non-photosynthetic organisms.

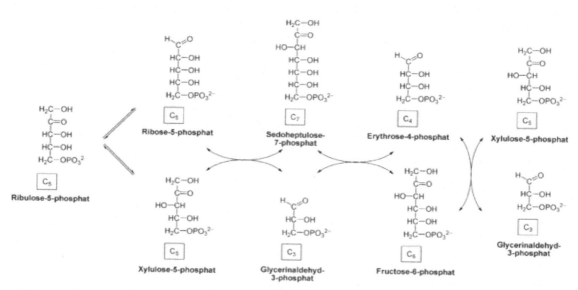

The oxidative phase of the pentose phosphate pathway

　　　1: glucose-6-phosphate

　　　2: 6-phosphogluconolactone

　　　3: 6-phosphogluconate

　　　4: ribulose 5-phosphate

The second phase of the pentose phosphate pathway with ribose-5-phosphate (nucleic acids) and erythrose-4-phosphate. Note the spelling of phosphate should include an -e ending

　　　The second phase of the pentose phosphate pathway is the non-oxidative phase, where other types of pentoses are synthesized from ribulose-5-phosphate. One of these pentoses is ribose-5-phosphate, which is used to synthesize nucleic acids. Another type of pentose is erythrose-4-

phosphate, which is used to generate aromatic amino acids. The pentose phosphate pathway is important, as it enables the biosynthesis of the biomolecules as the building blocks of life.

Net molecular and energetic results of respiration processes

From fermentation, the two ATP produced per glucose molecule is equivalent to 14.6 kcal. The ATP is produced via substrate-level phosphorylation as the direct enzymatic transfer of a phosphate to ADP, no extraneous carriers needed. Complete glucose breakdown to CO_2 and H_2O during aerobic cellular respiration represents a possible yield of 686 kcal of energy. Therefore, efficiency for fermentation is 14.6 / 686, or about 2.1%– much less efficient than aerobic respiration. Thus, the presence of an oxygen-rich atmosphere, which facilitated the evolution of aerobic respiration, is crucial in the complexity and diversification of life.

Krebs Cycle

Cellular respiration is the controlled release of energy from organic compounds in cells to produce ATP. Aerobic cellular respiration includes the anaerobic process of glycolysis, as well as the aerobic processes that occur in the mitochondria of eukaryotes. These three processes within the mitochondria are pyruvate decarboxylation, the Krebs cycle (the citric acid cycle or TCA) and oxidative phosphorylation via the electron transport chain (ETC).

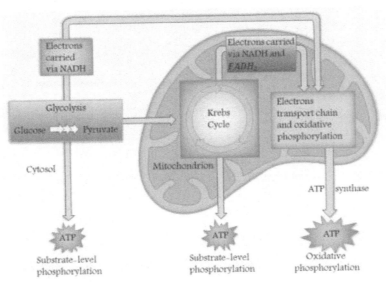

Cellular respiration outlining the sequence of glycolysis, Krebs cycle and electron transport chain

Acetyl-CoA production

Pyruvate decarboxylation follows glycolysis in aerobic respiration. Pyruvate decarboxylation is a link reaction as the intermediate step after glycolysis and before the Krebs cycle, thereby linking the two metabolic pathways. In eukaryotes, pyruvate decarboxylation occurs in the mitochondrial matrix (cytosol of mitochondria). In prokaryotes, it occurs in the cytoplasm and at the plasma membrane. It converts pyruvate into acetyl coenzyme A (acetyl-CoA) and is catalyzed by the pyruvate dehydrogenase complex (PDC), a complex of three enzymes. From two pyruvate molecules (one original molecule of glucose), two acetyl-CoA are created, and two NAD^+ are reduced to NADH. Additionally, two molecules of CO_2 are released.

Coenzyme A (CoA) is an important energy exchanger that contains adenosine, three phosphates, and a pantothenic acid-derived (vitamin B_5) portion. The two forms are acetyl-CoA (high energy) and CoA (low energy). Energy is released from acetyl-CoA when the C–S bond in the thioester group is hydrolyzed, producing an acetyl group and CoA.

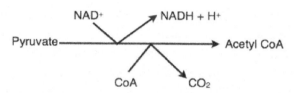

Coenzyme A exists in a low energy form as CoA or high energy form as acetyl-CoA

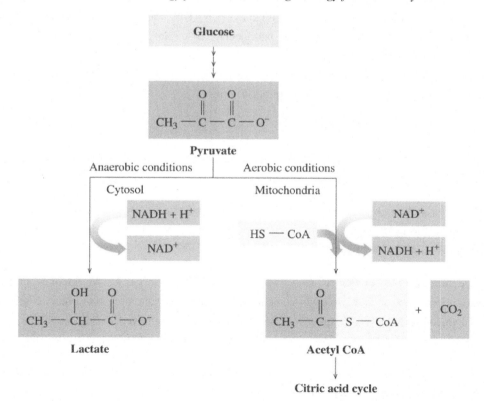

Pyruvate is the product of glycolysis that proceeds into the citric acid (Krebs) cycle or fermentation when oxygen is absent

In addition to glucose breakdown, there are other ways that acetyl-CoA can be produced for use in the Krebs cycle. Fat breaks down into glycerol and fatty acids. Glycerol is converted to PGAL, a metabolite in glycolysis. Beta oxidation is when the fatty acids can be broken down to generate acetyl-CoA. An 18-carbon fatty acid can be converted to nine acetyl-CoA molecules that each enter the Krebs cycle. Acetyl-CoA can be produced by degrading the carbon skeletons of ketogenic amino acids. However, regardless of whether acetyl-CoA is produced from carbohydrates, fats or proteins, the subsequent step for this molecule is the same: the acetyl-CoA proceeds to the Krebs cycle as the next phase of aerobic respiration.

Reactions of the Krebs cycle, substrates, and products

The *Krebs cycle* takes place in the fluid matrix of the cristae compartments of the mitochondria. The cycle is named after Sir Hans Krebs, who received the Nobel Prize for identifying these reactions. It is called the citric acid cycle or the tricarboxylic acid cycle (TCA) cycle because of the intermediate acids in the cycle. The Krebs cycle removes energy, carbon dioxide, and hydrogen from acetyl-CoA via enzyme-mediated reactions of organic acids. It begins by combining acetyl-CoA (two carbons) with oxaloacetate (four carbons), producing citric acid (six carbons). Citric acid undergoes several oxidations, decarboxylation, dehydrogenation, and hydration, yielding two CO_2, one GTP, three NADH, and one $FADH_2$ for each turn of the cycle. One original glucose is split into two pyruvates during glycolysis. Therefore, one glucose undergoes two turns of the Krebs cycle. The products for one glucose are four CO_2, two GTP, six NADH, and two $FADH_2$. GTP readily converts to ATP in the cell. The cycle regenerates oxaloacetate to begin again. The oxidations release energy, which is stored by the nucleotide carriers (NADH and $FADH_2$) when they accept the hydrogen electrons. This stored energy in NADH and $FADH_2$ is used by cytochromes in the electron transport chain to produce ATP by oxidative phosphorylation.

Steps of the Krebs cycle:

Reaction 1, Formation of Citrate: The acetyl group from acetyl-CoA (two carbons) combines with oxaloacetate (four carbons), forming citrate (six carbons) and CoA.

Reaction 2, Isomerization to Isocitrate: The OH and one of the H atoms are exchanged in citrate to form isocitrate. This rearrangement is necessary because isocitrate is oxidized in the next reaction.

Reaction 3, First Oxidative Decarboxylation (Release of CO_2): An alcohol undergoes oxidation (electron and two hydrogens are removed) to the ketone α-ketoglutarate and NAD^+ is reduced to NADH, accepting the proton and electrons removed during the oxidation. The six-carbon isocitrate is decarboxylated (release of CO_2) to the five-carbon α-ketoglutarate.

Reaction 4, Second Oxidative Decarboxylation: The thiol group of CoA is oxidized (loses an electron), and another NAD^+ is reduced (gains electron) to NADH. α-ketoglutarate (five carbons) is decarboxylated into a succinyl group (four carbons). The CoA is bonded to the succinyl group, thus producing succinyl CoA.

Reaction 5, Hydrolysis of Succinyl CoA: Succinyl CoA undergoes hydrolysis to yield succinate and CoA. The resulting energy produces the high-energy nucleotide GTP (guanosine triphosphate) from GDP and P_i. The GTP is converted to ATP in the cell.

Reaction 6, Dehydrogenation of Succinate: One hydrogen is eliminated from each of the two central carbons of succinate, forming a *trans* C=C bond, thus producing fumarate. The coenzyme FAD is reduced to $FADH_2$.

Reaction 7, Hydration of Fumarate: Water adds to the *trans* double bond of fumarate as H and OH to form malate.

Reaction 8, Oxidation of Malate: As in reaction 3, the secondary alcohol of malate is oxidized to a ketone, forming oxaloacetate and providing protons and electrons for reducing the coenzyme NAD^+ to NADH.

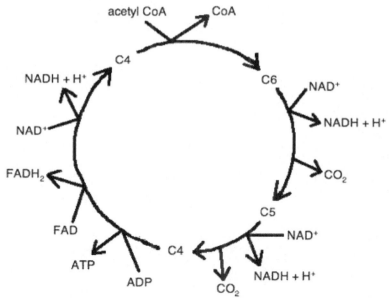

Krebs cycle showing the reactants and products. GTP is equivalent to ATP.

Net molecular and energetic results of respiration processes

The two pyruvate molecules produced during glycolysis are converted into two acetyl-CoA molecules in the link reaction. The acetyl-CoA then enters the Krebs cycle, which turns twice because two acetyl-CoA molecules enter the cycle per original glucose molecule. The final products of the Krebs cycle are oxaloacetic acid (to further drive the cycle), 2 ATP (converted from GTP), 6 NADH, 2 $FADH_2$, and 4 CO_2. The high-energy molecules NADH and $FADH_2$ are used in the final step of aerobic respiration, oxidative phosphorylation.

The equation for one turn of the Krebs cycle:

$$\text{Acetyl-CoA} + 3\ NAD^+ + FAD + GDP + P_i + 2\ H_2O \rightarrow 2\ CO_2 + 3\ NADH + 2\ H^+ + FADH_2 + CoA + GTP$$

The net production in cellular respiration from glycolysis through the Krebs cycle is 8 NADH, 2 $FADH_2$, 2 ATP, and 6 CO_2.

To summarize, the Krebs cycle degrades two carbon acetyl groups from acetyl CoA into CO_2, and the carbon dioxide is released in two reactions (as a waste product exhaled in animals). GTP is produced in one of the reactions and is converted into ATP. Hydrogen is removed in four reactions. NAD^+ accepts two electrons and the proton in three of these reactions, creating three NADH. FAD accepts two electrons and the proton in one reaction, creating one $FADH_2$.

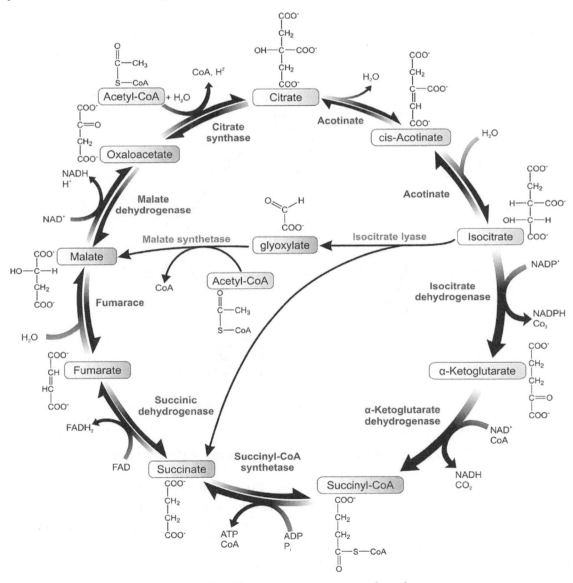

Krebs cycle with structures, enzymes and produces

Regulation of the cycle

The Krebs cycle must be carefully regulated so that the proper amount of ATP is generated. Although oxygen is not directly used in the Krebs cycle, the cycle can only occur under aerobic conditions, because FAD and NAD^+ can only be regenerated when oxygen is present. If oxygen is drastically reduced, respiration may drop to the point where it may ultimately lead to death. Low oxygen concentrations divert aerobic cellular respiration from the Krebs cycle to anaerobic fermentation.

The Krebs cycle is mainly regulated by substrate availability, product inhibition, and competitive feedback inhibition. During the cycle, ADP (a substrate) is converted to ATP, and a decreased amount of ADP reduces the rate of the cycle. This leads to an accumulation of NADH, decreasing the amount of NAD^+ available for the cycle. NADH can allosterically inhibit the enzymes in the Krebs cycle, including the pyruvate dehydrogenase complex (PDC), which catalyzes the link reaction (i.e., pyruvate decarboxylation to produce acetyl-CoA). Acetyl-CoA inhibits the PDC as an example of end product inhibition (or feedback inhibition) because acetyl-CoA is the product of the reaction catalyzed by the PDC.

Acetyl-CoA can enter the Krebs cycle from sources other than glycolysis, such as the breakdown of fatty acids. Calcium is an activator of the PDC and activates other dehydrogenase enzymes that catalyze reactions in the Krebs cycle. AMP (adenosine monophosphate) also activates the PDC. Citrate, the first compound in the Krebs cycle (formed from acetyl-CoA joining oxaloacetate), is used for feedback inhibition. Citrate inhibits phosphofructokinase, an important enzyme used in glycolysis. When high quantities of citrate accumulate, the rate of the respiration pathway is reduced to prevent the overproduction of ATP.

Oxidative Phosphorylation

Electron transport chain and oxidative phosphorylation, substrates and products and general features of the pathway

Oxidative phosphorylation is the next step in aerobic cell respiration, which includes the electron transport chain (ETC). This final step produces the majority of ATP. This is done by the oxidation (loss of electrons) by high-energy intermediates. NADH and $FADH_2$, which causes H^+ to be pumped into the intermembrane space (between inner and outer membranes) of the mitochondria. The reactions of the electron transport chain occur at the *cristae*, which are the folds of the inner mitochondrial membrane that increase the surface area for the electron transport chain. During this process, a gradient across the mitochondrial membrane is created to drive the production of ATP. Depending on cell conditions and if the prokaryotic or eukaryotic organism is present, about 32 to 34 molecules of ATP are produced by oxidative phosphorylation per glucose molecule.

The *electron transport chain* uses the NADH and $FADH_2$ from the Krebs cycle for a series of protein complexes that extract energy and pump protons across the inner mitochondrial membrane.

Energy is released as the hydrogen and electrons from the NAD^+, and FAD^+ carrier molecules flow into the system. When the electrons reach the end of the chain, they are accepted by oxygen (the final electron acceptor), and water is released when oxygen combines with the electrons and protons.

Chemiosmosis is the mechanism of ATP generation that occurs when energy is stored in the form of a proton concentration gradient across a membrane. ATP is produced by oxidative phosphorylation during the action of the electron transport chain. When H^+ molecules from the carrier molecules are transported from the matrix to the intermembrane space, a pH and electric charge gradient are created. The ATP synthase enzyme uses the potential energy on this gradient to create ATP when the protons flow through the ATPase channel in the membrane back into the matrix.

Note: A common question about this topic concerns pH changes from these processes. Remember that an *increase* in H^+ concentration means a *decrease* in pH.

Within the inner membrane of the mitochondria, there are four enzyme complexes (numbered I – IV). ATP synthase is sometimes referred to as complex V. The electron carriers' coenzyme Q (CoQ), and cytochrome c are not firmly attached to any complex and shuttle electrons between the complexes. CoQ (known as ubiquinone) is a fat-soluble carrier dissolved in the membrane. It can be fully reduced, fully oxidized or somewhere in between. This property enables it to perform in the electron transport chain. CoQ carries electrons from Complex I and Complex II to Complex III. Like all coenzymes, CoQ has a vitamin-like structure. *Cytochrome c* is a water-soluble protein that transfers electrons between Complex III and Complex IV. There is a central iron atom in this molecule; it is surrounded by heme protein. Cytochrome c is highly conserved across species, and it is often used to study evolutionary relationships between organisms.

The details of the electron transport chain are outlined below.

Complex I, NADH Dehydrogenase: NADH enters the electron transport chain. During its oxidation, two electrons and two protons are transferred to the electron transporter CoQ, reducing its two ketone groups to alcohols. NAD^+ is regenerated and returns to a catabolic pathway, as in the Krebs cycle. The reaction at Complex I is $NADH + H^+ + Q \rightarrow NAD^+ + QH$.

Complex II, Succinate Dehydrogenase: $FADH_2$ enters electron transport after the reduced nucleotide is produced in the conversion of succinate to fumarate in the Krebs cycle. Two electrons and two protons from $FADH_2$ are transferred to CoQ to yield QH_2. The reaction at Complex II is $FADH_2 + Q \rightarrow FAD + QH_2$.

Complex III, Coenzyme Q–Cytochrome c Reductase: The reduced coenzyme Q (QH_2) molecules are reoxidized to ubiquinone (Q). The electrons pass through a series of electron acceptors until they arrive at cytochrome c, which moves the electrons from Complex III to Complex IV.

Complex IV, Cytochrome c Oxidase: Single electrons are transferred from cytochrome c through another set of electron acceptors to combine with hydrogen ions and oxygen as the final electron accepts to form water. The reaction is $4 H^+ + 4 e^- + O_2 \rightarrow 2 H_2O$.

Three of the complexes (I, III and IV) span the inner membrane and pump protons out of the matrix and into the intermembrane space as electrons are shuttled through the complexes. The only complex that does not pump protons is Complex II.

The formation of the chemiosmotic (proton) gradient across the inner mitochondrial membrane provides the energy for ATP synthesis. Protons move back into the matrix through the transmembrane ATP synthase enzyme, and the resulting release of potential energy from the chemiosmotic gradient drives the synthesis of ATP by oxidative phosphorylation as protons and electrons join $\frac{1}{2}O_2$.

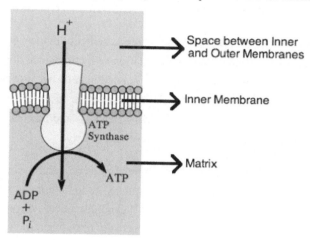

Oxidative phosphorylation with protons passing through the ATP synthase

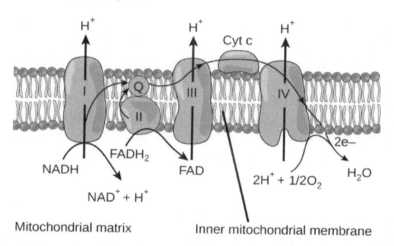

*Electron transport chain showing the membrane-bound cytochromes
as electrons move down their reduction potential towards O_2 as the final electron acceptor*

Electron transfer in mitochondria:
NADH, NADPH, flavoproteins, and cytochromes

The coenzymes nicotinamide adenine dinucleotide (NAD^+) and flavin adenine dinucleotide (FAD) are energy-transferring compounds that exist in different redox states. NAD^+ (low-energy oxidized form) accepts two electrons and proton to become NADH (high-energy reduced form). FAD (low-energy oxidized form) accepts two electrons and two protons to become $FADH_2$ (high-energy reduced form). The active end of each coenzyme contains a vitamin component. Nicotinamide is derived from the niacin (B_3) vitamin, and riboflavin (B_2) vitamin is in FAD. The FAD is, therefore, a type of flavoprotein.

Electrons received by NAD^+ and FAD can then be carried to cytochrome protein of the electron transport chain. NAD^+ and FAD are coenzymes of oxidation-reduction since they both accept and donate electrons. Only a small amount of these coenzymes are needed in cells because the exchange of electrons regenerates each molecule. For example, once NADH delivers electrons to the electron transport chain, it becomes oxidized to NAD^+ and then can be reduced with electrons and protons. Recycling NAD^+, FAD, and ADP eliminate the need to synthesize them *de novo* continuously.

NAD^+ is converted into nicotinamide adenine dinucleotide phosphate ($NADP^+$). $NADP^+$ is similar in structure to NAD^+ but has an additional phosphate group, and it can be reduced to NADPH. In non-photosynthetic organisms, the production of $NADP^+$ generally occurs through the pentose phosphate pathway. In plants, however, it is produced during the last step of the electron transport chain (across the thylakoid membrane) in the light reactions of photosynthesis. In general, NAD^+, $NADP^+$, and FAD are reduced during catabolic processes; their reduced forms (NADH, NADPH, and $FADH_2$) are oxidized during anabolic processes.

For each NADH formed within the mitochondrion (during the link reaction and the Krebs cycle), three ATP are produced. For each $FADH_2$ formed by the Krebs cycle, two ATP are produced. This is because $FADH_2$ delivers electrons after NADH (Complex II vs. Complex I). Therefore, NADH yields more energy than $FADH_2$, because more H^+ is pumped across the membrane per NADH (3:2 ratio). However, an exception occurs for NADH formed outside the mitochondrion (by glycolysis) in the cytoplasm. The NADH formed by glycolysis is the same molecule as the NADH formed in the mitochondria, but yields two ATP (equivalent to ATP production by FADH during the Krebs cycle) instead of three ATP. One ATP is consumed to transport the NADH of glycolysis from the cytoplasm into the mitochondrion, resulting in a net gain of two ATP.

ATP synthase and chemiosmotic coupling; proton motive force

The movement of protons from the mitochondrial matrix into the intermembrane space creates a concentration gradient. The *proton motive force* is the energy used to pump these protons across the inner membrane, and it comes from the energy released by the electrons passing through the electron transport chain. Remember that only three out of four enzyme complexes in the electron transport chain can pump protons into the intermembrane space: complexes I, III and IV. The protons create a concentration gradient (lowering the pH) as they move into the intermembrane space.

Chemiosmosis is ATP production tied to an electrochemical (H^+) gradient across a membrane. There is now a high concentration of protons in the intermembrane space and a low concentration of protons in the mitochondrial matrix. The protons then move down the concentration gradient from the intermembrane space back into the matrix. However, the inner membrane is impervious to protons, and they can move back into the matrix via the ATP synthase, an enzyme embedded in the inner membrane. ATP synthase complexes span the membrane and are channel proteins that serve as enzymes for ATP synthesis.

As the protons are transported back into the matrix through the channels of ATP synthase, they release energy, which is then used by ATP synthase to convert ADP into ATP by the addition of an inorganic phosphate (P_i). This process is oxidative phosphorylation, because the electrons come from previous oxidation reactions of cell respiration, and the ATP synthase uses $\frac{1}{2}O_2$ to accept the electrons to catalyze the phosphorylation of ADP into ATP. Once formed, ATP molecules diffuse out of the mitochondrial matrix through channel proteins and become available for use by the cell.

Tests confirmed the chemiosmotic nature of ATP synthesis with respiratory poisons (i.e., poisons that inhibit ATP synthesis), where these poisons caused the H^+ concentration gradient to increase. British biochemist Peter Mitchell received the 1978 Nobel Prize for his chemiosmotic theory of ATP production, which occurs in the membranes of both mitochondria in animal cells and chloroplasts in plant cells.

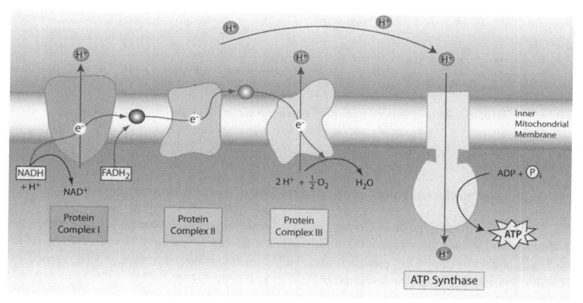

Inner membrane contains cytochromes that pass electrons. ATP synthase produces ATP by oxidative phosphorylation when protons pass down the concentration gradient into the matrix

The exergonic flow of electrons is generally coupled with the endergonic pumping of protons across the cristae membrane of the mitochondria. When electron flow and proton transport are uncoupled, the energy in the proton gradient is released as heat. This assists in maintaining body temperature through *thermogenesis*, which occurs primarily in warm-blooded animals.

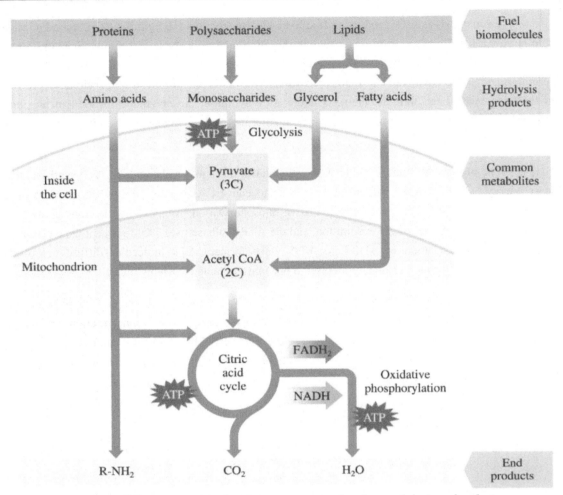

Aerobic cellular respiration showing precursor molecules used during glycolysis, citric acid cycle (Krebs) and electron transport chain

Net molecular and energetic results of respiration processes

During aerobic cellular respiration, glucose breakdown provides energy for a hydrogen ion gradient across the inner membrane of the mitochondria, which couples proton flow with ATP formation. At the end of cellular respiration, glucose is oxidized to carbon dioxide and water, and ATP is produced. The overall equation for aerobic respiration is:

$$C_6H_{12}O_6 + 6\ O_2 \rightarrow 6\ CO_2 + 6\ H_2O + energy$$

It is the reverse of the equation for photosynthesis, which uses carbon dioxide, water, and energy from sunlight to create glucose and oxygen.

The primary purpose of cellular respiration is the production of ATP for the cell's energy needs. Glycolysis produces 2 ATP, the Krebs cycle produces 2 ATP, and the electron transport chain (i.e., oxidative phosphorylation) produces approximately 30 to 32 ATP. In total, aerobic cellular respiration results in the production of approximately 34 to 36 ATP.

The energy yield can be calculated using the energy content of glucose and ATP. Glucose contains 686 kcal/molecule, while ATP contains 7.5 kcal/molecule.

$$7.5 \times 34 = 255 \text{ kcal/mol for all ATP produced}$$

$$255 / 686 = 37.2\% \text{ energy recovered from aerobic respiration}$$

Therefore, the ~36 resulting ATP molecules represent approximately 37% of the energy in one molecule of glucose. Aerobic respiration is almost twenty times as efficient as anaerobic respiration, which has a net of 2 ATP and efficiency of approximately 2%.

Overall, cellular respiration is an exergonic process ($\Delta G = -686$ kcal/mole).

Regulation of oxidative phosphorylation

The movement of electrons through the electron transport chain is regulated by the proton motive force (PMF). When the magnitude of the PMF is high (i.e., a high H^+ concentration gradient), the rate of electron flow through the electron transport chain is lower. When the magnitude of the PMF is low, the rate of electron flow is greater to increase the H^+ concentration gradient.

The energy demands of the cell directly influence the regulation of oxidative phosphorylation. During resting conditions, the demand for ATP is low, and there is a low rate of proton movement from the intermembrane space through the ATP synthase in the inner mitochondrial membrane. However, when the demand for energy is high (e.g., during vigorous activity), protons flow more quickly through the ATP synthase due to an increased concentration of ADP in the mitochondria. The level of ADP is the primary factor in determining the rate of oxidative phosphorylation. If protons are not flowing through the ATP synthases and thus allowing ADP to be phosphorylated into ATP, electrons do not flow through the cytochromes of the electron transport chain to oxygen, the terminal electron acceptor.

Oxygen is another important regulator of oxidative phosphorylation. If there is insufficient oxygen, then electrons cannot pass through the electron transport chain and NADH, and $FADH_2$ cannot be oxidized to NAD^+ and FAD. Eventually, NAD^+ and FAD are depleted, and the link reaction and Krebs cycle are suspended. By measuring oxygen consumption, the rate of the electron transport chain can be calculated (i.e., elevated oxygen consumption indicates faster electron transport). The rate of electron transport is the cellular respiratory rate.

Several compounds inhibit oxidative phosphorylation by preventing electron transport. One example is antimycin A, which inhibits cytochrome b in the coenzyme Q–cytochrome c Reductase (Complex III). Other inhibitors, such as Rotenone and Amytal, block the use of NADH as a substrate in NADH dehydrogenase (Complex I). The inhibitors cyanide, carbon monoxide, and azide prevent electron flow in Cytochrome c Oxidase (Complex IV).

There are compounds that can inhibit oxidative phosphorylation by acting as uncoupling agents. These include 2,4-dinitrophenol (DNP) and pentachlorophenol (acidic aromatic compounds). The protons in the intermembrane space of the mitochondria pass through the membrane with the protons attached, and then the uncoupling agent releases the protein once inside the mitochondrial matrix. This dissipates the proton motive force and allows protons to bypass the ATP synthases embedded in the membrane, preventing the production of ATP. This uncoupling of oxidative phosphorylation is a way for organisms to generate heat, (i.e., thermogenesis).

Additionally, the enzyme inhibitors oligomycin and dicyclohexylcarbodiimide can directly inhibit the ATP synthase by preventing the influx of protons.

Mitochondria, apoptosis and oxidative stress

Mitochondria are the organelles involved in energy production. A mitochondrion (powerhouse of the cell) has a double membrane with an intermembrane space between the outer and inner membrane. The matrix is a watery substance that contains ribosomes and many enzymes. Enzymes in the matrix and inner membrane catalyze the oxidation of carbohydrates, fats, and amino acids. These enzymes are vital for the link reaction and the Krebs cycle.

The inner membrane is where the electron transport chain and ATP synthase are located, and where oxidative phosphorylation takes place. The space between the inner and outer membranes is a small volume space which protons are pumped into. Due to its small volume, a high concentration gradient can be reached quickly, which is vital for chemiosmosis. The outer membrane is a membrane that separates the contents of the mitochondrion from the rest of the cell, creating the ideal environment for cell respiration. The cristae are tubular projections of the inner membrane that increase the surface area for oxidative phosphorylation.

Three parts of aerobic cellular respiration take place in the mitochondria: the link reaction, the Krebs cycle, and oxidative phosphorylation. During these processes, the pyruvate from glycolysis is broken down completely to CO_2 and H_2O, which is why aerobic respiration is thought of as glucose breakdown. CO_2 and ATP are transported out of the mitochondria into the cytoplasm. CO_2 enters the bloodstream for transport to the lungs and is expelled during expiration. H_2O can remain or enter the blood and be excreted by the kidneys.

Apoptosis is the programmed cell death that occurs in multicellular organisms. It occurs in both developing tissue (e.g., in embryonic development where parts of tissues are no longer needed) and adult tissue (where cell death balances cell division). When a cell dies by apoptosis, its cytoskeleton collapses, and its nucleus is fragmented, and its chromatin irreversibly condenses. After apoptosis, certain neighboring cells (e.g., macrophages) recognize alterations in the surface of the dead cell and phagocytize the cell before its contents are leaked, which allows for efficient recycling of the cell's biomolecules.

The machinery for apoptosis depends on an intracellular proteolytic cascade (which involves the breaking of peptide bonds within proteins). Target proteins (both enzymes and structural proteins) are cleaved by proteases, leading to the activation of more proteases. Eventually, this leads to the degradation of all the cell's organelles. However, before this cascade begins, signals must cause the initiation of the apoptosis pathway. Apoptosis usually begins in response to stress; nuclear receptors recognize many stress factors including nutrient deprivation, heat, viral infection, radiation and lack of oxygen. These stress factors lead to the release of intracellular apoptotic signals that cause proteins to initiate the apoptotic pathway.

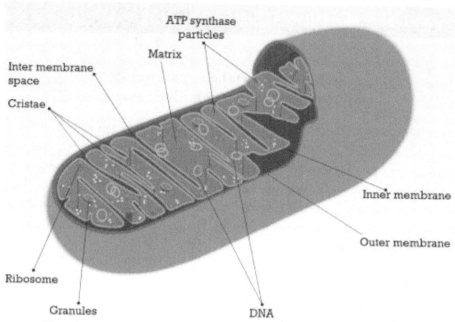

Mitochondrion with cellular compartments and select biomolecules

Mitochondria are the target of some apoptotic proteins. These proteins may create pores in the mitochondrial membrane and cause swelling, or they may change the permeability of the mitochondrial membrane. When the mitochondrial membrane is made more permeable, apoptotic effectors can leak out and activate the proteins of the apoptotic pathway.

Oxidative stress describes an imbalance between reactive oxygen species (i.e., free radicals) and the organism's ability to detoxify these harmful effects. *Free radicals* are atoms or molecules that have an unpaired electron, and thus an odd number of electrons. Free radicals, such as the hydroxyl radical (HO•), are highly reactive, and if they react with important cellular components such as the cell membrane or DNA, so the cell can be severely damaged. The extent of the damage can be extensive, because when a free radical abstracts an electron from a cell component, the cell component must then take an electron from another cell component, leading to a chain of free radical reactions.

The creation of a free radical could disrupt cellular signaling if the original molecule were a cellular messenger. Oxidative stress is involved in many diseases including cancer, Alzheimer's disease, and Parkinson's disease. However, oxidative stress can be useful, (e.g., when it attacks pathogens rather than the host's cells).

Antioxidants are molecules that donate electrons to free radicals without becoming destabilized, and they are the primary method by which the reactive oxygen species are counteracted. Oxidative stress can be caused by a decrease in antioxidant levels, an increase in free radical levels, or both. Sometimes, oxidative stress can trigger apoptosis or necrosis (cell injury leading to premature cell death).

Principles of Metabolic Regulation

Regulation of metabolic pathways: maintenance of a dynamic steady state

Metabolism refers to the biochemical reactions that occur in the cells of living organisms, including the reactions that allow the cell to grow, reproduce and function. Because organisms are usually in environments that are constantly changing, metabolism must be tightly regulated so that the internal conditions of the cell remain constant; this property is *homeostasis* and can be thought of as a "dynamic steady state." Homeostatic processes act at multiple levels, from the cell or the tissue to the whole organism. Examples include the regulation of pH, blood glucose and internal body temperature.

Both catabolic (breaking down) and anabolic (building up) pathways are extensively regulated. During the metabolic reactions that occur in steady-state conditions, the substrate is being converted to the product as efficiently as possible.

Regulation of glycolysis and gluconeogenesis

Glycolysis, the breakdown of glucose into two pyruvate molecules, is universal to biological organisms. Glycolysis is tightly regulated. The three most important steps in the pathway—the reactions catalyzed by phosphofructokinase (phosphorylation of fructose-6-phosphate), hexokinase (phosphorylation of glucose), and pyruvate kinase (phosphate transfer from phosphoenolpyruvate to ADP)—are the most tightly regulated because these three steps are essentially irreversible due to a large $-\Delta G$. The production of these three enzymes is regulated by hormones that control the rate of transcription (DNA→mRNA). For more immediate action, allosteric effectors bind to these enzymes reversibly, or the enzymes can be covalently modified to inhibit activity permanently.

Phosphofructokinase is the most important enzyme for glycolysis regulation. High ATP levels allosterically inhibit phosphofructokinase, meaning that the ATP binds to an allosteric site distinct from the catalytic active site. ATP inhibits the key reaction in glycolysis because if there are high levels of ATP, there is no need for more to be produced. Citrate, one of the early intermediates in the Krebs cycle, inhibits phosphofructokinase. It does this by enhancing ATP's inhibitory effect. When there are high levels of citrate, there is no need for additional glucose breakdown.

Some compounds activate phosphofructokinase. For example, the activator AMP reverses ATP's inhibitory action and allows the catalytic activity of phosphofructokinase to resume. Fructose 2,6-bisphosphate, another allosteric activator, reduces ATP's inhibitory effect and increases the enzyme's affinity for fructose-6-phosphate. The concentration of fructose 2,6-bisphosphate is controlled by two enzymes: phosphofructokinase 2 (different than phosphofructokinase) and fructose bisphosphate 2.

These *bifunctional enzymes* are contained in a single polypeptide chain. The activity of this bifunctional enzyme is controlled by the phosphorylation of a serine residue in the polypeptide chain.

The phosphorylation of this residue, which occurs when glucose levels are low, leads to a reduction in fructose 2,6-biphosphate production and thus a reduction in the rate of glycolysis, thereby conserving glucose. However, when glucose is abundant, dephosphorylation of the serine residue occurs, leading to an increase in fructose 2,6-biphosphate production and thus an increase in the rate of glycolysis. This regulation modulates the levels of glucose, so it is available when needed without being degraded.

Hexokinase, another important regulatory enzyme in glycolysis, is inhibited by its product, glucose-6-phosphate. Since glucose-6-phosphate is in equilibrium with fructose 6-phosphate (the reactant of phosphofructokinase), the inhibition of phosphofructokinase results in the inhibition of hexokinase. This further suggests that phosphofructokinase is the key enzyme in glycolysis regulation.

Pyruvate kinase controls outflow from the glycolysis pathway and is thus important in regulation. It is activated by fructose 1,6-bisphosphate, as an example of *feedforward stimulation*, since fructose 1,6-bisphosphate is an earlier intermediate in the pathway that enables pyruvate kinase to keep up with the oncoming flux of intermediates. Pyruvate kinase is allosterically inhibited by ATP (which signals that energy is abundant) and alanine (which signals that building blocks are abundant).

Transcription of the enzymes used in glycolysis is controlled by hormones, which allows for coordination between different organs and tissues. These hormones regulate the rate of gluconeogenesis (glucose from non-carbohydrate substrates). Gluconeogenesis and glycolysis are "reciprocally regulated," meaning that one pathway is active while the other is inactive. This prevents the occurrence of a *futile cycle*, where ATP is hydrolyzed without useful metabolic work, resulting in chemical energy dissipating as heat and being wasted.

As discussed earlier, gluconeogenesis uses the enzymes glucose-6-phosphatase, fructose-1,6-bisphosphatase and PEP carboxykinase/pyruvate carboxylase, replacing three strongly endergonic reactions with reactions that are more exergonic (and therefore more favorable). These replacements allow for the reciprocal regulation of these two pathways; i.e., the same compounds that activate one enzyme in glycolysis may inhibit the replacement enzyme in gluconeogenesis.

For example, phosphofructokinase in glycolysis is activated by AMP and inhibited by ATP and citrate; fructose-1,6-bisphosphatase, however, is inhibited by AMP and activated by ATP and citrate. Since gluconeogenesis consumes ATP while glycolysis produces ATP, high levels of ATP promote gluconeogenesis (by activation of fructose-1,6,-bisphosphate) and reduce the rate of glycolysis (by inhibition of phosphofructokinase).

Acetyl-CoA is an important activator of gluconeogenesis. It activates pyruvate carboxylase to catalyze the first reaction in gluconeogenesis. Conversely, it inhibits the enzyme pyruvate kinase in glycolysis, demonstrating the reciprocal control of the two pathways.

Metabolism of glycogen

Glycogen is a large, branched polysaccharide made of glucose residues, linked primarily by α-1,4-glycosidic bonds, although one of every ten bonds is an α-1,6-glycosidic bond.

Glycogen with α-1,4-glycosidic and α-1,6-glycosidic bonds

It is the main form of stored glucose in the body (mostly stored in the liver and skeletal muscles) because it can easily be broken down when energy is needed, allowing monomers of glucose to be released when levels are low. When glycogen is metabolized in the liver, it can be slowly released into the bloodstream (e.g., between meals). When it is metabolized in muscle, it can provide a quick burst of glucose for energy production of ATP for aerobic or anaerobic activity. Oxygen is not required for glycogen breakdown.

Glycogenolysis is the breakdown of glycogen in the following reaction:

$$\text{glycogen}_{(n \text{ residues})} + P_i \rightleftarrows \text{glycogen}_{(n-1 \text{ residues})} + \text{glucose 1-phosphate}$$

There are three main steps in glycogen metabolism. The first step is the release of a glucose-1-phosphate molecule from glycogen, which is catalyzed by glycogen phosphorylase. It is a phosphorolysis reaction (breakdown by the addition of a phosphate molecule). Glycogen phosphorylase only acts on non-reducing ends of a glycogen polymer that are five or more glucose residues away from a branch point (see figure above).

In the second step, the glycogen is remodeled for further breakdown of the polysaccharide. When there are only four glucose residues before a branching point, a glycogen debranching enzyme (a type of transferase) transfers three of the remaining four units of glucose (a trisaccharide) from the 1,6 branch to an adjacent 1,4 branch. This exposes the 1,6 branching point for hydrolysis by α-1,6-glucosidase. The final glucose molecule is removed, and the branch is eliminated, thus allowing the phosphorolysis of glycogen by glycogen phosphorylase to continue.

The third step of glycogen metabolism is the conversion of glucose-1-phosphate to glucose-6-phosphate by the enzyme phosphoglucomutase, which uses a phosphoserine to remove a phosphoryl group.

At this point, there are three possibilities for the glucose-6-phosphate molecule resulting from the glycogen breakdown: (1) enter the glycolysis pathway for the production of ATP, (2) enter the pentose phosphate pathway for the production of NADPH and ribose derivatives, or (3) become dephosphorylated and converted into free glucose for release into the bloodstream by the enzyme glucose 6-phosphatase (i.e., the final step in gluconeogenesis).

Regulation of glycogen synthesis and breakdown: allosteric and hormonal control

The regulation of glycogen anabolism and catabolism is complex. If both pathways were to occur at the same time, it would be a futile cycle, so the simultaneous action of these pathways must be avoided. Like glucose synthesis or breakdown, glycogen synthesis and breakdown are reciprocally regulated. Many of the enzymes involved in these processes are controlled allosterically, and they respond to metabolites that signal the cell's energy needs. The enzyme activity is adjusted so that the demands for energy are met. Glycogen phosphorylase is inhibited by high-energy molecules such as ATP, glucose-6-phosphate, and glucose, and is activated by low-energy molecules such as AMP.

In addition to allosteric control, hormones regulate glycogen synthesis and breakdown. Hormones trigger a cAMP (cyclic AMP) cascade, which is a signaling pathway that can act through the enzyme protein kinase A (PKA). PKA can activate phosphorylase kinase and deactivate glycogen synthase (through the addition of a phosphoryl group), which prevents glycogen synthesis and catabolism from occurring simultaneously. Conversely, protein phosphatase 1 reverses the regulatory effects of PKA. Insulin activates protein phosphatase 1, which stimulates the synthesis of glycogen. Hormonally-stimulated cascades allow the rate of glycogen synthesis or catabolism to be adjusted by the needs of both cells and the entire organism.

Analysis of metabolic control

Metabolic *regulation* refers to the changes that signaling molecules cause on the activity of enzymes that catalyze reactions in metabolic pathways, while metabolic *control* refers to how these changes in enzyme activity control the flux (i.e., the overall rate) of the pathway. These two concepts are linked.

There are multiple levels of metabolic control. *Intrinsic control* occurs when the reactions in metabolic pathways are self-regulated, meaning that they respond to changes in the levels of products and substrates. For example, the product of glycogen phosphorylase is glucose-6-phosphate, and it is inhibited by glucose-6-phosphate (feedback inhibition or negative feedback). Positive feedback occurs when the product of a reaction amplifies the reaction rate (e.g., blood clotting).

Extrinsic control occurs when a cell in a multicellular organism alters its metabolism in response to a signal sent from another cell. These signals are usually hormones or growth factors, and they are detected by receptors on the surface of the affected cell, which then leads to a transmission of signals within the cell (second messenger system). An example of extrinsic control is the hormone-triggered phosphorylation cascade that reduces the rate of glycogen synthesis (described earlier).

Extrinsic control systems allow for maintenance of homeostasis at the whole-organism level. There are three components to homeostatic control mechanisms that enable this to happen. The *receptor* is the first component, which senses environmental stimuli. The receptor sends a signal to the second component, the *control center* (e.g., brain). Then, the control center sends a signal to the third component, the *effector*, which responds to the stimuli.

Metabolism of Fatty Acids and Proteins

When glucose supply is low, the body uses other energy sources in the priority order of other carbohydrates, fats, and proteins. First, these molecules are converted to glucose or glucose intermediates; then they are degraded in glycolysis or by the Krebs cycle.

Monosaccharides and amino acids travel directly through the bloodstream to the cells for absorption. Triglycerides are packaged into lipoproteins (chylomicrons) for delivery.

Description of fatty acids

Fatty acids are a family of molecules classified as lipids, which are usually ingested and stored as triglycerides. A *triglyceride* (triacylglycerol) consists of a 3-carbon glycerol backbone with three fatty acids connected to it, as displayed below.

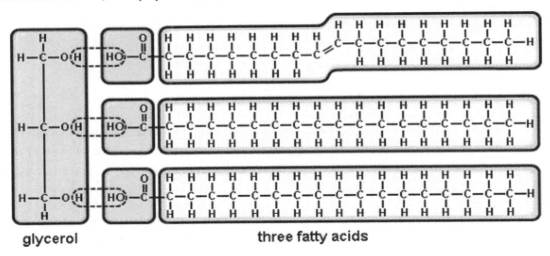

Glycerol and three fatty acids are each joined by a phosphodiester bond

Saturated fats have hydrogens that occupy all possible bonds and have only single bonds. *Unsaturated* fats do not have hydrogen atoms at all positions, and there is at least one double bond.

Fatty acids are an important source of energy, because they are reduced and anhydrous (hydrophobic), which allows for a higher energy yield. Carbohydrates are more hydrated, so the amount of energy that can be stored per unit mass is much lower than fatty acids. For this reason, it is ideal for an organism to store energy as fat in adipose tissue when a large quantity of energy needs to be kept in reserve for later use (e.g., hibernating bear).

trans-Oleic acid

cis-Oleic acid

Cholesterol is a four fused ring structure   *Cis and trans unsaturated fatty acids*

Digestion, mobilization, and transport of fats

The pancreatic enzyme lipase breaks down triglycerides into free fatty acids and monoglycerides (3-carbon glycerol and fatty acid) by hydrolyzing the ester bond, because triglycerides cannot be absorbed by the duodenum (first segment of the small intestine). Pancreatic lipase forms a complex with the protein colipase, which is essential for its activity, and this complex works at a water-fat interface because dietary fats are nonpolar molecules.

Bile is excreted from the gallbladder into the stomach to assist in digestion. Bile contains *bile salts*, which are *amphipathic* (i.e., having both hydrophilic and hydrophobic parts) to orient their nonpolar face toward the dietary fats and their polar face toward the water, forming *micelles*, which are spherical lipid droplets. *Emulsification* is the breaking of larger nonpolar globules into micelles. The micelles move the dietary fats closer to the intestinal cell wall so that cholesterol can be absorbed across the intestinal wall and triglycerides can be hydrolyzed.

Once across the intestinal wall, free fatty acids and monoglycerides are reassembled as triglycerides, while the cholesterol is linked to another free fatty acid, forming a cholesterol ester. These are repackaged as chylomicrons, lipoproteins that transport triglycerides in the bloodstream and then the tissues, where they are used for energy production or are stored. The liver is an important organ for fatty acid metabolism.

Adipocytes are cells where the majority of body fat is stored, as the entire cytoplasm is filled with a single fat droplet. Adipocytes synthesize and store triglycerides during food uptake and cluster to form adipose tissue, most of which is located beneath the skin and around internal organs.

Triacylglycerides $\xrightarrow[\substack{\text{lumen of} \\ \text{small} \\ \text{intestine}}]{\text{lipases}}$ Monoacylglycerides and fatty acids (transported into enterocytes) $\longrightarrow$ Triacylglycerides $\xrightarrow[\substack{\text{(packaged with} \\ \text{cholesterol)}}]{\text{chylomicrons}}$ adipose tissue (storage)

Oxidation of fatty acids (saturated and unsaturated fats)

Fatty acids can be degraded to produce ATP when glucose supplies are low. For fatty acids to undergo catabolism, they first must be activated. Catalysis by the enzyme fatty acyl-CoA synthase links fatty acids to coenzyme A, forming activated fatty acyl-CoA. Then, the fatty acyl-CoA is transported to the mitochondrial matrix with the help of the enzymes: carnitine acyltransferase I (which conjugates fatty acyl-CoA to carnitine), carnitine-acylcarnitine translocase (which shuttles carnitine and acylcarnitine inside the mitochondria), and carnitine acyltransferase II (which liberates the carnitine and turns the molecule back into fatty acyl-CoA). However, this applies to long-chain fatty acids; short-chain fatty acids can diffuse directly into the mitochondrial membrane with no need for the carnitine carrier system.

The molecule then undergoes β oxidation (beta carbon of the fatty acid is oxidized to a carbonyl group). In this process, fatty acyl-CoA undergoes oxidation by repeatedly cleaving two-carbon molecules, each time producing a new fatty acyl-CoA that is two carbons shorter, along with one molecule of acetyl-CoA. The details of β oxidation, including the enzymes that catalyze each step of the process, are in the figure below.

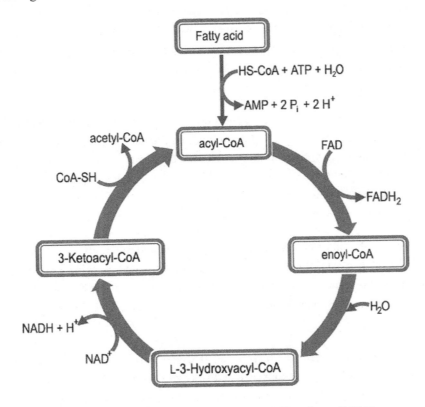

Each turn of β oxidation produces one NADH and one FADH₂,
which are later used in the electron transport chain for the production of ATP

β oxidation works best for even-numbered saturated fatty acids due to the repetitive cleaving of two carbons from the chain. Saturated fatty acids produce one NADH and one FADH₂ for every cut into two carbons.

The acetyl-CoA produced during β oxidation enters the Krebs cycle, and the typical progression of aerobic cellular respiration occurs. A short eight-carbon fatty acid can produce four acetyl-CoAs. Each acetyl-CoA yields 12 ATP (3 NADP, 1 FADH$_2$, and 1 ATP). Therefore, this eight-carbon fatty acid nets 48 ATP, and fat with three chains of this length produces 144 ATP, illustrating why fats are such a good source of energy. Fats provide 9 calories per gram, while carbohydrates and proteins each provide 4 calories per gram.

For odd-numbered saturated fatty acids in the lipids of some marine organisms and plants, the products of β oxidation are propionyl-CoA and acetyl-CoA. With the help of three enzymes and the vitamins cobalamin and biotin, the propionyl-CoA undergoes carboxylation and molecular rearrangement to form succinyl-CoA. The succinyl-CoA then enters the Krebs cycle.

Unsaturated fatty acids require additional enzymes for β oxidation because the double bonds in the fatty acid are usually in a *cis* configuration, which causes steric hindrance. First, β oxidation of the unsaturated fatty acid occurs normally, but when the presence of a *cis* bond interferes with the ability of acyl-CoA dehydrogenase or enoyl-CoA hydratase to catalyze the necessary reaction, the *cis* bond must be converted into a *trans* bond. Odd-numbered *cis* bonds are transformed by enoyl-CoA isomerase, which changes the *cis* bond to a *trans* bond. Even-numbered *cis* bonds are transformed by 2,4-dienoyl-CoA reductase, which creates an odd-numbered bond, and the enoyl CoA isomerase can then act. Unsaturated fatty acids yield one less FADH$_2$ per double bond compared to saturated fatty acids because they are already partially oxidized, thus reducing total ATP production.

The glycerol that was originally part of the stored triglyceride is converted to glyceraldehyde phosphate and enters glycolysis or gluconeogenesis (depending on the cell's needs at the time). However, it must be converted to glyceraldehyde-3-phosphate (PGAL) before it can enter either of these pathways.

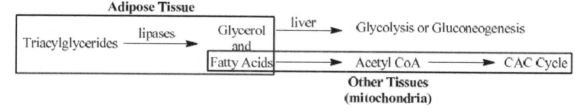

Ketone bodies

Sometimes there is an insufficient supply of glucose in the body (e.g., during periods of low food intake), and once the cellular carbohydrate stores have been depleted, it is necessary for the organism to find another way to obtain energy. *Ketogenesis* is the production of ketone derivatives of acetyl-CoA groups, as another form of fatty acid breakdown and occurs in the mitochondria of the liver.

The oxidation of large amounts of fatty acids can cause acetyl-CoA to accumulate in the liver, and if the quantities of acetyl-CoA are high, it may exceed the processing capacity of the Krebs cycle. This can occur when excess deaminated ketogenic amino acids are degraded, causing acetyl-CoA buildup. In cases of high acetyl-CoA accumulation, the ketogenesis pathway is initiated. In ketogenesis, the two carbon acetyl units condense in the liver, forming the following four carbon ketone molecules: β-hydroxybutyrate, acetoacetate, and acetone. These are collectively referred to as *ketone bodies* and the figure below illustrates how ketone bodies are produced.

These ketone bodies are transferred from the liver to the heart, brain, and muscles. In these organs, β-hydroxybutyrate and acetoacetate can be reconverted to acetyl-CoA for entry into the Krebs cycle. Acetone is converted to pyruvate, lactate, and acetate, or excreted as waste if it is not used quickly.

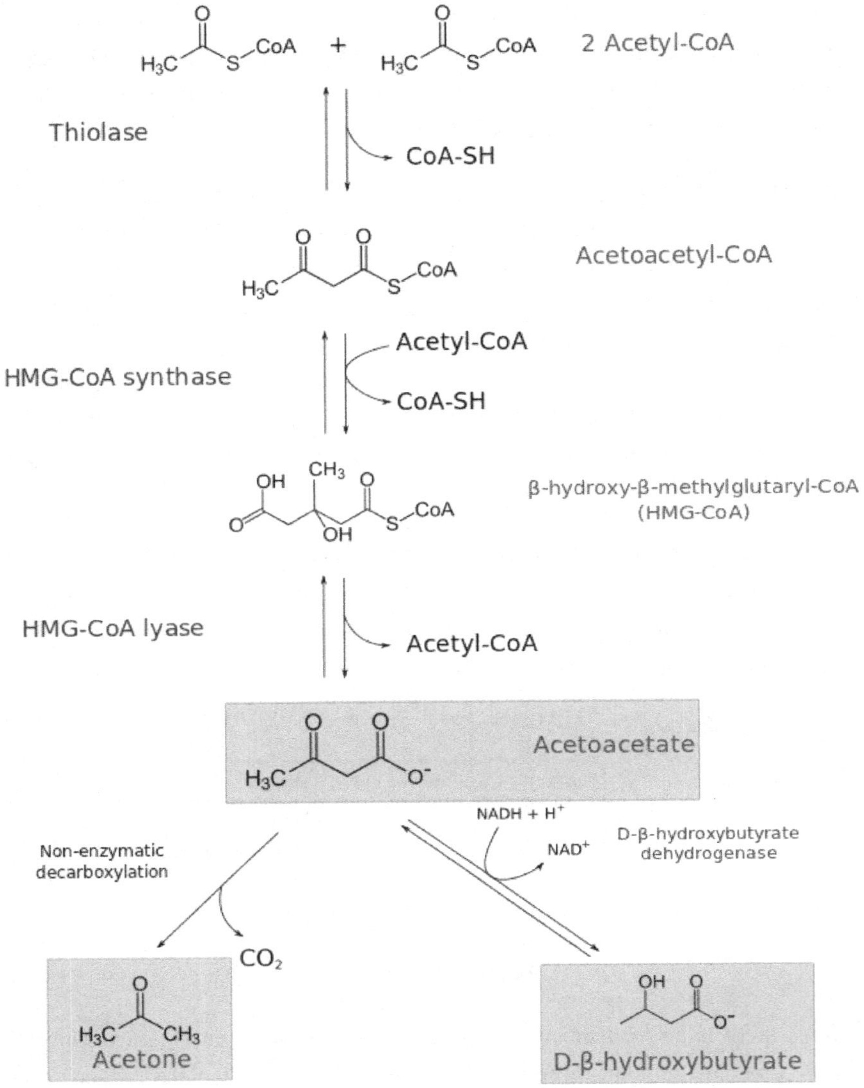

Conversion of acetyl-CoA to ketones (acetone) and β-hydroxybutyrate

Ketosis occurs when an excessive amount of ketone bodies is present in the body. In this metabolic state, the majority of energy comes from ketone bodies in the blood. It can occur during fasting or low-carbohydrate diets and is "fat-burning mode." Because acetoacetate and *β*-hydroxybutyrate are acidic, the excessive formation of ketone bodies can cause *ketoacidosis* or *metabolic acidosis,* a condition where blood pH is drastically decreased, sometimes to fatal levels. This occurs if the body fails to regulate ketone production properly.

Anabolism of fats

When there are excess dietary carbohydrates, the molecules must be converted to fat for storage. The anabolism of fats primarily occurs in the cytoplasm and the endoplasmic reticulum of liver cells, in contrast with fatty acid breakdown, which occurs in the mitochondria. *Lipogenesis* is the process whereby fats are produced from acetyl-CoA and malonyl-CoA precursors by fatty acid synthases that polymerize and reduce acetyl groups.

Lipogenesis is stimulated by insulin because insulin activates pyruvate dehydrogenase (an enzyme that catalyzes the formation of acetyl-CoA) and acetyl-CoA carboxylase (an enzyme that catalyzes the formation of malonyl-CoA). Then the synthesized fatty acids are esterified with glycerol to create triglycerides, and the triglycerides are packaged into lipoproteins and secreted from the liver.

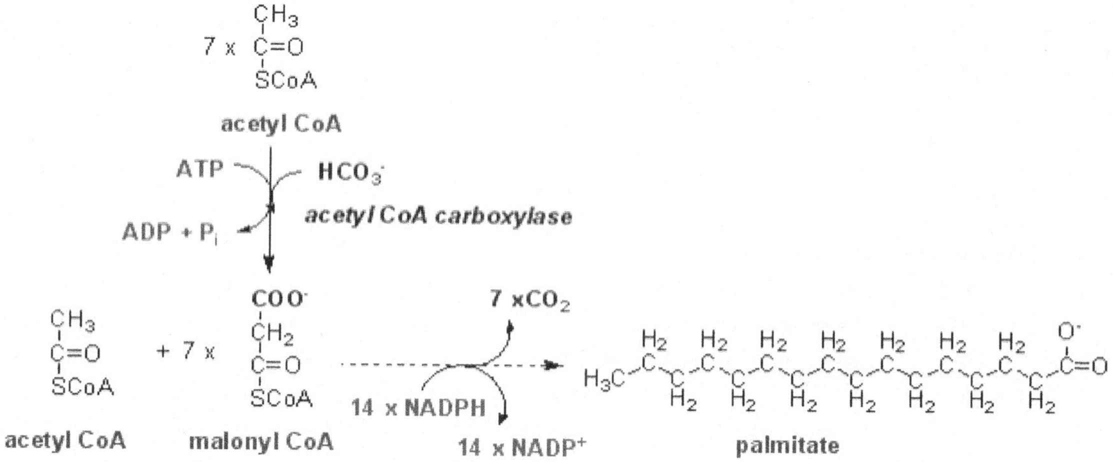

Synthesis of palmitate, the primary fatty acid synthesized in the human body

As seen in the diagram above, ATP and NADPH are required for fatty acid synthesis. During the process, these molecules are oxidized to ADP and NADP$^+$, respectively.

For the synthesis of an unsaturated fatty acid, a desaturation reaction introduces a double bond into the fatty acyl chain (usually requiring a desaturase enzyme).

Essential fatty acids cannot be synthesized in mammalian tissue and are required in the diet. The essential fatty acids for humans are linolenic acid (omega-3 fatty acid) and linoleic acid (omega-6 fatty acid). If these are not consumed in the diet, health problems develop from a deficiency of these essential fatty acids (from diet).

Non-template synthesis: biosynthesis of lipids and polysaccharides

Lipids are a group of molecules that include triglycerides, fat-soluble vitamins, waxes, sterols, and phospholipids. Polysaccharides (carbohydrate polymers) are synthesized by non-template *de novo* synthesis. For *de novo* synthesis, there is no template as there is in the synthesis of nucleic acids or proteins; rather, the synthesis of carbohydrate polymers is based solely on gene expression and enzyme specificity.

Additionally, other molecules are synthesized *de novo*. Terpenes, a class of lipids that exists in all organisms are created by the assembly and modification of isoprene units. These molecules can then be used to create sterols (e.g., cholesterol).

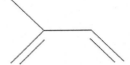

Isoprene as a component of carotenoids, steroids, etc.

Glycogen is a polysaccharide that is synthesized *de novo*, constructed by joining units of uracil-diphosphate-glucose by the enzyme glycogenin, which acts as a primer. Glycogen synthase adds glucose monomers to the existing chain, and the glycogen branching enzyme creates branches.

Metabolism of proteins

When carbohydrates and fats are unavailable, proteins are then used as an energy source. Proteins are the least desirable form of energy for the body because a large amount of energy is required for protein breakdown. *Proteolysis* (protein catabolism) requires proteases to break the peptide bonds of proteins via hydrolysis for the release of smaller polypeptides or amino acids. Amino acids can provide intermediates for a variety of other molecules. Most amino acids are deaminated (the amino group is removed) in the liver and are then converted to pyruvate, acetyl-CoA or other Krebs cycle intermediates. Depending on the specific amino acid, these intermediates enter cellular respiration at various points.

From dietary intake, protein digestion begins in the stomach, where proteins are denatured (unfolded) by the acidic digestive juices. Digestive enzymes like pepsin, trypsin, and chymotrypsin hydrolyze covalent peptide bonds. Amino acids are absorbed in the small intestine into the bloodstream for delivery to the tissues. Amino acids can produce ATP when other fuel supplies are low, and the cell does not require other nitrogen-containing compounds. For this process, the amino group of amino acid must either be removed by oxidative deamination to produce a keto acid and NH_3, or transferred to a keto acid by *transamination* (transfer of an amino acid group to an organic acid).

The keto acid enters the glycolytic pathway or the synthetic pathways for glucose and fat. The nitrogen from the amino group can be used to synthesize important nitrogen-containing molecules such as purines and pyrimidines. The toxic NH_3 (ammonia) passes into the bloodstream through the plasma membrane and is transported to the liver, where it is linked with CO_2 to form, in mammals, the relatively non-toxic urea, which is excreted by the kidneys. In fish, insects, and birds, the ammonia is converted to uric acid rather than urea.

Amino acids may replenish the intermediates in the Krebs cycle. Three-carbon amino acids (e.g., alanine) enter the pathways as pyruvate. Four-carbon amino acids (e.g., aspartate) are converted to oxaloacetate. Five-carbon amino acids (e.g., valine) are converted to α-ketoglutarate. Some amino acids can enter at more than one point, depending on cellular requirements.

Plants synthesize all the amino acids they need; animals lack some enzymes needed to make some amino acids. Humans synthesize eleven of the twenty necessary amino acids. The diet must provide the remaining nine amino acids as *essential amino acids*.

Total free amino acid pool in the body is derived from (1) ingested protein degraded to amino acids during digestion, (2) synthesis of non-essential amino acids from keto acids and (3) the breakdown of body proteins. The amino acids in these pools can be used for protein biosynthesis.

Enzyme Structure and Function

The function of enzymes in catalyzing biological reactions

Enzymes assist almost all chemical reactions that occur in living organisms. Enzymes are biological molecules that act as catalysts by increasing the rate of chemical reactions. The vast majority of enzymes are proteins, although RNA molecules (ribozymes) can catalyze reactions. An enzyme cannot force a reaction to occur if it would not normally occur (i.e., products are less stable than reactants); it simply makes a reaction occur faster. Enzymes are highly specific in their action and catalyze a single reaction or class of reactions. Enzymes are not consumed in a reaction; so only small amounts of enzymes are needed in a cell. The overall 3D shape (tertiary structure) of an enzyme plays an important role in the enzyme's function.

Enzymes are often named for their substrates by adding the suffix "-ase." For example, ribonuclease, abbreviated as RNase, is an enzyme that catalyzes the degradation of ribonucleic acid (RNA) into smaller components.

For example, the enzyme hexokinase is written above or below the reaction arrow. The reactants (starting material) upon which an enzyme acts, written to the left of the reaction arrow, are substrates. An enzyme may bind to one or more substrates (a molecule that enzymes act upon).

In the example below, there are two substrates: glucose and adenosine triphosphate (ATP). The products (to the right of the reaction arrow) are glucose-6-phosphate and adenosine diphosphate (ADP). Intermediates (not pictured in this example) are compounds temporarily formed between initial reactants and final products.

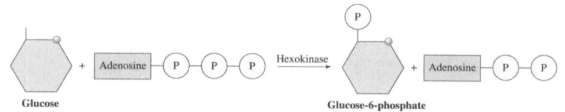

The substrates glucose and ATP are catalyzed by hexokinase to form glucose-6-phosphate and ADP. The hexokinase is not consumed in the reaction and continues the process with new substrates.

Reduction of activation energy

Reactants must reach a certain energy level before the reaction can proceed to convert the substrate(s) to the product(s). This energy is the *activation energy* (E_a) and is needed to break bonds within the reactants and thus enable them to react to form products. At a later stage in the reaction, energy is released as new bonds form within the product.

Enzymes decrease the activation energy of a reaction by lowering the energy of the transition state. The *transition state* (bond making and bond breaking state) is where the reactants are in an activated complex. As shown in the image below, the transition state corresponds to the highest energy level—the activation energy.

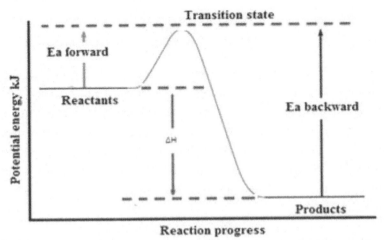

The energy is on the y-axis and time is shown on the x-axis. The products are lower in energy than reactants, and the reaction is spontaneous. The enzyme speeds the rate of spontaneous reactions.

Activation energy is an energy barrier to the reaction, in which the reactants must overcome before they can be converted into products. This can be accomplished non-enzymatically by increasing the temperature, which increases molecular collisions between reactants. However, homeostasis means the body temperature is maintained within narrow ranges in many organisms. The same enzyme may lower the activation energy barrier for both the forward and reverse reaction, increasing both rates of reaction, or another enzyme catalyzes each reaction separately.

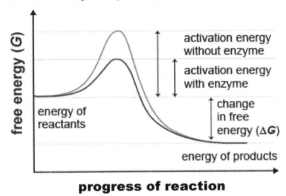

In the graph, the enzyme makes it easier for reactions to occur by lowering the required activation energy, but it does not alter the net energy release, ΔG (Gibbs free energy).

The enzyme provides an alternate pathway for reactants to form products. The lowering of the activation energy, which is accomplished during the formation of the enzyme-substrate complex, occurs in several ways:

Proximity: When the enzyme-substrate complex forms, the substrates are nearby, and therefore do not have to find each other as in a solution, thereby lowering the entropy (i.e., the disorder) of the reactants.

Optimizing orientation: The enzyme holds the substrates in the correct alignment and at the appropriate distance, usually by aligning active chemical groups. This lowers the entropy of the reactants.

Modifying bond energy: While binding to the substrate, the enzyme may stretch or distort a bond and weaken them so that less energy is needed to break the bond.

Electrostatic catalysis. Acidic or basic amino acids in the active site of the enzyme (a portion of the enzyme where the substrate binds) may form ionic bonds with the intermediate, which stabilize the transition state and lower the activation energy.

Substrates and enzyme specificity

When an enzyme binds to one or more substrates, it forms an enzyme-substrate complex, which is held together by hydrogen or ionic bonds. While bound, the catalytic action of the enzyme converts the substrate(s) to the product(s).

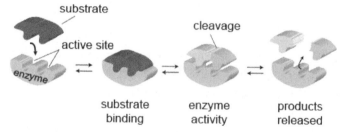

The substrate binds to active site of the enzyme with hydrogen on ionic bonds

An enzyme can distinguish its substrate from similar molecules and even isomers (same molecular formula but different molecules) of the same molecule. This enzyme-substrate *stereospecificity* refers to an enzyme's ability to distinguish between stereoisomers (different shapes) of the same molecule. For example, hexokinase binds D-glucose but not its stereoisomer L-glucose.

Enzyme classification by reaction type

While the names of some enzymes such as RNase describe their function, many enzymes have common names that do not refer to the substrate that they bind. Scientists created an enzyme classification system based on reaction type, with six main classes of enzymes. Note that within these six categories there are many subcategories. Knowing the reaction that a particular enzyme catalyzes is essential to categorization.

Oxidoreductases catalyze reactions that transfer electrons (oxidation-reduction or redox reactions). Common names for enzymes in this category are dehydrogenase, reductase, and oxidase.

Transferases catalyze group transfer reactions, whereby a group is transferred from a donor to an acceptor molecule. Hexokinase is a transferase that catalyzes the transfer of a phosphate group from ATP to glucose.

Hydrolases catalyze hydrolysis, in which water is used to break a bond (hydrolysis). Hydrolases transfer a functional group from a donor molecule to water (acceptor) and therefore are a specific type of transferase. An example is the digestive enzyme chymotrypsin, which can hydrolyze amide (amino acids) and ester (fatty acids) bonds.

Lyases catalyze the breakage of bonds using mechanisms other than hydrolysis (water) and oxidation (electron transfer). Lyases cleave double bonds by the addition of functional groups along with the reverse reaction (i.e., double bond formation via the removal of functional groups). Examples of lyases include fumarase (removes water), and adenosine deaminase (removes ammonia).

Isomerases produce isomeric forms (same atoms but the different connection between atoms) by transferring functional groups within a molecule, which allows for geometric or structural changes. For example, phosphoglucose isomerase converts glucose 6-phosphate to fructose 6-phosphate.

Ligases form covalent bonds by joining two molecules. A linkage reaction is usually coupled with an energy-producing reaction, such as the breakdown of ATP to ADP. An example is DNA ligase, which catalyzes the formation of a phosphodiester bond by joining DNA nucleotides together during replication of the genetic material.

Active Site Model

Every enzyme has an active site, which is a region on the surface of the enzyme where it binds to the substrate, orients it and facilitates the reaction. There are two prominent hypotheses on how substrates bind to the enzyme's active site. The *lock and key model* proposes that there is only one rigid active site that precisely fits the reactants, and the substrate fits inside its active site as a key fits into a lock. In this model, there is no modification of the enzyme after the substrate binds to the active site.

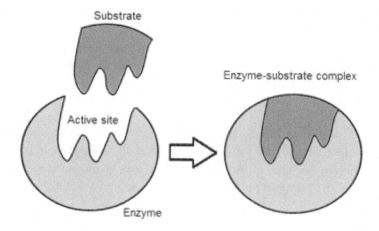

Induced Fit Model

As an alternative to the lock and key model, the more recently developed *induced fit model* proposes that there not be a perfect match between the active site of an enzyme and the substrate that it binds. In this model, a small conformation (rotation) change occurs as the enzyme and substrate come together, which allows the active site to bind more precisely with the substrate. The induced fit model hypothesizes that certain amino acid residues (i.e., side-chains) in the active site facilitate the enzyme finding the proper substrate. The initial interaction between the enzyme and substrate is weak, but these weak interactions cause conformational changes in the enzyme that strengthen binding. The energy barrier is lower in the "closed" form of the enzyme, where the active site fits snugly around the substrate.

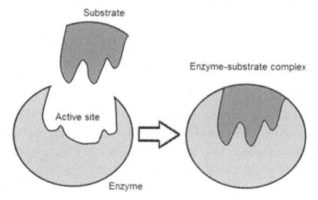

Think of a handshake where the fingers move to more tightly grasp the other hand. The induced fit is the favored model of enzyme-substrate binding.

Mechanism of catalysis (cofactors, coenzymes, and water-soluble vitamins)

When the enzyme and the substrate form the enzyme-substrate complex, the R-groups (side chains) of the amino acids in the active site catalyze the reaction. They pull or contort the substrate, temporarily weakening bonds or altering the substrate's conformation. In reactions with two or more substrates, the side chains form a template to guide the substrates into the most energy-efficient conformation.

Some enzymes require *cofactors*, which are non-protein molecules that assist in chemical reactions that cannot be performed by the active site alone. Not all organisms can produce the cofactors needed, so they are required (essential) in the diet. Cofactors may be inorganic trace elements or metal ions, such as Cu^{2+}, Fe^{3+} or Zn^{2+} (electron carriers), or organic molecules as coenzymes.

Coenzymes are small, complex organic molecules, usually derived from vitamins, which facilitate enzymatic reactions. Vitamins can be fat-soluble (A, D, E and K) or water-soluble (B and C), and it is the water-soluble vitamins that generally act as precursors to coenzymes. Vitamin deficiency may lead to the lack of a specific coenzyme and thus a lack of enzymatic action. Coenzymes may be loosely or tightly bound to the enzyme. Loosely-bound coenzymes are *cosubstrates*. Some coenzymes are so tightly bound to their enzyme that they cannot be removed without denaturing the enzyme and a coenzyme is a *prosthetic group*. Prosthetic groups may be attached to enzymes with covalent bonds. However, the division between loosely and tightly-bound coenzymes is not always clear. For example, NAD^+ can be loosely bound in some enzymes but tightly bound in other enzymes.

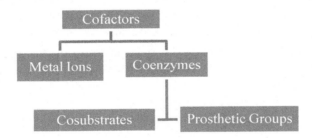

The relationship among molecules required by some enzymes

An *apoenzyme* is an enzyme missing the cofactor that it requires to function, and the enzyme is inactive. A *holoenzyme* is an active form in which the cofactor and enzyme are bound together. Cofactors must be regenerated (i.e., returned to their original state) to complete the catalytic cycle. Cofactors, unlike enzymes, may change during the course of the reaction and therefore need to be replenished.

Effects of local conditions on enzyme activity

Many factors can affect enzyme activity, including substrate concentration, pH, temperature and the presence of modulators. At a constant enzyme concentration, an increase in the substrate concentration increases the enzyme's activity by increasing the number of random collisions between the substrate and the active site on the enzyme. However, at some point, all available active sites are bound, and increasing the substrate concentration has no further effect on enzyme activity, a state of *saturation*. The substrate in a cell is regulated by an organism's diet, rates of absorption in the intestine, the permeability of the plasma membrane or changes in intracellular breakdown and synthesis of the substrate.

Enzymes have an optimum pH at which they work most efficiently. At this pH, the enzyme maintains its tertiary structure and, therefore, its active site. As the pH diverges from the optimum, enzyme activity decreases and may cease altogether. Additionally, changes in pH may change the nature of an amino acid side chain. If an enzyme requires a carboxylate ion (COO^-), lowering the pH could convert the carboxylate ion to a carboxylic acid (COOH), which would cause enzyme activity to decrease.

An enzyme's pH optimum depends on the location of the enzyme. Enzymes in the stomach function at a much lower pH because of the stomach's acidic environment.

Enzyme	Location	Substrate	pH Optimum
Pepsin	Stomach	Peptide bonds	2
Sucrase	Small intestine	Sucrose	6.2
Urease	Liver	Urea	7.4
Hexokinase	All tissues	Glucose	7.5
Trypsin	Small intestine	Peptide bonds	8
Arginase	Liver	Arginine	9.7

Sometimes, the active site of an enzyme functions as a microenvironment more conducible to the reaction, such as providing a pocket of low pH in an otherwise neutral cell.

There is usually a temperature optimum at which an enzyme exhibits peak activity. Like the pH optimum, it is dependent on the environment in which the enzyme normally operates. For example, a DNA polymerase of a human has a lower temperature optimum than a DNA polymerase of a thermophilic bacterium. For most human enzymes, the temperature optimum is body temperature (37 °C).

As temperature increases, molecules move faster, with more random collisions between enzymes and substrates. Within the upper-temperature limit range, enzyme activity generally doubles with every 10-degree increase.

However, as with pH, at a certain point, the temperature gets too high, and bonds maintaining the 2°, 3° and 4° structure of the protein dissociate. This causes the enzyme's active site to become unstable, leading to a loss of function. A *denatured* protein changes its three-dimensional shape. Denaturation alters the structure of an enzyme or another organic molecule so that the enzyme cannot carry out its intended function. The enzymes in bacteria are often denatured by high temperatures in processes like boiling contaminated drinking water and heat-sterilizing medical and scientific equipment. However, enzyme activity is decreased by low temperatures, as when food is stored in a refrigerator or freezer. Enzymes are major participants in food spoilage, but low temperatures can significantly slow the spoilage process.

Modulators are compounds that modify the binding site of an enzyme, either inhibiting or enhancing the enzyme's activity. Modulators can have a strong effect on enzyme activity through either covalent (irreversible) or non-covalent (reversible) interactions between modulators and an enzyme. Modulators are products of other chemical reactions or are activated by chemical signals. If modulators are required for enzymes to catalyze the reaction, they are cofactors. Modulators that increase an enzyme's catalytic activity are enzyme activators, enhancers or inducers; modulators that decrease or eliminate an enzyme's catalytic activity are enzyme inhibitors.

Notes

Chapter 3

Prokaryotic Cell

- **Bacteria Structure**

- **Growth and Physiology**

- **Genetics**

Bacteria Structure

Prokaryotic domains: Bacteria and Archaea

Prokaryotes include the Bacteria and Archaea domains of life. They were likely the first cells in evolutionary history; fossils of prokaryotes that date 3.5 billion years ago have been found. These fossils indicate that prokaryotes were alone on Earth for 2 billion years, during which time they evolved diverse metabolic capabilities and pathways.

Bacteria are encountered every day—they live inside and on humans and animals. Bacteria were discovered in the seventeenth century when Dutch naturalist Antonie van Leeuwenhoek examined scrapings from his teeth under a microscope. He called these organisms "little animals." At the time, it was believed that organisms could arise spontaneously from inanimate matter. Around 1850, Pasteur refuted the theory of spontaneous generation by showing that contamination was necessary for the growth of microbes. A single spoonful of soil can contain 10^{10} prokaryotic organisms; bacteria are the most numerous life form on Earth.

Classification of bacteria was originally based on metabolism and nutrition, among other characteristics. However, work done by Carl Woese since 1980 has revised the bacterial taxonomy based on similarity with 16S rRNA (genes encoding for the 30S small subunit of the prokaryotic ribosome). Bacteria were initially classified into 12 lineages based on these 16S rRNA sequences, but today the number of phyla has increased to around 52. Bacteria display a wide range of morphologies and live almost everywhere on the planet, including soil, water, hot springs and deep portions of the Earth's crust, and are vital for the recycling of many nutrients. They often live in symbiotic relationships with plants and animals, but some are pathogens that cause disease and even death.

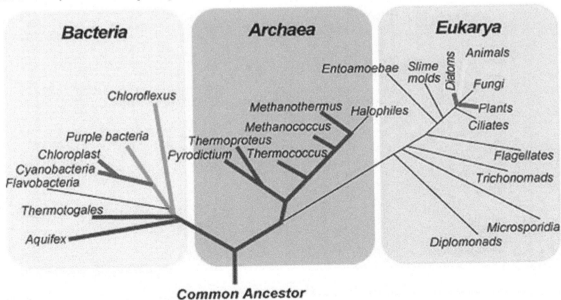

Three domains of life: Bacteria, Archaea and Eukarya and the lineage of their relationships

Archaea are prokaryotes with molecular characteristics that distinguish them from bacteria and eukaryotes. Archaea inhabit extreme environments, such as those with high salt, high temperature or harsh chemicals. Archaea and bacteria are believed to have diverged from a common ancestor about 3.7 billion years ago. The eukaryotes most likely split from archaea at some later time, as suggested by the fact that archaea and eukaryotes share some ribosomal proteins that are not in bacteria. Archaea and eukaryotes initiate transcription in the same manner and have similar types of tRNA.

Archaea come in many shapes, such as spherical, rod-shaped, spiraled, lobed, plate-shaped or irregular. DNA and RNA sequences in archaea are closer to eukaryotes than bacteria. The archaeal cell wall has various polysaccharides but no peptidoglycan, as in bacteria. Additionally, archaea may have unusual lipids in their plasma membranes (glycerol linked to hydrocarbons rather than fatty acids) that allow them to function at high temperatures. Most archaea are chemoautotrophs, and none are photosynthetic, suggesting that chemoautotrophs evolved first. Some archaea exhibit mutualism or commensalism, but none are parasitic or known to cause disease.

There are many types of archaea, including methanogens, halophiles, and thermoacidophiles. *Methanogens* live in anaerobic environments (e.g., marshes), where they produce methane. Methane is produced from hydrogen gas and carbon dioxide, and its production is coupled with ATP formation. Methane released into the atmosphere contributes to the greenhouse effect. Methanogenic archaea produce about 65% of methane in our atmosphere. *Halophiles* require high salt concentrations (e.g., Great Salt Lake). Their proteins use halorhodopsin (light-gated chloride pump) to synthesize ATP in the presence of light. They usually require salt concentrations of 12-15%; in comparison, ocean water is only 3.5% salt. *Thermoacidophiles* thrive in hot, acidic environments (e.g., geysers). They survive best at temperatures around 80 °C, although this varies depending on the species. The metabolism of sulfides forms acidic sulfates, and they generally thrive at a pH between 2−3.

Major classifications:
bacilli (rod-shaped), spirilla (spiral-shaped) and cocci (spherical)

Most bacteria, on average, are 1–1.4 μm wide and 2–6 μm long, making them just visible with light microscopes. They have three basic shapes:

- A *bacillus* is elongated, or rod-shaped.

- A *spirillum* is spiral-shaped.

- A *coccus* is spherical.

Cocci and bacilliform clusters and chains that vary in size between species.

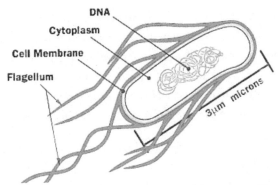

Escherichia coli (E. coli) bacterium with peptidoglycan cell wall and circular DNA

As cells get larger in volume, the ratio of surface area relative to volume decreases, which imposes a limit to how large an actively metabolizing cells can become. The adequate surface area is required for vital processes, such as nutrient exchange across the plasma membrane. If there is too much cell volume and not enough surface area, the exchange rate does not support metabolism. Cells that need a greater surface area-to-volume ratio use modifications, such as folding or microvilli (small projections of the membrane) that increase the surface area while minimizing the additional volume associated with these modifications.

Lack of nuclear membrane and mitotic apparatus

Unlike eukaryotes, all prokaryotic cells lack a nucleus (prokaryote means "before a nucleus"). However, prokaryotes have a dense area where their DNA is concentrated, the *nucleoid region*. It is irregularly shaped, dense with nucleic acid and not enclosed by any membrane. Prokaryotic DNA is a single, circular, double-stranded DNA chromosome. They do not have histones or associated proteins like eukaryotes. Bacterial cells may have additional *plasmids* or accessory circular DNA molecules. These plasmids, which are smaller than the genomic DNA, may confer additional abilities or characteristics (e.g., genes to encode proteins for antibiotic resistance).

Prokaryotes lack a mitotic apparatus, which is used in eukaryotes to separate chromosomes during mitosis. Instead, prokaryotes use their cytoskeleton to pull the replicated DNA apart during binary fission resulting in cell division.

Lack of typical eukaryotic organelles

Prokaryotic cells lack the membranous organelles of eukaryotic cells. They do not have the Golgi apparatus or endoplasmic reticulum, nor do they have mitochondria or chloroplasts. Their cytoplasm is a semifluid solution with the enzymes needed for essential chemical reactions. Some may have *inclusion bodies*, which are granules that store various substances.

Prokaryotes have thousands of ribosomes for protein synthesis, but they are smaller (the 30S, 50S subunits; 70S assembled) than eukaryotic (40S, 60S subunits; 80S assembled) ribosomes.

Prokaryotes have a cell wall

Like eukaryotes, prokaryotes have a typical plasma membrane with a phospholipid bilayer. The plasma membrane of prokaryotes can form internal pouches, *mesosomes*, that increase surface area for metabolic processes.

Outside the plasma membrane of bacteria, fungi and eukaryotic plant cells is a rigid *cell wall* that keeps the cell from bursting or collapsing due to osmotic changes. The bacterial cell wall is made of peptidoglycan, a polysaccharide-protein molecule that gives the wall much of its strength (archaea have cell wall polysaccharides, but no peptidoglycan). Plant cell walls are made of cellulose (polysaccharide of glucose monomers), and fungi cell walls are made of chitin. The cell wall protects the cell from the outside environment and maintains the shape of the cell. Additionally, some prokaryotes (e.g., Gram-negative bacteria) may have an *outer membrane* (lipopolysaccharide) outside their peptidoglycan cell wall.

Some prokaryotes have another layer of polysaccharides and proteins outside of their cell wall, a *glycocalyx* or *capsule*. The capsule is not easily washed off during experiments. Some prokaryotes may have a *slime layer*, a loose gelatinous sheath conferring some protection from environmental assaults (e.g., antibiotics or dehydration). The slime layer or capsule coverings are especially beneficial for parasitic prokaryotes as protection from host defenses.

Bacteria continuously remodel their peptidoglycan cell walls as they divide, which makes the cell wall a good drug target. β-lactam antibiotics (e.g., penicillin) prevent peptidoglycan cross-links in a bacterium's cell wall by inhibiting the enzyme that catalyzes the formation of these cross-links. This creates an imbalance in cell wall degradation and production, eventually causing the bacterium to lose its cell wall and become susceptible to rupture from osmotic pressure.

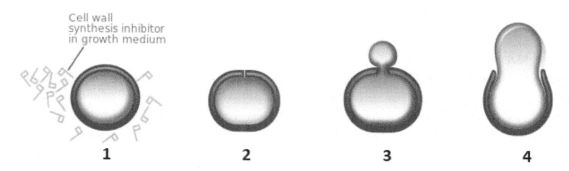

Penicillin's mechanism of action on peptidoglycan cell wall during cell division

The Gram stain procedure (developed in the late 1880s by Hans Christian Gram) differentiates bacteria based on their cell wall. The procedure involves first staining cells with crystal violet dye, washing away the dye, and then recoloring with a pink counterstain dye. Gram-positive bacteria stain purple, while Gram-negative bacteria stain pink.

The different results from the test are due to the differences in their structure. Gram-positive bacteria have a thick peptidoglycan layer and no outer membrane. Their thick peptidoglycan readily takes up the purple dye.

Gram-negative bacteria have a thin peptidoglycan layer surrounded by an outer membrane. They do take up the purple dye, but in the washing step (typically alcohol) their outer membranes are degraded, removing the purple color.

This washing step does not affect the thick peptidoglycan of the Gram-positive bacteria, leaving them purple. The recoloring with a pink dye is done to visualize the Gram-negative bacteria. The pink dye does not affect the purple color of the Gram-positive bacteria.

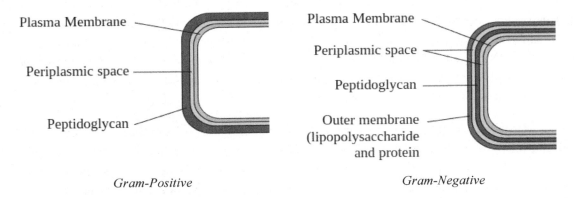

Gram-Positive   *Gram-Negative*

Gram-negative bacteria have a special *lipopolysaccharide* (LPS) in their outer membrane absent in Gram-positive bacteria. This is an *endotoxin*, and it invokes an immune response in humans. This outer membrane provides Gram-negative bacteria with some protection from antibiotics, while Gram-positive bacteria are generally more vulnerable to antibiotics.

Some bacteria in the *Firmicute* phylum (mostly Gram-positive) form resistant *endospores* in response to unfavorable environmental conditions. During spore formation, the bacterium's DNA, ribosomes, dipicolinic acid are encased by several protective spore layers: the exosporium, the spore coat, the spore cortex, and the core wall. The bacterial cell deteriorates, and the endospore is released.

Endospores can survive in the harshest of environments, including desert heat and dehydration, boiling temperatures, polar ice, and extreme ultraviolet radiation. Endospores survive for long periods of time (e.g., 1,300-year-old anthrax spores can cause disease). When environmental conditions are again suitable, the endospore absorbs water and grows out of its spore coat. In a few hours, newly emerged cells become typical bacteria that are capable of reproducing by binary fission. Endospore formation is not reproduction; rather, it is a dormant structure that aids in survival and dispersal to favorable locations.

Flagellar propulsion in bacteria

Some bacteria have *flagella* and are motile. The flagellum is a filament that is composed of three strands of the flagellin protein wound in a helix and inserted into a hook that is anchored by a *basal body*. A basal body is an organelle formed from a centriole and an array of microtubules. It is capable of 360° rotation, which causes the cell to spin and move forward. A motor powered by a proton (or sodium) gradient provides the energy. In contrast, flagella in eukaryotes are made of microtubules and are powered directly by ATP hydrolysis.

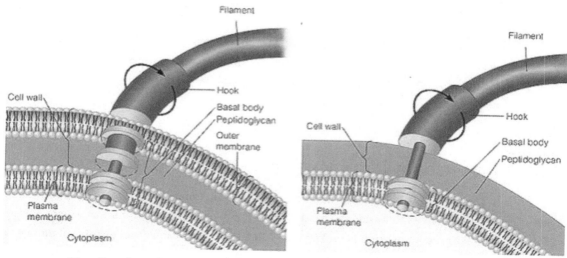

Flagellum for a Gram-negative bacteria (left) and Gram-positive bacteria (right)

Prokaryotes may have short, hair-like filaments or appendages of *fimbriae* that extend from their surface, which they can use to attach to another cell or inanimate object. The fimbriae of *Neisseria gonorrhoeae* allow it to attach to host cells and cause the disease gonorrhea.

Pili are similar appendages to fimbriae, but they are thicker and fewer in number. They help bacteria adhere to one another for motility or the exchange of genetic material.

Growth and Physiology

Reproduction by fission

Prokaryotes reproduce by *binary fission*, a form of *asexual reproduction*, which results in offspring with genetically identical chromosomes.

Binary fission in prokaryotes is similar to mitosis in eukaryotes because both result in daughter cells that are identical to the parent cell. However, in multicellular fungi, plants, and animals, cell division is a routine part of the growth process that produces and repairs cells within the organism. In contrast, binary fission is simply the method of prokaryotic reproduction. Eukaryotic cells have a spindle apparatus, which is required for distributing chromosomes to daughter cells during mitosis, while prokaryotes divide without microtubules, spindles or centrioles. *E. coli* bacteria can divide in about 20 minutes, while eukaryotic cells may require between an hour and a day to divide.

Binary fission follows the sequence:

1. Before division, the single, circular DNA chromosome is replicated and attached to a special site tethered to the plasma membrane.

2. The two chromosomes separate as the cell enlarges and pulls them apart.

3. When the cell is approximately twice its original length, the plasma membrane grows inward, and a new polysaccharide cell wall (cell plate) bisects the cell and then the plasma membrane forms, dividing the cell into two roughly equal-sized daughter cells.

High degree of genetic adaptability and antibiotic resistance

Prokaryotes have evolved to live in a vast variety of environments because they have a high degree of genetic adaptability, allowing species to differ in how they acquire and utilize energy. There are many sources for genetic variation in prokaryotes. Mutations are generated and distributed through a population more rapidly because prokaryotes have a short generation time. Also, prokaryotes are haploid (single copy of a gene), so mutations are immediately subjected to natural selection. Additionally, plasmids (extrachromosomal circular pieces of DNA) can carry genes for resistance to antibiotics and transfer them between bacteria (e.g., transformation and conjugation), which are described in the chapter.

Exponential growth

The growth of bacteria can be modeled with phases. In the *lag phase*, bacteria adapt to the conditions of their particular environment; the cells are maturing but not yet dividing. In the *log phase*, bacteria grow exponentially in a medium with adequate space and nutrients because of binary fission. The rate of population increase doubles with each consecutive period (e.g., 20-minute intervals). However, this cannot continue indefinitely. The *stationary phase* is when food and space become scarce, growth slows and eventually plateaus. *Death* or *decline phase* is when there is a lack of nutrients or improper conditions, where the bacteria die.

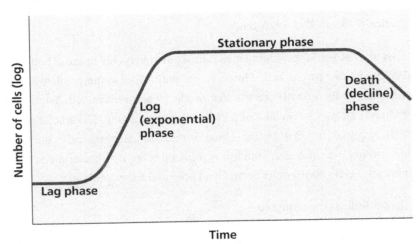

Bacterial growth with a log phase before reaching a stationary phase

Bacteria as anaerobic and aerobic organisms

Bacteria differ in their need for and tolerance of oxygen (O_2). *Obligate anaerobes* are unable to grow in the presence of O_2; this includes the species of anaerobic bacteria of the *Clostridium* genus that cause botulism, gas gangrene, and tetanus. *Obligate aerobes* must have O_2 for growth and die without it. *Facultative anaerobes* can grow in either the presence or absence of gaseous O_2, although they grow better with the O_2 present.

An autotroph (i.e., "producer") is an organism that is capable of self-nourishment; autotrophic prokaryotes include both photoautotrophs and chemoautotrophs. *Photoautotrophs* are photosynthetic and use light energy to assemble the organic molecules they require. Although the most well-recognized photoautotrophs are plants, many bacteria are in this category. *Primitive photosynthesizing bacteria*, such as green sulfur bacteria and purple sulfur bacteria, use bacteriochlorophyll and hydrogen sulfide (H_2S) as a proton and electron donor instead of H_2O, so they do not release O_2. *Advanced photosynthesizing bacteria*, such as cyanobacteria, use bacteriochlorophyll and chlorophylls in plants. H_2O is used as the proton and electron donor, so they do release O_2. *Chemoautotrophs* make organic molecules by using energy derived from the oxidation of inorganic compounds in the environment. Deep ocean hydrothermal vents provide H_2S and allow for the growth of chemosynthetic bacteria. The methanogens (previously mentioned in the context of archaea) include chemosynthetic bacteria that produce methane (CH_4) from hydrogen gas and CO_2. ATP synthesis and CO_2 reduction are linked to this reaction, and methanogens can even decompose animal wastes to produce electricity as an environmentally-friendly energy source. *Nitrifying bacteria* oxidize ammonia to nitrites (NH_3 to NO_2) and nitrites to nitrates (NO_2 to NO_3).

Heterotrophs cannot synthesize their food. Most free-living bacteria are *chemoheterotrophs* that take in pre-formed organic nutrients. With the existence of numerous *aerobic saprotrophs* (organisms that feed on the dead organic matter and use oxygen), there is probably no organic molecule that cannot be broken down by some prokaryotic species. *Decomposers* are critical in recycling materials in the ecosystem by decomposing dead organic matter and making it available to photosynthesizers.

Parasitic and symbiotic bacteria

Some bacteria are *symbiotic*, forming close, long-term relationships with members of other species, including *mutualistic* (both organisms benefit), *commensalism* (one organism benefits and the other is unaffected), and *parasitic* (one organism benefits while the other is harmed) relationships.

- Mutualistic nitrogen-fixing *Rhizobium* bacteria live in nodules on the roots of soybean, clover, and alfalfa, where they reduce nitrogen to ammonia, which the plant requires. *Rhizobium* bacteria use some of the plant's photosynthetically-produced organic molecules in return.

- Mutualistic bacteria that live in the intestines of humans benefit from partially-digested material and release vitamins K and B_{12}, which humans use to produce blood components.

- Mutualistic prokaryotes, in the stomachs of cows and goats, digest cellulose, which the animal cannot do by itself, and release nutrients that the cow or goat can use. In return, the bacteria get a warm, moist environment and a constant supply of food.

- Mutualistic cyanobacteria provide organic nutrients to fungi, and the fungus protects and supplies inorganic nutrients to the bacteria. This composite symbiotic organism is a *lichen*.

- Commensalistic bacteria live in (or on) organisms of other species and cause them no harm and no benefit, such as some bacterial species that live on the skin of humans.

- Parasitic bacteria (e.g., chlamydia or Cryptosporidium) are responsible for a wide variety of infectious plant and animal diseases.

Chemotaxis

Cells can engage in mechanical activities: they can move, and the organelles within them can move. *Cell migration* is the movement of cells from one location to another and is often the response to stimuli. *Chemotaxis* is movement in response to chemicals. For example, bacteria may swim toward the highest concentration of food molecules (positive chemotaxis) and may flee from poisons that they detect in their environment (negative chemotaxis). They can do this by sensing chemical gradients through transmembrane receptors that bind attractants (or repellents), and these receptors stimulate rotation of flagella, causing the bacteria to move.

Prokaryotes may engage in *phototaxis*, which is movable in response to light. Additionally, some prokaryotes can move in response to physical forces in their environment. This ability to respond to force is *mechanotaxis.*

Genetics

Plasmids as extragenomic DNA

Plasmids are pieces of DNA that exist apart from the genomic DNA and are in some prokaryotes. They range in size from less than one kilobase pair to several megabase pairs. Plasmids are generally circular, although examples of linear plasmids are known. They usually carry genes that are beneficial but are not always essential for growth and survival. Plasmids replicate independently of the genomic DNA, are inherited, and can be extracted in genetic engineering procedures to be used as vectors to carry foreign DNA into other bacteria. *Episomes* are plasmids that can incorporate themselves into the bacterial chromosome.

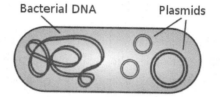

Bacterium with chromosomal DNA and plasmids

Conjugation

Conjugation transfers genetic material between bacteria via a sex pilus that temporarily joins a recipient bacterium. *Conjugative pili,* (sex pili) are the tubes specifically used by bacteria during conjugation to pass replicated DNA from one cell to another.

A plasmid is sent from the donor to the recipient through the sex pilus, where it can be incorporated through recombination. Plasmids can contain antibiotic resistance genes so that conjugation may allow the transfer of that resistance. The most well-studied example is the sex pilus of *E. coli* that possesses an F plasmid, which is an episome that contains the "fertility factor." Since an F plasmid is an episome, genomic DNA may be transferred in some instances. If the F plasmid is transferred to an F⁻ recipient and integrated successfully, the recipient becomes F⁺ and can then create its sex pili. Conjugation can occur between bacteria of the same species, closely or distantly related species. It is an important mechanism of horizontal gene transfer, which contributes to the genetic diversity of prokaryotes.

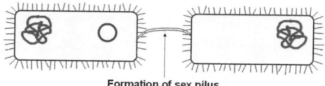

Formation of sex pilus that attaches to recipient

Bacterial conjugation: donor (with plasmid) is on the left, the recipient is on the right

Transformation
(incorporation into the bacterial genome
of DNA fragments from external medium)

Transformation is the process of bacteria taking up DNA fragments from outside the cell and incorporating them into their genomes. There are two major sources of these fragments: DNA secreted by live bacteria and DNA released from a bacterium that dies. When a cell lyses (i.e., breaks open), it spills its DNA into the environment. Successful transformation of DNA fragments containing an antibiotic resistant gene (i.e., plasmid) confers antibiotic resistance to the recipient bacterium.

Transposons

Transposons (transposable elements or "jumping genes") are pieces of DNA that can insert themselves into another place in the genome. They exist in both prokaryotes and eukaryotes and can move within or between chromosomes. The changes in the genome caused by transposon relocation in the genome can be advantageous, disadvantageous or neutral.

Some transposons make copies of themselves, and the copies are then inserted into other locations in the genome ("copy and paste"). Other transposons are removed from their original location in the DNA and insert themselves directly into another location ("cut and paste").

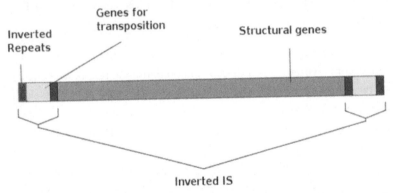

Bacterial DNA transposon flanked by inverted repeats as breakpoints on the DNA

Chapter 4

DNA
Protein Synthesis
Gene Expression

- **DNA Structure and Function**

- **DNA Replication**

- **Genetic Code**

- **Transcription**

- **Translation**

- **Eukaryotic Chromosome Organization**

- **Control of Gene Expression in Eukaryotes**

DNA Structure and Function

Description

Deoxyribonucleic acid (DNA) is the sequence of paired nucleotides that stores the genetic code, necessary for its replication and determining the sequence of amino acids in proteins. The process of transcription uses DNA as a template to form a *ribonucleic acid* (RNA), which serves as a temporary transcript of the hereditary genetic information. *Ribosomes* translate the information from mRNA into a sequence of amino acids to form polypeptides, which fold into proteins.

DNA contains four different nucleotides (adenine, cytosine, guanine, and thymine). *Nucleotides* consist of 1) at least one phosphate group, 2) a pentose sugar, and 3) a one or two-ringed structure containing carbon and nitrogen (i.e., nitrogenous base). The term "base" relates to its ability to accept hydrogen ions (protons). First, glycosidic bonds link the sugar to the nitrogenous base, creating a *nucleoside* (sugar and base only), as shown below.

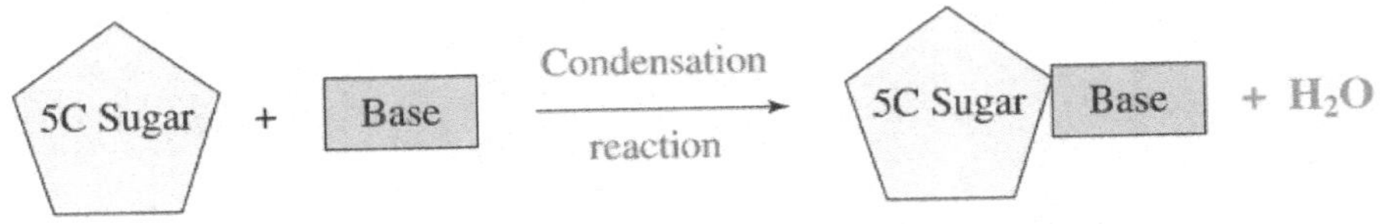

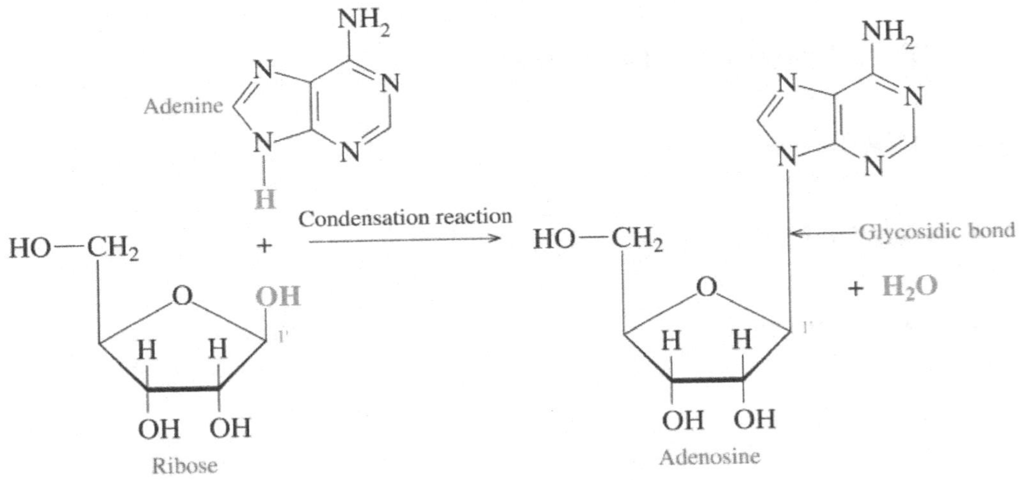

Nucleoside formation by a condensation reaction that joins the ribose to the adenine to form adenosine (lacking the phosphate group of a nucleotide)

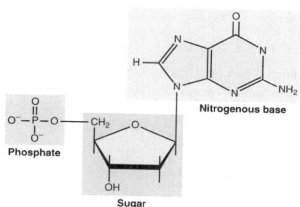

Nucleotides are linked by phosphodiester bonds
between the phosphate of one nucleotide and the sugar of another

The phosphate group then links to the nucleoside similarly through a condensation reaction, forming a nucleotide (sugar, base, phosphate)

These nucleotides form a long strand and the chain of phosphodiester bonds in the sugar-phosphate backbone for DNA. A polymer of nucleotides comprise a *nucleic acid*, and its acidity is due to the abundant phosphate groups. Nucleic acids (i.e., RNA and DNA) store, transmit and express hereditary, genetic information in cells.

DNA has two nucleic acid strands combined to form the double-stranded DNA molecule. Strands are joined when the bases of each strand (e.g., A = T and C ≡ G) form *base pairs*. The DNA in a single human cell contains about 3 billion base pairs. Because DNA has a helical twist, further coiling of the DNA strand makes the DNA more compact. The 3 billion base pairs in one human cell would stretch to about 6 feet in length. The sequence of base pairs encodes the genetic information of the cell. This genetic code, in the form of nucleic acids, determines the sequence of amino acids in the proteins synthesized by the cell.

DNA composition (purine and pyrimidine bases, deoxyribose and phosphate)

Early researchers knew that the genetic material must have a few necessary characteristics. It must store information used to control both the development and the metabolic activities of cells, it must be stable to be accurately replicated during cell division and be transmitted for many cell cycles and between generations of offspring (i.e., progeny), and it must be able to undergo mutations, providing the genetic variability required for evolution.

In 1869, Swiss chemist Friedrich Miescher removed nuclei from pus cells and isolated "nuclein." It was rich in phosphorus and lacked sulfur. Nuclein was analyzed by other scientists who found that it contained an acid—specifically, nucleic acid. The two types of nucleic acids were soon discovered: deoxyribonucleic acid (DNA) and ribonucleic acid (RNA). In the early twentieth century, researchers discovered that nucleic acids contain four types of nucleotides, the repeating units that make up the long DNA molecule.

Purines or *pyrimidines* are the two types of nucleotides within DNA.

Purine   Pyrimidine

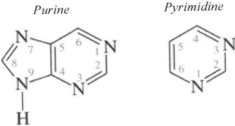

There are four different bases in DNA nucleotides. *Adenine* (A) and *guanine* (G) are purine bases and consist of two nitrogen-containing rings, while *thymine* (T) and *cytosine* (C) are pyrimidine bases and consist of one nitrogen-containing ring.

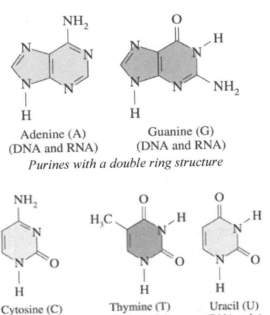

Adenine (A)   Guanine (G)
(DNA and RNA)   (DNA and RNA)

Purines with a double ring structure

Cytosine (C)   Thymine (T)   Uracil (U)
(DNA and RNA)   (DNA only)   (RNA only)

Pyrimidines with a single ring structure

Deoxyribose is a pentose (5-carbon) sugar in DNA. By convention, the carbons of the sugar are numbered. The phosphate group of DNA is attached to the 5' carbon. Deoxyribose has a hydroxyl group (OH) at the 3' carbon (as does the ribose sugar in RNA).

Comparison of the pentose sugars in RNA (left) and DNA (right). Ribose has a 2' hydroxyl, while DNA has no oxygen (deoxy) at the 2' position

RNA has a similar but not identical structure to DNA. The pentose sugar in RNA is *ribose*, which has the same structure as deoxyribose, except RNA has a hydroxyl group instead of hydrogen at the 2' position. Because deoxyribose is missing this hydroxyl group, it is *deoxy* (without oxygen). RNA contains the *uracil* pyrimidine instead of thymine (uracil replaces thymine in RNA). These two pyrimidine bases have similar structures, except for a methyl group (CH_3) present in thymine, but not in uracil.

Chain elongation of DNA with the nucleophilic attack of the 5' carbon phosphate by the 3' OH of the sugar. Elongation always proceeds 3'→5'.

By convention, the 5' phosphate end of a nucleic acid strand on the left,
and the 3' hydroxyl end on the right

Although the primary structures of DNA and RNA are similar, their structures in three-dimensional space (tertiary structure) are particularly distinct. RNA molecules are single-stranded, so base pairs can form between different parts of the same molecule, resulting in shapes such as stem-loops. Double-stranded DNA is a double helix with two strands bonded in an anti-parallel orientation and held together by hydrogen bonds between the bases (C≡G and A=T).

Single-stranded DNA with negatively-charged sugar-phosphate backbone and bases projecting inwards when the second strand of DNA hydrogen bonds

Base pairing specificity: A with T, G with C

In the 1940s, biochemist Erwin Chargaff analyzed the base content of DNA using chemical techniques. Chargaff discovered that for a species, DNA has the *constancy* required of genetic material. This constancy is *Chargaff's rule*, which states that the number of pyrimidine bases (T and C) equals the number of purine bases (A and G). Additionally, the bases always make hydrogen bond base pairs in the same way: the purine A uses a double bond with the pyrimidine T, and the purine G uses a triple bond with the pyrimidine C. This is a *complementary base pairing*. Therefore, Chargaff's rule states that the number of adenine in a DNA molecule equals the number of its base pair, thymidine, and the number of guanine equals the number of its base pair, cytosine. Hence, A = T and G = C.

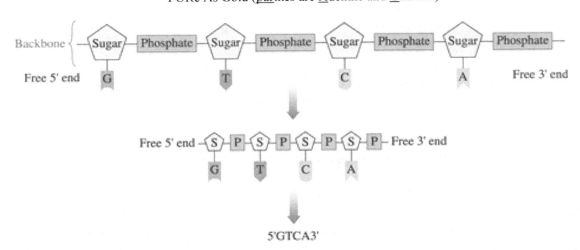

Adenine has 2 hydrogen bonds to Thymine   *Guanine has 3 hydrogen bonds to Cytosine*

Despite the restriction of base pair bonding, the G/C content relative to A/T content differs among species while still adhering to Chargaff's rules. Furthermore, because G/C pairs have three hydrogen bonds as opposed to the two hydrogen bonds in A/T pairs, the strands in DNA molecules with higher G/C content are more tightly bound than those with higher A/T content. Therefore, G/C base pairs have a higher T_m (melting temperature when ½ bonds are broken) than A/T base pairs.

Although there are four bases (A, C, G, T) and two types of base pairs (A/T, C/G) in DNA, the variability in the base sequence is enormous. A human chromosome contains about 140 million base pairs on average, and since any of the four possible nucleotides can be present at each nucleotide position, the total number of possible nucleotide sequences in a human chromosome is $4^{140,000,000}$, or 4 raised to 140,000,000.

Use this pneumonic to remember which bases are purines or pyrimidines:

CUT the PIE (Cytosine, Uracil, and Thymine are pyrimidines)

PURe As Gold (purines are Adenine and Guanine)

Chargaff's rule identifies the amount of all bases when the amount of one purine (A or G) and one pyrimidine (C or T) as a complementary-paired double-stranded DNA molecule

DNA: double helix, Watson–Crick model of DNA structure; RNA Structure

During the 1950s, English chemist Rosalind Franklin produced X-ray diffraction photographs of DNA molecules. Franklin's work provided evidence that DNA has a helical conformation, or more specifically, that the two strands of DNA wind together in a double helix. A double helix can be envisioned as a twisted ladder.

An American, James Watson, and an Englishman, Francis H. C. Crick, received the Nobel Prize in 1962 for their model of DNA. Using information gathered by Chargaff and Franklin, Watson and Crick built a model of DNA in a double helix secondary structure. Sugar-phosphate molecules form a backbone on the outside of the helix, while bases point toward the middle and form base pairs with the complementary strand. Their model was consistent with both Chargaff's rules and the dimensions of the DNA polymer provided by Franklin's x-ray diffraction photographs of DNA.

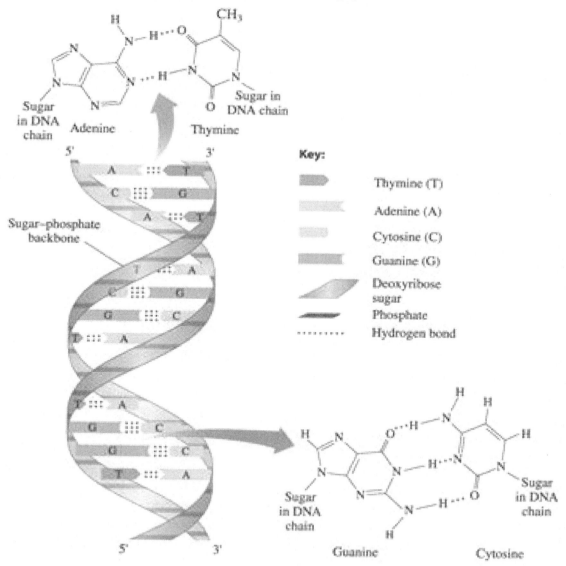

Antiparallel strands of DNA with phosphodiester bonds between the sugar-phosphate backbone.
Hydrogen bonds hold the complementary base pairs together

Each strand in DNA has a direction relative to the numbering on the pentose ring. In a free nucleotide, the phosphate is attached to the 5'− phosphate of deoxyribose, while the 3'− OH of deoxyribose is exposed. When phosphodiester bonds form, a 3' hydroxyl of one deoxyribose sugar attaches to a 5' phosphate of an incoming sugar. Thus, DNA strands have a distinct polarity, with a 5' end and a 3' end. Two strands bound in a double helix orient in opposite directions and the two strands are *antiparallel*.

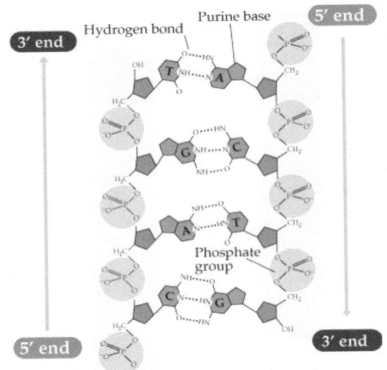

DNA as two antiparallel strands with hydrogen bonds between complementary base pairs

Function in the transmission of genetic information

In 1931, bacteriologist Frederick Griffith experimented with *Streptococcus pneumonia,* a pneumococcus bacterium that causes pneumonia in mammals. He first injected two sets of mice with different strains of pneumococcus: a virulent strain with a mucous capsule (S strain) due to the colonies' smooth appearance, and a non-virulent strain without a capsule (R strain) due to the colonies' rough appearance. Mice injected with the S strain died, while mice injected with the R strain survived.

To determine if the capsule alone was responsible for the virulence of the S strain, Griffith performed two more sets of injections. In one set of mice, he injected S strain bacteria that had been first subjected to heat ("heat-killed bacteria"). These mice survived. In another set of mice, he injected a mixture of the heat-killed S strain and the live R strain. These mice died, and Griffith was even able to recover living S strain pneumococcus from the mice's bodies, despite only heat-killed S strain being injected into the live mice.

Griffith concluded that the R strain had been "transformed" by the heat-killed S strain bacteria, allowing the R strain to synthesize a capsule and become virulent. The phenotype (virulent capsule) of the R strain bacteria must have been due to a change in their genotype (genetic material), which suggested that the transforming substance must have passed from the heat-killed S strain to the R strain. This passing of this unknown substance is *transformation.*

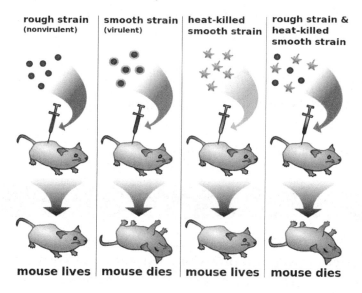

In 1944, molecular biologists Oswald Avery, Colin MacLeod, and Maclyn McCarty reported that the transforming substance in the heat-killed S strain was DNA. This conclusion was supported with several pieces of evidence that showed that purified DNA is capable of bringing about the transformation. In one experiment, they showed that enzymes that degrade proteins (proteases) and RNA (RNase) do not prevent a transformation. However, using enzymes that digest DNA (DNase) does prevent transformation. Additionally, the molecular weight of the transforming substance appeared great enough for some genetic variability. These results supported that DNA is the genetic material and controls the biosynthetic properties of a cell.

In 1952, researchers Alfred Hershey and Martha Chase performed experiments with bacteriophages (the virus that infects bacteria) to confirm that DNA was the genetic material. A *bacteriophage* (phage) is a virus that infects bacteria and consists only of a protein coat surrounding a nucleic acid core. They used the T2 bacteriophage to infect the bacterium *Escherichia coli*, a species of intensely studied bacteria that lives within the human gut.

The purpose of their experiments was to observe which of the bacteriophage components—the protein coat or the DNA—entered the bacterial cells and directed reproduction of the virus. In two separate experiments, they radiolabeled the bacteriophage protein coat with ^{35}S and then the DNA with ^{32}P, and allowed each aliquot of phages to infect bacterial cells. The separate populations of bacterial progeny were then lysed (blender experiment) and analyzed for the presence of isotope-labeled sulfur or phosphorous. The progeny became labeled with ^{32}P, while the sulfur of the progeny was unlabeled, confirming that DNA (contains P), not protein (contains S), is the transmissible genetic material.

Genes are sequences of DNA nucleotides that contain and transmit the information specifying amino acid sequences for protein synthesis. Each DNA molecule contains many genes. The *genome* refers collectively to the total genetic information encoded in a cell. Except for reproductive cells and red blood cells, human cells contain 23 pairs of bundled DNA as *chromosomes* in each cell nucleus, totaling 46 chromosomes per cell.

RNA molecules transfer information from DNA in the nucleus to the site of protein synthesis in the cytoplasm. RNA molecules are synthesized during transcription according to template information encoded in the hereditary molecule of DNA. These RNA molecules are then processed, and mRNA is translated by ribosomes to synthesize proteins.

DNA→ replication during S phase → DNA chromosome with sister chromatids

DNA → transcription → mRNA → translation → protein

DNA Replication

Mechanism of replication: separation of strands, the specific coupling of free nucleic acids, DNA polymerase and primer required

DNA replication is the copying of a DNA molecule. During the S (synthesis) phase of the cell cycle, DNA replicates when the strands of the double helix separate, and each exposed strand acts as a template for DNA synthesis. Free deoxyribonucleoside triphosphates (dNTPs) are base-paired to form new, complementary strands. Errors in the base sequence during replication may be corrected by a mechanism of *proofreading* or DNA repair.

DNA Replication in Prokaryotes and Eukaryotes	
Prokaryotes	Eukaryotes
Five polymerases (I, II, III, IV, V)	Five polymerases ($\alpha, \beta, \gamma, \delta, \varepsilon$)
Functions of polymerase:	Functions of polymerase:
I is involved in synthesis, proofreading, repair, and removal of RNA primers	α: a polymerizing enzyme
II is also a repair enzyme	β: a repair enzyme
III is main polymerizing enzyme	γ: mitochondrial DNA synthesis
IV, V are repair enzymes under unusual conditions	δ: main polymerizing enzyme
	ε: function unknown
Polymerase are also exonucleases	Not all polymerases are exonucleases
One origin of replication	Several origins of replication
Okazaki fragments 1000-2000 residues long	Okazaki fragments 150-200 residues long
No proteins complexed to DNA	Histones complexed to DNA

This text references polymerase III and I; substitute the corresponding polymerases when considering eukaryotic cells. The exact polymerase for a eukaryote's function is still actively investigated by researchers

Replication of linear DNA in eukaryotes starts at multiple points of origin (circular DNA in prokaryotes have a single origin). Once replication is initiated, the DNA strands separate at each of these points of origin as *replication bubbles*. The two V-shaped separating ends of the replication bubble are the sites of DNA replication or *replication forks*. Once a strand of DNA is exposed, the enzyme *DNA polymerase III* for prokaryotes (pol γ for eukaryotes) incorporates free deoxyribonucleoside triphosphates (dNTPs), which are nucleotides with three phosphate groups, into the complementary strand by catalyzing the exergonic loss of phosphate. The two phosphate groups cleaved in the process become nucleotides and release energy ($-\Delta G$), making the overall polymerization reaction thermodynamically favorable, thus driving the reaction forward.

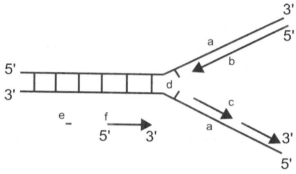

DNA replication: a: template strands, b: leading strand, c: lagging strand, d: replication fork, e: RNA primer, f: Okazaki fragment

Polymerization occurs on both strands and both ends of the replication bubbles until the entire DNA is replicated, a process that results in two complementary DNA molecules. Eukaryotes replicate their DNA at a relatively slow pace of 500 to 5,000 base pairs per minute, taking hours to complete replication, while prokaryotes can replicate their DNA at a much faster rate of 500 base pairs per second.

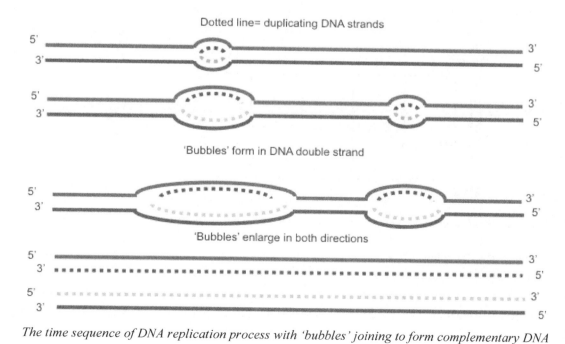

The time sequence of DNA replication process with 'bubbles' joining to form complementary DNA

The process of DNA replication can be divided into three steps:

1. Unwinding—the enzyme *DNA helicase* unwinds the double helix, pulling the DNA strands apart and breaking hydrogen bonds between base pairs. Each separated strand is now a template for the synthesis of a new (daughter) strand of DNA.

2. Complementary base pairing—free dNTPs form hydrogen bonds with their complementary base pair. Adenine pairs with thymine, and guanine pairs with cytosine.

3. Joining—DNA polymerase (III or γ) catalyzes the incorporation of nucleotides into the new strand. Incoming dNTPs cleave two phosphate groups, becoming nucleotides (only one phosphate group remains) as they are incorporated in a 5' to 3' direction, and the deoxyribose sugar and phosphate are covalently added to the new backbone.

Semiconservative nature of replication

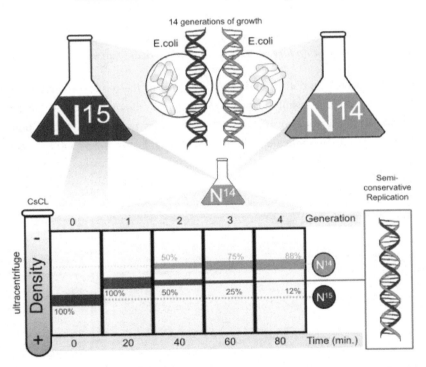

Meselson-Stahl experiment during replication with heavy nitrogen as the original growth medium

In 1958, Matthew Meselson and Franklin Stahl provided evidence for the model of DNA replication. They first grew bacteria in a medium with heavy nitrogen (^{15}N) and then switched the bacteria to light nitrogen (^{14}N) for further divisions. When they measured the density of the replicated DNA using centrifugation, they observed that the density of the replicated DNA was intermediate—less dense than a molecule made entirely with ^{15}N, but denser than a molecule made entirely with ^{14}N.

After one division, only these hybrid DNA molecules (1 light and 1 heavy strand) were present in the cells. After two divisions, half the DNA molecules were light, and half were hybrid. These results support the *semiconservative model,* one of three main theoretical models originally proposed in DNA replication.

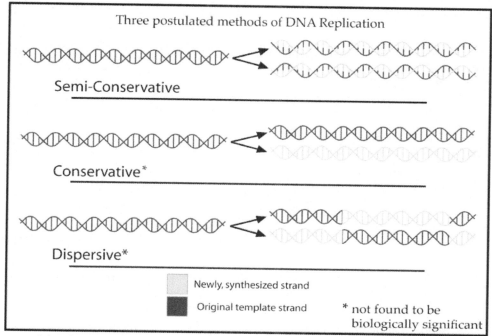

Three models of DNA replication with semi-conservative replication supported by experimentation

DNA replication is semiconservative because each daughter double helix consists of one parental strand and one new strand, meaning that half of the double helix is conserved from material in the parent generation.

In a *conservative model*, one entire double strand acts as the template, while the new double strand is composed *de novo* (i.e., starting from the beginning). If this model were true, Meselson and Stahl's experiment would produce only heavy and light DNA molecules in the daughter cells, with no hybrid strands of intermediate density.

In a *dispersive model*, the parent double-strand is made into two new strands, with each daughter containing a mixture of old and newly-incorporated nucleotides. The two different densities in the DNA of the cells in the second generation of the Meselson and Stahl experiment were inconsistent with the dispersive model because that model would have resulted in DNA of a single density.

Specific enzymes involved in replication

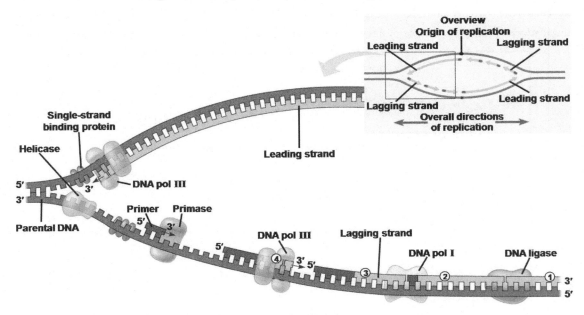

DNA replication with the associated protein involved in strand synthesis.
Pol III is used by prokaryotes and is functionally equivalent to γ in eukaryotes

DNA replication involves a complex of many proteins. The enzyme DNA *helicase* unwinds DNA to expose each template strand for DNA synthesis, forming two Y-shaped replication forks on the two sides of each replication bubble. ATP hydrolysis is required for helicase to break the hydrogen bonds between the complementary strands of DNA. *Single-stranded binding proteins* (SSBP) attach to exposed single strands to keep them from reforming base pair hydrogen bonds, as well as to prevent degradation by DNase. *Topoisomerase* relieves tension in the DNA supercoiling and prevents knots by breaking and rejoining strands. This is essential in allowing helicase to unwind the DNA without causing tension elsewhere along the DNA strand.

DNA polymerase III in prokaryotes (γ in eukaryotes), the primary polymerase responsible for replication, cannot attach directly to an exposed single strand of DNA; it can only start polymerizing from an existing strand, or more specifically from an existing nucleotide 3'−OH. The enzyme *primase* first creates a small sequence of complementary *RNA primer* of approximately 18−22 ribose bases. DNA polymerase begins synthesis after binding to this RNA primer. Primase can begin synthesis *de novo*, unlike DNA polymerase.

DNA polymerase III moves only from 3' to 5' along the template strand, synthesizing a new antiparallel strand; thus the new strand is formed from 5' to 3'. Since the original double strand is antiparallel, polymerization occurs in opposite directions on each strand at a replication fork. Nucleotides are incorporated as they form hydrogen bonds with their complementary base pairs and covalent phosphodiester bonds with their adjacent nucleotides.

The strand synthesized in the same direction as the movement of the replication fork is the *leading strand*, and the strand synthesized in a direction opposite the movement of the replication fork is the *lagging strand*. DNA polymerase (III or γ) can continuously polymerize DNA as the template unzips without detaching from the template, hence the continuous nature of the leading strand. However, DNA polymerase must repeatedly detach and reattach to remain in the proximity of the moving replication fork on the lagging strand. This action produces distinct fragments of new DNA on the lagging strand each time the polymerase detaches and reattaches, known as *Okazaki fragments*. In prokaryotes, Okazaki fragments are 1,000−2,000 nucleotides, while there are 150−200 nucleotides in eukaryotes.

Okazaki fragments each have their RNA primer, which is later replaced with DNA by *DNA polymerase I*. This polymerase can remove the ribonucleotides ahead of it while it synthesizes DNA to replace them. Since polymerases cannot connect separate fragments, *DNA ligase* connects the DNA sugar-phosphate backbones of adjacent Okazaki fragments. This enzyme is necessary because other polymerases can only add a free dNTP to a 3'−OH end, but cannot connect the ends of nucleotides that have already been incorporated. Ligase seals the DNA backbone between Okazaki fragments or when a repair mechanism replaces any nucleotides.

In summary:

1. Helicase uncoils and separates the DNA strands.

2. Primase adds RNA primers to which DNA polymerase III can bind.

3. DNA polymerase III begins polymerizing new DNA strands 5' to 3'.

4. Deoxyribonucleoside triphosphates lose two phosphate groups during incorporation, becoming typical nucleotides (sugar + base + phosphate).

5. The leading DNA strand is synthesized continuously.

6. The lagging DNA strand is synthesized in fragments (Okazaki fragments).

7. DNA polymerase I replaces the RNA primer with DNA.

8. DNA ligase joins Okazaki fragments together into a continuous DNA strand.

Origins of replication, multiple origins in eukaryotes

The average human chromosome contains 140 million nucleotide pairs, and each replication fork moves at a rate of about 50 base pairs per second. At this rate, the replication process would take about a month, but since there are many replication origins on the eukaryotic chromosome, the process takes hours. Replication begins at some origins earlier than others, but as replication nears completion, the replication bubbles meet and fuse to form two new DNA molecules. DNA replication occurs in S phase of interphase and must be completed before a cell can divide (e.g., mitosis or meiosis). Drugs with molecules similar to the four nucleotides (i.e., nucleotide analogs) can be used by patients to inhibit cell division of rapidly dividing cancer cells.

There are many origins of replication in linear DNA of eukaryotes, and several replication forks are formed simultaneously, forming several replication bubbles. In prokaryotes, there is just one origin of replication in each circular DNA molecule and replication occurs at one point. Accordingly, the Okazaki fragments in eukaryotes are shorter (100 – 200 nucleotides), while Okazaki fragments in prokaryotes are longer (1,000 – 2,000 nucleotides). Replication occurs about twenty times faster in prokaryotes than it does in eukaryotes, which undergoes more proofreading during DNA replication.

Replicating the ends of DNA molecules

DNA polymerase only adds nucleotides to a 3' –OH end of a preexisting polynucleotide, i.e., it cannot synthesize *de novo*. This is not an issue for circular DNA in prokaryotes, but it is a problem for the synthesis of the lagging strand at the ends of linear DNA in eukaryotes. Although primase can add an RNA primer to the end of the DNA molecule on the lagging strand, DNA polymerase I cannot replace DNA without RNA primers, since it cannot perform *de novo* synthesis. Without any special process, the RNA segment, as well as its complementary DNA on the opposite strand, would be degraded, since chromosomes are regulated to consist only of complementary DNA. This would lead to degradation of the RNA nucleotides of about 8-12 nucleotides at the ends of DNA strands after each round of replication, eventually encroaching on important genes on chromosomes and leading to cell death.

Telomeres are the end pieces of each chromosome. There are two telomeres on each of the 46 human chromosomes, which adds up to 92 telomeres in total. Their repetitive sequences and associated proteins protect the ends from degradation and provide a way for DNA ends to be replicated without loss of important sequence when the RNA primer initiates replication along the leading strand. In the 1980's, telomeres were proposed for the creation of special segments of DNA that are synthesized by the telomerase enzyme. This enzyme essentially lengthens the ends of DNA with repeating sequences, usually TTAGGG in humans and other vertebrates. The problem with terminal degradation of DNA still occurs, but since extra sequences have been added during embryogenesis and in stem cells, no essential information is lost. In healthy adult cells, telomerase is off.

The telomerase enzyme carries an internal RNA template. It attaches to the end of the DNA molecule and extends the 3' end with additional DNA. The new DNA that is added is complementary to the internal RNA template located on the enzyme. This new DNA is added during embryogenesis, leading and lagging strand synthesis takes place as normal, and a portion of the telomere is lost with each replication cycle during the organism's lifetime.

Since a double-stranded break is often indicative of DNA damage, telomeres have developed associated proteins that inhibit the cell's ability to recognize DNA damage, thereby preventing the unwanted activation of repair mechanisms, cell-cycle arrest or apoptosis.

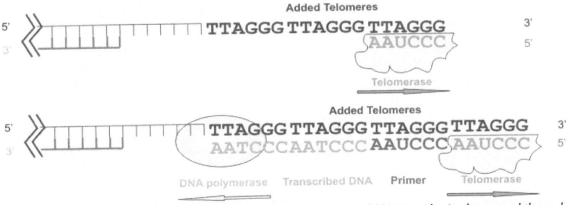

Telomerase has an internal RNA template to which complementary DNA is synthesized to extend the ends of the chromosomes

Repair during replication

The accuracy of DNA replication cannot be attributed solely to the specificity of base pairing, which has an error rate of 1 out of 10^5 base pairs. This rate is not consistent with the observed total error rate of 1 in 10^9 base pairs. To arrive at this fidelity, errors during replication must be repaired, the first of which is proofreading and performed by DNA polymerases.

DNA polymerase is an aggregate of subunits that combine to form an active *holoenzyme* complex. These aggregates often catalyze more than one type of reaction. When polymerases inevitably make errors in polymerization, DNA polymerase I and III use proofreading to make corrections.

Both DNA polymerase I and III have 3' to 5' exonuclease activity. When these polymerases incorporate an incorrect nucleotide into the strand (does not base pair correctly with the complementary strand), the exonuclease subunit breaks the phosphodiester bond at the 5' end, excises the nucleotide, and the polymerase subunit inserts the proper nucleotide. Since polymerases synthesize in the 5' to 3' direction, this excision is named for the complementary strand's 3' to 5' direction.

In addition to 3' to 5' exonuclease activity, DNA polymerase I (but not III) has a 5' to 3' exonuclease. This enzymatic activity allows DNA polymerase I to remove nucleotides ahead of it while synthesizing a new strand at the same time (5' to 3' polymerization). This is the basis by which DNA polymerase I excises ribonucleotides in the RNA primers and replaces them with DNA. This coupling of 5' to 3' exonuclease activity with 5' to 3' polymerization is *nick translation* since a single-stranded cut (nick) essentially translates along the strand as the sequence is replaced with new nucleotides. Ligase must seal the nicks with phosphodiester bonds in the backbone when repairs are made.

Repair of mutations

In addition to proofreading, *mismatch repair* and *excision repair* are two common systems that correct errors in DNA. In mismatch repair, a different group of enzymes detects a mismatched base pair in a double-stranded DNA molecule that has been missed by the DNA polymerase proofreading mechanism. The repair enzyme decides which DNA strand is the template (parent) strand of the new (daughter) DNA molecule by recognizing methylation sites. Newly synthesized DNA is unmethylated; therefore, the base located on the unmethylated strand must be the mismatched base. *Hemimethylation* is the temporary pattern where only one strand is methylated. Since full methylation is eventually reached after a period, mismatch repair is most accurate immediately after DNA synthesis.

Two types of excision repair, *base-excision repair* (BER) and *nucleotide-excision repair* (NER), act on bases with a mutated structure, rather than simply mismatched base pairs. BER generally works on small mutations such as deamination, alkylation or oxidation of the base that does not distort the helical structure. The excision of the damaged base occurs through breakage of the phosphodiester backbone at the resulting abasic site, and gap filling by DNA polymerase replaces the base while ligase seals the backbone. BER is accomplished through the concerted effort of a collection of many enzymes (DNA glycosylases, apurinic/apyrimidinic endonucleases, phosphatases, phosphodiesterases, kinases, polymerases, and ligases).

NER is a similar process, but it is used for mutations that more seriously affect the helical structure. For example, DNA exposure to UV light may cause the dimerization of adjacent thymines and sometimes the dimerization of adjacent cytosines (pyrimidine dimers). These dimers distort the DNA structure, potentially causing problems during replication, resulting in a pre-cancerous state (which is why UV exposure is linked to higher occurrences of skin cancer). NER recognizes the damage, removes the offending stretch of single-stranded DNA, and polymerizes new DNA using the remaining sequence as a template. NER usually replaces a larger region of DNA, rather than a single nucleotide or a small patch as in BER.

Sometimes during replication in bacteria, damage can accumulate so extensively that the NER system cannot keep up, and the *SOS repair system* is activated. SOS repair involves the induction of low-fidelity polymerases to prevent the normal high-fidelity polymerases from getting stuck along the DNA strand during synthesis.

Genetic Code

Central Dogma: DNA → RNA → protein

The *central dogma* of molecular biology describes the flow of genetic information in living systems. It states that information flows from DNA to mRNA to protein. DNA is *transcribed* by RNA polymerase to create mRNA molecules, and mRNA is *translated* by ribosomes to produce the polypeptide chains comprising proteins. The central dogma identifies DNA and RNA as information intermediates; information can flow back and forth between DNA and RNA (exemplified by retroviruses, where reverse transcriptase catalyzes the formation of DNA from RNA), but identifies the protein as an information sink. A protein sequence does not act as a template for the synthesis of DNA or RNA.

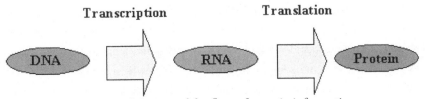

The central dogma of the flow of genetic information

Classical geneticists classify a gene as any of the particles of inheritance on a chromosome, while molecular biologists describe a gene as a sequence of DNA nucleotide bases that encodes for a protein. DNA is responsible for the genotype, or genetic makeup of an organism. Protein is responsible for the phenotype, or the observable characteristics (i.e., physical, physiological, developmental or behavioral traits) of the organism. Phenotypes result from the actions of enzymes, which are biological catalysts (mostly proteins) that regulate biological functions. Any alteration of the processes in the central dogma of molecular biology may affect the formation of proteins and thus affect the phenotype.

In the early 1900s, English physician Sir Archibald Garrod introduced the phrase *inborn error of metabolism*, in which inherited defects could be caused by the lack of an enzyme in a metabolic pathway. Knowing that enzymes are proteins, Garrod suggested a link between genes and proteins. In 1940, George Beadle and Edward Tatum x-rayed spores of red bread mold *Neurospora crassa*. They discovered that some cultures lacked a particular enzyme for growth on the medium, and soon after they found that a single gene was mutated, which resulted in the lack of this single enzyme. From their experiments, they proposed the *one gene-one enzyme hypothesis*, which states that one gene specifies the synthesis of one enzyme.

Later experiments built on this hypothesis. In 1949, biochemists Linus Pauling and Harvey Itano compared hemoglobin in red blood cells of persons with sickle-cell anemia with those of unafflicted individuals. By using electrophoresis to separate molecules by weight and charge, they discovered that the chemical properties of the chain of sickle-cell hemoglobin protein differed from the normal hemoglobin. Years later, biologist Vernon Ingram showed that the biochemical change to the sickle-cell hemoglobin chain was due to the substitution of a nonpolar valine amino acid for the negatively charged glutamate amino acid. Pauling and Itano proposed the *one gene-one polypeptide hypothesis*, which states that each gene specifies one polypeptide of a protein (proteins may contain multiple polypeptide chains). This hypothesis clarified the earlier one gene-one enzyme hypothesis.

The two strands of DNA are named by their relationship to the RNA and protein that their sequences lead to. The *template strand* is used as the template for RNA synthesis and has a complementary sequence to the coding strand. The *coding strand* has an identical sequence to the transcribed RNA but substitutes thymine for uracil. The template strand is the *antisense strand* or *anticoding strand*, and the coding strand is the *sense strand*.

Messenger RNA (mRNA)

RNA (ribonucleic acid) is a nucleic acid that carries a complementary copy of the genetic code of DNA and is translated into the amino acid sequence of proteins. Unlike DNA, RNA a 2'−OH sugar on the ribose instead of deoxyribose (2'− H) and the pyrimidine base uracil replacing thymine. RNA generally does not form helices. *Messenger RNA* (mRNA) is a single-stranded piece of RNA containing the bases complementary to the original DNA strand. The mRNA transcript is synthesized in the nucleus, but after *processing* it is transported into the cytoplasm where ribosomes are located. The ribosomes *translate* the mRNA sequence into amino acids and synthesize the polypeptide chains of proteins.

RNA, like DNA, can be replicated in special cases. However, a single-stranded RNA template must be used to synthesize a complementary strand, and then the new strand must serve as a template for another round of synthesis to create an additional RNA molecule identical to the first template.

Codon-anticodon relationship and degenerate code

Ribosomes read the RNA containing information for the synthesis of protein as a series of base triplets as *codons*. The three bases that make up each codon determine the amino acid that is incorporated into the growing polypeptide as the ribosome moves along the mRNA transcript. Each amino acid is added to the polypeptide chain, linked by peptide bonds between the amino acids. After the completed polypeptide dissociates from the ribosome, special modifications and three-dimensional folding take place (in the ribosome for eukaryotic cells) to form the final protein.

From the four types of ribonucleotide bases (adenine, cytosine, guanine, uracil), there are $4^3 = 64$ different codons possible from a series of three ribonucleotides. However, with few exceptions, there are only 20 naturally-occurring amino acids that make up an organism's proteins. In this way, the code is *degenerate*, meaning that more than one codon may specify the same amino acid.

In 1961, Marshall Nirenberg and J. Heinrich Matthaei assembled the initial relationships between naturally occurring amino acids and the codons that specify them. They found that an enzyme can be used to construct synthetic RNA in a cell-free system. By translating just three ribonucleotides at a time and observing the amino acid incorporated, they began to decipher the triplet code. This demonstrated the role of nucleotides in protein synthesis. Three nucleotides (codon) specify a single amino acid.

Initiation and termination codons (function, codon sequences)

The 64 codons also include 4 special codons. AUG (start codon) codes for methionine and signals the start of translation on an RNA transcript, forming the first amino acid in the nascent polypeptide. Three other codons, UAG, UGA, and UAA, do not encode for any amino acid but signal a ribosome to terminate translation. Although the code is degenerate (a single amino acid is specified by more than one codon), it is *unambiguous*: each nucleotide triplet encodes a single amino acid.

Second letter

First letter	U	C	A	G	Third letter
U	UUU UUC } Phe UUA UUG } Leu	UCU UCC UCA UCG } Ser	UAU UAC } Tyr **UAA Stop** **UAG Stop**	UGU UGC } Cys **UGA Stop** UGG Trp	U C A G
C	CUU CUC CUA CUG } Leu	CCU CCC CCA CCG } Pro	CAU CAC } His CAA CAG } Gln	CGU CGC CGA CGG } Arg	U C A G
A	AUU AUC AUA } Ile AUG Met	ACU ACC ACA ACG } Thr	AAU AAC } Asn AAA AAG } Lys	AGU AGC } Ser AGA AGG } Arg	U C A G
G	GUU GUC GUA GUG } Val	GCU GCC GCA GCG } Ala	GAU GAC } Asp GAA GAG } Glu	GGU GGC GGA GGG } Gly	U C A G

Genetic code with 3 nucleotides specifying an amino acid

From the genetic code, the amino acid sequence of a peptide encoded for by a specific nucleotide sequence can be determined. For example, AUG–CAU–UAC–UAA encodes for: Met–His–Tyr–Stop. Additionally, the degeneracy of the genetic code can be observed in the table. For example, CCC, CCU, CCA, and CCG all encode for the amino acid proline.

Mutations

A genetic mutation is a permanent change in the sequence of DNA nucleotide bases, evading proofreading and repair mechanisms. Major types of mutations include *point mutations*, in which a single base is replaced; *additions*, in which sections of DNA are added; and *deletions*, in which sections of DNA are deleted. The result is the potential for a misread in the DNA nucleotide code or the loss of a gene. Mutation can be categorized by their effect; these categories include *nonsense* mutations, *missense* mutations, *silent* mutations, *neutral* mutations, and *frameshift* mutations. Mutations have consequences that range from no effect to total inactivation of a protein's function.

At some point mutations, the corresponding change in the RNA may cause a change in the resulting polypeptide. For example, a DNA sequence C̲CA mutated to T̲CA would cause the RNA codon G̲GU for glycine to be changed to the A̲GU codon for serine. In this case, a single nucleotide change has caused a single amino acid change. *Missense mutations* cause codons to specify different amino acids. *Nonsense mutations* cause a codon that specifies for an amino acid to change to a stop codon, which results in a truncated (often nonfunctional) protein.

Not all mutations in DNA lead to changes in the resulting protein. For example, a DNA sequence G̲AT mutated to G̲AA would cause the RNA codon C̲UA to change to C̲UU; both codons translate the same amino acid, leucine. This is a *silent mutation* because it cannot be observed in the phenotype (i.e., protein sequence) of the organism.

A *neutral mutation* neither benefits nor inhibits the function of an organism. For example, a mutation leading to an amino acid change from aspartic acid to glutamic acid, both of which have negatively-charged side chains, may not cause a major structural or functional change in a protein. In this example, the organism may be unaffected by the amino acid change. Silent mutations are essentially neutral mutations unless there is some mechanism affected that depends specifically on the DNA or corresponding amino acid sequence.

There are several ways in which a mutation may not have a negative effect. For example, mutations within introns (segments excised when mRNA is processed), may not affect the final protein. Proteins may be unaffected by a particular amino acid change, especially if the new amino acid has similar properties. If a gene is damaged, there may be no negative effect on the organism if there is an intact homologous gene on the paired chromosome able to produce an intact protein. Damage to genes that synthesize amino acids may not affect if that amino acid can be obtained from an external source, such as the medium a bacterium is growing on.

Mutations can result in nonfunctional proteins, and even a single nonfunctioning protein can have dramatic effects. For example, phenylketonuria is a disease that results when the enzyme that breaks down phenylalanine is nonfunctional, causing phenylalanine to build up in the system. Albinism is caused by a faulty enzyme elsewhere in the same pathway. Cystic fibrosis results from the inheritance of a change in a chloride transport protein in the plasma membrane. A faulty receptor for male sex hormones causes androgen insensitivity in men, where the body's cells cannot respond to testosterone and instead develop like a female, even though all the cells have XY sex chromosomes.

Sickle cell anemia is the result of a single base change in the DNA: in the hemoglobin polypeptide chain, a glutamate amino acid is changed to valine at the sixth residue of the β-chain, distorting the structure of red blood cells into a sickle shape. The malformed red blood cells break down at a faster rate, causing anemia (low red blood cell count), and can become lodged within small vessels (capillaries), causing ischemia (restriction in blood supply).

A *frameshift mutation* alters the normal triplet reading frame so that codons downstream from the mutation are out of register and not read properly. They occur when one or more nucleotides are inserted or deleted, resulting in a new sequence of codons and nonfunctional proteins; it may affect the position of the stop codon. For example, if there is a mRNA sequence GAC CCG UAU corresponding to aspartic acid, proline, and tyrosine, a deletion of the first amino acid would cause a frameshift mutation. The mutated mRNA sequence would be ACC CGU AU, and it now encodes for threonine and arginine. The human transposon *Alu* causes hemophilia when a frameshift mutation leads to a premature stop codon in the gene for clotting factor IX.

Spontaneous mutations occur randomly in the cell, but they are rare and due largely to imperfections in the replication machinery. *Mutagens* are environmental agents that produce changes in DNA. Proofreading and other repair mechanisms lower the likelihood of mutation to one out of every billion base pairs replicated, but high exposure to mutagens may increase this rate. Many mutagens are *carcinogens* or cancer-causing agents.

If a mutation occurs in a somatic cell (cell other than an egg or sperm), it affects only the individual organism and can cause conditions like cancer. Future generations can inherit mutations that occur in germ cells (sperm or egg cells) and cause genetic diseases; more than 4,000 genetic diseases have been identified.

Radiation is a common mutagen. X-rays and gamma rays are forms of ionizing radiation that create dangerous free radicals (atoms with unpaired electrons), and ultraviolet (UV) radiation can cause pyrimidines (thymine or cytosine) to form covalent linkages as pyrimidine dimers, cellular repair enzymes must remove that. A lack of repair enzymes can cause xeroderma pigmentosum, a condition leading to a higher incidence of skin cancer.

Organic chemicals can act directly on DNA. The mutagen 5-bromouracil pairs with thymine, so the A–T base pair becomes a G–C base pair. Other chemicals may add hydrocarbon groups or remove amino groups from DNA bases. Tobacco smoke contains some chemical carcinogens. One common chemical mutagen is sodium nitrite ($NaNO_2$), used as a preservative in processed meats. In the presence of amines, sodium nitrite forms nitrosamines, which assist in the conversion of the base cytosine into uracil.

Transposons are DNA sequences that can move within and between chromosomes and can cause mutations when they change the DNA sequence. These "jumping genes" were proposed by Barbara McClintock and first detected in maize (corn). Now transposons have been observed in bacteria, fruit flies and other organisms. Charcot-Marie-Tooth disease is a rare human disorder where muscles and nerves of legs and feet wither away. It is believed to be the result of a transposon that caused the partial duplication of a chromosome, giving the patient three copies of a series of genes, leading to the breakdown of myelin (insulating sheath around neurons) and thus affects nerve transmission. Viruses can cause mutations when they integrate their DNA into the host genome.

Transcription

mRNA composition and structure:
RNA nucleotides, 5' cap, 3' poly-A tail

Messenger RNA (mRNA) is the single-stranded RNA transcript comprising the nucleotide sequence information for the synthesis of polypeptides. It is different from tRNA or rRNA because it has a modified guanine (7-methylguanosine) base 5' cap and a series of adenine nucleotides as a 3' poly-A tail. 5' capping occurs early, often even before the RNA polymerase is finished transcribing RNA. The cap is used by a ribosome for attachment to begin translation; it provides some stability for the mRNA molecule. The polyadenylation at the 3' end of the transcript creates the poly-A tail of approximately 150-200 adenine (A), nucleotides. This is *template-independent* because it does not require a template strand for the polymerization to occur. This tail inhibits degradation of mRNA in the cytoplasm by hydrolytic enzymes. Delay of degradation allows the mRNA to remain in the cell cytoplasm for longer, leading to more polypeptides translated from the same transcript. Polyadenylation can occur in prokaryotes, but it instead promotes degradation.

The region between the 5' cap and the mRNA start codon is the 5' untranslated region (5'–UTR). Similarly, the region between the mRNA stop codon and the poly-A tail is the 3'–UTR. Although UTRs are not translated, they function for stability and localization of the pre-processed mRNA (pre-mRNA, heterogeneous nuclear RNA or hnRNA).

mRNA structure including the untranslated regions (UTRs)

mRNA processing in eukaryotes, introns, and exons

In eukaryotes, the newly-formed RNA (*primary transcript RNA* or *pre-RNA*) are processed before leaving the nucleus to yield a *mature RNA product*. Processing involves capping and polyadenylation (described above). Additional processing that must occur during the formation of the mature RNA molecule is the splicing of the *introns*, the non-coding RNA regions that must be excised. The *exons* are the coding regions that remain in the final transcript. After the introns are excised, the exons are joined in the process of *RNA splicing*. The mRNA exons are eventually translated into polypeptides after leaving the nucleus. The organization of genes into introns and exons is of particular evolutionary importance. Exons generally represent functional protein domains, so splicing and changes in exon composition allow the easy shuffling of protein domains to create new proteins. Splicing differs between developmental stages (i.e., fetus or adult) or tissue-specific differences (e.g., lung or heart).

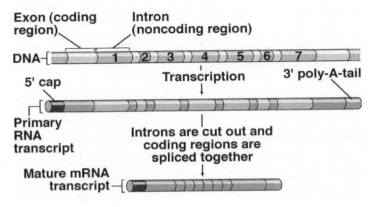

mRNA with introns removed and exons ligated together

An exception for RNA processing involves mitochondria, which have the complete set of machinery needed to produce their proteins, and their circular DNA molecules (like prokaryotes) lack introns.

DNA is not always transcribed into mRNA; it produces other types of RNA such as *transfer RNA* (tRNA) or *ribosomal RNA* (rRNA). tRNA molecules travel out of the nucleus after transcription, where they are "activated" when a tRNA synthetase enzyme attaches the corresponding amino acid. rRNAs (synthesized in the nucleolus within the nucleus), along with a variety of proteins, form the subunits of ribosomes before the subunits migrate out of the nucleus into the cytoplasm.

Mechanism of transcription:
RNA polymerase, promoters, primer not required

Transcription (DNA → RNA) transforms the information in stable DNA into dynamic mRNA, a necessary component for the production of protein.

Transcription takes place in the nucleus, where DNA to be transcribed adopts an "open" conformation, uncoiling and thus exposing the template strand. DNA is usually tightly bound to histones, but the binding of histone deacetylases causes a conformational change in the DNA-histone complex, allowing the association to become loose and open. In an open conformation, exposed DNA promoter regions are likely to be recognized by transcription factors. The *promoter* is regions on DNA where the RNA polymerase binds to begin transcription of mRNA. Promoters are often at 30, 75, and 90 nucleotide base pairs upstream (towards the 5' end) from the *transcription start site* (TSS) where transcription begins.

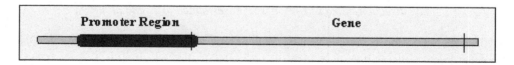

A segment of DNA with upstream promoter region where RNA polymerase binds to initiate transcription

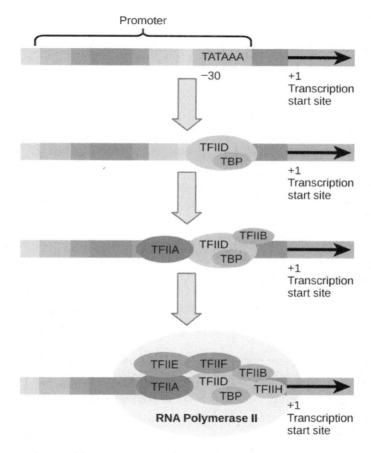

RNA transcription is influenced by transcription factors that increase the affinity of the RNA polymerase and the rate of mRNA synthesis as enhancers or decrease the rate as inhibitors

RNA polymerase uncoils DNA to expose the template strand and processes from 3' to 5' along the DNA, incorporating free ribonucleotide triphosphates (NTPs) into the growing mRNA strand, where these NTPs (ATP, CTP, GTP or UTP) become ribonucleotides. Since RNA polymerase moves from the 3' to the 5' end of a particular DNA template sequence, the RNA transcript is synthesized in a 5' to 3' direction (same as in DNA synthesis), which require a free 3'−OH end.

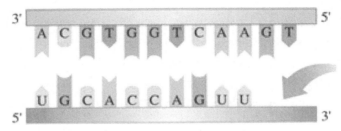

DNA template for transcription 3' A C G T G G T C A A G T 5'

U G C A C C A G U U 5' 3'

The direction of transcription with bases added to 3' end of mRNA

The mRNA strand is complementary to the DNA template and the same as the DNA coding strand (except uracil replaces thymine). Phosphodiester bonds link ribonucleotides as is DNA, but with ribose sugars instead of deoxyribose as for DNA.

RNA polymerization creates transcripts that must dissociate from the template DNA. A short portion of the RNA is base paired with the DNA for the correct sequence to be polymerized, but otherwise, the 5' end of the RNA transcript is not hydrogen bonded to the template strand of the DNA double strand. The strands of DNA then reform their double helix once the newly synthesized RNA dissociates from the DNA.

Unlike DNA polymerase III, RNA polymerase does not require a primer to initiate synthesis; the stretch of DNA template strand that encodes for a single RNA transcript is the *transcription unit* (i.e., promoter, RNA-coding sequence, and terminator).

In addition to the promoter and TSS, eukaryotes can contain a *TATA box* (or *Hogness box*) in their promoter, which is a specialized sequence of thymine and adenine nucleotides. Eukaryotic transcription requires special proteins, or *transcription factors*, to control and enable transcription. Many of these transcription factors bind at the TATA box to regulate transcription. Not all genes have a TATA box. The *transcription initiation complex* is the complete assembly of transcription factors and RNA polymerase bound to the DNA.

Prokaryotes do not require transcription factors; the RNA polymerase recognizes the promoter and begins transcription. *Pribnow box* is a sequence in their promoter similar to the TATA box.

Termination of transcription occurs when RNA polymerization ends, and the RNA transcript is released from the DNA coding strand. Termination in prokaryotes occurs when the RNA polymerase transcribes the *terminator* sequence. Termination is not well understood in eukaryotes, but it includes various protein factors that interact with the DNA strand and the RNA polymerase. Transcription in eukaryotes usually proceeds at least 30 base pairs after the RNA stop codon, and termination usually occurs in one of two ways. *Intrinsic termination* is where specific sequences known as *termination sites* create a stem-loop in the RNA that causes the RNA to dissociate from the DNA template strand. The second mechanism is *rho (ρ) dependent termination*, where the ρ protein factor travels along the synthesized RNA and dislodges the RNA polymerase off the DNA template strand.

Cells produce thousands of copies of the same RNA transcripts. Since there are so many transcripts available for translation, protein synthesis occurs more quickly than if translation were to occur along a single mRNA. In both prokaryotes and eukaryotes, multiple RNA polymerases can transcribe the same template simultaneously.

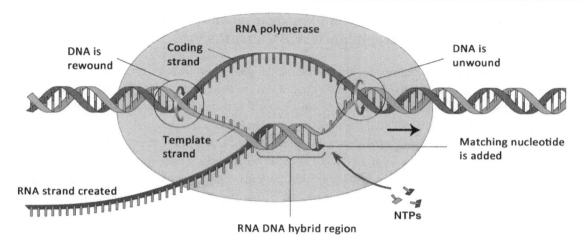

RNA polymerase synthesizes a single strand of RNA complementary to DNA template strand

Summary of transcription:

1. RNA polymerase binds to the promoter region on the DNA, initiating transcription.

2. RNA polymerase uncoils and separates the DNA strands as it synthesizes the RNA strand in a 5' to 3' direction.

3. One DNA strand is used as a template (antisense strand). The other strand (coding strand) is not used, but has the same sequence as the RNA transcript (sense strand), except with thymine (in DNA) instead of uracil (in RNA).

4. Free nucleoside triphosphates (NTP) are incorporated, losing two of their phosphates. The new RNA nucleotides temporarily form base pairs with the template DNA.

5. The terminator sequence causes the RNA polymerase to stop transcription and separate from the DNA, and the DNA rewinds.

6. RNA destined for protein translation is messenger RNA (mRNA).

7. mRNA processes in the nucleus with splicing (remove introns and join exons), adding 5' G−cap and 3' poly−A tail.

8. Processed mRNA migrates through nuclear pores to enter the nucleus for translation into protein.

Ribozymes, spliceosomes, small nuclear ribonucleoproteins (snRNP), small nuclear RNA (snRNA)

RNA splicing is necessary for the production of processed mRNA. Originally it was thought that eukaryotic genomes are completely continuous (as bacterial genomes), but in the late 1970s, sequencing technology became sophisticated enough to allow the comparison of vertebrate mRNA and DNA sequences. The results replaced the hypothesis that mRNA is a direct copy of DNA, and marked the first real conception of differential gene expression. A remarkable facet of eukaryotic biology is that every cell has the same genetic information but is capable of making completely different proteins. For example, a duct cell in the kidney makes certain proteins that allow the passage of ions across a membrane to direct the flow of water, but a neuron (using the same genome) synthesizes completely different proteins to maintain and harness a membrane potential.

Splicing does not often decide cell fate, but it does allow for the synthesis of different proteins from the same DNA template by ligating (joining) exons and removing introns. *Small nuclear RNA* (snRNA) are special types of RNA that combine with proteins to form *small nuclear ribonucleoproteins* (snRNPs), which make up *spliceosomes*, the complexes that perform RNA splicing. The spliceosome recognizes splice sites (exon and intron boundaries) and enzymatically forms stretches of RNA as *lariats* from the intron sequences. The lariat is then cleaved from the transcript, and the remaining exons are ligated together. Splicing, like all mRNA processing, occurs in the nucleus, separate from the ribosomes. After the RNA is fully processed, it is exported from the nucleus via nuclear pores into the cytoplasm to be translated. Splicing does not occur in bacteria since they do not have introns; instead, translation often occurs immediately after transcription since both processes are in the nucleus.

Alternative splicing allows an array of unique mRNAs to be generated from the same primary RNA transcript. A change in the splice recognition site can cause a putative exon in a different transcript to become an intron in another transcript (i.e., cell type or stage of development). In this way, different proteins can be constructed by shuffling exons. This is done by selectively removing parts of a primary RNA transcript and arranging different combinations; as a result, each mRNA encodes for a different protein product.

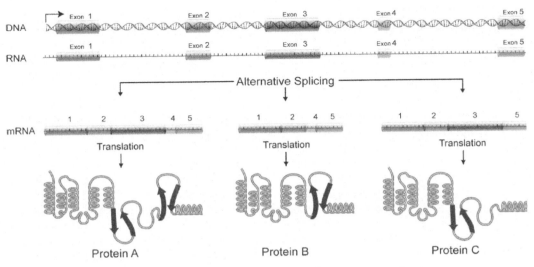

DNA alternative splicing allows the same genome to transcribe tissue or developmentally-specific proteins for specific functions

---

In 1989, American scientists Sidney Altman and Thomas Cech were awarded the Nobel Prize for their discovery that some RNA molecules have an enzymatic function. *Ribozymes* are RNA molecules that include the snRNA involved in RNA splicing and the RNA molecules in the protozoan *Tetrahymena*, which catalyze condensation and hydrolysis of phosphodiester bonds. It is postulated that RNA has served as both the genetic material and as enzymes in early life forms. The ribozyme suggests that RNA is the answer to the persistent uncertainty about whether DNA or RNA came first in evolutionary history.

Functional and evolutionary importance of introns

The role of RNA non-coding (intronic) regions of the genome is still contested. Molecular biology is moving toward understanding that these regions are important in the regulation of gene products. Intron sequences often contain short stretches of RNA, known as small interfering RNA (siRNA), which have important effects in regulating gene expression. The evolutionary importance of the noncontinuity of the genome is controversial. Some researchers assert that because spliceosome-splicing is not conserved in prokaryotes, it can have only limited importance to the origin of species. However, others hypothesize that the ability to shuffle genes is important to the evolution of unique phenotypes within a population.

Introns can provide important evolutionary advantages, particularly because they can enable alternative splicing. For example, the thyroid and pituitary glands use the same primary mRNA transcript, but via alternative splicing produce different proteins. Investigators have found that the simpler the eukaryote, the less likely is the presence of introns. Though introns are mostly restricted to eukaryotes, an intron has been discovered in the gene for a tRNA molecule in the cyanobacterium *Anabaena*; this particular intron is *self-splicing* (similar to a ribozyme) and capable of splicing itself out of an RNA transcript.

Translation

Roles of mRNA, tRNA, and rRNA; RNA base-pairing specificity

Messenger RNA (mRNA) molecules containing information transcribed from DNA are transported into the cytosol from the nucleus after processing (e.g., 5' cap, 3' poly-A tail and splicing – removal of introns and ligation of exons). The mRNA contains the information necessary for ribosomes to assemble amino acids into polypeptides, the building blocks of proteins.

Translational control occurs in the cytoplasm after mRNA leaves the nucleus, but before there is a protein product. The life expectancy of mRNA molecules and their ability to bind ribosomes can vary. The longer an active mRNA molecule remains in the cytoplasm; the more proteins are synthesized. Some mRNAs may need additional changes before they are translated.

Ribonucleases are enzymes that degrade RNA (e.g., mRNA). Mature mRNA molecules contain a 5'− cap and 3'− poly-A tail, non-coding segments that influence how long the mRNA can avoid being degraded by ribonucleases. An example of translational control is mature mammalian red blood cells that eject their nucleus, but synthesize hemoglobin protein for several months, so the mRNAs in red blood cells must persist during this time since no additional RNA is transcribed. Another example of translational control involves frog eggs that contain mRNA as "masked messengers" that are not translated until fertilization occurs. When fertilization of the frog egg occurs, the mRNA is "unmasked," and there is a rapid synthesis of proteins.

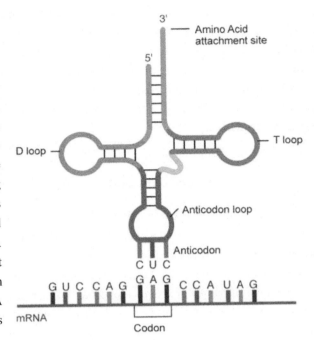

Transfer RNA (tRNA) is the interpreter of the codons contained in a mRNA. tRNA associate each three-nucleotide anticodon with a particular amino acid and transfer the corresponding amino acids to growing polypeptides. tRNA carries a single amino acid on its 3' end and have an anticodon segment. An *anticodon* (contained within the *anticodon loop*) is a special three-nucleotide sequence on the tRNA molecule that base-pairs with a complementary three-nucleotide codon on the mRNA. After a tRNA activated with an amino acid base pairs (via hydrogen bonds) with a codon, the ribosome incorporates the amino acid into the growing polypeptide. The tRNA, now free from the amino acid which is part of the growing polypeptide, dissociates and returns to the cytoplasm, ready to become charged by binding another specific amino acid at its 3'− end.

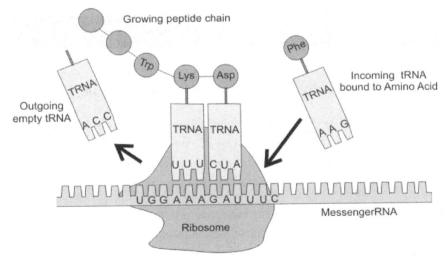

Peptide synthesis using codons on mRNA and anticodons on tRNA with associated amino acids

There are many tRNA with specific anticodons which are complementary to codons. A tRNA-activating enzyme (aminoacyl tRNA synthetase) charges the 3'−end of a tRNA with the correct amino acid, a process of *aminoacylation*. There are twenty tRNA-activating enzymes, corresponding to each of the 20 unique amino acids. The different chemical properties and three-dimensional structures of each type of tRNA allow the tRNA-activating enzymes to recognize their specific tRNA. When the amino acid is attached to the tRNA molecule, a high-energy bond is created using ATP. The energy stored from this high energy bond is used to transfer and bind the amino acids to the growing polypeptide chain during translation.

A recent analysis of entire genomes revealed that some organisms do not have genes for all twenty aminoacyl-tRNA synthetases. They do, however, use all twenty amino acids to construct their proteins. The solution, as is often the case in living cells, is that more complex mechanisms are used. For instance, some bacteria do not have an enzyme for charging glutamine onto its tRNA. Instead, a single enzyme adds glutamic acid to the glutamic acid tRNA molecules and the glutamine tRNA molecules. A second enzyme then converts the glutamic acid into glutamine on the latter tRNA molecules, forming the proper pairing of the tRNA with the amino acid.

The third base of a mRNA codon is the *wobble position* because there is some "flexibility" in the third position of codon-anticodon (hydrogen bonding) base pairing. For example, the base U of a tRNA anticodon in the third position can base pair with either A or G. However, the most versatile tRNA have the modified base inosine (I) in the wobble position, because inosine can form hydrogen bonds with U, C or A. Wobble explains why degenerate (synonymous) codons for a given amino acid only differ by the third position. It allows a tRNA holding a particular amino acid to potentially bind to multiple codons with a different third base, that each encodes for the same amino acid specific to that tRNA.

Ribosomal RNA (rRNA) is synthesized from a DNA template in the nucleolus (organelle within the nucleus). Many proteins are transported from their site of synthesis in the cytoplasm into the nucleus, where rRNA and proteins form the small subunit (the 30S for prokaryotes and 40S for eukaryotes) and the large subunit (50S for prokaryotes and 60S for eukaryotes) of the complete ribosome. These two subunits travel out to the cytoplasm through the nuclear pores, where they join to form the complete ribosome (the 70S for prokaryotes and 80S for eukaryotes) when translation occurs. Each ribosome is composed of dozens of associated proteins. The "S" stands for Svedberg units, which measures density and corresponds to the sedimentation value during configuration of particles.

tRNA and rRNA composition and structure (e.g., RNA nucleotides)

Each tRNA molecule is a single strand containing 75 to 95 nucleotides, and particular sequences on this single strand form base pairs with other parts of the molecule, forming a tightly-compacted T-shaped structure. Most of the ribonucleotides in tRNA are the normally-occurring RNA bases (A, C, G, and U), but there are some variant bases, such as pseudouridine, that result from modifications (alkylation, methylation, and glycosylation) to the typical bases that occur after RNA transcription. These modified bases, which usually occur in restricted sites of the tRNA molecule, allow for the formation of unusual base pairs.

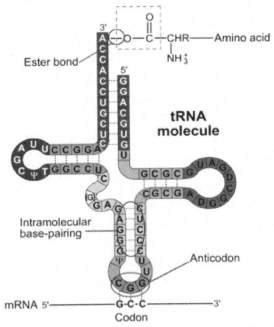

tRNA with amino acid joined at the 3'−OH end, creating a phosphodiester bond

The secondary (two-dimensional) structure of a tRNA molecule resembles a cloverleaf while the tertiary (three-dimensional) structure is L-shaped. Five regions in tRNA are not base-paired: the CCA acceptor stem, the D-loop, the TΨC loop, the anticodon loop, and the extra arm. The nucleotide sequence CCA is at the 3'−OH end of the tRNA and allows for the attachment of an amino acid by a phosphodiester bond, creating a charged tRNA. Which amino acid is attached depends on the anticodon, the three-base sequence binding to a complementary triplet codon on the mRNA according to the base-pairing rules. Each tRNA has a slightly different chemical property and three-dimensional structure, which allows the tRNA-activating enzyme to attach the correct amino acid to the 3'−OH end of the tRNA. The cell's cytoplasm contains all twenty amino acids either by synthesizing them or importing them into the cell.

Ribosomal RNAs, which are structural components of the ribosome, perform critical functions for protein synthesis. rRNAs are synthesized in the nucleolus, while ribosomal proteins are synthesized in the cytoplasm and are brought to the nucleolus to be joined with the rRNAs for the assembly of the two ribosomal subunits: the *large subunit* and the *small subunit*. The rRNA in the large subunit has ribozyme activity and catalyzes the formation of peptide bonds between adjacent amino acids. The secondary structure of rRNA is extensive, and it plays an important role in recognizing tRNA and mRNA that bind to the ribosome. Secondary rRNA structure has been conserved throughout evolution. Prokaryotes contain 16S (in small subunit), 23S and 5S (both in the large subunit) rRNA, while eukaryotes contain 18S (in small subunit) and 5S, 5.8S, 28S (in large subunit).

Role and structure of ribosomes

The ribosome is composed of a few rRNA molecules and many proteins. Each ribosome consists of a small subunit and a large subunit. Prokaryotic ribosomes are the 70S (30S small subunit + 50S large subunit), while eukaryotic ribosomes are 80S (40S small subunit + 60S large subunit). Note that the Svedberg units for the subunits do not add up to the total sedimentation coefficient for the whole ribosome due to differences in density of the complete structure compared to the individual subunits.

In ribosomes, the ribonucleotide sequence of mRNA is interpreted and synthesized into an amino acid sequence. The mRNA strand fits into a groove on the small subunit with the bases pointing toward the large subunit. The ribosome acts as a "reader," and when it reaches a termination sequence in the mRNA, the link between the synthesized polypeptide chain and tRNA is broken. Then, the completed polypeptide is released from the ribosome.

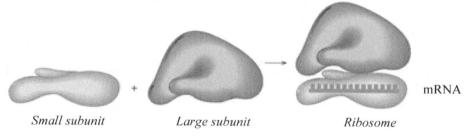

Small subunit   *Large subunit*   *Ribosome*   mRNA

Ribosomes can float freely in the cytosol or attach to the endoplasmic reticulum (ER); the *rough ER* is due to its appearance caused by the ribosomes studding its surface. Prokaryotic cells contain about 10,000 ribosomes, and eukaryotic cells contain many times more. Ribosomes have binding sites for mRNA and tRNA molecules.

Free ribosomes synthesize proteins that are primarily used within the cytosol of the cell. As small proteins emerge from the ribosome, they undergo folding. Larger proteins fold within the recess of small, hollow chambers in *chaperones* proteins. *Bound ribosomes* on the rough endoplasmic reticulum usually synthesize proteins used for the secretory pathway (e.g., secreted from the cell, embedded into membranes or targeted for organelles). Multiple ribosomes can simultaneously translate the same mRNA with elongating polypeptides of different lengths about the ribosomes' progress. A group of ribosomes on a single RNA is a *polyribosome* (polysome). However, mRNA molecules cannot be translated indefinitely and are eventually degraded into ribonucleotides by cytoplasmic enzymes.

A ribosome has three sites that can each hold a tRNA: the *E* (exit) *site, P* (peptidyl) *site* and *A* (aminoacyl) *site*. The sites are in the order of the E site on the 5' of the mRNA held by the ribosome and the A site on the 3'−side. The E site holds a discharged tRNA that is ready for dissociation, the P site tRNA holds the growing polypeptide, and the A site tRNA holds the next amino acid to be added to the polypeptide chain.

Initiation and termination co-factors

Translation, the process by which a mRNA nucleotide sequence is read as triplet codons to assemble a polypeptide chain, involves three steps: chain initiation, elongation, and termination. The mRNA codons base pair with the tRNA anticodons, which represent specific amino acids based on the nucleotide sequence. Enzymes are required for all three steps, and energy is needed for the first two steps. Once the polypeptide is fully formed and translation has terminated, the ribosome then dissociates into its two subunits.

Chain initiation is the first step in translation. Before translation begins, the two subunits of the ribosome are not yet combined. They are assembled separately in the nucleus and travel through the pores of the nuclear envelope and combine in the cytoplasm to form a functional ribosome when translation starts. Since translation moves 5' to 3' along the mRNA strand, the small subunit first binds the 5' end of the mRNA. The subunit moves along the transcript until it finds the start codon, AUG. The tRNA for AUG (attached amino acid is methionine) hybridizes to the start codon using its anticodon, and the large subunit binds the small subunit, forming the complete ribosome. This first tRNA is the *initiator tRNA*.

Initiation factor proteins drive this initial binding. The activity of these factors regulates the rate of protein synthesis. The initiation phase is the slowest of the three phases of the assembly process.

Chain elongation is the process of adding new amino acids to the growing polypeptide chain. After initiation, the tRNA holding the first amino acid (methionine) is bound to the P site.

The A site is now empty and has an exposed codon for the next amino acid. The tRNA with the appropriate anticodon then binds the A site, which requires hydrolysis of a GTP. Elongation factor proteins facilitate this base pairing. With the two amino acids on the tRNAs now nearby, the methionine forms a peptide bond with the amino acid held in the A site, dissociating from the tRNA in the P site at the same time. A ribozyme catalyzes this transfer in the large subunit. The tRNA in the A site now holds two amino acids, with the methionine at the N-terminus (the end with the free amino group) of the growing polypeptide and the newest amino acid attached to the tRNA.

The mRNA transcript now slides through the ribosome, 3 ribonucleotides forward. The tRNA anticodons are still associated with their matching codons, so this movement causes the tRNA to move sites, a process of *translocation*. The discharged tRNA in the P site moves to the E site, and the tRNA holding the growing polypeptide moves to the P site. The A site is now empty and ready for a charged tRNA to recognize the next codon. As the tRNA holding the next amino acid binds the A site, the discharged tRNA in the E site dissociates away from the ribosome, where it is free to associate to a new amino acid in the cytoplasm.

A peptide bond attaches the newly-arrived amino acid. After translocation, the tRNA attached to the recent amino acid moves into the P site, and the tRNA formerly attached to A site moves to the E site and is released. An amino acid–tRNA complex is in the P site.

This process of elongation continues as more codons on the mRNA move through the ribosome. The growing polypeptide chain elongates and is passed from tRNA to incoming tRNA, and the N-terminus of the polypeptide emerges. This elongation step is rapid and occurs about 15 times per second in *E. coli*.

The elongation repeats until the ribosome eventually reaches one of three stop codons on the mRNA, which leads to *chain termination*. No tRNA anticodons recognize the stop codons; instead, *release factors* bind to the stop codon, causing translation to stop. This binding causes the addition of a water molecule instead of an amino acid to the polypeptide chain. The polypeptide chain is released, the uncharged tRNA dissociates from the ribosome, and the ribosomal subunits dissociate.

Summary of translation:

1. mRNA attaches to a small subunit of the ribosome and binds to a charged tRNA molecule (with an attached amino acid) based on its specific codon sequence.

2. The large subunit of ribosome joins the complex.

3. Initiator tRNA resides in P site of the ribosome, and a new tRNA recognizes the next codon sequence on mRNA and attaches to the A site of the ribosome.

4. A peptide bond is formed between the amino acid attached to the tRNA in the P site and the amino acid attached to the tRNA in the A site.

5. The uncharged tRNA in the P site moves to the E site, where it is then released, and the tRNA in the A site moves to the P site.

6. Another tRNA binds to the A site, and the pattern continues, creating a growing polypeptide chain until a termination codon is reached.

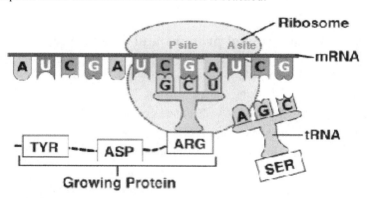

Post-translational modification of protein

Post-translational control regulates the activity of the protein in the cell after translation. For a polypeptide product of translation to become a functional protein, post-translational modifications are made. These modifications include bending and twisting the chain into the correct three-dimensional shape (protein folding), sometimes facilitated by chaperone molecules. The growing polypeptide folds into its tertiary structure, forming disulfide links, salt bridges, or other interactions that make the polypeptide a biologically-active protein. Other changes include additions to the polypeptide chain, such as carbohydrate or lipid derivatives that may be covalently attached when the functional protein is folded. The initial amino acid methionine is often removed from the beginning of the polypeptide. Some molecules are composed of multiple polypeptide chains that must be joined to achieve the final protein (quaternary structure).

Post-translational control may involve degradation to "activate" a protein. For example, the bovine protein proinsulin is inactive when first translated, but after a sequence of 30 amino acids is removed from the middle of the chain and disulfide bonds join the two pieces, the protein becomes active. In other cases, proteins are degraded to cause deactivation. *Proteasomes* (enzymes that target proteins) are large protein complexes that carry out this task. For example, cyclins that control the cell cycle are only present temporarily and must be degraded.

Proteins to be secreted from a cell have a signal sequence that binds to a specific membrane protein on the surface of the rough endoplasmic reticulum. Early during translation, the protein is fed into the lumen of the rough ER; the signal sequence is removed. Once the protein is properly folded in the rough endoplasmic reticulum, portions of the endoplasmic reticulum bud off, forming vesicles that contain the properly folded protein. The vesicles migrate to the Golgi apparatus and fuse with the Golgi membrane. Within the Golgi, carbohydrates and other groups may be added or removed according to final destinations of the proteins (almost all secreted proteins are glycoproteins). The proteins are then again packaged into vesicles that bud off the surface of the Golgi membrane and may travel to the plasma membrane (secretory pathway), where they fuse and release their contents in the extracellular fluid through *exocytosis*.

Eukaryotic Chromosome Organization

Chromosomal proteins and supercoiling

Histones are positively-charged chromosomal proteins responsible for the compact packing and winding of chromosomal negatively-charged DNA. A histone protein octamer and a histone H1 protein (nucleosome) form a protein core around which DNA winds to achieve a compact state. Nonhistone chromosomal proteins are associated with the chromosomes. They have various functions, such as regulatory and enzymatic roles. An active area of research is chromatin remodeling via histones to regulate gene expression within cells.

octamer of core histones:
H2A, H2B, H3, H4 (each one ×2)

core DNA

histone H1

linker DNA

The nucleosome consists of an octamer and a histone H1 between the linked regions of the chromosome

A chromosome consists of a single DNA molecule wound tightly around thousands of *histone* proteins. The basic unit of compact DNA is a *nucleosome*, which consists of negatively-charged DNA wound around a positively-charged histone octamer core and held in place by an additional histone H1. A nucleosome consists of two H2A, two H2B, two H3, two H4 and one H1 histone.

Chromatin is a strand of nucleic acid and associated protein. A nucleosome is a bead-like unit made of DNA wound around a complex of histone proteins. When DNA is wrapped around several of these nucleosomes in sequence, the resulting structure looks like beads on a string.

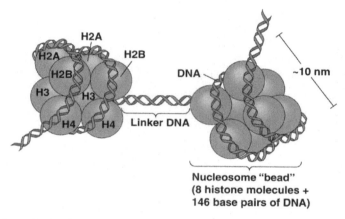

Nucleosomes form the basic unit of coiling in DNA. In turn, these nucleosomes then form higher-order coils, as *supercoils*. The level of supercoiling influences transcription, with a decreasing level of transcription for more compacted DNA. Human DNA is separated into 46 compact, supercoiled pieces (organized into 23 pairs) with the help of nucleosomes and other proteins. These separate pieces of nucleic acid comprise the chromosomes.

Single copy vs. repetitive DNA

Highly repetitive base sequences in DNA are between 5 and 300 nucleotide bases long and may be repeated up to 10,000 times. They are not translated into proteins. Highly repetitive base sequences constitute 5–45% of eukaryotic DNA. Single-copy genes (unique genes) are transcribed and translated to constitute a small proportion of eukaryotic DNA.

Centromeres

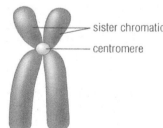

A *centromere* is a region of heterochromatin on the chromosome that can be at the center (metacentric) or close to one of the ends (telocentric). After replication, sister chromatids are attached at the centromere. During mitosis, spindle fibers (comprised of tubulin) are attached to the centromere (via the kinetochore) and pull the sister chromatids apart during anaphase. During anaphase of mitosis, the centromere splits and the sister chromatids become chromosomes in the daughter cell during cell division.

Control of Gene Expression in Eukaryotes

Transcription regulation

Transcriptional control in the nucleus determines which structural genes are transcribed as well as the rate of their transcription. It includes the organization of chromatin and the protein transcription factors initiating transcription. Regulatory proteins include repressors and activators, and they influence the attachment of RNA polymerase to the promoter region on the DNA.

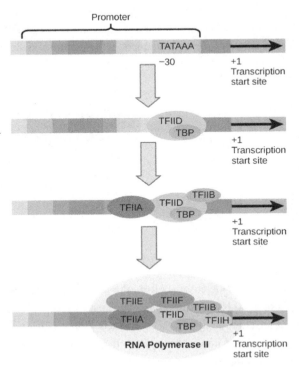

RNA transcription is influenced by transcription factor proteins that increase the affinity of the RNA polymerase and the rate of mRNA synthesis as enhancers or inhibitors

DNA binding proteins and transcription factors

Transcription factors are positively charged proteins that have DNA-binding domains that allow them to bind to the promoter, enhancer and silencer regions of DNA to regulate transcription. *Enhancers* increase transcription when bound, while *silencers* decrease it.

Transcription factors can be influenced by intracellular or extracellular signals, accounting for the wide variation in gene expression in different cell types. Many pathways and types of signals exist to influence the activity of transcription factors, including the allosteric regulation of proteins as well as covalent modifications by kinases, phosphatases, and other enzymes. The DNA-binding domains themselves are varied in the way they interact with the DNA double helix. Some common domains include the helix-turn-helix (HTH), the zinc finger, and the basic-region leucine zipper (bZIP).

Pre-initiation complex forms when transcription factors gather at the promoter region (segment of DNA where RNA polymerase binds) adjacent to a structural gene. The transcription factor complex leads to activation (or repression) of the gene. The complex may attract and bind RNA polymerase, or even promote the separation of DNA strands, but transcription may or may not begin at this point, depending on which transcription factors (activators or repressors) are bound.

While promoters are generally close to the affected gene in both prokaryotes and eukaryotes, eukaryotic regulatory elements (i.e., enhancers and silencers) can be far from the promoter – even thousands of nucleotides away along the DNA strand that bends to stabilize the structure. This is not true for prokaryotic regulatory elements. Since the enhancers and silencers must interact with the promoter to influence transcription, eukaryotic DNA can loop back on itself so that the transcription factor bound to the enhancer or silencer can make contact with the promoter or RNA polymerase. Intermediate proteins between the transcription factors and RNA polymerase are often involved in the process.

Eukaryotes differ in that they lack certain special transcription regulation mechanisms in bacteria, such as the operon (except in rare cases) and attenuation. The *operon* (e.g., *lac* or *trp* operon) is a cluster of tandem genes in bacterial DNA under the control of a single promoter. Transcription of the operon results in several genes being transcribed simultaneously. *Attenuation* is a process where transcription and mRNA structure can influence ribosome translation. It is only possible in prokaryotes because transcription and translation can occur simultaneously. In eukaryotes, the two processes are separate; transcription occurs in the nucleus and translation occurs in the cytoplasm.

Gene amplification and duplication

Gene duplication (gene amplification or chromosomal duplication) is a mechanism by which genetic material is duplicated, and can serve as a source for molecular evolution. There are many ways in which gene duplication can occur. *Ectopic recombination* occurs during unequal crossing over between homologous chromosomes (during meiosis) due to the DNA sequence similarity at duplication breakpoints.

Replication slippage arises from an error in DNA replication; the DNA polymerase dissociates and then reattaches to the DNA at an incorrect position and mistakenly duplicates a section. *Aneuploidy* (an abnormal number of chromosomes, often harmful) is another example of gene duplication, as is *polyploidy* (whole genome duplications), which are both due to *nondisjunction*, the failure of sister chromatids (i.e., mitosis) or homologous chromosomes (i.e., meiosis) to separate properly during cell division.

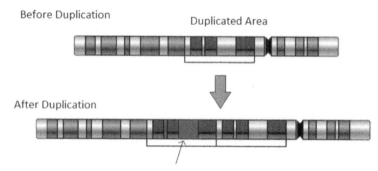

Point mutations are common in duplicated regions and may accelerate the evolution of new proteins

Gene duplication may be evolutionarily advantageous because it creates genetic redundancy, and a mutation in the second copy of a gene may not have harmful effects on the organism because the original gene can still function to encode functional protein products. Since mutations of the second copy of a gene are not directly harmful, mutations accumulate more rapidly (in the duplicated region) than usual, and the second copy of the gene can even develop a new function. Therefore, gene duplication is believed to have played an important role in evolution.

Post-transcriptional control, the basic concept of splicing
(introns, exons)

Post-transcriptional control occurs in the nucleus after DNA has been transcribed and mRNA has formed. In this regulation, the RNA strands are processed before they leave the nucleus with certain variations for a particular effect in the cell.

Timing is one form of control. The speed at which mRNAs leave the nucleus affects the ultimate amount of gene product available per unit of time. Various mRNA molecules may differ in the rate they travel through the nuclear pores. Additionally, mRNA must eventually be degraded, and the rate and timing of mRNA degradation, controlled by the cell, affect how much protein is translated. Some modifications, such as the addition of a 5' cap and a 3' poly-A tail, affect control by protecting the mRNA from ribonuclease degradation.

Post-transcriptional control affects the sequences present in the final RNA products. Alternative splicing is the process by which introns are removed (cut from the transcript), and exons are ligated (rejoined) in different ways, forming different mRNA products from the same initial hnRNA transcript. Both the hypothalamus and the thyroid gland contain the gene that encodes for the peptide hormone calcitonin, but the mRNA that leaves the nucleus, and therefore the translated protein, are not the same in both types of cells.

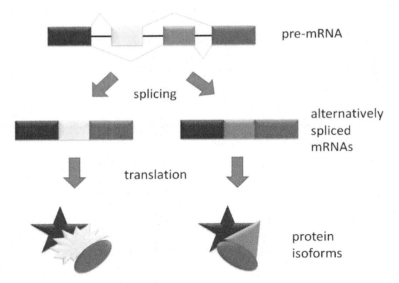

Different types of RNA are subject to post-transcriptional control. For example, special modifications to nucleotides are made to control the structure of tRNA and rRNA.

Alternative splicing of the calcitonin gene occurs in the hypothalamus, leading to the production of a distantly-related peptide, the calcitonin-gene-related peptide (CGRP), while the thyroid gland produces regular calcitonin. Radioactive labeling experiments show different splicing in these strands. Evidence of alternative splicing has been found in cells that produce neurotransmitters, muscle regulatory proteins, and antibodies. Additionally, experiments indicate that alternative splicing occurs at different stages of development (i.e., embryogenic vs. adult cells).

Cancer as a failure of normal cellular controls, oncogenes and tumor suppressor genes

Cancer is a disorder that arises from mutations in the somatic cells and results from the failure of the control system that regulates cell division, which leads to uncontrolled growth. Cancer may develop in different tissues and has special terminology depending on the location and the way it develops. *Carcinomas* are cancers in epithelial cells, *sarcomas* are cancers in muscle cells, and *lymphomas* are cancers involving white blood cells. The lungs, colon, and breasts are the organs most commonly affected by cancer. The incidence of cancer increases with age due to the accumulation of defective mutations (i.e., mutagen exposure or errors during DNA replication).

Oncogenes are dominant cancer-producing genes that encode for abnormal forms of cell surface receptors that bind growth factors, producing a continuous growth signal. Oncogenes cause cancer when they are activated. The products of many oncogenes are involved in increasing cell division. Before an oncogene is activated, it may be a harmless *proto-oncogene*. Researchers have identified many proto-oncogenes whose mutation to an oncogene cause increased growth and leads to tumor formation. The *Ras* family of genes is the most common group of oncogenes implicated in human cancers. The alteration of one nucleotide pair converts a normal functioning *ras* proto-oncogene to an oncogene.

Tumor suppressor genes are recessive cancer-producing genes with mutated forms. Tumor suppressor genes protect cells from becoming cancerous by inhibiting tumor formation through the control of cell division. Mutations in tumor suppressor genes alter the protective proteins encoded by these genes and disrupt their function and thus can lead to cancer. A *tumor* is an abnormal replication of cells that form a mass of tissue. If the cells remain localized, the tumor is *benign* (i.e., remains localized), but if the tumor invades the surrounding tissue because it undergoes metastasis, it is *malignant.*

A major tumor-suppressor gene, *p53*, is more frequently mutated in human cancers than any other known gene. The p53 protein acts as a transcription factor to turn on the expression of genes whose products are cell cycle inhibitors. The p53 can stimulate *apoptosis* (i.e., programmed cell death), the ability of cells to self-destruct by autodigestion with endogenous enzymes. In apoptosis, the plasma membrane is kept intact, and the digested contents are not released; instead, phagocytic cells engulf the whole cell to eliminate these undesirable cells.

Cancer cells continue to grow and divide in situations where normal cells would not (lack contact inhibition); they fail to respond to cellular controls and signals that would halt growth in normal cells. Cancer cells avoid the apoptosis (self-destruction) that normal cells undergo when extensive DNA damage is present. Cancer cells stimulate angiogenesis (the formation of new blood vessels) to nourish the cancer cell, and they are immortal (continue to divide for more generations than a normal cell), while normal cells die after some divisions. Cancer cells can *metastasize* (relocate) and then grow in another location.

Chromatin structure: heterochromatin vs. euchromatin

DNA exists as euchromatin and heterochromatin within the cell. *Euchromatin* is a looser conformation of DNA and histones, compared to the tightly-condensed *heterochromatin*. Euchromatin appears lighter than the darker heterochromatin when viewed under an electron microscope. DNA sequences in heterochromatin are generally repressed, while those in euchromatin are available and actively transcribed when the RNA polymerase binds to the single-stranded DNA. Much of the satellite DNA (large, tandem repeats of noncoding DNA) appears in heterochromatin.

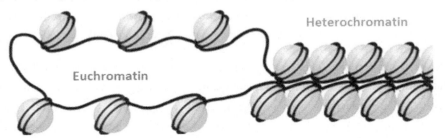

Euchromatin is transcribed while heterochromatin is not transcribed

When DNA is transcribed, activators known as remodeling proteins can push aside the histone portion of the chromatin, allowing transcription to begin. During interphase (G_1, S and G_2 phase), chromatin exists as either of the two types, but during mitosis, it condenses to supercoiled heterochromatin.

The form of compactness that the DNA adopts depends on the cellular needs of the cell and can be regulated by covalent *histone modifications* by specific enzymes. Examples of modifications are *histone methylation*, which causes tighter packing that prevents transcription, and *histone acetylation*, which involves uncoiling of the DNA and promotes transcription. There are many other types of histone modifications, such as ubiquitination and phosphorylation. As an active area of investigation, the *histone code hypothesis* states that DNA transcription is partly regulated by these histone modifications, especially on the unstructured ends of histones.

Chromatin remodeling complexes is another mechanism for regulating chromatin structure. These protein complexes are ATP-dependent and thus have a common ATPase domain. ATP hydrolysis gives these domains the energy to reposition nucleosomes and move histones, which creates uncoiled DNA regions available for transcription.

DNA methylation

DNA methylation, which reduces the rate of transcription, is another method that the cell uses to regulate gene expression. DNA methyltransferase enzymes add a methyl ($-CH_3$) group to the cytosine bases of DNA, converting these bases to 5-methylcytosine. The methylated cytosine residues are usually adjacent to guanine, which results in methylated cytosines that are diagonal from each other. The patterns of DNA methylation are heritable to daughter cells, as it is passed on during cell division. *Epigenetics* is the study of changes in transcriptional potential (such as DNA methylation and histone modification) that do not involve changes in DNA sequence.

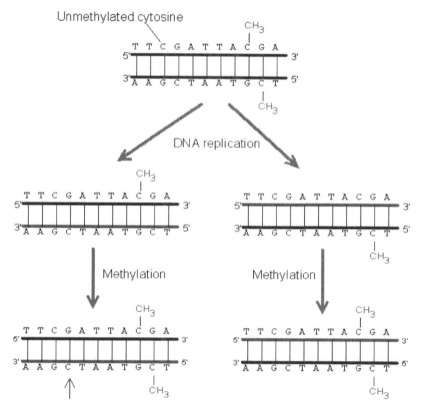

Modifications such as methylation facilitate proper DNA repair

Role of non-coding RNAs

Non-coding RNA (ncRNA) is a functional RNA that is not translated into proteins. They have an important role in many cellular processes, including RNA splicing, DNA replication, and the regulation of gene expression. ncRNA can participate in histone modification, DNA methylation, and heterochromatin formation. The majority of ncRNA are *long ncRNA* (over 200 nucleotides) that form a complex with chromatin-modifying proteins, and function in chromatin remodeling.

There are three classes of *short ncRNA* (less than 30 nucleotides), including microRNA (miRNA), short interfering RNAs and piwi-interacting RNA.

microRNAs (miRNA) are folded RNA molecules with hairpin loops that bind to target mRNA sequences through complementary base pairing. miRNA can induce degradation of the mRNA by shortening its poly-A tail, which destabilizes the mRNA, or they can cleave the mRNA into pieces. This silences the mRNA and prevents translation from occurring. A single miRNA molecule can target and repress several different mRNAs.

Short interfering RNAs (siRNAs) are double-stranded RNA molecules that are often created through catalysis by the Dicer enzyme, which produces siRNAs from longer double-stranded RNAs. siRNAs function similarly to miRNAs, as they interfere with the expression of genes that have complementary sequences. siRNA can degrade mRNA, blocking translation, and can induce heterochromatin formation, blocking transcription.

Piwi-interacting RNAs (piRNA) form RNA-protein complexes with the piwi family of proteins, and they are the largest class of short ncRNA in animal cells. They suppress transposon activity in germline cells through the formation of an *RNA-induced silencing complex*. piRNAs do not have any known secondary structure motifs.

Chapter 5

Genetics

- **Mendelian Concepts**

- **Meiosis and Genetic Variability**

- **Analytic Methods**

Mendelian Concepts

Gregor Mendel (1822-1884) was an Austrian monk whose experiments with plant breeding in the mid-1800s formed the basis of modern genetics. Mendel was a friar at the monastery of St. Thomas in Brunn, Austria, while teaching part-time at a local secondary school. Mendel's goal was to have firm scientific evidence for how genetic information is passed from parents to offspring. He focused on how plant offspring acquired traits from their parents. He traced the inheritance of individual traits and kept careful records of numbers, then used his understanding of probability to interpret the results.

Unfortunately, Mendel's results, published in 1865 and 1869 in the *Proceedings of the Society of Natural History of Brunn*, went mostly unnoticed until 1900. Three independent investigators, Carl Correns (Germany), Hugo De Vries (Holland) and Erich Von Tschermak (Austria), conducted similar experiments and reached the same conclusions as Mendel. After researching the literature, the three researchers gave credit to Mendel. This rediscovery and confirmation of Mendel's work launched the field of modern genetics.

Mendel chose to study the *Pisum sativum* (garden pea), as it was easy to cultivate, had a short generation time, and could be cross-pollinated by hand. Mendel chose 22 of the many pea varieties for his experiments. Concerning certain traits, all the plants used in the experiment had to be *true breeding* (progeny identical to parents). He chose this approach to ensure accuracy and simplicity in his studies.

True-bred organisms are obtained by inbreeding genetically similar individuals for many generations. Pea plants can be *self-pollinated*, an effective form of inbreeding. The process eliminates genetic variation from the gene pool and results in a strain with certain identical traits in all individuals.

Mendel studied seven simple traits: seed shape, seed color, flower position, flower color, pod shape, pod color, and plant height. He correctly hypothesized that the pattern of inheritance of these traits from generation to generation was because the genetic information was being passed from their parents. He terms this *hereditary factors*. The *particulate theory of inheritance* proposed by Mendel stated that these factors, or "discrete particles," do not blend. This is in contrast to the *blending theory of inheritance*, now discredited, which stated that offspring have traits that are a blend of the parents. For example, red and white flowers would produce pink flowers. This is observed in nature, but the pink offspring then self-fertilize and produce several all-red and all-white offspring. Proponents of the blending theory dismissed this as mere instability in genetic material. Charles Darwin wanted to develop a theory of evolution based on hereditary principles, but the blending theory did not account for genetic variation and species diversity.

Mendel's particulate theory properly accounts for Charles Darwin's theory of evolution, although he never knew this because he did not live to see Mendel's work rediscovered. A *particulate inheritance theory* is essential for Darwin's theory of *natural selection*. Otherwise, any selectively favored trait would be blended away as soon as the selected individual reproduced with one of a different trait.

Mendel's explanation of heredity was remarkable, since it was based solely on the interpretation of breeding experiments, and it was done long before scientists understood cell division and molecular biology. During the early part of the 20th century, the processes of mitosis and meiosis were discovered by the microscopic examination of cells. Researchers found that Mendel's hereditary factors were not free-floating particles, but rather were located on chromosomes in the nucleus. The hereditary factors were given the name genes.

Gene

A *gene* is a stretch of DNA that encodes for a specific trait (or characteristic). On a molecular level, each gene encodes for a protein which produces a particular trait. An organism's *genome* includes its entire set of genes.

The *chromosomal theory of inheritance* states that genes are located on chromosomes and that inheritance patterns can be explained by the specific locations of genes on chromosomes. Chromosomes can be classified as autosomes and allosomes. *Autosomes* are non-sex chromosomes that are the same number and kind in all sexes. *Allosomes* are sex chromosomes (e.g., X or Y). Humans have 22 pairs of autosomes (44 total), and 1 pair of allosomes (2 total). *Diploidy* is the characteristic of having pairs of chromosomes. One member of each pair is from the mother, and the other is from the father. *Homologous chromosomes* are similar but not identical copies.

Locus

A *locus* is a specific location of a gene or segment of DNA on a chromosome. Loci can be mapped by their physical position on the chromosome, or by their relative distance from each other.

The loci of two homologous chromosomes are identical in their placement and align during meiosis.

Allele: single and multiple

Loci are identical, but the gene at each locus may be in different forms (e.g., brown eyes or blue eyes) as *alleles.* An allele is an alternative form of the gene and differs from another allele by one or more nucleotide bases that encode a different protein. Diploid organisms have two similar versions of each chromosome, so an individual can have a maximum of two alleles for each gene, one on each chromosome. One allele comes from the mother, and one comes from the father. Traits exhibited by an organism are determined based on the alleles an individual has inherited.

Although each diploid organism inherits only two alleles, there are usually more than two possible alleles of each gene in the population. *Multiple alleles* are when more than two types of alleles can encode for a particular characteristic. A classic example is blood type in humans. There are three blood group alleles: I^A, I^B and i. Each can have only two alleles, so the possible combinations are I^AI^A, I^Ai, I^BI^B, I^Bi, I^AI^B and ii. An organism's combination of alleles is its *genotype.*

Blood type comparison of genotype and phenotype:

Genotype	Blood type (phenotype)
$I^A I^A$ or $I^A i$	A
$I^B I^B$ or $I^B i$	B
$I^A I^B$	AB
ii	O

Homozygosity and heterozygosity

Zygosity is the similarity of the alleles at any given locus. An individual is *homozygous* for a gene when the two alleles that the individual carries are identical. In the blood type example, individuals with the genotype $I^A I^A$, $I^B I^B$ and ii are homozygous, since they have two identical alleles. A *heterozygous* individual has two different alleles, as the genotypes $I^A i$, $I^B i$, and $I^A I^B$.

Wild-type

The *wild-type allele* is the most prevalent and most dominant version of the allele, with some exceptions. *Mutant alleles* are the product of changes in the nucleotide sequence within DNA (mutation) and are generally less common in a population. For example, red eyes are a wild-type trait in Drosophila (fruit fly), while white eyes are the mutant trait. However, alleles often do not fit into either of these categories with additional variation.

Recessiveness

When an allele is *recessive,* the individual must inherit two copies of this allele to express the trait. The *dominant* allele is when only a single copy is needed for it to be observed. A dominant allele masks or hides expression of a recessive allele. For example, the allele for attached earlobes in humans is recessive, and the allele for unattached earlobes is dominant. An individual must inherit two attached earlobe alleles to express this trait. If she inherits one attached allele and one unattached allele, the unattached allele dominates the other, leading to unattached earlobes.

By convention, the dominant allele is an uppercase letter, and the recessive allele is a lower case letter. The letter may be the first letter of either the dominant or recessive allele (by convention). For example, in Mendel's pea plants the alleles for seed shape are named R for the dominant round allele. The recessive allele, wrinkled, is denoted *r*. However, in fruit flies, the alleles for wing size are named for the recessive allele, vestigial wings. Therefore, the dominant allele (normal wings) is denoted V and the recessive allele as v. In many cases, the locus is given one letter, and the alleles are superscripts of the letter. For example, with sex chromosomes (e.g., $X^A X^a$). Other, more complex naming schemes are often seen in the case of multiple alleles (e.g., blood types) and codominant or incompletely dominant alleles, as discussed later.

Genotype and phenotype:
definitions, probability calculations, and pedigree analysis

Genotype is the alleles received when an organism is conceived. The *phenotype* describes the observable traits which are expressed. Due to recessiveness and dominance, zygosity cannot be deduced by observation of phenotype. Two organisms with different allele combinations can have the same phenotype. For example, if a homozygous individual has the alleles I^AI^A and a heterozygous individual has the alleles I^Ai, they have different genotypes, but the same phenotype and both exhibit blood type A since I^A is a dominant allele. Its presence overshadows the i allele in the heterozygous individual.

Mendel inferred genotype from observable phenotype by performing breeding experiments that revealed how traits emerge, disappear and re-emerge over generations. Traits are *characters,* and their expression in an individual is a *character state.* Mendel was able to streamline his experiments by using character states that were *discontinuous* (no intermediates). Discontinuous traits have discrete, distinct categories; the trait is either there or not. Mendel could quantify his results by simply counting the two different phenotypes among the offspring of each experimental cross. Mendel's knowledge of statistics enabled him to recognize that his results followed theoretical probability calculations, even though his small sample size resulted in slight deviations from the expected ratios.

Probability is the likelihood that a given event occurs by random chance. With each coin flip, there is a 50% chance of heads and a 50% chance of tails. In Mendelian genetics, the chance of inheriting one of two alleles from the parent is 50%.

The *multiplicative law of probability* states that the chance of two or more independent events occurring together is the product of the probability of the events occurring separately. If two parents that are heterozygous for unattached earlobes (genotype Ee) have a child, the probability of the child's genotype can be calculated using probability.

In half (½) of cases, the mother passes an E allele to the child, and in the other half of the cases, she passes an e allele. The same is true for the father. The possible combinations for the child's genotype, where one allele is inherited from the mother and another from the father:

$$EE = ½ \times ½ = ¼, \quad eE = ½ \times ½ = ¼ \quad Ee = ½ \times ½ = ¼ \quad ee = ½ \times ½ = ¼$$

The *additive law of probability* calculates the probability of an event that occurs in two or more independent ways; it is the sum of individual probabilities of each way an event can occur. In the example, the chance that the child is a heterozygote is the sum of the probability of having the genotype eE (¼) or Ee (¼) = ½, or a 50% chance. The probability of having unattached earlobes is the sum of the probabilities of inheriting at least one E (dominant) allele. This sum is ¼ + ¼ + ¼ = ¾, or a 75% chance.

These laws of probability were used in creating a method that predicts the genotypic results of genetic crosses: the *Punnett square,* introduced by R. C. Punnett. For example, in a Punnett square, all possible types of alleles from the father's gametes may be aligned vertically, and all possible alleles from the mother's gametes may be aligned horizontally; possible combinations of offspring are placed in squares. The Punnett square predicts the chance of each child's genotype and corresponding phenotype.

In genetics, probability depends on independent, mutually exclusive events. For example, the odds of a couple having a boy or a girl is always 50%. Using the multiplicative law of probability, it is overall unlikely that a couple has 5 boys: $\frac{1}{2} \times \frac{1}{2} \times \frac{1}{2} \times \frac{1}{2} \times \frac{1}{2} = 1/32$, or 3.125%. However, this probability has no bearing on each event. Even if the couple has four boys, there is still a 50% chance their fifth child is a boy or a girl. Each fertilization is an independent event. In the Punnett square below, note that the mother has two copies of the X allele, while the father has one X allele and one Y allele. In 50% of cases, the child inherits an X from the mother and an X from the father, while in the other 50% of cases the child inherits an X from the mother and a Y from the father.

Mother

		X	X
Father	X	XX (Girl)	XX (Girl)
	Y	XY (Boy)	XY (Boy)

x

Punnett square for the probability that parents have a girl (XX) or boy (XY). Probability is 50%.

Mendel began his experiments by creating true-breeding plants and then *crossing* (mating), two strains that were true breeding for different alleles of the same trait. An example is to cross a true breeding pea plant with round seeds with another plant with wrinkled seeds; these two individuals constitute the *parental (P) generation*. Crossbreeding is accomplished by removing the pollen-producing male organs from a "father" plant and using them to fertilize the ovary of another. Because pea plants are *monoecious* (both male and female reproductive organs). Mendel had to remove the male organs from the "mother" plant to prevent it from self-pollinating.

The *first filial* (F1) generation is the hybrid offspring produced by this cross. These individuals breed with one another to produce the *second filial* (F2) generation. This is a *monohybrid cross* because it is between two individuals that are heterozygous for a single trait, e.g., Rr × Rr. This produces F2 offspring in a phenotypic ratio of 3:1 round to wrinkled seeds.

Example using true-breeding plants with round seeds and true-breeding plants with wrinkled seeds. Compare results to genotype to observed phenotype ratios. The proportions of a test cross with a homozygous dominant:

P generation:	homozygous round (RR) × homozygous wrinkled (rr)
F1 generation:	100% heterozygous round (Rr)
F2 generation:	25% homozygous round (RR)
	50% heterozygous round (Rr)
	25% homozygous wrinkled (rr)
	Genotypic ratio: 1:2:1 RR to Rr to rr
	Phenotypic ratio: 3:1 round-seeded to wrinkle-seeded

The F2 offspring ratio is conceptualized using a Punnett square of the F1 generation cross, Rr × Rr:

Parent 1 gametes

		R	r
Parent 2 gametes	**R**	RR	Rr
	r	Rr	rr

When Mendel performed this experiment with the six other traits of pea plants, he obtained similar results. He recognized a pattern of 3:1 phenotypic ratio in the F2 generation. Mendel extended his experimental results through inductive reasoning to claim that all discontinuous traits, no matter what animal or plant species is studied, would follow the same pattern. This was how he developed the notion that each distinct phenotypic trait in an individual is controlled by two "hereditary factors" (now called alleles). The alternative hypothesis that each trait is controlled by one factor was not a viable explanation, because it could not explain the reappearance of wrinkled seeds in the F2 generation. These observations led him to develop the principles of dominance and recessiveness since he understood that the factor for wrinkled seeds was masked by the factor for round seeds.

Notice that he could have hypothesized that there are more than two hereditary factors in each, but he used the principle of *parsimony*, which states that the simplest explanation for observation is likely the most accurate.

From these conclusions, Mendel was able to deduce his first law of inheritance: *the law of segregation*. Each organism contains two alleles for each trait, which segregate during the formation of gametes for only one allele in each gamete. During fertilization, gametes from two individuals are united, giving the offspring a complete set of alleles for each trait.

Although Mendel did not understand meiosis when he formulated his theories, he correctly outlined its principles in this law. He understood that if two alleles control each distinct phenotype, it then follows that only one is passed on to an offspring by each parent. Otherwise, the number of alleles would double with each generation. Mendel's law of segregation is consistent with the particulate theory of inheritance because many individual "discrete particles" of inheritance are passed on from generation to generation.

Mendel then performed a *dihybrid cross* between two organisms that are heterozygous for two traits rather than just one. A dihybrid cross is achieved by first crossing parent organisms that are true breeding for different forms of two traits; it produces F1 offspring that are heterozygous for both traits (dihybrids). Mendel performed a dihybrid cross of the F1 individuals with one another. He expected that the dihybrids would produce two types of gametes: dominant (RY) gametes and recessive (ry) gametes. He thought he would see a phenotypic ratio of 3:1 in the F2 generation, as with his monohybrid crosses. This would mean that 75% of the F2 individuals would be round and yellow and 25% would be wrinkled and green.

Mendel's expected Punnett square for his F1 dihybrid cross:

	RY	**ry**
RY	RRYY	RrYy
ry	RrYy	rryy

P generation:	homozygous round yellow (RRYY) × homozygous wrinkled green (rryy)
F1 generation:	100% heterozygous round yellow offspring (RrYy)
F2 generation:	25% homozygous round, homozygous yellow (RRYY)
	50% heterozygous round, heterozygous yellow (RrYy)
	25% homozygous wrinkled, homozygous green (rryy)
	Genotypic ratio: 1:2:1 RRYY to RrYy to rryy
	Phenotypic ratio: 3:1 round yellow to wrinkled green offspring

However, Mendel observed the following results from the dihybrid cross.

Parents' genotype	RRYY × rryy
Parents' gametes	RY ry

F1 generation	RrYy
F1 gametes	RY rY Ry ry

Punnett square of gametes produced

	RY	Ry	rY	ry
RY	RRYY	RRYy	RrYY	RrYy
Ry	RRYy	RRyy	RrYy	Rryy
rY	RrYY	RrYy	rrYY	rrYy
ry	RrYy	Rryy	rrYy	rryy

	Round and yellow phenotype		Wrinkled and yellow phenotype
	Round and green phenotype		Wrinkled and green phenotype

The offspring from that cross produced a phenotypic ratio of 9:3:3:1 (of round and yellow, round and green, wrinkled and yellow, and wrinkled and green). It was not predicted that the offspring could have the dominant form of one phenotype and the recessive form of the other. He deduced that dihybrids produce not two types of gametes (RY and ry) but *four:* RY, Ry, rY, and ry. He realized that dominant alleles do not have to be shuffled into the same gametes as other dominant alleles, nor do recessive alleles.

Mendel's law of independent assortment states that alleles assort independently from other alleles and that a parent's gametes contain all possible combinations of alleles. This leads to a phenotypic ratio of 9:3:3:1 in dihybrid crosses. However, the ratio often breaks down in more complex cases, patterns of *non-Mendelian inheritance* (do not follow Mendel's laws), for reasons that would not be understood until the 20th century.

Humans cannot be bred like pea plants, so their genetic relationships are studied using pedigrees. *Pedigrees* are charts that portray family histories by including phenotypes and family relationships. In a pedigree chart, squares represent males, circles represent females, and diamonds represent individuals of unspecified sex. If an individual displays a trait studied, their shape is filled. Heterozygotes are half-filled (if the disorder is autosomal) or have a shaded dot inside the symbol (if the disorder is sex-linked). If the trait of interest is a disease, heterozygotes are *carriers*. However, it is often not known if an individual is heterozygous, since this may not be visible phenotypically.

Lines between individuals denote relationships. Horizontal lines connect mating couples, and a vertical line connects to their offspring. Siblings are grouped under a horizontal line that branches from this vertical one, with the oldest sibling on the left and the youngest on the right. If the offspring are twins, they are connected by a triangle. If an individual dies, its symbol is crossed out. If it is stillborn or aborted, it is indicated by a small circle.

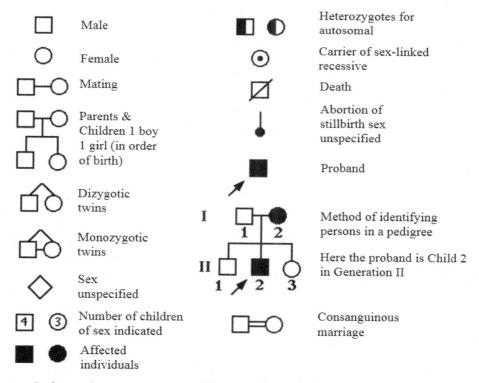

Pedigree chart convention specifying relationships and observed phenotypes

The generations are shown using Roman numerals (I, II, III, etc.), and each from the same generation is indicated by Arabic numbers (1, 2, 3, etc.). The pattern of inheritance of a particular trait can be determined by analyzing pedigrees. Often, the trait in question is a genetic disease. These are typically recessive traits, which can be seen by the way the disorder skips generations, passed by carriers. The first family member known to seek treatment for the disease is the *proband* and may be indicated by an arrow. The proband serves as the starting point for the pedigree, and researchers may work both backward and forward from there.

Complete dominance

Dominance is a relationship between alleles of one gene, in which one allele is expressed over a second allele at the same locus. Thus far, the only *complete dominance* of alleles has been discussed. This occurs when only one dominant allele is needed to express a trait since it masks the recessive allele. The traits of Mendel's pea plants had complete dominance, with a round seed shape completely dominates a wrinkled seed shape.

Genotype	Phenotype
RR	Dominant - round
Rr	Dominant - round
rr	Recessive - wrinkled

The homozygous dominant and heterozygous individual exhibit the "dominant phenotype," but the heterozygote has a recessive allele, though not observed.

Codominance

In *codominance*, two or more alleles are equally dominant, so a heterozygote expresses the phenotypes associated with both alleles. The most famous example in humans is the ABO blood type system, a multiple allele system. The locus is named "I." A person with genotype $I^A I^A$ or $I^A i$ expresses the A blood type; a person with $I^B I^B$ or $I^B i$ expresses the B blood type; and a person with $I^A I^B$ expresses roughly equal amounts of both "A" and "B" antigens, giving the blood type AB. Thus, I^A and I^B are codominant. Note that allele i represents the absence of any antigens. Therefore it is recessive and must be homozygous to express type O blood.

Blood groups with genotype and phenotype:

Genotype	Phenotype
$I^A I^A$	*A*
$I^A I^O$	*A*
$I^A I^B$	Both *A* and *B*
$I^B I^B$	*B*
$I^B I^O$	*B*
$I^O I^O$	*O*

Incomplete dominance, expressivity, and penetrance

Although Mendel completely dismissed the notion that traits "blend." *Incomplete dominance* (partial dominance) occurs when the phenotype of a heterozygote is an intermediate of the phenotypes of the homozygotes. For example, true-breeding red and white-flowered four-o'clock plants produce pink-flowered offspring. The red allele is only partially dominant to the white allele, so its effect is weakened and it appears pink. This does not support a blending theory of inheritance since the red and white parental phenotypes reappear in the F2 generation.

P   true breeding red × true breeding white; $(C^R C^R) \times (C^W C^W)$

F1   All pink offspring; $(C^R C^W)$

If these F1 individuals are crossed, the results are:

pink × pink; $(C^R C^W) \times (C^R C^W)$

F2   1 red: 2 pink: 1 white; $1(C^R C^R) \times 2(C^R C^W) \times (C^W C^W)$

Another example of incomplete dominance is *sickle-cell anemia,* a blood disorder controlled by incompletely dominant alleles. Homozygous dominant individuals $(Hb^A Hb^A)$ are asymptomatic and healthy. Homozygous recessive individuals $(Hb^S Hb^S)$ are afflicted with sickle-cell anemia. Their red blood cells are irregular in shape (sickle-shaped) rather than biconcave, due to abnormal hemoglobin. Sickle-shaped red blood cells clog vessels and break down, which results in poor circulation, anemia, low resistance to infection, hemorrhaging, damage to organs, jaundice and pain in the abdomen and the joints. This is an example of *pleiotropy,* where one gene affects many traits.

Incomplete dominance is heterozygous individuals $(Hb^A Hb^S)$ since they do not have the full-blown disease but have some sickled cells and minor health problems. This is the *sickle-cell trait.* In regions prone to malaria (e.g., Africa) being heterozygous for the sickle-cell allele confers an advantage, because the malaria parasite dies as potassium leaks from sickle cells. *Heterozygote advantage* describes the case where the heterozygous condition bears a greater advantage than either homozygous condition.

Penetrance is the frequency by which a particular genotype results in a corresponding phenotype. For example, Mendel's pea plants had 100% penetrance for seed color. A plant with genotype Yy or YY always had yellow seeds, and a plant with genotype yy always had green seeds. However, many genes have *incomplete penetrance.* For example, the BRCA gene (breast cancer) for women with this mutation have an 80% lifetime risk of developing cancer. Certain people may have a gene that predisposes them to lung cancer, but due to their lifestyle habits and random chance, they may never develop cancer. This means that in some cases, a given genotype does not guarantee that the expected phenotype is expressed.

Phenotypes can exhibit *expressivity* (variation in the presentation). This is different from penetrance, which is simply a question of whether or not the phenotype is expressed. Expressivity refers to individuals that express their phenotype, but to varying degrees. In the case of *constant expressivity,* individuals with the same genotype have the same phenotype. For example, if pea plants are homozygous recessive for flower color (pp), they are all approximately the same shade of white. Discontinuous traits

163

typically exhibit fairly constant expressivity. However, a trait like a polydactyly (extra digits) in humans is prone to *variable expressivity,* when a phenotype is expressed in different degrees from the same genotype. Although a single gene controls polydactyly, afflicted individuals may have extra fingers or toes. The expressivity varies from person to person.

Crossing individuals with different alleles of a continuously varying phenotype do not produce the discrete ratios that enabled Mendel to discover the laws of inheritance. The genetics of *continuous variation* is far more complex than that of discontinuous variation (e.g., pea plants). Such traits follow a bell-shaped curve when the number of individuals is plotted against the range of the variable trait.

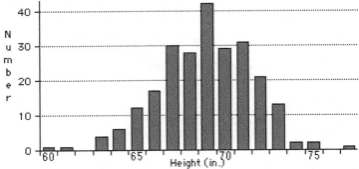

Continuous variation for height distribution in a population

With continuous variation, there is a phenotype in between any two chosen for comparison, because the distribution of phenotypes in the population varies along a continuum; individuals differ by small degrees. Mendel's theory explains both discontinuous and continuous patterns of individual variation. With continuous variation, however, many different genes typically influence the same trait, not just one, as Mendel proposed.

Traits such as height and weight are not due to variable expressivity of a single gene but due to *polygenic inheritance.* Polygenic inheritance is when two or more genes govern one trait. Often, the genes have an additive effect, with each dominant allele adding to the "intensity" of the phenotype. The more genes involved, the more continuous the variation in phenotypes, resulting in a bell-shaped curve.

A human example of polygenic inheritance is skin color. Many genes control skin pigment and the more dominant alleles that an individual has, the darker their skin. Parents with intermediate skin color can have offspring with light or dark skin.

Albinism, a condition where the eyes, skin, and hair have little to no color due to lack of pigment production, is an example of *epistasis,* a phenomenon where one gene interferes with other genes in the expression of a phenotype. It does not matter if an individual has other genes which would otherwise give her dark skin; if the individual is homozygous recessive for the albinism gene, this shuts off pigment production and prevents the other genes from being expressed.

Traits such as height and weight have a polygenic component but are heavily influenced by outside factors, such as environment, exercise, and nutrition (nature vs. nurture).

Hybridization and viability of offspring

Leakage is gene flow from one species to another, which occurs when individuals of two related species mate and produce *hybrid* offspring. The hybrid now has genetic information from both species and may mate with either one, causing genes to "leak" from each gene pool and flow into the other.

A hybrid is the product of parents that are true breeding for different forms of one trait. The parents may be of different breeds of a single species or a different species altogether. When hybrids are created from two different species, they may not always be *fertile* (capable of producing offspring). This may be because the hybrid is infertile or it dies before reproductive age. Even if the hybrid is fertile and able to reproduce, the offspring may prove not to be viable (able to survive until reproduction), a *hybrid breakdown*.

Hybrid vigor (heterosis) is when the hybrid has the best qualities of both species or strains. The hybrid's superior quality is due to the suppression of recessive alleles and the increase in heterozygotic traits, leading to heterozygote advantage. Hybrids are bred with the intention to create new breeds that are healthier or more desirable than either parent breed.

Outbreeding is the mating of genetically dissimilar individuals and is a powerful agent of genetic diversity. Inbreeding tends to promote harmful recessive alleles and decrease the number of alleles in the population.

Gene pool

Population genetics studies the variation of alleles within a population. The *gene pool* is the total of all genetic information in the population, described by gene frequencies.

Demes are local gene pools, consisting of individuals who are likely to breed with each other. A great deal of evolutionary change occurs in these groups.

Modern biologists view each individual as a temporary vessel housing a small fraction of the gene pool. Thus, the concept of a gene pool is an abstract pooling of all genetic variation in the population; it does not exist apart from the individuals themselves.

Meiosis and Genetic Variability

There are several sources of genetic variability in nature, some of which are random, while others are the result of selective processes. These sources include mutation, sexual reproduction, diploidy, outbreeding and balanced polymorphisms.

Significance of meiosis

Meiosis is essential to sexual reproduction, diploidy, and genetic diversity. Asexual organisms produce offspring that are genetic clones of themselves, but sexual reproduction creates offspring that are both similar to and unique from either parent. Meiosis creates haploid gametes that fuse and form a diploid zygote. Meiosis provides several opportunities to promote variability in the gene pool and allow for new combinations of alleles.

Important differences between mitosis and meiosis

Mitosis and meiosis begin similarly with a somatic cell in G1 of interphase, which has 46 total chromosomes. Each chromosome has a homologous structure that originated from the other parent; there are 23 homologous pairs, and each chromosome is scattered randomly within the nucleus. During S phase, each chromosome duplicates. Each duplicate is renamed a sister chromatid, and together, two sister chromatids comprise one chromosome. Sister chromatids are chromatin strands attached at the centromere, which are identical copies (except for low-frequency mutations of nucleotides) of the same chromosome (not homologous). One chromosome (a pair of sister chromatids) has another chromosome, its homolog.

During prophase / metaphase of mitosis, the chromosomes do not attempt to locate their homolog. Each chromosome lines up along the metaphase plate individually. During anaphase, the sister chromatids of each chromosome are separated so that one chromatid (resulting as a single strand separated at centromere) is partitioned into a daughter cell. The result is two diploid daughter cells with identical genetic makeup.

However, during prophase I and metaphase I of meiosis, homologs do pair into a *tetrad* (bivalent). The tetrad consists of two homologous chromosomes or four sister chromatids in total. Tetrads line up on the metaphase plate so that one homologous chromosome is on one side, and the other homolog is on the other side of the midline. During anaphase I, the tetrad is separated, and one chromosome is pulled into one pole, while the second is pulled toward the other pole. The result of meiosis I is that two daughter cells have a different genetic makeup; one has half of the organism's genome and the second cell has the other half.

A key difference is that mitosis involves one set of cell divisions, in which the chromatids of each chromosome separate. In meiosis, there are two sets of cell divisions. During the first division, the chromosome of each tetrad separates (chromosome has 2 strands attached at the centromere). In the second division, the centromere splits and the sister chromatids of each chromosome separate.

Mitosis	*Meiosis*
One set of divisions	Two sets of divisions
Occurs in the body (somatic cells)	Occurs in the testes or ovaries (gametes)
Two identical daughter cells	Four unique daughter cells (four sperm cells or one egg with up to three polar bodies)
Daughter cells are diploid 2n → 2 cells with 2n	Daughter cells are haploid 2n → 4 cells with 1n
No crossing over occurs	Crossing over (tetrad) during prophase I create genetic variability

During meiosis, when the homologous pairs are arranged into tetrads, their ends overlap and cause *recombination* (exchange of genetic material). Homologous chromosomes exchange some DNA so that once they are separated, the maternal chromosome has some paternal DNA, and vice versa. Gametes are unique from the organism's genome.

Segregation of genes

Mendel understood that different genes are segregated into different gametes despite being unaware of meiosis. His first *law of segregation* supports the conclusion that each diploid individual has two alleles, but passes only one to its offspring. Segregation of genes and diploidy are important concepts for explaining the genetic variability.

Independent assortment

Mendel's second law, the *law of independent assortment*, states that any gamete may have any combination of alleles. In the era of molecular biology, it is known that this occurs during meiosis. During metaphase I of meiosis, homologous chromosomes pair as tetrads along the metaphase plate in a random orientation. They have then pulled apart so that some homologous chromosomes from each parent end up in one daughter cell and some in another. The law of independent assortment requires that homologous chromosomes line up randomly during metaphase. The segregation of alleles of one gene does not affect the segregation of alleles of another gene. Mendel initially thought that a dihybrid (e.g., AaBb) would produce the gametes AB and ab. However, he realized that the A alleles and the B alleles are not linked; they can segregate into gametes in four combinations: AB, aB, Ab and ab. He noted that these gametes are produced in equal numbers; no one combination is favored over another.

Mendel's research was based on simplistic traits of pea plants. He sometimes encountered puzzling phenotypic ratios, which suggested that some allele combinations were more frequent than others. If Mendel chose two phenotypic traits controlled by two tightly linked loci for his dihybrid cross, he would have only obtained two types of gametes due to the *linkage.*

Linkage

Mendel's law of independent assortment does not apply to the linkage, whereby some genes are located on the same chromosome and inherited together. Because of this, his law is applied to chromosomes rather than genes; chromosomes assort independently, but alleles do not. *Linked genes* are close on the same chromosome, while those far apart or on different chromosomes altogether are *unlinked genes*.

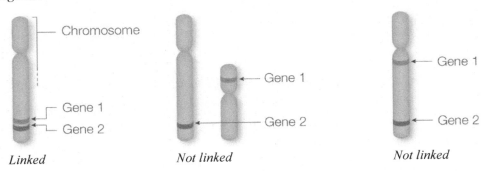

Genes on separate chromosomes are unlinked and can segregate independently, but genes on the same chromosome must often be inherited together as a *linkage group*. However, this is not always the case. *Crossing over* allows genes on the same chromosome to become unlinked due to the probability that the genes undergo recombination in the region between them.

Recombination

Genetic recombination introduces genetic diversity into the gametes during meiosis. It includes both independent assortment and crossing over. Crossing over occurs during prophase I of meiosis when homologous chromosomes are paired together into tetrads. The pairing of tetrads is *synapsis* and is facilitated by a protein structure of the *synaptonemal complex*. It is thought that the synaptonemal complex functions primarily as a scaffold to allow interacting chromatids to complete their crossover activities.

During prophase I, synapsis links the four sister chromatids (4 strands with 2 chromosomes) of a pair of homologous chromosomes. The ends of different chromatids often overlap and make contact at sites, as *chiasmata*, between loci. Both chromatids cut at the same locus, allowing them to bind where they are cut (chiasma) and swap their DNA. In this way, an individual can create gametes with some chromatids that are different from the chromatids they inherited. *Recombinant* are chromatids which have undergone recombination, and those which did not are *parental*. Once crossing over is finished, the homologous chromosomes are no longer tightly linked, and the homologous chromosomes are independent. However, the recombinant chromatids remain connected by their chiasma until anaphase, when the centromeres divide and liberate each sister chromatid strand, which becomes a chromosome.

Sex-linked characteristics

In a human, the normal chromosome complement is 46, two of which are sex chromosomes. A human female has two X chromosomes, while a human male has an X and a Y chromosome. The sex chromosomes carry genes that determine the sex of an organism and various unrelated traits that are *sex-linked*. Because the Y chromosome is small, only the X chromosome essentially carries all sex-linked traits as X-linked traits. The alleles are designated as superscripts to the letter X.

In females, one of the duplicate X chromosomes is deactivated during embryonic development, resulting in an unused chromosome, a *Barr body*. This is random for each cell, so in one cell the paternal X chromosome may become the Barr body, while in the other the maternal X chromosome becomes the Barr body.

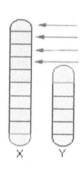

In males, chromosomes X and Y do not make a homologous pair since they are of different sizes and contain different genes. The X chromosome in humans is much longer than the Y chromosome and contains many more genes. In a male, the smaller Y chromosome has no opposing alleles to those on the X chromosome. Males are *hemizygous* for X-linked traits, meaning they have only one allele since they have only one X chromosome. In the pairing seen below, all genes on the X chromosome with an arrow pointing to them are dominant, because they have no opposition from the Y chromosome. Note that there are some genes with loci on both X and Y chromosomes.

Males are more prone to certain genetic diseases due to this characteristic. A female who inherits a sex-linked recessive allele that encodes for a genetic disease may inherit the dominant allele on the other X chromosome. She then is unafflicted (in the case of complete dominance) or has minor afflictions (incomplete or codominance). A male, however, has no such mitigating protection.

The number of genes on the Y chromosome

The Y chromosome is significantly smaller than the X chromosome and contains a much lower density of genes, perhaps 50 to 60. However, these few genes do have an important impact on sex determination and male characteristics. Genes on the Y chromosomes that do not recombine are passed from father to son and are not present in females. The lack of recombination weakens the effectiveness of natural selection to reduce the frequency of disadvantaged variants and select for favorable ones.

Sex determination

Humans have an *XY sex-determination system.* Most other mammals, along with some insects and fish, follow this as well. Every female produces gametes with a single X chromosome, while sperm from the male can contain either a Y chromosome or an X chromosome. There is a 50% chance that a gamete contains either chromosome. Therefore, when the gametes fuse during fertilization, there is a 50% chance of a female (or male) sex.

The maternal gamete is *homogametic* because all its cells possess the XX sex chromosomes. Sperm gametes are the variable factor and are thus *heterogametic* because around half contain the X chromosome, and the other half possess the Y chromosome.

In the absence of a Y chromosome, genes on the X chromosomes direct a fetus to produce female sex hormones and develop internal and external female sex organs. However, if a Y chromosome is present, its *SRY gene* inhibits the development of female sex organs and promotes male sex organs. The SRY gene is so powerful since it controls the activity of many other genes, which direct the development of internal and external male characteristics. Mutations in this gene can cause the fetus to develop nonfunctional testes or even ovaries. In some cases, errors during the male production of gametes attach the SRY gene to an X chromosome instead. This results in a female fetus developing male characteristics.

Note that many other organisms have different sex determination schemes. For example, in an *X0 sex determination system* females have two X chromosomes, and males have one. In a *ZW sex-determination system* (e.g., birds, reptiles, and many other organisms), the males have two Z chromosomes, while females have a Z and a W chromosome. Several other sex-determination systems involve different chromosomes, often more than two; for example, platypuses each have ten sex chromosomes.

In many reptiles and invertebrates, sex is determined not by genetics but by environmental factors (e.g., temperature). Some species can change sex throughout their lifetime.

Cytoplasmic and extranuclear inheritance

Extranuclear inheritance, known as *cytoplasmic inheritance,* is the inheritance of genetic material outside the nucleus. This was discovered by Carl Correns in 1908. The two most prevalent examples are the inheritance of mitochondria and chloroplasts in many eukaryotes. Along with the nuclear chromosomes, DNA from the mitochondria (and chloroplasts for plants) are transferred in the cytoplasm of the maternal gamete. Sperm contains these organelles as well, but they either do not enter the egg during fertilization or do enter but are destroyed by the egg. Therefore, extranuclear DNA is always passed along the maternal line, making it useful for certain genetic testing.

Mutation

Mutations are changes in the DNA nucleotide sequence that arise by means other than recombination. Mutant genes may produce abnormalities in structure and function, leading to disease. Cystic fibrosis, sickle-cell anemia, hemophilia, and muscular dystrophy are *single gene diseases* because they arise from mutations that occur in a single gene. *Polygenic diseases,* such as diabetes, cancer, cleft lip and schizophrenia, result from several defective genes that have little effect on their own, but collectively can have significant effects. The environment greatly influences many genetic disorders, physical features, and behavioral traits.

Genetic mutations can be beneficial, neutral or harmful. Beneficial or *advantageous mutations* provide an improvement to the fitness of the organism. *Deleterious mutations* disrupt gene function and result in a harmful effect to the fitness of the organism. It is possible for mutations to be harmful in one situation but advantageous in another.

Neutral mutations have a negligible effect on fitness, considered neither harmful nor beneficial. Mutations may not affect the phenotype as *silent mutations,* or the effect does not arise until later generations.

Types of mutations: random, transcription error, translation error, base substitution, deletion, insertion, frameshift and mispairing

Random mutations are changes in DNA sequence that occur at any time, due to radiation, chemicals, replication errors or other chance events. *Transcription errors* occur specifically during transcription of DNA into mRNA. This results in mRNA with some RNA nucleotide sequences that do not accurately correspond to the original DNA code. *Translation errors* occur during translation of mRNA into a protein with a mutant amino acid sequence.

Mutations may be classified by their structural effects. *Base substitutions* are when another replaces one or more nucleotides. They range from advantageous to fatal (as with many mutations), but most base substitutions are minor. Nucleotide base substitutions may cause a stop codon to halt transcription, a *nonsense mutation*. Substitutions could cause a different amino acid to be transcribed in the final protein, causing a *missense mutation*. Even a single amino acid change may change the protein's function or render it inoperable. However, since some mRNA codons can encode for the same amino acid (i.e., code is degenerative), sometimes base substitutions may not result in a different amino acid sequence. *Silent mutations* do not affect the phenotype.

Deletions involve a base (or several bases) being removed from the DNA or mRNA sequence, while *insertions* involve the addition of one (or more) bases. One or two insertions (or deletions) result in *frameshift mutations* since they shift the reading frame of the three-base codons.

For example, the mRNA sequence AUGUUGACUGCCAAU is meant to be read:

AUG - UU<u>G</u> - ACU - GCC – A …

Met - Leu - Thr - Ala - …

If the 6th base (guanine) is deleted, a frameshift mutation changes the transcribed amino acids.

AUG - UUA - CUG - CCA - …

Met - Leu - Leu - Pro - ...

The first codon is unaffected and still encodes for methionine. The second codon *is* changed, but since this codon encodes for the same amino acid (leucine), it is a silent mutation. However, the third and fourth codons are changed so that they encode for entirely different amino acids. It is assumed that many other amino acids in the sequence are changed. This is a serious mutation that often renders the protein inoperable. However, deletions and insertions of nucleotides that involve multiples of threes do not cause a frameshift, since they remove an entire codon (i.e., three nucleotides that encode an amino acid). The reading frame remains intact, but the absence of the single amino acid may be a serious issue.

Slipped-strand mispairing is a mutation that occurs during transcription, translation or DNA replication. After DNA strands are denatured during replication, the template or replicated strand may slip (become temporarily dislodged) and cause incorrect pairing of complementary bases. This is believed to have led to the evolution of many repetitive DNA sequences in the human genome.

Portions of the chromosome are subject to deletion, especially during meiosis. Chromosomal deletion may be severe and even render the gamete incapable of fertilization or spontaneous abortion.

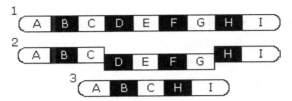

Chromosomal deletion where sections D, E, F, and G are absent after deletion

Chromosomal rearrangements: inversion and translocation

Along with base mutations, the entire structure of the chromosome is subject to rearrangement, especially during meiosis. Chromosomal mutations may be severe and even fatal, terminating the fetus before birth.

Inversions involve a stretch of DNA breaking and reattaching in the opposite orientation. There are two types of inversions: paracentric and pericentric. *Paracentric inversions* do not involve the centromere; they occur when a piece of the arm of chromosome breaks, inverts and reattaches. A *pericentric inversion* includes the centromere; the breakpoint is on either arm of the chromosome.

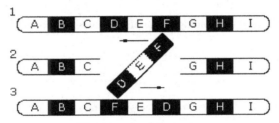

Chromosomal inversion

The organism experiencing the newly inverted sequence may not be viable if it includes a necessary region of the chromosome. The mutation could be advantageous. However, inversions usually do not affect the phenotype of the organism and go undetected. Their greatest impact is on the production of gametes since they are often marked by loops in the chromosome that affects recombination. An individual with a chromosomal inversion may generate some gametes with altered linkage relationships or abnormal chromatids. The former case is likely harmless, while the latter case usually yields inviable gametes.

Translocation is when a segment of one chromosome separates and binds to the other. This is a drastic rearrangement that is often lethal. An individual inheriting a chromosome that has been altered due to translocation has extra alleles or too few alleles, leading to a variety of defects. This may occur in both autosomal and sex chromosomes, where it leads to infertility issues and other genetic disorders. *XX male syndrome* occurs when the portion of the Y chromosome containing the SRY gene is translocated to an X chromosome during a male's production of gametes. If the sperm cell containing the mutant X chromosome fertilizes an egg, a female fetus develops that has male secondary sex characteristics and genitalia.

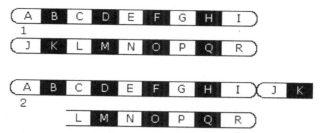

Chromosomal translocation where section J and K join the other chromosome

Although there are safeguards in place to ensure chromosomes properly separate during meiosis, these checks sometimes fail. Failure of proper chromosome separation causes *nondisjunction errors,* in which three chromosomes of a tetrad are pulled to one side of the spindle, and only one is pulled to the other side. This results in some gametes having an extra chromosome and some gametes lack a chromosome. A chromosome with three copies is a *trisomy*, while a chromosome with only one copy is *monosomy*. Both examples of gamete formation by nondisjunction usually lead to an inviable zygote.

Down syndrome is a common nonlethal autosomal trisomy, involving chromosome 21. In general, the chance of a woman having a Down syndrome child increases with age. This disorder leads to faster aging, moderate to severe developmental issues and a higher risk of health complications in the child.

Many nonlethal nondisjunction errors involve the sex chromosomes. Females with *Turner syndrome* have only one X sex chromosome. This results in nonfunctional ovaries and the absence of puberty. Afflicted females have somewhat masculine characteristics and are sterile, but they usually have no cognitive issues and use hormone therapy.

Klinefelter syndrome occurs when a zygote receives one Y chromosome and two (or more) X chromosomes. Presence of the Y chromosome makes affected individuals identifiably male, but they have underdeveloped sex organs and are sterile. The extra X chromosomes cause the development of breasts, lack of facial hair, and may result in developmental delay. Males with Klinefelter syndrome have one or more Barr bodies due to the extra X chromosomes.

Females with *Poly-X syndrome* have extra X chromosomes and therefore extra Barr bodies. They do not exhibit enhanced feminine characteristics and usually appear physically normal. Some experience menstrual irregularities but most have regular menstruation and are fertile. Females with three X chromosomes are not developmentally delayed but have some impaired cognitive skills. However, four X chromosomes cause severe cognitive impairment.

Jacob's syndrome is the condition of one X chromosome and two Y chromosomes. Although it was previously believed that individuals with Jacob's syndrome were more likely to be aggressive, this claim is now refuted. Males with Jacob's syndrome are usually taller than average, suffer from persistent acne and tend to have speech and reading problems.

Many genetic disorders, especially those due to nondisjunction, can be detected during pregnancy. Chorionic villi sampling testing, amniocentesis and karyotyping are prenatal testing methods. *Karyotyping* (visual examination of chromosomes) may be used to determine the gender of a fetus and look for chromosomal abnormalities.

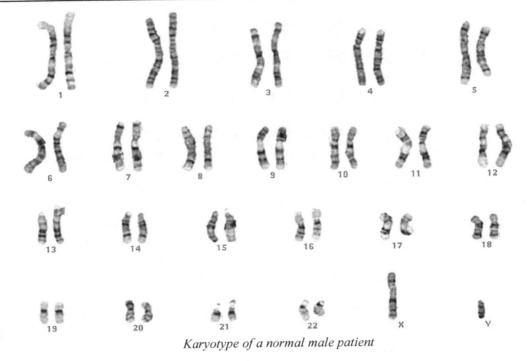

Karyotype of a normal male patient

The structure of the sex chromosomes can deduce the gender since Y chromosomes are markedly smaller than X chromosomes. Missing or extra chromosomes can visualize nondisjunction. If there are two of each chromosome, the 23 chromosomes pairs should result in 46 chromosomes in total. Deviations from this, such as three copies of chromosome 21, signifies a nondisjunction.

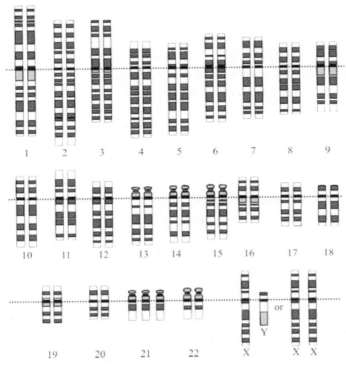

Karyotype of a male patient with Down syndrome (three chromosomes, 21) and an extra set of X chromosomes (XXXY) as a variant of Klinefelter syndrome (XXY)

Inborn errors of metabolism

Inborn errors of metabolism are genetic disorders which cause mild to severe metabolic issues. These diseases are caused by a mutant gene that results in abnormal enzyme production, which may affect the gastrointestinal system, the circulatory system, the nervous system or any area of the body. They are typically rare and have severe health implications. However, there are more benign inborn errors of metabolism, such as lactose intolerance, which arises from an inability to produce the digestive enzyme lactase. Unlike many genetic disorders, inborn errors of metabolism are caused by the mutation of only a single gene.

Autosomal recessive disorders can be discerned from pedigrees by establishing the presence of certain patterns of inheritance. The first observation is that affected children usually have unaffected carrier parents since recessive alleles are statistically rare in the general population. Two homozygous dominant parents produce all unaffected children. However, if one parent is affected, the children are unaffected carriers. If one parent is affected and the other is a carrier, they have a 50% chance of producing a carrier child or affected child. This is in the pedigree chart below. Note the many carriers (half-filled circles or squares), due to their affected parents.

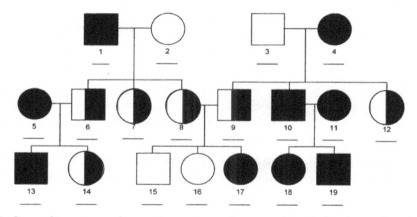

Pedigree for autosomal recessive: squares denote males, circles denote females; completely shaded figures are afflicted, while half-filled figures are carriers.

It must be noted that individuals 15 and 16 (third generation) may be homozygous dominant or carriers. Only genetic testing reveals if they received two dominant alleles from each carrier parent or a dominant allele from one and a recessive allele from the other parent.

Tay-Sachs disease is an autosomal recessive disorder that is rare in the general population but afflicts approximately 1 in 3,600 Ashkenazi Jews at birth. Tay-Sachs results in death by about age three due to progressive neurological degeneration. A genetic mutation prevents the production of the enzyme hexosaminidase A (Hex A), leading to accumulations of its substrate, glycosphingolipid, in lysosomes of brain cells. Tay-Sachs is one of many *lysosomal storage disorders,* caused by abnormal lysosomal function.

Cystic fibrosis is among the most common lethal genetic diseases in Northern European ancestry. About 1 in 20 Caucasians in the U.S. is a carrier for cystic fibrosis, and about 1 in 3,000 is afflicted. The disease is caused by a mutation of chromosome 7 that prevents chloride ions from passing into some cells. Since water normally follows Cl⁻, lack of water in lung cells causes the production of viscous mucus and subsequent respiratory issues. This disease has gastrointestinal, kidney and fertility effects.

Phenylketonuria (PKU) is the most common inherited disease of the nervous system, occurring once in about every 5,000 births. A mutant gene on chromosome 21 results in a lack of the enzyme that metabolizes the amino acid phenylalanine. The absence of the enzyme causes accumulation of phenylalanine in nerve cells and impairs the CNS. From neonatal diagnosis, children are placed on low-phenylalanine diets that prevent severe neural degeneration.

Sex-linked recessive disorders, like autosomal recessive disorders, results in all children affected if the parents are both affected. Furthermore, two unaffected parents can bear affected offspring (male only), if the mother is a carrier. She has a 50% chance of donating a recessive allele to a son, afflicting him with the disease. If the mother is homozygous recessive, this results in 100% of male offspring as affected. Female offspring are all unaffected because the father donates a dominant X-linked allele; however, all are carriers.

If an affected male mates with a homozygous dominant female, all offspring are unaffected. However, if the father is affected and the mother is a carrier, 50% of their children, regardless of sex, are affected.

Color blindness can be an X-linked recessive disorder involving mutations of genes coding for green-sensitive pigment or red-sensitive pigment; the gene for blue-sensitive pigment is autosomal. About 8% of Caucasian men have red-green color blindness.

Duchenne muscular dystrophy is the most common form of muscular dystrophy and is characterized by the wasting (atrophy) of muscles, eventually leading to death. It affects 1 in 3,600 male births. This X-linked recessive disease involves a mutant gene that fails to produce the protein dystrophin. The lack of dystrophin promotes the action of an enzyme that dissolves muscle fibers. Affected males rarely live to be fathers; the allele survives in the population due to transmission by carrier females.

About 1 in 10,000 males is a hemophiliac with an impaired ability of blood clotting. This is a classic example of an X-linked recessive disorder, famously seen in the royal families of Europe throughout the 19th and 20th century. Queen Victoria was a carrier of the disease and passed it on to many of her descendants. Due to diplomatic marriages within a small subpopulation, it was transmitted to royal families in Russia, Germany, and Spain. The issue was exacerbated by incest, not uncommon at the time to keep power and assets within the family.

Autosomal dominant disorders can be discerned from pedigrees by establishing the presence of certain patterns. The first is that all affected children must have an affected parent. Two affected parents can have a child that is unaffected if both parents are heterozygous. However, it is impossible for two unaffected parents to have a child that is affected. As with all autosomal disorders, both males and females are affected with equal frequency.

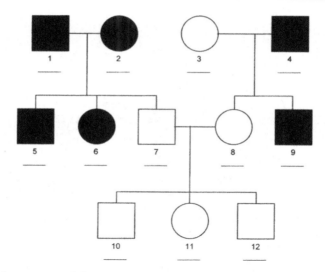

*Pedigree for autosomal dominant: circles denote females, squares denote males;
filled shapes are affected individuals, while unfilled shapes are not carriers of the disorder.*

Neurofibromatosis is an autosomal dominant disorder in about 1 in 3,500 people. It is caused by an altered gene on chromosome 17 that controls the production of the neurofibromin protein, which normally inhibits cell division. When this gene is mutated, neurofibromin is nonfunctional and affected individuals develop neurofibromas, benign skin tumors. In most cases, symptoms are mild, and patients live healthy lives but can be severe. Since the severity of symptoms varies, this is an example of *variable expressivity.*

Huntington's disease is an autosomal dominant disorder that, while fatal, usually does not onset until middle age, after an afflicted individual may already have children. Therefore, the disease continues to pass through the generations. The gene for Huntington's disease is on chromosome 4. This gene encodes for the *huntingtin protein* as extra glutamine in the amino acid sequence, which causes the mutant huntingtin protein to form clumps inside neurons.

Achondroplasia is a form of dwarfism caused by a defective bone growth that occurs in about 1 in 25,000 people. Individuals with achondroplasia have short limbs, a deformed spine and a normal torso and head. Like many genetic disorders, being homozygous dominant for achondroplasia is lethal; afflicted individuals are heterozygotes.

Sex-linked dominant disorders have certain characteristics distinguished from other forms of inherited genetic disorders, which can be discerned by analyzing a pedigree. If a male is affected, then all of his female offspring are affected because the male donates his affected X chromosome to his female offspring, and since the allele is dominant, all offspring are affected. By contrast, male offspring can only be affected if the mother has the disease because the father donates a Y chromosome (does not carry the gene) to all of his male offspring. Unlike autosomal genetic disorders, sex-linked genetic disorders affect the sexes disproportionately, as in the pedigree below. Note that there are no carriers, since having one dominant allele guarantees affliction. Like autosomal dominant disorders, dominant sex-linked disorders are not common in the population, since they lead to health problems which reduce the reproduction frequency. This makes the alleles rarer in the gene pool, a self-fulfilling cycle that keeps them at a low frequency.

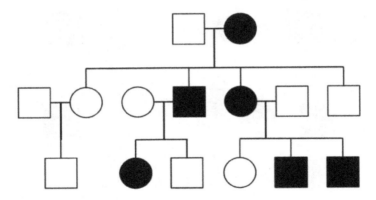

Pedigree for dominant sex-linked disorder: squares denote males, circles denote females;
filled shapes are afflicted individuals, while unfilled are healthy individuals

The *fragile X syndrome* is an X-linked dominant disorder in which the *FMR1* gene is mutated. This causes a deficiency of the protein FMRP protein, affecting neural and physical development. The mutation is among both sexes, though it is more prevalent in females. Affected children often have hyperactive behaviors, intellectual disabilities and autism spectrum disorders. These symptoms are more severe in males, although about one-fifth are unaffected due to incomplete penetrance.

Pedigrees must determine sex-linked traits in humans since it is not ethical to orchestrate human mating, while other organisms can be bred to determine patterns of inheritance. In non-humans, *reciprocal cross* involves two sets of parents, in which the male and female of each set have opposite traits. For example, in one set a female pea plant has white flowers, and the male has purple flowers, and in another, the female has purple flowers, and the male has white flowers. This determines the inheritance of sex-linked traits. If the trait is autosomal, both crosses produce the same results, but if not, they produce different results.

Test cross uses the homozygous recessive to mate with an organism of known phenotype but unknown genotype. If all progeny resemble the known phenotype parent, the parent is homozygous dominant. A 50:50 ratio means that the parent is heterozygous for the phenotypic trait. For example, AA × aa = Aa and Aa × aa = Aa and aa.

Relationship of mutagens to carcinogens

The genetic component of cancers is the leading cause of death in developed nations and the second leading cause of death in developing nations. Most cancers are caused by both a genetic predisposition and by carcinogens. *Carcinogens* are any physical, chemical or biological agents that cause cancer. The majority are *mutagens,* harmful agents which cause DNA mutations. Toxic chemicals, radiation, free radicals, viruses, and bacteria are all possible mutagens. *Exogenous mutagens* come from an external event, like smoking a cigarette or exposing one's skin to UV radiation. *Endogenous mutagens* arise internally as byproducts from metabolic processes. *Reactive oxygen species* (ROS) are a class of endogenous mutagens that contain oxygen and are highly reactive (e.g., H_2O_2 and O^{-2}).

Mitogens are another class of carcinogens that trigger an increase in the rate of mitosis. While all carcinogens are either mutagens or mitogens, there are mutagens and mitogens that do not lead to cancer.

Genetic drift, gene flow, and balanced polymorphisms

Genetic drift refers to random changes in allele frequencies of a gene pool over time. This occurs in both large and small populations, but the effect is magnified in small populations. Isolated gene pools can quickly diverge from the parent population. Over time, large populations may speciate (organisms are unable to reproduce to produce viable and fertile offspring). Genetic drift causes some alleles to be lost and others to become *fixed*, meaning they are the only allele for a particular gene in the population. Variation among populations can often be attributed to the random effects of genetic drift.

The *founder effect* is an example of genetic drift, whereby a handful of founders leave a source population and establish a colony. The new population contains a fraction of the total genetic diversity of the original population. Over time, the founders' alleles may occur at high frequencies in the new population, even if they are rare in the original population. For example, cases of dwarfism are much higher in the Pennsylvania Amish community because a few German founders were dwarfs.

The *bottleneck effect* may occur after excessive predation, habitat destruction or a natural disaster, rather than a founding event. After the catastrophe, there is a major decrease in the total genetic diversity of the original gene pool. Purely by chance, some alleles may be lost, and this affects the future genetic makeup of the population. Today, relative infertility is in cheetahs due to a bottleneck in earlier times. Small populations suffer low genetic variation due to the high rates of inbreeding.

Genetic drift is a random process, greatly enhanced when other agents of genetic diversity are random. Random mating, in which individuals pair by chance and not by any selection, is one example. However, the majority of populations practice some sort of nonrandom mating which inhibits genetic diversity. *Inbreeding*, where relatives mate, can occur as a form of nonrandom mating because these individuals are near others. *Assortative mating* occurs when individuals mate with those that have similar phenotypes. This may divide a population into two phenotypic classes with reduced gene exchange. Phenotypes can be selected due to *sexual selection* when males compete for the right to reproduce, and (often) the females select males of a particular phenotype.

Gene flow is the introduction (or removal) of alleles from populations when individuals leave (emigration) or enter the population (immigration). Gene flow increases variation within a population by introducing novel alleles from another population. Continued gene flow decreases variation between populations, causing their gene pools to become similar. Because of this, gene flow can be a powerful opposing force in speciation.

Balanced polymorphisms add to genetic variability by maintaining two different alleles in the population rather than encouraging homozygosity. Balanced polymorphisms are maintained by heterozygote advantage, hybrid vigor, and frequency-dependent selection.

Frequency-dependent selection is when the frequency of one phenotype affects the frequency of another. For example, a prey animal population may have several coloring phenotypes in the population (i.e., gray, brown and black fur). When the gray phenotype becomes the most frequent in the population, predators become familiar with this phenotype and can identify prey by their gray fur. Natural selection makes it so that the rarer phenotypes (brown and black fur) have an advantage. More prey evolves these phenotypes until one becomes most common, at which point frequency-dependent selection changes. This is *minority advantage,* in which the rarest phenotype has the highest fitness (i.e., chance of survival).

Phenotypes can have a positive frequency dependent relationship from safety in numbers. For example, some individuals of a poisonous species may evolve coloring that signals their toxicity. Predators learn this signal and avoid individuals with this phenotype. In this example, it is a disadvantage to have a unique phenotype that is not identifiable as poisonous, because these individuals have an increased probability of being eaten by predators.

Synapsis (crossing-over) for increasing genetic diversity

It is hypothesized that meiosis evolved from either the bacterial analog of sexual reproduction (transformation) or from mitosis. It is ubiquitous in eukaryotes and is the source of genetic variation in sexual reproduction. Recombination creates a recombinant chromosome, and then meiosis independently sorts chromosomes into different gametes.

Suppose a woman inherited the alleles A and b from her mother and a and B from her father. She has genotype AaBb. Her alleles are not necessarily arranged in a *cis* configuration:

$$\frac{\text{A B}}{\text{a b}}$$

The top alleles (AB) represent one chromosome, and the bottom (ab) represent its homolog inherited from the other parent.

She could have inherited the alleles on the chromosomes as a *trans* configuration:

$$\frac{\text{A b}}{\text{a B}}$$

It is not possible to know from her phenotype alone what configuration (*cis* or *trans*) she has, since both results in the same phenotype. One must know her parents' genotypes or perform statistical analysis of her children's phenotypes. However, a sample size in the hundreds would be required to make the latter strategy accurate, which is improbable. Her configuration is determined by her parents' gametes and influences how she produces gametes.

Suppose she is *cis* (AB / ab), and the genes are closely linked. When she produces gametes, the crossover is unlikely because the genes are close together. Each somatic cell divides into the four gametes: AB, AB, ab, and ab. She has produced all parental chromosomes since they are identical to herself (parental). However, if the alleles are farther apart, there is the increased probability of crossover.

Crossover allows a *cis* configuration to become a *trans* configuration or vice versa. A *single crossover* is a crossing-over between two adjacent non-sister chromatids of a tetrad, at only a single chiasma. The two chromatids which crossed over are recombinant, while the other two remain parental (nonrecombinant). The following image illustrates a tetrad that has undergone recombination to produce some recombinant chromatids.

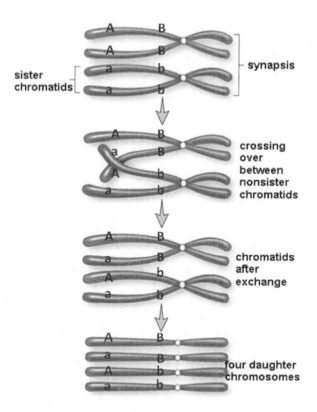

Crossing over with exchange at tetrad with A for a.
From AB and ab, the four gametes produced are AB, aB, Ab and ab

A *double crossover* is more complicated and can have a few different outcomes. A *two-stranded* double crossover involves two chromatids that overlap each other at two points. They exchange alleles at first, but then exchange them back, resulting in no net recombination. When a crossover occurs between alleles, the following summaries are accurate. Although each chromatid ends up with a region of the other, these are noncoding regions between the loci, so the crossing-over has no observable difference. This results in four parental chromatids.

Two-strand double crossing-over

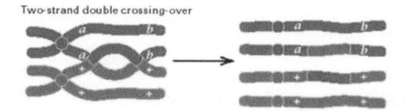

Double crossover results from two exchanges and preserves the parental genotype chromosomes

A *three-stranded* double crossover involves three chromatids of two chromosomes. Like a single crossover, this results in two recombinants and two parentals.

A *four-stranded* double crossover involves all four chromatids in the tetrad. It results in four recombinants and no parentals.

If the crossover event flanks an allele on each side, the following is observed:

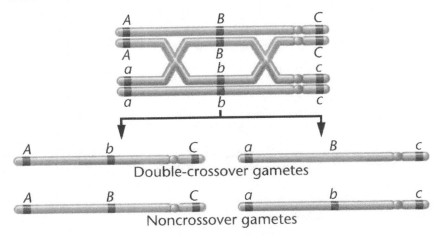

Double-crossover gametes

Noncrossover gametes

The result is two noncrossover chromosomes and two recombinant chromosomes.

The random orientation of homologous chromosomes during metaphase I and their independent assortment into separate daughter cells is another important mechanism of recombination. The number of possible orientations at the metaphase plate is equal to 2^n, where n is the haploid number of chromosomes.

Since humans have a haploid number of 23, they have 2^{23} possible outcomes, yielding over 8,388,000 possible combinations of gametes from the division of a single cell. Compounded with the effects of exchange of genetic material on a chromosome during crossing over, the number of combinations increases even further to virtually infinite genetic variety.

Summary of important terms in genetics

Carrier - an individual that has one copy of a recessive allele that causes a genetic disease in individuals that are homozygous for this allele.

Codominant alleles - pairs of alleles that both affect the phenotype when present in a heterozygote.

Dominant allele - an allele that has the same effect on the phenotype, whether it is present in the homozygous or heterozygous state.

Genotype - the alleles of an organism.

Heterozygous - having two different alleles of a gene.

Homozygous - having two identical alleles of a gene.

Locus - the particular position of a gene on a chromosome.

Phenotype - the characteristics of an organism.

Recessive allele - an allele that only has an effect on the phenotype when present in the homozygous state.

Analytic Methods

Hardy-Weinberg principle

In the 1930s, scientists were able to apply genetics to populations and observe the small-scale evolution of a gene pool.

A population in stasis has constant gene frequencies and is in *Hardy-Weinberg equilibrium*. This provides a baseline by which to judge whether evolution has occurred. Populations can be in Hardy-Weinberg equilibrium if they meet the requirements of the *Hardy-Weinberg principle*. These assumptions describe a population without changes in allelic and genotypic frequencies, due to the absence of evolutionary pressures. The conditions are as follows:

1) Mutation does not occur at any significant rate.

2) Gene flow is absent as individuals do not migrate among populations.

3) Nonrandom mating does not occur; individuals pair by chance and do not engage in mate selection.

4) Genetic drift is minimal; the population is large, so changes in allele frequencies due to chance are insignificant.

5) Natural selection does not occur; the population does not experience competition, and no traits have a selective advantage.

In the natural world, the conditions of the Hardy-Weinberg principle are rarely, if ever, met, so allelic and genotypic frequencies change, and thus evolution occurs.

The Hardy-Weinberg principle includes mathematical models to predict frequencies.

1) Allele frequencies: $p + q = 1$, represented by p and q, sum to 100% in the gene pool.

2) Genotype frequencies: $p^2 + 2pq + q^2 = 1$, which arise from the two alleles, sum to 100% in the gene pool. The homozygous dominant genotype (*pp*) equals the product of $p \times p$ (or p^2). The homozygous recessive genotype (*qq*) equals the product of $q \times q$ (or q^2). The heterozygous genotype is represented by two possibilities, *pq,* and *qp*, and therefore is the sum of both their products (or *2pq*).

Example: A plant population has two phenotypes for flower color: the wild-type red and the white mutant phenotype. The wild-type red flowers are inherited from at least one dominant allele (R), and the mutant white flowers are inherited from two recessive alleles (r). In this gene pool, 84% of the flowers have the red phenotype, and 16% have the white phenotype. What are the frequencies of the alleles? What are the frequencies of the genotypes?

Solution:

Assume that the red allele (R) is "*p*" and the white allele (r) is "*q*."

Since the frequency of the white phenotype is 16%, then $q^2 = 0.16$. This is the frequency of the homozygous recessive genotype.

Since the frequency of the red phenotype is 84%, then $p^2 + 2pq = 0.84$. Remember that p^2 and $2pq$ are the homozygous dominant and heterozygous genotypes, respectively. Together, they represent the wild-type (red) phenotype.

Solving for the white allele, *q*:

$$q = \sqrt{q^2} = \sqrt{0.16} = 0.4$$

Therefore, the frequency of the white allele in the population is 0.4 (or 40%).

Solving for the red allele, *p*:

$$p + q = 1$$
$$p = 1 - q$$
$$p = 1 - 0.4 = 0.6$$

Therefore, the frequency of the red allele in the population is 0.6 (or 60%).

Solving for p^2:

$$p^2 = 0.6^2 = 0.36.$$

The frequency of the homozygous dominant genotype is 36%.

Solving for $2pq$:

$$2pq = 0.84 - p^2$$
$$2pq = 0.84 - 0.36$$
$$2pq = 0.48.$$

The frequency of the heterozygous genotype is 48%.

Testcross for parental, F1, and F2 generations

A *test cross* is mating of an individual of unknown genotype against an individual with a homozygous recessive genotype. This may be done for one or more traits. Mendel used this to ensure his plants were true breeding. For example, he may have had a plant with white flowers (a recessive trait), so he knew it was homozygous recessive for white flowers; genotype pp. Another plant may have had purple flowers, but since this is the dominant phenotype, he could not be sure whether it was Pp or PP. He would cross the purple (unknown genotype) plant with the white plant. The phenotypic results of a test cross indicate the presence of the different genotypes even though the genotypes (PP or Pp) could not be observed because only phenotype can be observed.

Example:

A white-flowered pea plant (genotype pp) and an unknown purple-flowered pea plant (genotype P_) are test-crossed to determine the genotype (PP or Pp) of the purple parent.

Solution:

To analyze a test cross between a white plant and an unknown purple plant, first, establish the possible gametes.

Since the white parent is a homozygote (pp), there is a 100% chance it contributes a p allele and a 0% chance it contributes a P allele; every gamete contains a p allele.

If the purple parent is a homozygote (PP), there is a 100% chance it contributes a P allele to its offspring and a 0% chance that it contributes a p allele. Therefore, by the laws of probability, the chance that their offspring are genotype Pp is 1.0 × 1.0 = 1.0 (or 100%). The chance of genotype PP is 0%, and the chance of genotype pp is 0%; all offspring are Pp to exhibit the phenotype of a purple flower.

However, if the purple parent is a heterozygote (Pp), there is a 50% chance it contributes a p allele and a 50% chance it contributes a P allele. Therefore, by the laws of probability, the chance that they have an offspring with genotype Pp is 1.0 × 0.5 = 50%. The chance of genotype pp is 1.0 × 0.5 = 50%. The chance of genotype PP is 0%. Half of their offspring will be Pp and exhibit a purple flower phenotype, and half will be pp and exhibit a white flower phenotype.

The purple plant genotype can be determined by its offspring with the white plant. If the offspring are all purple, the unknown parent must be genotype PP. If the offspring are half purple, half white, the unknown parent must be genotype Pp.

This test cross (homozygous recessive with unknown genotype) is summarized:

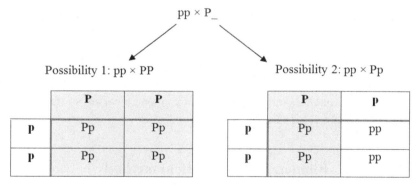

100% purple offspring indicated a purple
a parent that is homozygous

50% purple offspring and 50% white offspring
indicates a purple parent that is heterozygous

For a two-trait test cross (e.g., seed color and seed shape) the results are simply an extension of the single trait test cross. If a double homozygous recessive parent (rryy) is crossed with a round-seeded yellow parent of unknown genotype, the unknown parent has four possible genotypes: RRYY, RrYY, RRYy, RrYy.

Known parent genotype	Possible parent genotype	Possible parent gametes	Offspring phenotype ratio
rryy	RRYY	RY, RY, RY, RY	1 round yellow
rryy	RrYY	RY, RY, rY, rY	1 round yellow : 1 wrinkled yellow
rryy	RRYy	RY, RY, Ry, Ry	1 round yellow : 1 round green
rryy	RrYy	RY, rY, Ry, ry	1 round yellow : 1 wrinkled yellow : 1 round green : 1 wrinkled green

After creating hybrid F1 individuals from a parental generation, a *backcross* can be performed by breeding the offspring with the parents (or an individual with the same genotype as the parents). This is used to conserve desirable traits in the F2 plants and animals or to "knock out" (remove) a certain gene from an organism. Backcrossing occurs naturally in small populations, especially among plants. Artificial backcrossing is done in genetic research to study the function of certain genes by eliminating them and then observing the effect when absent.

Gene mapping from crossover frequencies

The rate of "unlinking" of genes is used to map the physical distances between two genes on the same chromosome. The further apart two genes are, the more likely they become unlinked during crossing over. The frequency of recombination is an estimate of linkage because it indicates the distance between loci. The maximum frequency of recombination is 50%. For example, genes undergoing recombination 50% of the time are completely unlinked. Genes undergoing recombination 10% of the time must be close together since it is apparently difficult for them to be separated.

Frequencies are found by dividing the number of recombinant offspring by the number of total offspring. Recombinants are identified by their rarity amongst the offspring since parental type offspring are always more frequent. If more than two genes are studied, different types of crossing over can occur, and the recombinants can be further divided into two-, three-, or four-stranded crossovers. Two-stranded crossovers occur more frequently than three-stranded crossovers; four-stranded crossovers are rarest. 1% recombination frequency equals 1 map unit, measured in *centimorgans* (cM) as an arbitrary, relative unit and not physical distances. If crosses are performed for three genes, and their recombination frequencies are calculated, the genes can order (arrange in the relative distance) since only one map relationship explains the distances between them. Biochemical methods are used to map the physical distances between loci by the number of DNA nucleotide bases. Human chromosomes must be mapped this way since it is not possible to measure recombination frequency among a person's offspring due to the small sample size and lack of control over reproduction.

Biometry and statistical methods

Mendel's knowledge of statistics, a new branch of mathematics at the time, greatly aided his work. Today, statistics is an integral field of life sciences. *Biometry,* or *biostatistics,* is the application of mathematical models and statistics to a vast array of biological fields. Biometry is used for biological experiments, the collection, summarization, and analysis of data, and interpretation of and inferences from the data. Problem-solving in genetics usually relies on statistics. Allelic, genotypic and phenotypic frequencies are calculated using statistics to determine the characteristics of a family (or population). Models can be applied that describe genetic inheritance, even for complex non-Mendelian patterns.

Chapter 6

Evolution

- Background

- Natural Selection

- Speciation

- Evolutionary Time Measured by Changes in Genome

- Origin of Life

Background

The pre-Darwinian worldview was defined by deep-seated beliefs held to be intractable truths, namely that each species was specially designed and has not changed since Earth's creation. Variations among organisms of the same species were explained as circumstantial imperfections in a perfectly-adapted creation. However, Charles Darwin (1809-1882) lived during a time of great change in the scientific and social realms. His ideas were part of a larger change in thought and perspective already underway among scientists.

Georges Louis Leclerc (1707–1788), known by his title Count Buffon, was a French naturalist who wrote the 44-volume *Natural History of All Known Plants and Animals*. Buffon provided evidence of descent with modification and speculated on how environment, migration, geographical isolation and competition influence the traits of organisms. His work was essential to the development of the theory of evolution and natural selection, as he was among the first to assert that traits are inherited from earlier descendants. Paradoxically, Buffon's beliefs often directly contradicted his work, as he believed in a young Earth and the fixity of species.

Erasmus Darwin (1731–1802), the grandfather of Charles Darwin, was a physician and a naturalist whose writings on both botany and zoology suggested the possibility of *common descent*, the theory that organisms descended from common ancestors. He based his conclusions on observations of embryonic development and the changes incurred in domestic plants and animals due to selective breeding by people. However, Erasmus Darwin offered no mechanism by which descent with modification might occur.

Jean-Baptiste Lamarck (1744–1829) was the first credited with definitively stating that descent with modification occurs and that organisms adapt to their environments. Lamarck was an invertebrate zoologist who went against the thinking of many of his contemporaries by embracing evolution. His assertion that organisms change over time would prove to be correct, but his proposed mechanism for this phenomenon was flawed. *Inheritance of acquired characteristics* was the Lamarckian belief that organisms become adapted to their environment during their lifetime and pass on these adaptations to their offspring. Lamarck suggested that body parts are enhanced by increased usage, while unused parts are weakened. Under this theory, a giraffe's neck would lengthen over its lifetime as it continually stretched to reach leaves, and the slightly-lengthened neck would be passed onto the giraffe's offspring. Over time, these adaptations would accumulate in the long-necked giraffes of modern times. Today, Lamarck's theory of inheritance of acquired characteristics has been dismissed in favor of Darwin's theory of natural selection, along with modern additions to the Darwinian theory of evolution.

At the time, however, Lamarck's ideas were fervently held by some scientists, while others, including the naturalist Georges Cuvier, adamantly opposed his theory. The term "evolution" was not used; Lamarck referred to the changes of species as *transmutation*.

In 1831, at the age of 22, an English naturalist named Charles Darwin (1809-1882) accepted a position aboard the ship *HMS Beagle*; this worldwide voyage would provide Darwin with a wealth of observations that shaped his theories. He spent several weeks of his journey on the Galapagos Islands, volcanic islands off the South American coast. He was astonished by the wide variety of island species, which often varied drastically from the mainland species and from island to island. The finches of the Galapagos are one such example. Darwin noted that they resembled the mainland finch, but varied in their nesting sites, beak size and eating habits. These variations led Darwin to ruminate about the descent of these finches from the mainland species, and how isolation on the islands may have contributed to their different traits.

After the *HMS Beagle* returned to England in 1836, Darwin waited over 20 years to publish his findings, as he knew the publication would be controversial. He used the time to publish other works and to gather further evidence on his grand theory of how life forms arise by descent from a common ancestor and change over time. In 1859, Darwin was compelled to publish *On the Origin of Species*, when the English naturalist Alfred Russel Wallace (1823-1913) published a similar theory.

At the same time that Darwin was developing his theories in the mid-1800s, Gregor Mendel was quietly studying genetics. It is unclear whether Darwin ever read Mendel's work, but he did not include any mention of genetics in *On the Origin of Species.* Mendel's research went largely unnoticed until after his death when early 20th century scientists realized the profound implications of his work. Working from the foundations he had laid, they gradually came to understand that organisms have a genetic code, the *genotype*. The observable expression of the code, the *phenotype*, depends on gene variants as *alleles*. The newly emerging wealth of knowledge about genetics provided an explanation and elaboration on natural selection that Darwin had not been able to describe.

On the Origin of the Species popularized the term *evolution*, as the cumulative change in the heritable characteristics of a population, species or group over time. It occurs when the *Hardy-Weinberg Law,* a central tenet of population genetics, is violated. This law states that phenotypic (allelic) and genotypic frequencies do not change from generation to generation, provided the following: genes are unaffected by evolutionary forces (no mutation), the population is infinitely large and is not affected by migration, and sexual selection does not occur (i.e., mating is random). In nature, these assumptions are often violated, resulting in evolution. Darwin described one mechanism of evolution, natural selection, but genetic drift, gene flow, and mutation are violations of the Hardy-Weinberg law.

Genetic drift is random changes in an allele's frequency in a population and occurs by chance. Some organisms, even if they are not the fittest of the population, may pass on more of their genes simply by the probability of chance.

Gene flow is the transfer of genes between two populations. This may occur by migration when one population moves into a new habitat and encounters another population. Gene flow may be within a species or between two different species for hybridization. *Hybrids* are offspring from two genetically dissimilar parents. Parents may be different from the same species or different species altogether. When two species hybridize, genes flow between the groups, introducing genetic diversity.

Mutations are changes in an organism's DNA genetic code from damage or replication errors. Mutations can be fatal, harmless, beneficial or in between. When mutations introduce new traits that allow an organism to be more successful, that organism may survive and pass on its mutated code. Mutations are powerful agents of evolution.

Evolution can be regarded from two perspectives. *Microevolution* describes a change in allele frequency within a population. The population, known as the gene pool or deme, is the arena of microevolution. Microevolution can be observed over short periods of time, especially in organisms such as bacteria.

Macroevolution is a pattern of change in groups of related species over long periods of geologic time. In a sense, macroevolution is the sum of all microevolution. The patterns of macroevolution determine *phylogeny*, the evolutionary relationships among species and groups of species. There are two models of macroevolution: phyletic gradualism and punctuated equilibrium.

Phyletic gradualism describes the constant, uniform accumulation of small changes, resulting in the gradual transformation of one species into a new species. Phyletic gradualism is generally disregarded as a model of macroevolution due to fossil evidence which indicates sudden, drastic speciation. *Punctuated equilibrium,* which describes geologically long periods of stasis with little to no evolution, is the model supported by the fossil record where geologically short periods show rapid evolution.

Natural Selection

Both Wallace and Darwin proposed *natural selection* as a driving mechanism of evolution. They theorized that environmental pressures select for organisms adapted in ways that make them most fit to reproduce. There are three pre-conditions for natural selection. The first condition is that the members of a population have random but heritable variations. Individuals in a population differ due to mutations and chromosomal recombination; new adaptations arise.

The second condition is that in the population more individuals are being produced in each generation than the environment can support. This creates selective pressure to adapt to survive in the face of scarce resources.

The third condition relies on both the first and second; it states that some individuals have adaptive characteristics that enable them to survive and reproduce.

Natural selection results in the better-adapted individuals passing on their adaptations to their offspring, while the less-adapted individuals have their alleles eliminated from the gene pool, as they are more likely to die before they can reproduce. An increasing proportion of individuals in succeeding generations have these adaptive characteristics. With time, the population may become a new species altogether. Since the environment is always changing, there is no perfectly-adapted organism. Over time, the fittest organisms may be poorly adapted to their changing environment. Extinction occurs when the fitness of a population (or species) has declined to the point that in each generation more individuals die than are born. This leads to a dwindling population and eventually to the complete elimination of the group.

Fitness

Fitness is the ability of an organism to pass genetic information onto future generations. Fitness is a measure of an organism's reproductive success. Evolutionary fitness differs from the colloquial term used to refer to a strong, healthy individual. Even the healthiest organism is considered to be "unfit" if it fails to reproduce. *Relative fitness* compares the fitness of one phenotype to another.

Organisms whose traits enable them to reproduce to a greater degree have greater fitness. For example, black western diamondback rattlesnakes are more likely to survive on lava flows, while lighter-colored rattlesnakes are more likely to survive on desert soil. Therefore, each species has adapted to maximize their fitness to survive in their habitat. *Survival of the fittest* is a component of the theory of natural selection that describes the process by which fitter individuals survive and pass on their traits, while fewer fit individuals die out.

Selection by differential reproduction

The term "survival of the fittest" is a bit misleading as it implies that mere survival of the organism is the driving force behind natural selection. However, organisms may be well-suited for survival but still fail to reproduce for some reasons. Evolutionary success relies on reproduction so that adaptations may be passed to subsequent generations. Because of this consideration, survival of the fittest may be more accurately termed reproduction of the fittest or *differential reproduction*. Differential reproduction is related to the survival of the fittest and describes the way some individuals have a reproductive advantage over others.

Differential reproduction occurs naturally in the wild, but humans have manipulated natural selection for thousands of years. *Artificial selection* refers to the selection of certain traits in plants and animals that humans deem desirable, otherwise known as breeding. For example, humans have bred dogs from wolves by selecting for wolves with friendlier, more domestic traits. Over time, this artificial selection compounded, eventually producing the domestic dog; further selection for specific traits resulted in the wide variety of breeds seen today. Other domesticated animals have been produced by artificial selection, as well as many crop plants. Breeders of plants and animals are trying to produce organisms that possess desirable characteristics, such as high crop yields, resistance to disease, high growth rate and many other phenotypical characteristics that benefit people who consume or make use of these organisms.

Selective breeding is often performed with the goal of producing a hybrid from two parents that have different traits which are both desirable. Usually, the hybrid offspring possess the desirable traits of both parents. For example, one parent may possess dominant alleles for long life, while the other parent possesses dominant alleles for fast growth. When *crossed*, or bred, the hybrid offspring ought to be both long-lived and fast-growing. The high fitness characteristic of hybrids is *hybrid vigor*. Hybridization often occurs in nature when two groups with overlapping habitats mate at a geographic boundary in the *hybrid zone*.

Concepts of natural and group selection

Natural selection can utilize variations that are randomly provided; therefore, there is no directedness or anticipation of future needs. The organism is incapable of consciously picking and choosing the genome which it desires.

Only alleles that cause phenotypic differences are subject to natural selection. In diploid organisms, recessive alleles may only be expressed when there are two copies (i.e., homozygosity). Heterozygosity is responsible for the preservation of recessive alleles over time, as heterozygotes carry rare recessive alleles that may otherwise be selected against.

An example of this preservation is two alleles that encode for hemoglobin: a dominant, normal allele and a recessive, defective allele. Heterozygous expression of one abnormal hemoglobin allele results in the sickle cell trait, while homozygous expression of both recessive alleles results in full-blown anemia. Homozygotes with both dominant alleles have normal hemoglobin production.

There is a high frequency of the sickle cell allele in Africa, where there is a high incidence of malaria. A substantiated link has been made between those with the sickle cell trait and immunity to the effects of malaria. The sickle cell trait proves to be advantageous in areas where malaria is a greater threat to survival. Therefore, the higher the frequency of the sickle cell allele in Africa is evidence of natural selection. There is, of course, a tradeoff, as a higher frequency of the allele results in a higher frequency of regressive heterozygosity, causing sickle cell anemia. Theoretically, anemia is an excellent defense against malaria, but this is offset by the severe health issues which it can inflict.

Natural selection can be described in a variety of ways. *Directional selection* occurs when a trait at one extreme of a spectrum is favored, while traits on the opposite end of the spectrum are selected against. Over time, directional selection results in a shift of the distribution curve of allele frequency towards the favored allele. Natural selection leading to drug resistance in bacteria represents this type of selection. Another example is in trees in the rainforest, where trees compete for sunlight and selection favors the taller spectrum for trees.

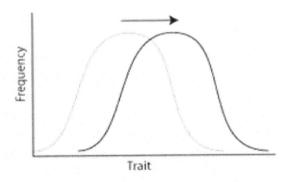

Directional selection shifts the distribution toward favored traits

Artificial selection is mostly directional selection as humans progressively select one trait over another. It has allowed humans to increase the efficiency of livestock animals and crop plants, such as increasing milk yield from cows by continuously breeding cows with high milk production.

Stabilizing selection occurs when both extreme phenotypes are eliminated, and the intermediate phenotype is favored. For example, the optimum number of eggs laid by Swiss starlings is four or five. If the female lays more or less than this number, fewer survive. Stabilizing selection results in the selection of those alleles that produce intermediate clutch sizes.

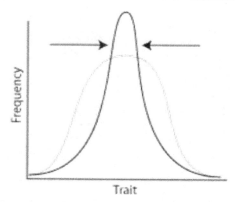

Stabilizing selection favors a narrow range of phenotypes

Disruptive selection occurs when the phenotypes at both extremes are favored. Birds occupying an area with two distinct niches, such as eating berries and eating seeds, may exhibit disruptive selection. Small beaks are selected for eating berries, while large beaks are selected for cracking seeds. The intermediate phenotype (medium beaks) is selected against because a medium beak is not useful for either berries or seeds.

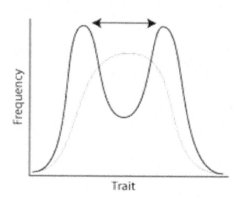

Disruptive selection favors traits at the extreme

Group selection is a natural selection which acts upon the group and not the individual. Group selection is often provided as an explanation for the existence of altruism. Altruism is when the fitness of an individual is sacrificed to benefit the group, usually the family, as the family shares similar genes. If altruism enables another family member to survive so that the genes can be passed on, the individual may exhibit altruism even if it means sacrificing his survival.

Sexual selection is the differential mating of males (or females) in a population. In most species, the females select for superior males, which increases the fitness of their offspring. Females invest more energy in reproduction and thus attempt to maximize the quality of their mates. Males primarily attempt to maximize the number of their mates. Male competition for mates leads to fights, with mating opportunities awarded to the strongest male. The genetically superior males exhibit traits which aid them in this competition and prove their strength (e.g., musculature, large stature or bigger horns). Sexual selection by females may lead to traits or behaviors in males that are not practical and do not increase the male's ability to survive. An example is colorful bird plumages, which may make certain species more easily identifiable by predators. The pressure for males to attract females usually leads to *sexual dimorphism*, in which males and females are observably different.

Adaptive radiation is the rapid development and evolution of many new species from a single ancestral species. It occurs when the ancestral species is introduced into an area where diverse geographic or ecological conditions are available for colonization. The observations by Darwin of finches illustrates the adaptive radiation of several species from one founder species of mainland finch.

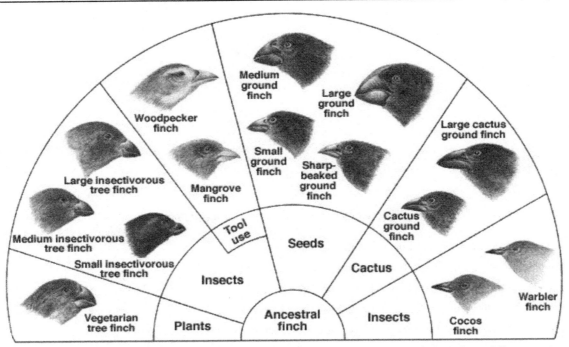

Darwin's theory of finches on the Galapagos Islands illustrates adaptive radiation from differences in their environments

Evolutionary success as an increase in percent representation in the gene pool of the next generation

An increase in allele frequency represents evolutionary success for that allele. An increase in an individual's allele frequency is an evolutionary success for that individual.

The peppered moth in European cities can illustrate the changes in allele frequency over time. Pre-industrialization, these moths were light-colored. However, as pollution increased in the 1800s, soot collected on the sides of buildings, making the light-colored peppered moth more visible to predators. Mutations resulted in a new phenotype, darker than that of the white peppered moth. These new, darker peppered moths were favored over the lighter moths by natural selection, specifically directional selection. Over time, the frequency of the dark allele increased in the population, reaching 95% in most industrial European cities. However, in more recent years, soot has been reduced, and most buildings are lighter than in the late 1900s. Therefore, the allelic frequency of dark-peppered moths has decreased since then, demonstrating directional selection back towards the light phenotype.

Speciation

At a certain stage of evolution, one population may become so genetically different from the other that it becomes a new species, the process of *speciation*. After enough changes in gene pool phenotypic and genotypic frequencies have amassed, speciation is an inevitable result. The definition of a species is controversial and inexact, but a good rule of thumb is that members of different species are unable to reproduce with another and produce viable and fertile offspring.

Biogeography is the study of the geographic distribution of life forms on Earth. Comparison of the animals of South America and the Galápagos Islands led Darwin to conclude that adaptation to the environment can cause diversification, including the origination of new species. His conclusions were greatly influenced by Charles Lyell (1797-1875), a geologist who first presented the idea that geological variations were formed over time by slow, continuous processes such as erosion. This notion was in direct contrast to the beliefs of the time, which stated that the contours of the Earth had been shaped by divine intervention and had only changed due to sudden, violent catastrophes. Lyell's theory of *uniformitarianism* supported what Darwin had observed in his study of geology and fossils.

Lyell and Darwin's conclusions were the precursors to tectonic plate theory, which describes the movement of Earth's crust. In 1920, German meteorologist Alfred Wegener presented data from across disciplines supporting continental drift. By the 1960s, the theory of continental drift was widely confirmed due to overwhelming geologic and evolutionary evidence. For example, many of the same fossils have been found on different continents, indicating there was a time when the continents were united.

Continental drift explains why the coastlines of several continents are mirror images of each other, as in the outlines of the west coast of Africa and the east coast of South America. The same geological features are found throughout the world in regions where the continents separated or collided.

The single, ancestral supercontinent of Pangea has separated over millions of years into the land masses seen today. As the continents drifted apart, organisms were separated, and this resulted in great speciation. One example is the marsupials of Australia, who were isolated from other mammals, and therefore are different from mammals on other continents.

Allopatric speciation is when a new species results from populations being separated by a geographic barrier. While geographically isolated, variations accumulate via natural selection, mutation, gene flow, and genetic drift until the two populations are reproductively isolated.

Sympatric speciation occurs when members of a single population develop genetic variations that prevent them from reproducing with the original population without geographic isolation. An example of sympatric speciation is polyploidy in plants. *Polyploidy* is the possession of more than two sets (2N) of chromosomes (e.g., 3N, 4N). The failure to reduce the chromosome number produces polyploid plants that reproduce successfully only with other polyploid plants. Backcrosses with diploid plants are sterile. Polyploidy leads to new species of plants.

Aside from the special case of polyploidy, evidence for sympatric speciation is sparse. Many researchers contend that sympatric speciation is, in fact, a poorly understood subtype of allopatric speciation. They assert that without a geographic barrier, interbreeding makes true sympatric speciation impossible. This issue is under intense research and debate in evolution.

Taxonomy

Taxonomy is the science of classifying organisms, an undertaking which humans have attempted for thousands of years. Carl Linnaeus (1707–1778), a Swedish naturalist, revolutionized the field of taxonomy when he introduced a streamlined, standardized taxonomic system which soon replaced the hundreds of disorganized systems previously used. Linnaeus' system featured binomial nomenclature, in which a two-part name is given for each species (e.g., *Homo sapiens*). Like other taxonomists of his time, Linnaeus believed that each species had an "ideal" structure and held a fixed place in the *scala naturae*, the divine hierarchy of life. Linnaeus thought that classification should describe the fixed features of species and reveal God's plan, and believed that the ideal form of each organism could be deduced and arranged according to the scala naturae. However, his later work with hybridization suggested species might change with time. He was the first prominent scientist to propose that humans were closely related to other primates. This was a bold declaration which clashed with the predominant belief that humans were fundamentally distinct from animals.

Classification establishes categories to assign species by their relationship to other species. A *taxon* is a group of organisms in a classification category; *Homo* or *Felis* are taxa at a genus level, describing the genera of hominids and cats, respectively. The term *species* is used for a taxonomic category below the rank of *genus*. When a species has a wide geographic range, variant types may interbreed where they overlap; these populations may be named as subspecies. For example, the wolf species *Canis lupus* contains around 40 subspecies, including *Canis lupus lupus* (Eurasian wolf) and *Canis lupus familiaris* (domestic dog). Addition of the subspecies makes for a trinomial, or three-part name.

Previous taxonomic classifications separated organisms into two categories, Animalia and Plantae. This proves to be incorrect, as many new organisms were discovered that appeared to fit in both or neither category. Currently, at least seven categories of classification are used. From broadest to narrowest, these taxa are a kingdom, phylum, class, order, family, genus, and species. Additional classification can be added to all taxonomic levels by adding the prefixes super-, sub-, or infra-.

The *domain* is a taxon above the level of the kingdom and is used to classify all organisms into Eukarya, Bacteria, and Archaea. These domains, as well as all the subsequent taxa, are under revision, controversy, and debate.

Definition of species

Most of the contention in taxonomy comes from trying to define a species. The controversy of what makes a species has sparked entire books and movements, and the discussion has been ongoing since the earliest days of taxonomy. Many of the arguments come down to philosophy, which makes consensus incredibly difficult.

One of the most common definitions in use is that a species is a group of organisms that could interbreed and produce fertile, viable offspring without outside intervention. This is the *biological definition of a species.*

Gene flow, via sexual reproduction, cannot occur between species which are reproductively isolated. A *reproductive isolating mechanism* is any structural, functional or behavioral characteristic that prevents successful reproduction. *Prezygotic isolating mechanisms* are barriers that prevent two species from successfully mating or undergoing fertilization. These barriers include habitat and temporal, behavioral and mechanical isolation.

Habitat isolation occurs when two species occupy different habitats so that they are less likely to meet and attempt to reproduce. *Temporal isolation* is another prezygotic barrier, where two species occupy the same habitat but mate at different cycles (i.e., times of the year).

The species may have certain behaviors which preclude them from mating, as *behavioral isolation.* Flowers exhibit behavioral isolation when one is bee-pollinated, and the other is bird-pollinated so that the two do not exchange gametes. Another example is two species with different mating rituals so that the females of one species do not recognize the males of the other species as a mating partner. Even if a male and female manage to mate, *mechanical isolation* can prevent successful mating due to incompatible reproductive structures or another anatomy.

Gamete isolation is prezygotic reproductive isolation involving incompatibility of the gametes from two different species so they cannot fuse to form a zygote (i.e., fertilized egg). For example, the egg may have receptors only for the sperm of its species.

Postzygotic isolating mechanisms prevent the development of a hybrid after fertilization. These barriers include sterility or inviability of the hybrid. Hybrids may be fertile in the first generation, but the second generation is infertile, as a *hybrid breakdown.* Sometimes populations are reunited before reproductive isolation is complete. In this scenario, they may still be able to reproduce fertile hybrids with increased genetic diversity, which is a benefit to the population.

Defining species based on reproductive isolation has limitations. Some groups are considered separate species based on other differences, even if they can hybridize where their ranges overlap. Furthermore, many species do not reproduce sexually and thus cannot be characterized by their sexual reproduction abilities. Finally, it is difficult to test and observe reproductive isolation.

Polymorphism

Early taxonomists such as Linnaeus classified organisms based on morphology, but this method has several limitations. This becomes apparent for variations among members of the same species, like sexual dimorphism. Sexual dimorphism is an example of morphological variations within a species or *polymorphism.*

Phenotypic polymorphism is the existence of two or more distinct phenotypic variations within a species. Sexual dimorphism falls under phenotypic polymorphism, as does blood type.

Polymorphism may be a result of sexual selection (e.g., sexual dimorphism) or due to different ecological niches. The two different forms of the peppered moth discussed earlier, are an example of polymorphism. As might be expected, it is sometimes difficult to determine whether two groups are polymorphs within the same species or different subspecies.

Genetic polymorphism applies the concept of polymorphism to the genotype. In the age of genomics, genetic polymorphism refers to the unique differences between the genomes of two individuals. In this regard, a species has an almost infinite number of genetic polymorphs.

Balanced polymorphism is a form of natural selection which favors the balance of different polymorphs in the population, meaning the frequencies for each allele are equal. The *heterozygote advantage,* in which the heterozygous expression of two alleles is favorable, is one mechanism of balanced polymorphism. For example, the conservation of the recessive sickle cell allele in populations afflicted by malaria illustrates this mechanism, as heterozygotes have an advantage over homozygous individuals.

Balanced polymorphism may be a consequence of *frequency-dependent selection.* In this mechanism, the fitness of one phenotype is dependent on the frequency of another phenotype, so that an increase in the frequency of either phenotype is eventually balanced by the other.

In some cases, polymorphism may be controlled by environmental effects. In contrast to genetic polymorphism, this is a *polyphyletic system* (i.e., phyletic polymorphism), in which the phenome of the individual, rather than the genome, is altered to produce a morph. For example, many organisms have polyphyletic sex determination, in which the sex of an individual is regulated by factors such as temperature, predator density or current male-to-female ratio in the population. When polyphenic forms are in balance, they can be hard to distinguish from genetic polymorphs.

Adaptation and specialization

An *adaptation* is any change that an organism develops to become more suited to its environment. Because of natural selection, adaptive traits accumulate in each succeeding generation. *Specialization* occurs as these adaptations allow the species to adapt to a niche.

Most environments change, and so organisms must adapt or become extinct. Organisms might adapt to an environmental change in their habitat (e.g., change in soil acidity or precipitation). They may adapt to a new habitat due to migration or an invasion from outside species. Organisms must adapt in response to the evolution of other species in their habitat. When two species evolve in response to one another, they exhibit *co-adaptation.* For example, pollinating insects and flowers continually co-adapt with one another to maximize their mutualistic relationship. Co-adaptation may be more threatening, as when a host species evolves a more hostile gut environment, and the parasitic species must then adjust to maintain its survival in the host species' gut.

Exaptation is when, over time, many traits are readapted for a new purpose. *Maladaptation* is when traits are selected against because they became a distinct disadvantage.

Concepts of ecological niche and competition

Organisms occupy an *ecological niche,* the environment in which it lives and the role which it performs. The ecological niche is the organism's habitat as well as its behavior and relationship with the habitat. Organisms attempt to fill their niche by maximizing resources. This is accomplished by the specialization of the organism, as it adapts for its unique niche.

Resources can be best used when organisms occupy distinct niches, as they do not have to compete with one another. When niches overlap, resources become scarce, and the organisms threaten each other's survival. This results in competition, which can be both interspecific and intraspecific. Since members of the same species tend to occupy similar niches, *intraspecific competition* is more common. Competition is a powerful selection pressure that drives organisms to evolve so that they can adapt to a different niche or better compete for resources in their current niche.

The concept of population growth through competition

Population growth is balanced by competition. At a low population density, competition is minimal, and population growth can occur rapidly. However, as the population grows the number of available resources dwindle. *Carrying capacity* describes the maximum population that can be supported by the environment. As a population approaches its carrying capacity, intraspecific competition increases and slows population growth. Competition within a species can force members within the species to occupy different niches, which drives speciation.

Individuals may directly fight for mates, food, water, space or other such advantages. They may compete indirectly by depleting a resource which other organisms utilize. The *competitive exclusion principle* says that when two species use the same resource in the same place at the same time, the species diverge into different niches. This process is *niche differentiation.* If this divergence does not occur quickly enough, one species outcompetes the other by pushing the species from the shared overlap of their niches.

Exponential population growth is rare due to the pressures of competition. However, humans exhibit exponential population growth due to our ability to find and exploit new resources constantly.

Evolutionary ecology places species on a spectrum that relates to their population growth strategy. *r-selected* species mature rapidly, reproduce early and produce many offspring. A classic example of an r-selected organism is the mouse. Mice populations can grow exponentially due to their early onset of reproductive capacity and large brood sizes. They reserve the energy that would otherwise be spent on caring for their young by minimizing parental investment. Of course, this results in a high offspring mortality rate. The mice population often quickly grows above the carrying capacity of the environment, exhausting their resources. This results in sudden and intense competition (along with disease, overcrowding, and other threats), and the population dwindles. This fluctuation in population growth and subsequent decline is characteristic of r-selected organisms.

K-selected organisms, such as primates, tend to mature slowly. Once at reproductive age, they produce a few offspring at a time but reproduce throughout their lifetimes. These organisms live longer and have a higher offspring survival rate, but at the expense of great energy investment. They usually maintain a steady population near their carrying capacity.

Most organisms lie somewhere between r-selection and K-selection and may even be able to shift strategy depending on the environment and population density. Furthermore, organisms frequently exhibit r-selected traits in some areas and K-selected traits in others.

Inbreeding

Inbreeding is mating between individuals that are closely related. It increases the frequency of homozygotes, decreases the frequency of heterozygotes and decreases genetic diversity. Some organisms avoid inbreeding because it results in increased rates of disease and other undesirable traits. However, inbreeding occurs in the wild often, being unavoidable in small populations.

Inbreeding is a common result of artificial selection by humans. Continuous inbreeding for selection of particular genes results in the loss of certain genes decreased genetic diversity and can cause the accumulation of genetic defects. *Inbreeding depression* is the loss of fitness that results from these effects. In the long term, it is generally more advantageous for organisms to promote genetic diversity, and most present-day breeders strive to avoid inbreeding.

Genetic diversity serves as a way for populations to adapt to changing environments. More variation in the gene pool may help a species be prepared for a wide range of scenarios such as food shortage or an epidemic of disease. For example, an extremely contagious disease may threaten 99% of a species, but the remaining 1% possess an allele that provides them with resistance to the disease. With this allele, the population has a chance of surviving the disease that would otherwise force their extinction.

Outbreeding

Outbreeding (opposite of inbreeding) is when genetically dissimilar individuals are mated to increase genetic diversity. Generally, outbreeding increases the fitness of a population. However, crossing individuals from different populations rarely result in offspring that are less fit than those from within the same population. This is *outbreeding depression.*

Outbreeding depression may be the result of disruptive selection when the heterozygous genotype is selected against in favor of the homozygous genotypes. Then, the outbred progeny display an intermediate phenotype that is not useful in either of the populations from which its parents originated.

Bottlenecks and genetic drift

A bottleneck is a severe reduction in population size. This can be caused, for example, by a natural disaster that eliminates a majority of the population. Bottlenecks can be deleterious to the population because the population is less able to adapt to environmental changes with a smaller, less diverse gene pool.

Bottlenecks increase the effect of genetic drift, because certain traits may be represented disproportionately when a population drastically decreases, especially if the decrease is random. For example, in a population of white and brown rabbits in equal numbers, a natural disaster may kill mostly brown rabbits by random chance. The bottlenecked population then displays a disproportionate representation of the white rabbits' alleles. In subsequent generations, the population may grow to its former size but still be made of mostly white rabbits.

The effects of genetic drift are amplified when a small group migrates out of a large population. As with a bottleneck, the small population has a small gene pool which may not accurately represent the original population. This is the *founder effect*. Founders can have a profound effect on the gene pool of the population after many generations. Genetic drift, along with mutations, gene flow, and natural selection, may cause the two populations to diverge over time.

Divergent, parallel and convergent evolution

Divergent evolution occurs when species that originate from the same lineage or common ancestor evolve to become increasingly different over time. An example can be seen in bats and horses. Both bats and horses share the same mammalian lineage, but the ancestral mammalian forelimbs became wings in bats, hooves in horses. *Homologous structures* are different structures that arose from the same common ancestor (e.g., bat's wings and horse's hooves).

Parallel evolution involves two related species that are of the same lineage and evolve similarly. An example of this is the feeding structure in different species of crustaceans. The feeding structure of these crustaceans came from a mutation of their ancestors' legs, evolving them into mouthparts. This is a prime example of parallel evolution: the same lineage, similar traits, evolving from similar mechanisms or mutations.

Convergent evolution is a phenomenon in which two unrelated species of different lineage, by different mechanisms, evolve closer together to become more similar. For example, bats and butterflies both have wings, but they came from entirely different lineages and evolved through different mechanisms or mutations. These similar structures which arose by convergent evolution are *analogous structures*. Analogous structures can be deceptive, as they often seem to suggest common lineage when in fact the structures are similar merely by chance.

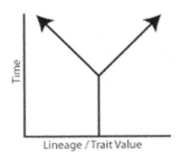

Divergent evolution giving rise to homologous structures

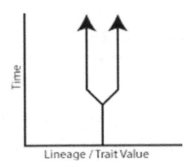

Parallel evolution for two related species undergoing a similar evolution

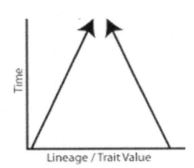

Convergent evolution from unrelated species that become more similar

Coevolution occurs when two species evolve in response to one another. It is essentially the accumulation of co-adaptations. For example, a predator may develop a new trait that aids in better hunting prey. In response to this newly developed trait in the predator, the prey may develop a trait that allows it to evade the predator more successfully.

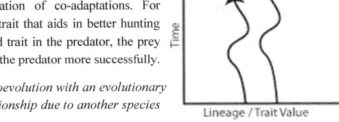

Coevolution with an evolutionary relationship due to another species

Symbiotic relationships: parasitism, commensalism, and mutualism

Symbiosis is a relationship between two species involving three types: parasitism, commensalism, and mutualism.

Parasitism is a relationship in which one organism, the parasite, benefits at the expense of the other organism, the host. An example of this is in tapeworms that live in the intestines of animals, causing illness. All pathogenic bacteria and viruses are parasites.

Commensalism occurs when one organism benefits while the other is not affected. Plants exhibit commensalism when they disperse their seeds by sticking them to an animal's fur. The animal is neither harmed nor benefited, but the plant has the advantage of wider seed distribution.

Mutualism is when both species benefit. A classic mutualistic relationship is that between fungi and algae. The fungus provides anchorage and absorption for the alga, while the alga provides photosynthesis; together they form a *lichen*.

Evolutionary Time as Measured by Gradual Random Changes in Genome

Random genetic mutations that are not acted on by natural selection occur at a constant rate. By measuring the amount of these neutral mutations, the amount of elapsed time can be determined. The genome differences between two species can be compared to determine how long ago they may have diverged. This concept of dating is the *molecular clock*.

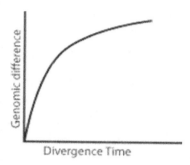

Molecular clock to estimate time-based on genomic differences

Genetic differences may be measured in the *phenome* (i.e., the total of phenotypic traits), or in the genome, by DNA nucleotide sequences.

The simple molecular clock may indicate relative periods of time between two organisms, but cannot be used to assign a numerical date. Calibration of the clock, via comparison against known fossil dates or other evidence, allows for this precision.

Origin of Life

About 13.8 billion years ago (b.y.a.), all material in the universe was created in a massive expansion of matter as the Big Bang. From a singularity of infinite density, all matter and space were suddenly expelled. The sudden cooling of the superheated ejecta facilitated the combination of atomic components into atoms and molecules. These clouds of gasses eventually cooled and formed the principal components of galaxies, including stars and planets.

The Earth came into being about 4.6 b.y.a. when dust and gasses leftover from the formation of the sun coalesced. Heat and gravity stratified the Earth into several layers: a molten core of iron and nickel, a semi-liquid mantle of silicate minerals and a solid crust created by the upwelling of lava from the mantle. The early Earth was so hot that water existed as a vapor in dense, thick clouds. As the Earth cooled over the first billion years, water vapor condensed into rain, which formed the primordial oceans.

The Earth's size provides a gravitational field strong enough to hold an atmosphere. Earth's primitive atmosphere differed from its current atmosphere, which consists of water vapor (H_2O), nitrogen

(N_2), carbon dioxide (CO_2), small amounts of hydrogen (H_2), and carbon monoxide (CO). The primitive atmosphere was likely formed by gases from volcanoes and celestial bodies which bombarded the earth.

The primitive atmosphere contained little free oxygen and was probably a reducing atmosphere, as opposed to the oxidizing atmosphere of today. An *oxidizing atmosphere* contains free O_2 and inhibits the formation of complex organic molecules. A *reducing atmosphere* lacks free O_2 and allows the formation of complex organic molecules (i.e., carbon-based compounds). Scientists believe that this characteristic of the primitive atmosphere was crucial to allow simple elements to form organic molecules that are the basis of life. Another important contribution may have come from the comets and meteorites which pelted the Earth, as they may have carried organic chemicals. Abiotic (non-living) factors are the physical and chemical components of a system (e.g., light, temperature, soil, pH, etc.)

Some evidence suggests that the primitive atmosphere was less reducing than previously thought, containing less ammonia. In this case, abiotic synthesis may have been carried out in the oceans at hydrothermal vents, which expelled large amounts of ammonia. It is possible that chemical evolution occurred both at these vents and in the atmosphere.

Monomers evolve

Chemical evolution is the increase in complexity of chemicals that led to the first cells. Chemical evolution gave way to biological evolution when the first self-replicating molecules arose and began to undergo natural selection. Like living organisms, biotic molecules are subject to variations and natural selection, enabling evolution.

In the early 1900s, a biochemist named Alexander Oparin suggested that organic molecules could be formed from the elements of the reducing atmosphere: methane (CH_4), ammonia (NH_3), hydrogen (H_2) and water (H_2O). In Oparin's vision, lack of oxidation and decay allowed organic molecules to form a thick, warm organic soup from which life arose.

His theory motivated chemists, such as Stanley Miller, to perform experiments that imitated the conditions of the early Earth. In 1953, Miller published the results of his experiment on the chemical origin of life, called the Miller Spark-Discharge Experiment, or the Urey-Miller Experiment. His experiment assumed a reducing atmosphere that had been bombarded by kinetic energy from lightning. This was simulated by a sealed flask of water vapor, methane, ammonia, and hydrogen being continually subjected to electrical sparks. Within a week, a deep red solution formed that contained hydrogen cyanide (HCN), formaldehyde (CH_2O) and other related compounds. These simple organic molecules then assembled into many different amino acids, the monomers of proteins.

Miller concluded that preliminary organic molecules could have been originated in a reducing atmosphere. Amino acids formed readily in Miller's spark-discharge experiment so that they could have formed in the conditions of early Earth. More recent evidence suggests that amino acids could have been seeded from comets and meteorites. Interstellar dust that constantly falls on Earth contains hydrogen cyanide and aldehydes, key reactants for forming amino acids.

Other evidence since Miller's experiment has indicated that Earth had a more oxidizing atmosphere than previously thought. Experiments based on this assumption have yielded more diverse molecules than those from the Urey-Miller experiment, including amino acids, nitrogenous bases, sugars, and hydroxy acids – all crucial for the existence of life. Current scientific thought assumes an atmosphere that is 60% hydrogen, 20% water vapor, 10% carbon dioxide and 5% hydrogen sulfide.

One puzzle that persists is that chemical evolution experiments have yielded a mixture of left-handed and right-handed enantiomers (non-superimposable mirror images). Although both forms of every amino acid exist in nature, the left-handed forms are in living organisms. Scientists have not yet been able to answer the question of how only the left-handed forms were utilized in living organisms.

Polymers evolve

As the Earth's primordial oceans began to collect the organic monomers formed in the atmosphere, amino acids and nucleotides polymerized to form proteins and nucleic acids, respectively. Polymerization occurred by condensation, the removal of a molecule of water from two monomers to form a bond between them. The opposite process, hydrolysis, depolymerizes macromolecules by the addition of water to break bonds.

There are a few challenges that come with the formation of polymers. The first is that hydrolysis dominates in the chemical equilibrium over condensation because hydrolysis increases entropy (randomness) while condensation decreases it. This means that the breaking of polymers by hydrolysis is a more thermodynamically (energetically) favorable reaction. Polymerization requires energy input, which is accomplished in living organisms by coupling the reaction with an energy-producing reaction (e.g., hydrolysis of ATP). However, scientists still do not know exactly how this was done in the pre-biotic era. It can be challenging to model polymerization reactions based on the presumed conditions of ancient Earth.

More recently, scientists showed that macromolecules could have polymerized in the muddy tide pools and beaches of ancient Earth. They mixed nucleotides and minerals from mud and collected the resulting polymers, finding nucleic acids up to 40 nucleotides long. The experiment was repeated with amino acids, and once again they found that the monomers had managed to polymerize into proteins up to 55 amino acids long.

Sidney Fox, who demonstrated that amino acids form peptides abiotically if exposed to dry heat, described the *Protein World Hypothesis*, or *Protein-first Hypothesis*. This hypothesis states that amino acids collected in shallow puddles along the rocky shore and the heat of the sun caused them to form *proteinoids*, small polypeptides that have some catalytic properties. Proteinoids form small *microspheres* in water due to their hydrophobic amino acids. These proteinoid microspheres have some similarity to lipid cell membranes, so the Protein World Hypothesis contends that polypeptides were the first precursors to living entities.

This hypothesis has been largely refuted since it was found that proteinoids are not proteins. Furthermore, the first living macromolecules would need to be able to self-replicate, something that neither proteinoids nor proteins are capable of.

The *RNA World Hypothesis* states that RNA was the precursor of the first cell. It has been proven that special RNA of *ribozymes* can act as enzymes. It is known that some viruses contain RNA genes with a protein enzyme of reverse transcriptase that uses RNA as a template to form DNA. This evidence shows that RNA could have possibly assumed the roles of both DNA and protein enzymes on early Earth, and eventually have led to the rise of both. Supporters of this hypothesis label this an "RNA world."

RNA is an excellent model for chemical evolution. When replicating, many inexact copies (mutations) are made randomly, which produces diversity in the population of self-replicating RNA.

Mutants that are more efficient self-replicators produce more copies of themselves, dominating the population. UV radiation or a chemical reaction may destroy less efficient self-replicators before they can copy themselves. Thus, chemical evolution results, changing the composition of the population over time. This is essentially natural selection, just directed toward abiotic systems rather than living organisms.

Most researchers subscribe to the idea of an RNA world, but this is not a definitive theory.

The Clay Hypothesis, first suggested by Graham Cairns-Smith in the 1960s, proposes that amino acids polymerized in clay, with radioactivity providing energy. Clay attracts small organic molecules and contains iron and zinc atoms, which serve as inorganic catalysts for polypeptide formation. In this hypothesis, clay would have collected energy from radioactive decay and discharged it whenever there were temperature or humidity changes. Nucleotides and amino acids may have polymerized in this environment and began to associate with one another, meaning they evolved simultaneously. An advantage of this theory is that it addresses the chicken-and-egg paradox, with both RNA and proteins arising at the same time.

DNA is not considered a possible candidate for the first living entity because it has low reactivity and therefore would not be an effective catalyst. Unlike RNA, DNA has never been shown to have catalytic properties in a living organism.

Evolution of a protocell

Before the first true cell arose, there would have been a *protocell* with a lipid-protein membrane and the ability to perform metabolic reactions. Although the concept of the proteinoid has been shelved, Fox's experiments proved that lipids and proteins could form complex membranes under the right conditions.

Oparin believed that the protocell could have developed from *coacervate droplets*. Coacervate droplets are spheres of lipids that spontaneously form under the right temperature, ionic environment, and pH. Coacervate droplets selectively absorb and incorporate various substances from the surrounding solution, similar to the semipermeable membranes of cells. Oparin's hypothesis still has relevance today, but some of his concepts have been superseded. Modern scientists cannot decide on a model for the protocell. Some scientists postulate that fatty acids organized into *micelles*, another form of a lipid sphere, in a mineral-rich clay environment. The clay would have allowed for the development of nucleotides that would spontaneously assemble into RNA.

There is evidence that abiotic amino acid synthesis and peptide polymerization could have occurred at hydrothermal vents. The first protocells may have used pre-formed ATP to drive reactions, but as supplies dwindled, natural selection would favor protocells that could synthesize ATP.

Several hypotheses now operate under the assumption that life began at hydrothermal vents. *Chemosynthetic* models claim that energy for peptide synthesis came from the electrochemical gradients within the microcompartments of the vents, while *thermosynthetic* models suggest thermal cycling as the primary energy source.

In 1998, Gunter Wachtershauser and Claudia Huber published a paper describing how they created peptides using iron-nickel sulfides under vent-like conditions. These minerals have a charged surface that attracts amino acids and provides electrons to polymerize them. Supporters of the *Iron-Sulfur World Hypothesis* (a chemosynthetic model) suggest that life arose in these hydrothermal vents, aided by the high pH, high temperature and the richness of iron and sulfur. Under this theory, lipids would be synthesized and anchor to the walls of the vent, before forming into a lipid membrane sphere and associating with proteins to form a protocell. Another hydrothermal vent hypothesis focuses on zinc rather than iron and sulfur (the *Zinc World Hypothesis*). Both hypotheses assume that early energy production was accomplished by fermentation, which generates ATP in the absence of oxygen. This would make glycolysis a crucial adaptation for the first protocells. The glycolysis pathway is prevalent in living organisms, indicating that it must have evolved early in the history of life.

Thermosynthetic models posit chemiosmosis rather than glycolysis as a method of ATP production. Chemiosmosis is an important part of cellular respiration and photosynthesis, describing the movement of hydrogen atoms across a membrane to drive the enzyme ATP synthase. Thus, thermosynthesis assumes ATP synthase was the first protein enzyme to evolve.

A *heterotroph* is an organism that cannot synthesize organic compounds from inorganic substances and therefore must take in preformed organic compounds. An *autotroph* is an organism that makes organic molecules from inorganic nutrients. If a protocell were a heterotrophic fermenter living on the organic molecules in the prebiotic soup, this would indicate heterotrophs preceded autotrophs. However, if the protocell evolved at hydrothermal vents in the oceans, then chemosynthetic autotrophs would have preceded heterotrophs.

A self-replication system evolves

The chemical evolution theory rests on the idea that the precursor to life was a self-replicating molecule, which many agree would have to be RNA. Supporters of the Protein-first Hypothesis argue that proteins could have self-replicating abilities that have not yet been discovered.

In most biological systems, genetic information flows from DNA to RNA to protein. It is possible that this path developed in stages. Once the protocell was capable of reproduction, it became a true cell and began the age of biological evolution.

After DNA formed, the genetic code still had to evolve to store information. Because the current code is subject to fewer errors than other possible codes, and because it minimizes mutations, it is likely to have undergone a natural selection process.

In summary, most biologists suspect life evolved in basic steps. Abiotic synthesis of organic molecules such as amino acids occurred in the atmosphere or at hydrothermal vents. Monomers then joined together to form polymers, perhaps in tidal pools, clay or hydrothermal vents. The first polymers were likely RNA, followed by proteins. A self-replicating RNA molecule appeared in the prebiotic soup due to chemical evolution and made copies of itself using free ribonucleotides in its environment. Lipids aggregated to form a protocell that had limited ability to grow; if it developed in the primordial soup, it was a heterotroph, but if it developed at a hydrothermal vent, it was a chemoautotroph. Once the protocell was able to replicate its genetic code, it became a true cell.

Development of species from protocell

The history of Earth is divided into *eons,* which are further subdivided into *eras,* then *periods*. Life first arose in the eons before the *Phanerozoic Eon,* the current eon. Collectively, these eons are the *Precambrian Supereon,* because they occurred before the first period of the Phanerozoic, the *Cambrian*. The Precambrian encompasses more than 80% of the geologic time scale. This time is when the chemical and biological evolution of life, the early development of prokaryotes, eukaryotes and even the first multicellular life.

After the development of the protocell, it would evolve for millions of years until it developed into a true unicellular organism as far back as 4 b.y.a. In all likelihood, these organisms were chemoautotrophs that resemble today's archaea as prokaryotic, lacking any membrane-bound organelles.

The unicellular chemoautotrophs found around 3.5 b.y.a. were the *last universal common ancestor* (LUCA) of all living organisms. LUCA diverged into two major domains: *Bacteria* and *Archaea*. Photosynthesis arose with the emergence of cyanobacteria around 3.4 b.y.a. These bacteria use water as a reducing agent during photosynthesis, producing oxygen as a by-product. *Stromatolites,* fossilized remains of the sediments excreted by cyanobacteria, provide fossil evidence of their age. From cyanobacteria, oxygen levels in the primitive oceans rose so high over the next billion years that it converted the reducing atmosphere into an oxidizing atmosphere. This *Great Oxygenation Event* resulted in the extinction of many obligate anaerobes, which could only survive in areas of low or nonexistent oxygen. Since that time, aerobic organisms have dominated the Earth.

Eukaryotic cells appeared later in the Precambrian, around 1.6 to 2.1 b.y.a. Their exact origin is unclear. Some scientists propose that membrane-bound organelles arose from invaginations of a prokaryote's cell membrane. Others propose that two prokaryotes fused, or one engulfed the other by phagocytosis. These may have been two bacteria, or perhaps a bacterium and an archaeon. They would have developed a symbiotic relationship, with the host cell becoming the eukaryote and the other evolving into an organelle. This is *symbiogenesis* or *endosymbiotic theory.* Under this theory, the anaerobic host cell would have taken in an aerobic bacterium which would share its genome with the host cell and eventually develop into a mitochondrion. Some host cells would also take in a cyanobacterium that would develop into a chloroplast. Endosymbiosis can still be observed in modern unicellular organisms, lending credence to the theory. Mitochondria are present in all eukaryotic cells, while chloroplasts exist only in plants and algae, suggesting that mitochondria arose before chloroplasts in the evolutionary timeline since they are almost universal.

At this time, some eukaryotes were developing cilia and flagella to facilitate movement. Motility was a huge advantage in the Precambrian Eon, allowing species to explore new niches. Organisms were continually diversifying to occupy and adapt to these previously sterile environments. Early prokaryotes and eukaryotes were exclusively asexual, but around 1.2 b.y.a. some of them evolved meiosis and sexual reproduction. It is not known exactly when multicellular organisms appeared, but they would have been microscopic. The variety of eukaryotic microorganisms that emerged around this time are *protists* (e.g., algae, plankton and amoebas). They took the form of large unicellular organisms or multicellular colonies. Their size must have been an advantage at the time, allowing them to out-compete other organisms.

At the time, all organisms existed in the sea, because on land the harsh UV radiation made life improbable. The oxygen content of the air continued to increase until 600 million years ago (m.y.a.) when it allowed for the formation of an ozone layer in the upper atmosphere. This ozone layer shielded most UV radiation from reaching Earth's surface, allowing organisms to live on land for the first time. Near the end of the Precambrian, complex multicellular life began to arise. Most data about this time comes from fossils of the Ediacaran Hills in South Australia, which date to 600 m.y.a. These organisms were soft-bodied primitive invertebrates, which likely inhabited the oceans and shallow mudflats.

Following the eons of the Precambrian came the Phanerozoic Eon. The first era of the Phanerozoic is the *Paleozoic Era* ("ancient life"), which includes the Cambrian Period along with five others: Ordovician, Silurian, Devonian, Carboniferous, and Permian. The Paleozoic Era lasted over 300 million years and was an active period with three major *mass extinctions*, events in which a large number of organisms went extinct in a small window of time. Tectonic, oceanic and climatic changes are often cited as possible triggers of mass extinction. These changes may be drastic (e.g., asteroid impact) or gradual (e.g., sea level change and global heating or cooling).

The *Cambrian Explosion*, about 542 m.y.a., marked the beginning of the Cambrian Period. This time included widespread evolution and diversification of existing prokaryotes and eukaryotes, along with the arrival of fungi and invertebrate animals. Nearly all modern phyla evolved at this time.

Animals are believed to have descended from flagellated colonial protists, making their closest relatives the modern choanoflagellates. Colonies of the choanoflagellate ancestor perhaps gave rise to the first animals: the sponges. Sponges could grow to one meter across, making them massive compared to existing species at the time. They have no nervous system and are asymmetrical.

Shortly after the sponges, the cnidarians (jellyfish, sea anemones, etc.) and the ctenophores (box jellies) evolved. These animals had symmetry in all planes, as *radial symmetry*, along with rudimentary sensory organs.

Animals higher than the sponges, cnidarians, and ctenophores are *bilaterians* for their bilateral symmetry. The first bilaterians were protostomes, which evolved throughout the Cambrian Period. *Protostomes* are defined by gastrulation during embryonic development. In protostomes, the first opening formed in the blastula, the blastopore, becomes the mouth. The earliest would have been small, primitive, parasitic organisms, such as flatworms and nematodes. Soon, mollusks overtook the Cambrian seas. At this time, marine arthropods (crustaceans) and early deuterostomes such as *echinoderms* (sea stars and kin) were evolving. *Deuterostomes* are characterized by a blastopore which first becomes an anus, with the mouth forming afterward.

In the early Cambrian Period, plants were simple, unicellular and aquatic. Throughout the Cambrian and the ensuing *Ordovician Period,* 490 m.y.a., marine algae expanded to freshwater. Fungi began symbiotic relationships with plant roots, allowing plants to colonize bare rocks. Back in the oceans, invertebrate fish diverged from the deuterostome ancestor, but protostomes such as mollusks and crustaceans still dominated.

At the end of the period, 450 m.y.a., the *Ordovician-Silurian extinction* eliminated 70% of all marine life, paving the way for fish to dominate in later periods. The marine invertebrates re-diversified, and crustaceans overtook mollusks as the rulers of the seas.

The *Silurian Period* 450 m.y.a. includes the establishment of terrestrial plants. More advanced adaptations such as vascular structures, stems, leaves, roots, and seeds evolved over the next 100 million years. The Silurian witnessed the first terrestrial arthropods, the arachnids. Jawless fish emerged at this time, evolving into jawed fishes some 395 to 420 m.y.a. Some jawed fish became "armored" with heavy, bony plates; these armored fish are all but extinct. The armored fishes diverged into bony and cartilaginous fish at the end of the Silurian Period. As competition increased and available habitats decreased, many cartilaginous fish adapted to be large, aggressive, and predatory, leading to shark species. Other cartilaginous fish include skates and rays.

In the ensuing *Devonian Period,* 420 m.y.a., the bony fish diversified into ray-finned and lobe-finned fish. This period is the "Age of Fishes," as they dominated the seas. However, arthropods were undergoing significant diversification, and an explosion of plant life occurred at this time, resulting in the first true trees and seeds. Near the end of the Devonian, as early as 395 m.y.a., lobe-finned fishes ventured onto land. These became amphibians, the first *tetrapods* (four-limbed vertebrates). The *Late Devonian extinction* 360 m.y.a. eliminated 70% of species worldwide.

The end of the Late Devonian extinction marked the beginning of the *Carboniferous Period,* which spawned the growth of forests and an abundance of seedless vascular plants such as horsetails and ferns. These ancient plants provided most of the carbon (in the form of coal) that is used today. Insects, which had by now ventured onto land, developed wings and exploded in diversity. Amphibians began to dominate land in the *Carboniferous Period,* marking this the "Age of Amphibians." Some amphibians became particularly well-adapted to land, evolving into reptiles at the end of the Carboniferous, around 300 m.y.a.

Reptiles quickly colonized land and specialized even further to their terrestrial environment. At this time, terrestrial prey was scarce because not many animals were able to survive on land. As a result, many reptiles were herbivores so that they could take advantage of the plants available on land and shorelines. The first reptiles and amphibians to reach land and die would have broken down into simpler organic compounds. This enriched the nutrients in the soil, allowing plants to grow and microorganisms to exist on a larger scale. Organisms that relied on these microorganisms would migrate to land, followed by predators of these organisms. This continued ecological succession would eventually allow land to support life on a scale that approached that of the sea.

After the Carboniferous, came the final period of the Paleozoic Era: the *Permian Period*. At the border of the Carboniferous and Permian, reptiles had diverged into sauropsids and synapsids. The *sauropsids* were the ancestors of dinosaurs, modern reptiles, and birds, while the *synapsids* were proto-mammals that resembled reptiles more than the mammals of today. These organisms are para mammals, proto-mammals or mammal-like reptiles. During the Permian, sauropsids and synapsids continued to flourish in a climate which was becoming increasingly dry; a massive desert developed in the interior of the supercontinent Pangea. Early conifers and seeded ferns largely replaced the swampy forests of the Carboniferous. The end of the Permian, 252 m.y.a., saw the greatest extinction in history, the *Permian-Triassic extinction*. This resulted in the decimation of over 95% of marine life and 70% of terrestrial vertebrates, as well as many insects. The cause of this event, like most other extinctions, is unknown, but it has been suggested that an excess of carbon dioxide due to a change in ocean circulation contributed to the extinction.

The Permian-Triassic extinction marks the beginning of the *Mesozoic Era*. Throughout the Mesozoic, gymnosperms flourished, but flowering plants had yet to evolve. Marine animals began to diversify to re-occupy all the niches that had been left empty by the extinction. Meanwhile, many of the surviving reptiles became larger and evolved into true dinosaurs and crocodilian species. The entire era is the "Age of Dinosaurs."

The first period of the Mesozoic, the *Triassic Period*, covered the evolution of dinosaurs into both avian (bird-like) and reptilian species. It included the splitting of *Pangea* (hypothetical supercontinent including all land masses) into separate continents. The mass *Triassic-Jurassic Extinction* at the end of the Triassic (around 200 m.y.a.) eliminated most synapsids and almost all large amphibians. This event allowed dinosaurs to dominate in the succeeding *Jurassic Period*. The diversification of dinosaurs was truly astonishing; they came to occupy a massive number of ecological niches. Dinosaurs ruled the land, sea, and air, and grew to mammoth proportions throughout the Jurassic. The surviving synapsids evolved into the first true mammals, but these were small and resembled rodents.

Dinosaurs were the most advanced tetrapods at the time, and some of them even moved towards bipedal motion. Many dinosaurs were huge compared to any previous organisms on Earth. Their sheer bulk and power had an obvious selective advantage. These herbivores were tall enough to graze on the top of trees, which other organisms were not able to reach. Many predators became bigger and stronger, such as the *Tyrannosaurus rex*, the largest carnivore on Earth at the time. The Jurassic Period witnessed the emergence of flying reptiles as *pterosaurs*, the most famous of which is the pterodactyl.

The Jurassic was succeeded by the *Cretaceous Period* some 145 m.y.a. This period was when avian dinosaurs evolved into true birds, some of which dominated the skies and began to outcompete flying dinosaurs. The first flowers emerged. Meanwhile, the continents from Pangea had come to resemble the continents of today. This geological change was a major factor that molded the evolution of species. For example, the collision of the supercontinent Eurasia with Africa allowed species from each region to interbreed, compete or predate on one another, greatly accelerating evolution.

From the Triassic Period until the dinosaurs' last moment of glory in the Cretaceous Period, a competitor was quietly emerging: the mammals. Throughout the Mesozoic Era, mammals gradually shed their reptilian features in favor of hair, live birth (viviparity) and other mammalian characteristics. By the Cretaceous Period, they had diverged into placentals and marsupials. Although the dinosaurs still dominated, the mammals became a competitor for land resources and even preyed on smaller dinosaurs.

No one knows exactly how the dinosaurs became extinct, but their disappearance at the end of the Cretaceous, 65 m.y.a., made many ecological niches available to other organisms. The consensus is that a major geological event as the *K-T* or *K-Pg extinction* extinguished nearly all the dinosaurs except for the avian dinosaurs, which would develop into birds. It spared many of the non-dinosaur reptiles and mammals. Many causes of this extinction have been proposed. One of the most popular theories is that an asteroid struck the Earth, resulting in catastrophic changes to Earth's climate. It has been suggested that mammals survived because they were smaller and required fewer resources to survive.

Furthermore, many lived underground and thus could have survived an event like an asteroid impact. Whatever the reason for their survival, the mammals flourished once relieved of competition and predation by dinosaurs. The K-T extinction 65 m.y.a. marks the end of the Mesozoic Era and the beginning of the *Cenozoic Era*, which extends to the present day. It is rife with the evolution and proliferation of mammals, as well as drastic climate changes. Although the entire era is the "Age of Mammals," birds dominated in the first period, the *Paleogene Period*. Many of these ancient birds were large and aggressive and posed a great threat to mammals in the early years of the Cenozoic.

In the Paleogene Period, mammals were small and inhabited overcrowded jungle environments. Flowering plants were diverse and plentiful by the Cenozoic era, and primates had adapted to living in flowering trees. The first primates were small squirrel-like animals, eventually evolving into monkeys and apes. Early whales were semi-terrestrial but fully returned to the oceans in later years as their limbs regressed.

Later in the Paleogene, global cooling resulted in the reduction of jungles and the proliferation of grasslands, allowing mammals to grow large. This trend continued throughout the *Neogene Period,* which began about 23 m.y.a. Many modern mammals evolved during this time, including the apes. At the end of the Neogene, it is believed that some of the great apes evolved into the *hominids,* the human-like ancestors of modern *Homo sapiens sapiens*. These apes would have moved from the jungles and ventured into the open savanna. They developed bipedal motion, an important advantage which allowed them to look across the tall grasses of the savanna for predators and prey alike. Bipedalism freed the hands and allowed for more efficient terrestrial travel in savannahs and scattered forests. They are *australopithecines,* which translates as "southern ape."

The previous scientific thought held Europe to be the birthplace of hominids, due to the Eurocentric outlook of the time and the discovery of some fossils in Europe. However, throughout the

20th century, archaeologists uncovered even older australopithecine remains in Africa, with Africa coming to be generally acknowledged as the origin of humans. The most famous of these discoveries was Lucy, found in Ethiopia in the 1980s. She represents the species *Australopithecus afarensis*, meaning "southern ape from afar," which roamed the Earth 3 to 5 m.y.a. While *A. afarensis* is generally cited as a direct ancestor to modern humans, that title may have belonged to other australopithecines of the time.

The paleoanthropologists who discovered Lucy and other fossilized remains used skeletal features to map the subtle evolutionary changes in hominids. The position of the pelvis and spine can be used to determine whether an organism was bipedal and upright, or walked on all fours. Lucy would have been bipedal, with short, stout stature and a relatively small cranium. The skull is one of the most important skeletal features in hominid fossils. Development of a larger brain, and thus a larger skull, is one of the most prominent markers of human evolution.

Today, the advent of DNA sequencing, genomics, and molecular biology has helped researchers map evolutionary relationships between humans and the great apes. There is evidence to suggest that the species *Australopithecus africanus* would have eventually superseded *A. afarensis* some 2 to 3 m.y.a. due to some key adaptations. They would have been taller and slimmer, and their hands better adapted for tool use. Most notably, they had a larger cranium with more human-like facial features. *Africanus* developed into *Australopithecus robustus* and *Australopithecus boisei,* two species once thought to be direct ancestors of man. However, it is now believed that they became extinct and that other descendants of *A. africanus* fathered the first members of genus *Homo.*

Homo habilis, dated to 2.8 m.y.a. at the earliest, may mark the transition from genus *Australopithecus* to genus *Homo.* These hominids exhibited the first evidence of intelligence comparable to that of modern man, most notably in their use of basic stone tools.

The *Quaternary Period,* which continues today, began 2.6 m.y.a. Early in the Quaternary Period, global cooling resulted in extensive glaciation, the Ice Ages. The Ice Ages spurred the evolution of mammals with significant capacity to retain heat, such as giant ground sloths, beavers, wolves, bison, woolly rhinoceroses, mastodons and mammoths. Many of these species are now extinct, likely due to hunting by humans. While the other latitudes iced over, deserts developed at the tropics.

During the Quaternary Period, *H. habilis* diverged into some groups, the classification of which is still debated. The prevailing view is that *Homo erectus* evolved from *H. habilis* around 1.3 to 1.8 m.y.a., and began to migrate out of Africa into Asia and Europe.

The earliest evidence of true humans has been dated to around 1.6 m.y.a. These individuals, *archaic humans,* included numerous groups, such as *Homo neanderthalensis* and *Homo heidelbergensis,* which are extinct today. It is often debated whether these groups were subspecies of *Homo sapiens* or a different species. There is still little consensus on the issue, and it is nearly impossible to draw clear lines between the various *Homo* species and subspecies.

Archaic humans spread throughout Africa, Asia, and Europe over the next 1 million years. They began to develop wooden tools, like spears and exhibited cranial capacity comparable to modern humans. Their cognitive abilities allowed them to survive in hostile environments and begin to dominate Earth. Humans would eventually learn to control their environment and protect themselves against many of the pressures of natural selection. *Homo sapiens sapiens* is the subspecies which represents modern humans, originating as far back as 500,000 years ago and persisting down to the present day. Intentional duplication of sapiens distinguishes them from their direct ancestor *Homo sapiens idaltu.*

The *Neanderthals* are well known for their hypothesized common ancestry and co-existence with early man. They became widespread across both Europe and Asia around 250,000 years ago. Once believed to be of a separate genus, it is now accepted that they were a species or subspecies of genus *Homo.* Neanderthals' robust and stocky nature made them well-suited to the cold, but *Homo sapiens* dominated the deserts. Eventually, *Homo sapiens* prevailed even in the tundra. Opinion is divided on whether modern man outcompeted the Neanderthals or incorporated them via interbreeding, or both. Whatever the reason, the Neanderthals vanished, and *Homo sapiens sapiens* became the sole humans on Earth around 40,000 years ago. They eventually inhabited all continents except Antarctica.

The advent of *Homo sapiens* characterizes the Quaternary Period as the "Age of Man." Many argue that a sixth mass extinction event, driven by humans, is occurring at the present moment. This would signal the end of the Quaternary within the unknown future.

Present day	Eon	Era	Period	Events
2.6 m.y.a.	Phanerozoic	Cenozoic	Quaternary	▪ Homo sapiens appears ▪ Modern mammals continue to evolve ▪ Ice ages
23 m.y.a.			Neogene	▪ Continents drift to current positions ▪ First australopithecines ▪ Modern mammals evolve and dominate ▪ Grasslands proliferate
65 m.y.a.			Paleogene	▪ Mammals diversify and grow ▪ Birds dominate land and skies
145 m.y.a.		Mesozoic	Cretaceous	▪ First flowering plants ▪ First true birds ▪ Mammals evolve and diversify ▪ Dinosaurs continue to dominate
200 m.y.a.			Jurassic	▪ Dinosaurs grow, diversify and dominate the land ▪ Mammals evolve and diversify
252 m.y.a.			Triassic	▪ Pangea separates ▪ First true mammals ▪ First true dinosaurs ▪ Life recovers from Permian-Triassic extinction

300 m.y.a.		Paleozoic	Permian	▪ Pangea forms ▪ Sauropsids and synapsids proliferate
360 m.y.a.			Carboniferous	▪ Trees proliferate ▪ Reptiles diverge into sauropsids and synapsids ▪ First reptiles (evolution of amniotic eggs)
420 m.y.a.			Devonian	▪ First amphibians ▪ Global heating ▪ First trees and seeds ▪ Arthropods diversify ▪ Bony fish diversify into ray-finned and lobe-finned fish
445 m.y.a.			Silurian	▪ Vascular plants colonize land ▪ Arthropods colonize land ▪ Fish evolve into jawed fish, then cartilaginous and bony fish
490 m.y.a.			Ordovician	▪ Glaciation ▪ First fish ▪ Invertebrate marine life proliferates
540 m.y.a.			Cambrian	▪ Cambrian explosion
2.5 b.y.a.	Proterozoic			▪ First fungi and invertebrates ▪ Ozone layer forms ▪ First protozoa ▪ First multicellular organisms ▪ First eukaryotes
4 b.y.a.	Archaean			▪ Cyanobacteria evolve and begin producing oxygen; Great Oxygenation Event ▪ LUCA diverges into Bacteria and Archaea ▪ First life
4.5 b.y.a.	Hadean			▪ Cooling of crust ▪ Water condenses into oceans ▪ Reducing atmosphere ▪ Volcanic activity ▪ Celestial bombardment ▪ Earth forms

Notes

Please, leave your Customer Review on Amazon

Chapter 7

Classification and Diversity

- **Common Ancestry: Shared Conserved Features**

- **Phylogenetic Trees and Cladograms**

- **Comparative Anatomy**

Common Ancestry: Shared Conserved Features

Structural and functional evidence of the relatedness of all domains

The theory of *common descent* is a central tenet of evolutionary biology. This theory states that any given group of living organisms share a common ancestor from which they have descended. It naturally follows that there is a common ancestor to every single organism on Earth, the *last universal common ancestor* (LUCA). Darwin and his contemporaries gathered much of the evidence that is the foundation of this theory, but modern advances in molecular biology have provided many new insights.

In contemporary biology, life is generally divided into the three domains of Archaea, Bacteria, and Eukarya. It is in relatively recent years that scientists have realized that Archaea constitutes a separate domain from Bacteria. Although both are unicellular prokaryotes, the archaea are too different from bacteria to be classified under the same domain. It is believed that the LUCA first diverged into bacteria and archaea, with eukaryotes later developing from one of these domains or a combination of both.

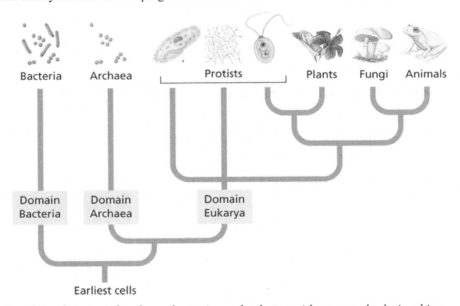

Three domains of archaea, bacteria, and eukarya with proposed relationships

Archaea live in a wide range of habitats, but the most well-known species are the extremophiles, which inhabit hostile environments such as hot springs, salt lakes, and anaerobic swamps. Bacteria and archaea exhibit common prokaryotic features, such as a cell wall, but they are different on a biochemical level.

Domain Eukarya comprises a diverse array of unicellular to multicellular organisms, the most well-studied on Earth. It includes the kingdoms Animalia, Plantae, Fungi, and Protista.

While all three domains have their unique differences, their underlying structure and function display undeniable similarity. Modern evidence comes from DNA and RNA sequencing, analysis of common enzymes and metabolic pathways, fundamental similarities in cell structure and observation of ongoing evolution. There is a large body of evidence gathered from fossils, comparative anatomy, and other investigations that have been underway since the 18th century.

Biochemical evidence is some of the most compelling. All living organisms essentially use the same fundamental building blocks, genetic code, and metabolic pathways. There are certain genes which are largely conserved across all domains of life, coding for proteins that can be remarkably similar or even identical. This all alludes to the fact that all three domains of life descend from a common ancestor. It is seen consistently in all forms of life that the same biomolecules are used in the same way, i.e., all organisms use DNA and RNA as carriers of genetic information, encoded in the same base nucleotides (A, T, G, C, and U). All organisms interpret these bases as triplet codes to produce proteins made of the same twenty amino acids. Furthermore, the code and not just the method of interpretation is widespread in its similarity. Many organisms even share the same introns (noncoding sequences of mRNA), which is notable because these units are inoperative and there is no functional reason why they need to be so similar.

Structural evidence of the relatedness of all eukaryotes

The eukaryotes are the most well-researched domain on the Earth, as they are the most readily observable and of the most interest to humans. Consequently, there is a wide array of evidence for the relatedness of all eukaryotes.

Common structural features of eukaryotes include an extensive cytoskeleton and organelles such as the mitochondrion, chloroplast, and nucleus. These organelles are all a part of the endomembrane system, a collective term for the membranous sacs that divide a cell into its constituent parts. Eukaryotes, in contrast to bacteria and archaea, are the only organisms with this system of membrane-bound organelles. Other notable similarities among eukaryotes include DNA bound linear chromosomes and common metabolic pathways, such as glycolysis.

The earliest evidence for the structural relatedness of all eukaryotes comes from fossils. The fossil record is the history of life as recorded by remains from the past. Fossils may be skeletons, shells, seeds, imprints, and even soft tissues. Unfortunately, the fossil record is often incomplete, because most organisms decay before they become fossilized, and soft-bodied organisms generally do not fossilize. The great majority of fossils are embedded in or recently eroded from sedimentary rock. Sedimentation has been going on since the Earth was formed; it is an accumulation of particles forming a stratum, a recognizable layer in a sequence of other strata. Its place in the sequence indicates the age of a fossil; a stratum is older than the one above and younger than the one below. This is relative dating.

Relative dating does not establish the absolute age of fossils. For this, radioactive dating must be used. The technique is based on radioactive isotopes that have a quantifiable half-life, the time it takes for half of a radioactive isotope to decay into a stable element. Carbon-14 (^{14}C) is a radioactive isotope contained within the organic matter. Half of the carbon-14 decays to nitrogen-14 every 5,730 years. Therefore, comparing carbon-14 radioactivity of a fossil to modern organic matter allows for calculating the age of the fossil. However, after 50,000 years, carbon-14 radioactivity is so low it can no longer be used to measure age accurately. For this, paleontologists turn to potassium-40 and uranium-238, which have half-lives in billions of years.

Paleontologists use strata and fossils to study the history of life. It is an essential principle of fossil study (paleontology) that a living organism most resembles a fossil that is the most recent in the line of descent. Underlying similarities become fewer the farther back is the lineage. Similar fossils can be used to construct a timeline showing changes from a common ancestor millions of years ago to a modern species. For example, transitional forms such as the bird-like dinosaur Archeopteryx establish that birds descended from reptiles.

Fossils can reveal a wealth of information about an extinct animal, as in the case of the horse ancestor Hyracotherium, which was small with cusped, low-crowned molars, four toes on each front foot, and three on each hind foot. These were all adaptations for forest living and were gradually replaced by larger size, grinding teeth and reduction of toes into hooves as the forests gave way to grasslands. Intermediate forms show this transition between Hyracotherium and the modern horse, genus Equus.

However, it can be difficult to identify the true patterns in the fossil record and ignore misleading ones. Fossils are continually reclassified and reevaluated. For example, some paleontologists have taken a new look at turtle fossils and proposed that they are closer to crocodiles than previously thought.

Biogeographical evidence provides a valuable frame of reference for the fossil record. *Biogeography* examines the distribution of organisms across the Earth, providing additional insight into evolutionary research. Physical factors, such as the location of continents, determine where a population can spread and greatly influence evolution. Related forms evolving in one locale and spreading to other accessible areas may explain the distribution of organisms. For example, in the 19th century, Darwin observed that the Galapagos Islands have a wide variety of finch species, but the mainland has one. He concluded that the Galapagos finches had originated on the mainland and diversified once they reached the islands. Biogeography, as applied to evolution, is sometimes known as *phylogeography*.

Phylogenetic Trees and Cladograms

Systematics (classification) is the study of the diversity of organisms using evidence from the molecular to the population level. It is often used synonymously with the term taxonomy, although *taxonomy* refers to nomenclature, a subset of the broader field of classification. Systematics may be applied only to modern species or extinct species, in which case it falls under the field of phylogeny. Studying the phylogeny of species reveals the evolutionary relationships between all organisms, past, and present.

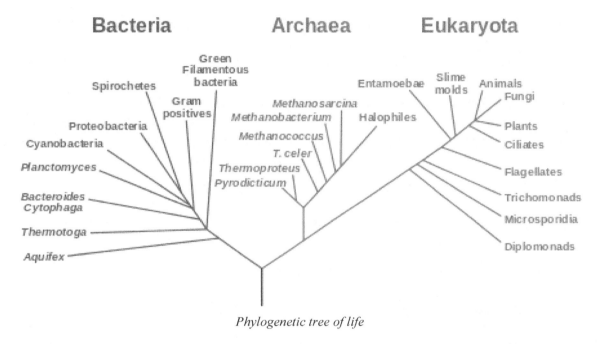

Phylogenetic tree of life

A phylogenetic tree visually indicates common ancestors and lines of descent. It is represented by a branching diagram that, naturally, resembles a tree. A common ancestor is wherever two branches diverge at the node, while the tip of each branch is labeled with an individual taxon, usually a species. Often the tree includes an outgroup, which does not share the most recent common ancestor of other groups on the tree. The outgroup provides a frame of reference for the taxa of interest. Phylogenetic trees can be constructed as broadly or as narrowly as desired, from all of life to the subforms of a single species. A traditional phylogenetic tree has branches which correspond to the relative length of evolutionary time.

Traits derived or lost due to evolution

From the diversity of life, each of the species on a phylogenetic tree must have lost some traits seen in the common ancestor and adopted new ones. For the most part, by natural selection, these changes are influenced by the viability of certain traits in a given environment.

In systematics, *traits* (characters) are any structural, chromosomal or molecular features that distinguish taxa. Phylogenetic trees are constructed by comparing which characters are shared between organisms. For example, members of a kingdom share general characters, whereas members of a species share specific characters. These general characters, *plesiomorphies* (primitive character or ancestral states), are traits present in a common ancestor and the members of the descending group. For example, a vertebral column is a plesiomorphy of all vertebrates.

Apomorphy (derivate character) are characters which are in one or a few descendants. Apomorphies are present in a specific line of descent. Different lineages that diverge from a common ancestor have different apomorphies, but share plesiomorphies.

Although the terms "primitive character" and "derived character" are still in use, phylogeneticists prefer to use plesiomorphy or apomorphy. This is because "primitive" implies that the ancestral character was simpler and more rudimentary, which is not always the case. Furthermore, "derived" implies that the character is a trait not in the ancestor, but arose in the descendant. This is not necessarily true; a derived character can represent the absence of a trait. For example, the common reptilian ancestor had four limbs. Some reptiles became snakes as they gradually lost their limbs. The characteristic of four limbs was a plesiomorphy of reptiles (and indeed all tetrapods), while the characteristic of having lost them is an apomorphy shared amongst snakes.

Speciation and common ancestry

A speciation event occurs when two species diverge from a node of a tree. Defining speciation is a difficult process, but it can be aided by examining the genetic code of two species for changes in nucleic acid and amino acid sequences over time. Advances in this analysis have made abundant data available to researchers.

Comparing homologous (similar) DNA and RNA sequences elucidate the degree of difference between the two organisms. A high degree of homology indicates highly related organisms, while sequences which are drastically different indicate distantly related organisms.

Biomolecular studies may show that organisms that physically resemble each other are not as related as once thought. For example, the red panda and the Chinese giant panda both have false thumbs, but the former has raccoon-like teeth and bones, while the latter does not. Results of DNA analysis confirm that the two pandas are distantly related, the red panda having originated from the same lineage as raccoons and the giant panda from the same lineage as bears.

The rate of mutations in the DNA nucleotide sequence is fairly constant and provides a "molecular clock." Molecular dating compares the relative age of two organisms, quantifying the number of base pair changes between their genomes. By comparing trace DNA from two fossils thought to be of the same lineage, it can be determined which organism the ancestor was, and how long it took the other organism to evolve from the ancestor.

Mitochondrial DNA (mtDNA) is often used for molecular dating because it changes about ten times faster than nuclear DNA, which makes it useful for comparing related species. Mitochondrial DNA is passed directly from mother to offspring. This removes the effect of recombination as for chromosomal DNA and gives a clearer picture of the timeline. The fossil record can then be used to calibrate the clock and confirm the hypothesis drawn from molecular data. For example, a comparison of mitochondrial DNA sequences equated a 5.1% nucleic acid difference among a songbird common ancestor and descendant. In regards to the molecular clock, this means that the descendant diverged from its common ancestor 2.5 million years ago. However, mtDNA is less useful for distantly related organisms. For those applications, RNA or nuclear DNA is used.

Protein comparison is another method for determining speciation or common ancestry, albeit less robust. The amino acid sequences of homologous proteins can be compared, as with cytochrome c, to a protein in all aerobic organisms. The 140 amino acid composition of cytochrome c differs between chickens and humans by twelve amino acids (8.6%), but by three (2.1%) between chickens and ducks. With this evidence, it can be reasonably assumed that humans, chickens, and ducks share a common ancestor, but that chickens and ducks are more related to each other than to humans. Protein analysis has limitations when it comes to distant relations since few proteins are across all domains of life.

Biomolecular evidence has become one of the most invaluable tools of the modern phylogeneticist, but it is most useful in conjunction with other evidence.

Morphological similarities of living and extinct species

The evolutionary relationship between living and extinct species is observable in comparisons of the function and development of their morphology, the structures of external and internal anatomy. A living species appear morphologically similar to a recent ancestor, with perhaps a few modifications. By contrast, a distant ancestor may appear entirely alien compared to the modern organism. However, further analysis would reveal the evolutionary link between them.

George Cuvier, a distinguished French vertebrate zoologist, was the first to use comparative anatomy to classify animals. He recognized that organisms often have morphological similarities when they are related because of common descent. *Homologous structures* are similar anatomical traits which have been derived from the same ancestor but have been modified over time. For example, vertebrate forelimbs contain the same sets of bones organized in similar ways, despite their dissimilar functions. The flippers of a whale and the forelimbs of a human are homologous structures, providing evidence of divergent evolution from a common tetrapod ancestor. *Analogous structures* are inherited from unique ancestors but have come to resemble each other because they happen to serve a similar function. Fish and marine mammals have the same sleek, hydrodynamic body plan, but evolved this trait separately. Although the shape is present in their common ancestor, a primitive fish, it was lost in mammals, then later regained by a select few. Therefore, streamlined shape is an analogous structure that has evolved separately many times in response to an aquatic environment.

Using morphology to classify organisms can be misleading due to analogous structures. Organisms may appear to be related due to their similar morphologies, but genetic or fossil analysis may reveal that these traits evolved separately.

Many traits that do not provide a present-day advantage are still present in organisms because they do not negatively affect their fitness. *Vestigial structures* are traits which are regressive and nonfunctional in the modern organism but served a purpose in an ancestor. For example, whales have vestigial hind limbs that harken back to their tetrapod origins.

The process of losing vestigial structures is an ongoing process that has been well-studied in humans. Most humans no longer eat the plant roughage, which requires extra molars to break down. As a result, human jaws are becoming smaller and forcing out the wisdom teeth. In modern times, hundreds of millions of people lack wisdom teeth altogether. Another example of a vestigial structure in humans is the coccyx, a bone at the end of the spine (i.e., tailbone), essentially a regressed tail.

Dynamic changes of phylogenetic trees and cladograms

The first unified form of systematics arose in the 18th century when Carl Linnaeus devised a taxonomic system of classification. Because taxonomy assigns a hierarchy, this system required each taxon to have a rank about other taxa. This seemed logical to scientists at the time, who mostly subscribed to the idea that species have a fixed place in the divine hierarchy of God's plan. However, assigning ranks to groups can be confusing and prone to bias. Countless modifiers were added to the simple taxa of kingdom, phylum, class, order, family, genus and species, leading to variations such as infraclass, subspecies, superphylum, etc. Another weakness of Linnaeus' method was that it did not acknowledge evolution and allow for the inclusion of extinct species.

Linnaeus' original scheme divided life into kingdom Animalia, kingdom Plantae and the third kingdom for minerals (since abandoned). In much of the 20th century, a five-kingdom classification was popular, which included kingdom Animalia and kingdom Plantae, in addition to kingdom Monera (for prokaryotes), Protista (for protists) and Fungi. As discussed, the prokaryotes have more recently been divided into archaea and bacteria. To reflect this, Monera was eliminated, and the kingdom Bacteria and kingdom Archaea were installed in its place. Because protists, animals, plants, and fungi are all eukaryotes, Eukarya was added to unify them under one domain. To represent bacteria and archaea as equal to eukaryotes, they were named as domains. This may seem pointless since each of their domains includes one kingdom of the same name. However, it would be misleading to say that life consists of domain Eukarya, kingdom Bacteria and kingdom Archaea because eukaryotes did not come any earlier than the other two and are not any more "overarching" of a group. This example reveals how taxonomy can become convoluted due to ranking.

Darwinian taxonomy, or *evolutionary taxonomy,* includes species past and present and relies heavily on the notion of common descent. At this time, taxonomy came to represent part of a larger whole: phylogeny. Darwinian taxonomy did not attempt to construct orderly lists, but to demonstrate phylogenetic relationships in their actuality. However, it was still too constricted in the types of relationships that were allowed, and there was no consensus as to how these relationships should be constructed in the first place.

Phenetic systematics is a more quantitative notion of classification which arose in the 20th century. Phenetic systematics constructs phylogenetic trees based on mathematics and statistics, counting the number of shared characteristics between two species and running various analyses to construct the most likely tree. This method clusters species merely based on the number of shared characteristics, making no distinction between plesiomorphies and apomorphies. It does not take into account the nuanced forms of evolution, such as parallel, divergent, convergent and co-evolution. This form of classification is vulnerable to misleading similarities, such as analogous structures or those which have evolved alongside one another. Systematists of this school do not believe that a classification that reflects phylogeny can be constructed; it is better to rely strictly on a method that does away with nuance. Results of their analysis are depicted in a *phenogram*. Phenograms vary for similar groups of organisms, depending on how the data is collected and processed.

Phenetic systematics has been mostly superseded in modern times by *cladistic systematics*, or *cladistics*. Cladistics requires more complex mathematics and processing power to take into account how evolution may influence characteristics. Analysis of characteristics classifies organisms into *clades*. A clade is any taxon with a single common ancestor; it may be broad or narrow. Clades can be organized into trees as *cladograms*, which show the relationships between species based on their shared characteristics. A cladogram, unlike a traditional phylogenetic tree, does not infer how much species have changed over time, but merely that they diverged at one point. In a cladogram, all branches are placed equidistant from one another and do not correspond to relative evolutionary time since the common ancestor.

Clades are usually *monophyletic*, meaning they include a given common ancestor and every of the ancestor's descendants. Even if one descendant is left out of the clade, it is no longer monophyletic, but *paraphyletic*. A paraphyletic clade has a common ancestor and many of its descendants, but not all. Monophyletic clades are the most useful in phylogeny, but paraphyletic clades have their place as well. However, there is seldom good cause to construct a *polyphyletic* clade. In this case, the group contains several descendants who do not share the same common ancestor. Polyphyletic clades are often accidentally created due to misleading evidence. With further investigation, researchers may be able to remedy them into monophyletic clades.

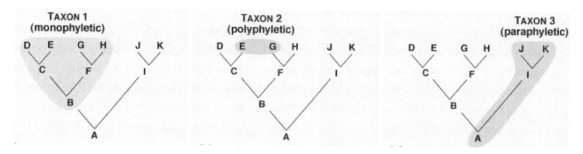

Cladistics is guided by the principle of *parsimony*, which states that the solution with the least amount of assumptions is the most logical. According to parsimony, the best cladogram is the one with a minimal number of assumptions made, generally the simplest one. This approach attempts to minimize bias but is still dependent on the knowledge and skill of the particular investigator, as well as the evidence available. Like earlier methodologies, cladistics is prone to ambiguity and confusion. It is no easy task to pinpoint a common ancestor or to articulate what exactly makes two organisms more evolutionarily related to one another. Cladistics cannot be improved until there is a consensus as to what exact characters should be used to distinguish different groups.

Comparative Anatomy

Although sequence analysis and molecular dating are integral to modern phylogenetics, morphology remains the simplest and most accessible way to describe the relationships between organisms. For this reason, this text uses comparative anatomy to present the diversity of the plant and animal kingdoms. Understand that underneath these anatomical traits is molecular evidence which strengthens the classifications presented.

The diversity of Kingdom Plantae

Kingdom Plantae is a clade of eukaryotes which are autotrophic by photosynthesis. This clade usually includes the multicellular green algae and terrestrial nonvascular plants ("lower plants") and terrestrial vascular plants ("higher plants"). However, many phylogeneticists would exclude multicellular green algae, since their inclusion arguably makes the clade paraphyletic; unicellular green algae can be found in kingdom Protista.

Algae and the *bryophytes* (terrestrial nonvascular plants) are green, low to the ground and generally found in moist environments. They do not have a vascular system or true stems, leaves, and roots, but are anchored to the ground by root-like *rhizoids*. Capsules at the end of stalks protrude from a bryophyte and release spores into the wind or water. Algae and bryophytes reproduce via these spores, rather than seeds.

The development of vascular tissues allowed the higher plants to reach unprecedented heights. Vascular plants can be divided into seed and seedless plants. The most well-known phylum of the vascular seedless plants are the ferns, also *filicophyta* or *polypodiophyta*. These plants have true roots, leaves, and stems, but reproduce using spores like the nonvascular plants. Spores are produced on the underside of long fronds divided into many leaflets. In the Paleozoic Era, ferns ruled the forests, becoming so large that they resembled trees. Today, most ferns are much smaller, with a maximum height of about 15 meters. Other seedless plants are the "fern allies" and are classified under phylum *Lycopodiophyta*. The lycopods include club mosses, spike mosses, and quillworts. These plants designated as "mosses" are not mosses. The lycopods are distinguished from the ferns by their *microphylls,* simple leaves with a single vein. Ferns have *megaphylls,* which have complex, branching veins.

The *spermatophytes* (seed plants) comprise a monophyletic clade of all plants reproducing with seeds. This group is the largest, most recent, and most successful taxon of plants on Earth. It can be divided into the *angiosperms* (flowering plants) and the *gymnosperms* (cone plants). Angiosperms have their seeds fully enclosed within the ovaries of flowers, while gymnosperms have their seeds partially exposed on the scale-like leaves of cones. Both groups include the tallest plants on Earth: angiosperms with Eucalyptus, which may reach 100 meters tall, and gymnosperms with the redwood, up to 115 meters.

General characteristics of angiosperms include vascular tissues, roots, stems, leaves, flowers, and fruit. Flowers contain ovaries that enclose *ovules.* When pollinated by a male flower, an ovule is fertilized and develops into the seed. The ovary ripens around the seed to form a fruit, which may be dispersed by wind, water, gravity or animals. Once dispersed, the seed germinates in the soil and begins to grow.

Gymnosperms have neither flowers nor fruit. They are typically woody; the most well-known gymnosperms are the conifers (e.g., pines and fir trees). Gymnosperms have their ovules on female cones, *ovule cones* or *seed cones.* Pollen from male cones, *pollen cones,* fertilizes an ovule cone and causes a seed to develop. The ovule cone then drops to the ground to allow the seed to germinate.

The diversity of Kingdom Animalia

Kingdom Animalia is a monophyletic clade of multicellular eukaryotes which are heterotrophic by ingestion, and generally motile. The most primitive animals are sponges (phylum Porifera, meaning "pore bearer"), which are asymmetrical and lack specialized tissues or organs. Sponges are diploblastic with two cell layers. The sponges intake water through pores lined with *choanocytes,* flagellated cells which filter food and then expel the remaining water through their mouth, the *osculum.* The choanocytes pass food to *amoebocytes* for digestion and distribution of the nutrients. Many sponges feature a calcium carbonate exoskeleton secreted by their cells. They may feature *spicules,* skeletal needles of silica or calcium carbonate that strengthen their body walls. Unlike other animals, sponges are sessile and lack even a rudimentary sensory system.

Phylum Cnidaria evolved shortly after the sponges. The cnidarians include jellyfish, hydra, sea anemones and corals. Like the sponges, they are diploblastic but exhibit more complex cell specialization. A gelatinous layer of *mesoglea* between their two cell layers develops into muscle-like tissue. Cnidarians may take either a sessile form, the *polyp,* or a swimming, bell-shaped form, the *medusa.* Polyps may live in colonies, as do coral, while medusae are free-living, as do jellyfish. Many cnidarians have a larval polyp stage and a medusa adult stage. Both forms have radial symmetry and exhibit symmetry in all planes. They have one opening, through which they ingest food, expel waste and carry out respiration. Cnidarians have a rudimentary nervous system and sensory organs, with nothing resembling a brain. They are noted for their *nematocysts,* stinging cells which they use to stun or kill prey.

Phylum Ctenophora, the comb jellies, are a group of diploblastic organisms with radial symmetry. They have traditionally been grouped with the cnidarians, but recent evidence suggests that they may be closer to the higher level *bilaterians.* Bilaterians are animals that are more complex than the sponges, ctenophores, and cnidarians. With a few exceptions, they exhibit bilateral symmetry and have two distinct, symmetrical sides. The bilaterians are triploblastic, having originated from three germ layers: ectoderm, mesoderm, and endoderm. Bilaterians exhibit *cephalization,* the concentration of nervous tissues at the anterior end of an organism.

Worms were some of the first bilaterians to evolve. "Worms" refer to a variety of phyla, including but not limited to the flatworms of phylum Platyhelminthes, the nematodes of phylum Nematoda and the segmented worms of phylum Annelida. Despite the umbrella term of the worm, these three phyla are more related to other organisms than to each other. The flatworms are the simplest because they have no body cavities (coeloms), making them *acoelomates*. They undergo respiration by simple diffusion and have one digestive opening. Flatworms may be free-living, as with the carnivorous *planarians,* or parasitic (e.g., flukes and tapeworms). Their closest relatives are the *rotifers* (phylum Rotifera), microscopic filter feeders. Rotifers are *pseudocoelomates;* they have a coelom, but it is not completely lined by mesoderm-derived tissue.

The nematodes are pseudocoelomates. These roundworms inhabit a wide range of habitats and have more complex respiratory and digestive systems. Often, they are free-living soil dwellers that help decompose and recycle nutrients.

Annelids, such as earthworms and leeches, are true *coelomates.* They have complex, segmented bodies; each segment contains a set of organs identical to those in other segments. They move by muscle contraction, often aided by bristles of *setae* and limbs as *parapodia* that push them along the soil. Their closest relatives are the mollusks.

Mollusks, of phylum Mollusca, are the largest marine phylum today. They are triploblastic, true coelomates, and have a rudimentary nervous system with some cephalization. In general, mollusks have a mantle, a radula, and a foot. The *mantle* encloses a cavity used for respiration and excretion, while the *radula* is essentially a raspy tongue. The foot is used for motility, sensation and for anchoring to surfaces. The evolution of these features in the Cambrian seas would have given early mollusks advantages in motility, protection, and feeding. Many classes are in the phylum Mollusca, but three are widely known. The bivalves (class Bivalvia), such as clams and oysters, secrete a hinged shell from the mantle. They are generally filter feeders which can be on the ocean floor, in tidal areas or even free swimming. The gastropods (class Gastropoda), the snails and slugs, have a single continuous shell and are noted for their large, developed feet. Most gastropods are herbivorous grazers. Finally, the cephalopods (class Cephalopoda) are the most advanced mollusks; these include octopus, squid, cuttlefish, and nautilus. They have well-developed cephalization and are primarily predatory, using their tentacles to attack prey.

Phylum Arthropoda includes the arthropods, an incredibly diverse phylum on land, sea, and air. At a basic level, they can be characterized by their jointed limbs, segmented body plan, and an exoskeleton. The exoskeleton is made of *chitin,* a polysaccharide. It undergoes continuous molting and replacement. Their small coelom surrounds only the reproductive and excretory system, with a special body cavity of a *hemocoel* facilitating their open circulatory system. Coelomic fluid and blood flow through the hemocoel, bathing surrounding tissues.

Arthropods have a relatively advanced nervous system, with rudimentary brains and sophisticated photo-, mechano- and chemoreceptors. They may exhibit one of two life cycles: the immature *nymph* gradually develops into an adult, or an immature *larva* enters a cocoon and undergoes *metamorphosis,* or sudden development, into an adult.

Phylum Arthropoda is divided into four extant subphyla. A fifth subphylum, Trilobita, now extinct, comprised the trilobites. These creatures were ubiquitous in the Cambrian and later periods.

Subphylum Crustacea, meaning "insects of the sea," includes the crustaceans: lobsters, crabs, shrimp, krill and barnacles, among others. Many crustaceans are herbivores and are of great importance to aquatic ecosystems. They feed on phytoplankton, making them a crucial link in the food chain. A few crustaceans have colonized land and freshwater environments, but the majority are marine, carrying on gas exchange with their gills.

By contrast, the species of subphylum Myriapoda are exclusively terrestrial. Myriapods include centipedes and millipedes, which use their many feet to their advantage. Centipedes have fewer legs and are fast, making them effective predators. They have venomous glands which assist in killing prey. Millipedes are slower, so they are mostly herbivores and detritivores. They have venomous glands, which they use as chemical defenses against predators.

Along with the myriapods, class, Arachnida (subphylum Chelicerata) were some of the first creatures to occupy the land. Only a few arachnids have returned to the water. The most famous arachnids are spiders, scorpions, ticks and mites. All arachnids have eight legs and advanced eyes, making them excellent predators. They exhibit unique "book lungs," stacks of alternating air pockets used for gas exchange.

Class Insecta (subphylum Hexapoda) is by far the most successful and diverse taxon on Earth. There are more species of insect than any other species combined. All insects are characterized by their three-part body plan, six legs, and antennae, but insect species has developed additional features.

The adaptation of wings allowed some insects to escape from predators and travel large distances. The more primitive insects (most likely the first insects) are wingless, which suggests that flying was an evolutionary adaptation over time. Most insects inhabit the land and air. Their respiratory systems have *spiracles* (openings), which allow air to enter their internal *tracheal tubes*. Insects are limited in their size because their anatomy cannot support a large body. Most insect species are relatively small, allowing them to occupy small areas and minimize their food requirements.

Collectively, the triploblasts (Platyhelminthes, nematodes, annelids, rotifers, mollusks, and arthropods, along with other smaller phyla and classes) are protostomes.

Taxa are continually debated, so only echinoderms and chordates have been definitively established as deuterostomes. For instance, scientists cannot decide whether the *bryozoans*, a phylum of marine invertebrates, are protostomes or deuterostomes. *Brachiopods*, bottom-dwelling animals which superficially resemble mollusks, are another such phylum which has come under question in recent years. Traditionally, the brachiopods have been classified as deuterostomes, but new evidence suggests that they may be protostomes.

It is inconclusive which deuterostomes evolved first, but some of the evidence points to phylum *Echinodermata*. The echinoderms include sea stars, sea urchins, sea cucumbers, sand dollars, brittle stars, and sea lilies. Like all organisms higher than cnidarians and ctenophores, the echinoderms are classified as bilaterians. However, this is a misnomer, as echinoderms evolved radial symmetry after descending from their bilateral ancestor, making this an apomorphy. This legacy can be seen in their larvae, which are bilateral.

Phylum	Common Names	Tissue Complexity	Germ Layers	Body Symmetry	Gut Openings	Coelom	Embryonic Development
Porifera	sponges	parazoa	-	none	0*	-	-
Cnidaria	jellyfish, corals	eumetazoa	2	radial	1	-	-
Platyhelminthes	flatworms	eumetazoa	3	bilateral	1	acoelomate	-
Nematoda	roundworms	eumetazoa	3	bilateral	2	pseudo-coelomate	-
Rotifera	rotifers	eumetazoa	3	bilateral	2	pseudo-coelomate	-
Mollusca	clams, snails, octopuses	eumetazoa	3	bilateral	2	coelomate	protosome
Annelida	segmented worms	eumetazoa	3	bilateral	2	coelomate	protosome
Arthropoda	insects, spiders, crustaceans	eumetazoa	3	bilateral	2	coelomate	protosome
Echinodermata	sea stars, sea urchins	eumetazoa	3	radial	2	coelomate	deuterostome
Chordata	vertebrates	eumetazoa	3	bilateral	2	coelomate	deuterostome

* *Amoebocytes carry out digestion*

- *Characteristic does not apply to this phylum.*

Echinoderms had strong exoskeletons and accomplished camouflaging abilities. While they appear sessile, echinoderms do move. Some are herbivores, but many sea stars are carnivorous and can move surprisingly quickly. Echinoderms are noted for their robust powers of regeneration and their complex coelom, which includes a water vascular system for water pressure to extend *tube feet*, allowing the animal to move. It functions in the respiratory, digestive and nervous systems.

The most advanced deuterostomes belong to phylum *Chordata* and are primarily defined by the development of a *notochord* in the early embryo. In most chordates, the notochord develops into a vertebral column; these organisms are classified under subphylum *Vertebrata*. Because nearly all chordates are vertebrates, the two taxa are often conflated. Non-vertebral chordates include tunicates and lancelets, both simple filter feeders. These species likely represent the earliest chordates.

Chordate features

Notochord

The notochord is a flexible rod that extends below the dorsal nerve cord along the length of all chordate embryos. It is derived from the mesoderm and develops during neurulation of the embryo. All chordates possess a notochord during the embryonic stage.

In non-vertebral chordates, the notochord persists after birth; muscles work against it to help the animal move. It may be lost in sexual maturity or remain throughout adulthood. However, for invertebral chordates, the notochord is replaced by the vertebral column later in embryonic development.

Pharyngeal pouches and branchial arches

Another key distinguishing feature of chordates includes pharyngeal pouches, present in all chordate embryos. *Pharyngeal pouches* are openings between the pharynx (a portion of the throat) and the environment. They develop into various structures depending on the organism. In non-vertebrate chordates, the pouches develop into *pharyngeal slits,* used for filter feeding and gas exchange.

In fish and some amphibians, the pouches are modified into the jaw, the hyoid bone, and the *branchial arches*. The branchial arches are *gill arches* because they support the gills. In all other vertebrates, the pouches form *pharyngeal arches,* which develop into the jaw, the hyoid bone, the thyroid, the larynx, the tonsils and other components of the neck and head.

Dorsal nerve cord

The *dorsal nerve cord*, derived from ectoderm, forms the central nervous system. Like the notochord, it develops during neurulation. This hollow chord extends the length of the embryo on top of the notochord. In vertebrates, most of the dorsal nerve cord develops into the spinal cord, and the anterior part enlarges into the brain. A *post-anal tail* extends at the posterior end and may be lost or retained in the adult organism.

Vertebrate phylogeny

Most scientists agree that the first vertebrates would have been jawless fishes similar to hagfish and lampreys. These primitive filter feeders diversified greatly in the periods after the Cambrian, some evolving into jawed fishes. Some of the jawed fish evolved a cartilaginous jaw and skeleton, today the cartilaginous fishes (class Chondrichthyes). In this class, which includes sharks, skates, and rays, the notochord is replaced by cartilage rather than bone. Cartilaginous fishes have tough skin and are typically predatory.

Shortly after the cartilaginous fish, the bony fishes (superclass Osteichthyes), would have split off from their common ancestor. They are the origin of nearly all other fish species. Bony fish have a bony jaw and skeleton and are divided into ray-finned fish (class Actinopterygii) and lobe-finned fish (class Sarcopterygii). Lobe-finned fish are more primitive, and few survive today, while ray-finned fish represent the majority of fish species on Earth.

Nearly all fish are *ectothermic* ("cold-blooded"), as their body temperature is dependent on the environment and cannot be internally regulated. A typical fish has a streamlined shape to optimize swimming efficiency, although many fish diverge significantly from this morphology. Other typical characteristics of fish include scaly skin, fins, and reproduction with eggs. Most breathe with gills, pumping blood past the gills with their two-chambered heart, although a few fish have rudimentary lungs.

Those ancient lobe-finned fish that developed lungs were able to venture onto land with the help of their fleshy, lobed fins. These are the ancestors to lungfish as well as all terrestrial vertebrates, known as tetrapods (superclass Tetrapoda) for their four limbs. Within this group are class Amphibia, class

Reptilia, class Aves, and class Mammalia. Many tetrapods lost most or all of their limbs over time, and some even returned to the water. Tetrapods were the origin of animals such as snakes and marine mammals.

The earliest tetrapods, the amphibians (class Amphibia), never fully left their aquatic origins. Most amphibians are adapted for both terrestrial and aquatic habitats, and many spend the entire first stage of their lives in water. Their larval form typically breathes with gills but undergoes metamorphosis into an adult form that breathes with lungs and through its skin. Their skin is moist and able to exchange gases and water with the environment. A two-chambered heart pumps blood throughout their closed circulatory system. Like fish, amphibians are ectotherms. The three main orders of amphibians are frogs and toads (order Anura), salamanders (order Caudata) and caecilians (order Gymnophiona). Caecilians are primitive, limbless amphibians which resemble a large worm.

As the ancient amphibians thrived, an important apomorphy was arising. The *amniotic egg*, a specialized egg with complex membranes that can sustain an embryo on land or, in the case of mammals, within the womb. Earlier organisms, such as the amphibians and fish, laid their eggs in water. These eggs are simple and gelatinous, and exchange materials with their aquatic environment. By contrast, amniotic eggs can be laid on land and carry on gas exchange with the air while protecting the embryo with their hard, leathery shells. The embryo fully develops within the egg before hatching, unlike fish and amphibian embryos, which are still in the larval stage when they exit the egg.

Those amphibians which evolved amniotic eggs, along with other adaptations such as scaly, waterproof skin, became the reptiles. Reptiles quickly colonized land and specialized even further to their terrestrial environment. Reptiles are widespread and diverse. Turtles, tortoises, and terrapins (order Testudines) inhabit the seas and land, protected by their large, tough shells. The lizards and snakes (order Squamata) are the most diverse reptile group, with nearly 10,000 known species.

By contrast, the crocodilians (order Crocodilia) have 25 living species, and the tuatara (order Sphenodontia) is the only reptile of its kind. Many reptiles are capable of maintaining a remarkably stable body temperature despite being ectotherms. This is due to adaptations such as low resting metabolism, muscle contractions and sunning. A three-chambered heart pumps blood throughout their system. All reptiles have lungs and must breathe air, even the aquatic and semi-aquatic species.

When the dinosaurs ruled the Earth millions of years ago, mammals began to evolve from a niche of reptiles deemed the synapsids. These organisms are para mammals, proto-mammals or mammal-like reptiles. They diverged from the true reptiles, the sauropsids, and became true mammals, gradually shedding their reptilian features for hair, internal amniotic eggs, and other mammalian characteristics. Although the dinosaurs still dominated for some time, the mammals became a competitor for land resources and even predated on some smaller dinosaurs.

Mammals (class Mammalia) are a diverse group of organisms, but they share common apomorphies. Some of these include hair, varied teeth (*heterodonty*), a warm-blooded metabolism (*endothermy*) and mammary glands which feed their young with milk. The majority of mammals are placental, with offspring developing in the uterus and are nourished by a placenta. The marsupials carry their offspring in an external womb, the pouch. Nearly all mammals give live birth, except for the monotremes, which lay eggs like their reptilian ancestors.

The other descendants of the reptiles were the birds (class Aves). They are a highly diverse class and represent thousands of species on Earth. Class Aves can be divided into *paleognaths* (flightless and weak-flying birds) and *neognaths* (all other birds). Birds have many unique apomorphies which separate them from their reptilian kin. Unlike reptiles, most birds have bills rather than teeth. They are adapted for flight, having lightweight bones, forelimbs which have been modified into wings, large flight muscles, and feathers.

To meet the high energy demands of flight, birds have a high metabolism and sophisticated respiratory and circulatory systems. A large amount of heat they generate makes them endotherms like the mammals. This is an analogous trait; endothermy was not present in their common ancestor but evolved separately in birds and mammals. A similar example is the ability of flight in both birds and bats.

Birds have even more powerful lungs and massive four-chambered hearts than the mammals, allowing them to circulate oxygen efficiently. Respiration has the additional benefit of filling the spaces within their bones, contributing to their buoyancy. Many birds are such strong fliers that they can migrate thousands of miles each year. Birds may be insectivorous, herbivorous, carnivorous, omnivorous or even eat carrion. The wide variation in bird morphology is a testament to their diversification into many different niches. However, their reptilian ancestry can be seen in their scales, feathers (scale derivatives) and their amniotic eggs.

The relationship between ontogeny and phylogeny

Ontogeny refers to the development of an organism throughout its lifetime, from the zygote to the adult form. *Phylogeny* is the development of a species over evolutionary time. Evolutionary biologists historically have subscribed to the idea that "ontogeny recapitulates phylogeny." This dogma, the *recapitulation theory*, states that the development seen in an organism (i.e., ontogeny) reflects the way in which the species evolved (i.e., phylogeny). Take for example the notochord. A vertebrate embryo has a notochord in early development, but vertebrae later replace it.

Similarly, the chordate ancestor had a notochord which was replaced by vertebrae during the evolution of vertebrates. In this case, the theory holds. However, the notion that ontogeny recapitulates phylogeny is an exaggeration that has largely been refuted. If it were true, a human would appear as a mature invertebrate (e.g., fish, reptile) and then an ancestral mammal until it developed into a mature human. This does not occur. Rather, the human embryo diverges from more generalized forms as it matures. It is not that it looks like a fish at any point during development, but merely that an early embryo cannot be visually identified as either fish or human.

Embryology can be separated from the recapitulation theory to provide valuable insights into embryos of different species. It cannot be concluded that a human embryo must have evolved from a bird ancestor merely because a human embryo and bird embryo look the same until several weeks after fertilization. However, it *can* be reasonably inferred that since they look more similar, bird embryos are more related to human embryos than to fish embryos. Ontogeny can reveal evolutionary relatedness but not the order of evolution or the exact characteristics that were derived.

Notes

Chapter 8

Plants

- **Evolution and Diversity of Plants**

- **Structure and Organization of Plants**

- **Nutrition and Transport in Plants**

- **Control of Growth and Responses in Plants**

- **Reproduction in Plants**

Evolution and Diversity of Plants

Evolutionary history of plants

Plants are multicellular, photosynthetic eukaryotes characterized under the kingdom *Plantae*. The most primitive plants are aquatic green algae, believed to be the ancestors of all land plants. Evidence for this ancestry includes many adaptations which terrestrial plants and algae have in common, including the use of certain photosynthetic pigments, the propensity to store nutrients as starch and cell walls which contain *cellulose,* a tough polysaccharide. Genetic analysis suggests that terrestrial plants' closest living relatives are green algae, commonly known as "stoneworts."

Some green algae began to colonize tidal areas and evolved into terrestrial plants in the Paleozoic Era as early as 600 million years ago. These algae and early terrestrial plants are *nonvascular,* lacking a sophisticated vascular system. They are more primitive and require a moist environment to maintain adequate water supply for their tissues.

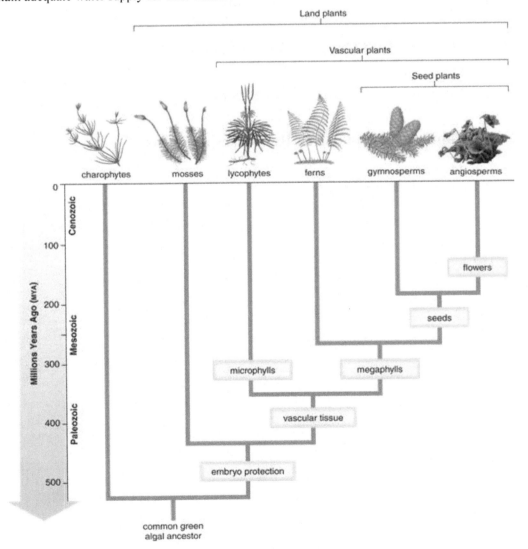

Many nonvascular plants adopted certain adaptations to avoid desiccation, such as a *cuticle*, a waxy, protective layer. However, the cuticle limits gas exchange, so plants have developed *stomata* (openings) to allow CO_2 and oxygen to enter and exit their tissues.

Around 430 million years ago, during the Silurian Period, those nonvascular plants, which were particularly well-adapted to drier environments, developed *vascular tissues* with specialized cells that conduct water and solutes throughout the plant. Vascular tissue marks a significant division in the evolutionary history of plants, as it allowed them to venture inland and grow larger. They began to grow upright and quickly evolved seeds and flowers as adaptations for life on land around 400 million years ago.

Angiosperms represent all vascular plants with *flowers,* useful reproductive structures which attract pollinators. *Gymnosperms* are the non-flowering vascular plants, which use *cones* as their reproductive structures. Both angiosperms and gymnosperms produce *seeds*, which contain food storage and a developing embryo within an outer coat. Seeds have some advantages, such as drought and predator resistance, which make it more likely for the embryo to successfully mature. The evolution of seeds catapulted gymnosperms and angiosperms to the forefront of the plant kingdom.

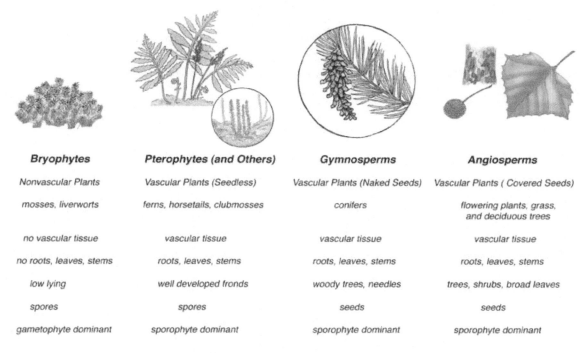

Bryophytes	*Pterophytes (and Others)*	*Gymnosperms*	*Angiosperms*
Nonvascular Plants	*Vascular Plants (Seedless)*	*Vascular Plants (Naked Seeds)*	*Vascular Plants (Covered Seeds)*
mosses, liverworts	*ferns, horsetails, clubmosses*	*conifers*	*flowering plants, grass, and deciduous trees*
no vascular tissue	*vascular tissue*	*vascular tissue*	*vascular tissue*
no roots, leaves, stems	*roots, leaves, stems*	*roots, leaves, stems*	*roots, leaves, stems*
low lying	*well developed fronds*	*woody trees, needles*	*trees, shrubs, broad leaves*
spores	*spores*	*seeds*	*seeds*
gametophyte dominant	*sporophyte dominant*	*sporophyte dominant*	*sporophyte dominant*

Plant divisions with distinctive characteristics between species

Plants exhibit a two-generation life cycle of *alternation of generations*, in which the morphology and reproductive strategy of the organism alternates. The two generations are named for the type of offspring they produce.

The *sporophyte generation* produces haploid *spores,* a unit of asexual reproduction, while the *gametophyte generation* produces haploid *gametes,* a unit of sexual reproduction. Sporophyte translates to "spore plant," while gametophyte translates to "gamete plant" ("phyte" is the Greek root for the plant, *phyton).* The gametophyte and sporophyte are both multicellular, while their gametes and spores are unicellular. Plant species differ in which generation, gametophyte or sporophyte, is dominant (i.e., more conspicuous), and the appearance of the plant may vary widely between generations.

In nonvascular plants, the gametophyte generation is generally dominant, while in vascular plants it is the sporophyte generation. Adaptations for water transport and conservation accompany sporophyte dominance. The shift to sporophyte generation dominance in vascular plants is an adaptation to life on land and resulted in the gametophyte becoming microscopic and dependent on the sporophyte.

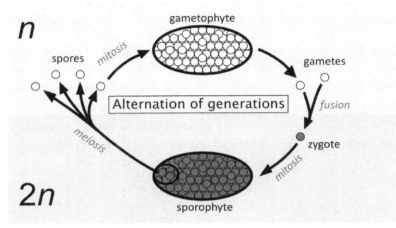

Alternation of generations with the 1N gametophyte compared to the 2N sporophyte

Nonvascular plants

Nonvascular plants lack true roots, stems, and leaves, but may have adaptations which resemble these structures. They include the green algae and the *bryophytes*. Nonvascular plants cannot grow large on land due to their lack of vascular tissue, and usually, require a moist habitat for fertilization. Algae are aquatic or semi-aquatic, and may be single-celled, colonial or multicellular (e.g., seaweed). Many algae species live symbiotically with fungi as *lichens*. Unlike most other plants, in which one generation is dependent on the other, algae generally have sporophytes and gametophytes which are both free-living. Their sporophyte and gametophyte form are practically indistinguishable.

Bryophytes are terrestrial nonvascular plants

The terrestrial nonvascular plants collectively known as bryophytes include three living phyla: hornworts (phylum *Anthocerotophyta*), liverworts (phylum *Hepatophyta*) and mosses (phylum *Bryophyta*). Bryophytes exhibit an alternation of generations in which the gametophyte is dominant to the sporophyte generation. Like many of their algal cousins, bryophytes are small, low to the ground and moist. They are anchored to the surface below by root-like *rhizoids*. Like algae, many species are capable of asexual reproduction.

The larger, dominant gametophyte form appears as a mat of leafy or filamentous tissue from which small sprouts project. Some of these are male gamete-producing (sperm) organs, and some are female (eggs). The gametes combine within a female reproductive organ and develop into a sporophyte, which then produces spores. Spores are dispersed from the plant and develop into new gametophytes elsewhere, where the cycle continues. Both algae and bryophytes are *homosporous* and one type of spore.

Vascular plants

Vascular plants may be homosporous or *heterosporous* (two types of spores). Vascular plants have a more complex spore strategy as a hallmark of the dominant sporophyte generation. The gametophytes of vascular plants are usually tiny or even microscopic, held within reproductive structures on the sporophyte. Other characteristics of vascular plants include a waxy cuticle, stomata for gas exchange and vascular tissue. There are two types of vascular tissues, xylem, and phloem. *Xylem* passively conducts water and mineral solutes, the *xylem sap,* from roots to leaves. Water in the xylem rises against gravity without the help of any mechanical pump to reach heights of more than 100 meters in the tallest trees. *Phloem* transports water, sucrose and other organic compounds, the *phloem sap,* throughout the plant. Unlike xylem transport, phloem transport is active and thus requires energy to move its sap.

Seedless vascular plants

The first vascular plants were seedless, dominating the plant world from the late Devonian period (360 m.y.a.) through the Carboniferous period (300 m.y.a.). Many of these plants were much larger than their modern counterparts and formed vast, expansive swamps. Like nonvascular plants, they reproduce via spores rather than seeds, and most species are homosporous.

The seedless vascular plants include the ferns, club mosses, whisk ferns and horsetails. The most well-known of the seedless vascular plants are undoubtedly the ferns (class *Pteriopsida*). They are the most diverse seedless land plant and are especially abundant in warm, moist, tropical regions. Ferns may range in size from low-growing, moss-like forms to tall trees. Ferns are the only group of seedless plants to have well-developed *megaphylls*, which are "true leaves" with complex, branching veins. Fern megaphylls are *fronds.* Other seedless plants, as well as the lower plants (algae and bryophytes), have *microphylls* with a single vein. Many fern species live in the tropics or subtropics as *epiphytes,* plants which live on host trees without harming them.

Ferns are similar to vascular seed plants in many ways but have a life cycle which exhibits characteristics of nonvascular plants. Their small gametophytes anchor to the soil with rhizoids and are free-living. In many ways, the fern gametophyte resembles a bryophyte, and it is unlikely that most people would recognize it as a fern. In a rainy season, gametes produced on the underside of the gametophyte combine to create a diploid zygote, which matures into the adult sporophyte. The sporophyte grows from the soil as the large, leafy plant that is typically recognized as the fern. It generates spores on the underside of its fronds.

Club mosses (lycopods or "ground pines") are not mosses but are another group of seedless vascular plants, common in temperate woodlands, with tightly packed, scale-like microphylls covering their stems and branches. Like ferns, club mosses have small, inconspicuous, free-living gametophytes and a more recognizable sporophyte. Although seedless vascular plants may have independent gametophytes, the sporophyte is much larger and is still considered dominant.

Seed plants

Seed plants developed from seedless plants that were heterosporous; all seed plants still exhibit heterospory, with small male spores and large female spores. The male microspores are produced in the male reproductive organs. Each microspore develops into a multicellular *microgametophyte* (male gametophyte), which produces sperm contained in *pollen grains.* Meanwhile, the female megaspores are produced within *ovules,* housed on the female plant. The megaspores produced in the ovules then divide by mitosis into a *megagametophyte* (female gametophyte).

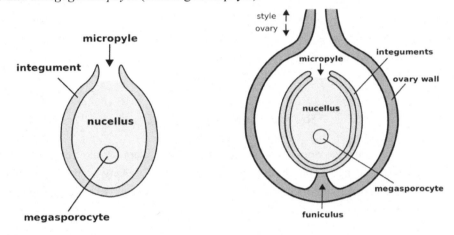

Gymnosperm (left) is a naked seed, and angiosperm (right) is an enclosed seed

When pollen fertilizes a macrogametophyte, it combines with the egg within to develop into a diploid zygote, which grows into an embryo contained in the seed. The seed initially develops in the ovule but is eventually ejected so that it may germinate in the soil. It is enclosed in a protective coat to make it resistant to adverse conditions such as dryness and extreme temperatures. Food is stored in the seed to nourish the embryo until it can exist on its own. The success of seed plants and their present dominance is due in large part to the utilization of seeds.

There are two divisions of seed plants: the gymnosperms and the angiosperms. Gymnosperms have ovules which are not completely enclosed by sporophyte tissue at pollination. In angiosperms, the ovules are completely enclosed within this tissue, the flower. The distinction is in the naming of these two groups: in ancient Greek, the term gymnosperm means "naked seed," while angiosperm translates as "enclosed seed."

Gymnosperms

The gymnosperms include conifers, cycads, ginkgo, and gnetophytes. All have their reproductive organs partially exposed on *cones,* clusters of modified leaves with a scale-like appearance. These spore-producing leaves are *sporophylls.* The male cones are *pollen cones* because they produce the male gametophyte (pollen), while the female cones are *ovule cones* (seed cones) because they contain the ovules which house the seeds upon fertilization.

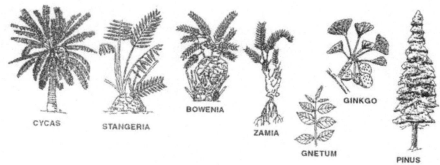

Gymnosperms

Conifers (Phylum *Coniferophyta*) contain about 575 species of cone-bearing trees and shrubs such as pine trees, hemlocks, and spruces. Usually, conifers have evergreen, needle-like leaves that are well-adapted to withstand extremes in climate due to their thick cuticles and recessed stomata. Conifer forests cover vast areas of northern temperate regions and are characterized by being the oldest and largest trees in existence.

Cycads (Phylum *Cycadophyta*) have a trunk that is stout and unbranched with large, compound leaves, giving a palm-like appearance. Cycads have pollen and seed cones on separate plants. The cycad life cycle is similar to that of pine trees, except they are pollinated by insects.

One species of ginkgo survives from the phylum *Ginkgophyta* with fork-veined, fan-shaped leaves. Ginkgo ovules reside at the end of short, paired stalks.

Gnetophytes (phylum *Gnetophyta*) consist of three living genera containing about 70 species. Gnetophytes are the gymnosperms most related to angiosperms. Some gnetophytes even produce nectar in their reproductive structures to recruit insects for pollination.

Angiosperms

The most numerous plant group, with six times the number of species of all other plant groups combined, is the "flowering plants" (angiosperms). There are over 275,000 known species of angiosperms living in a range of habitats, from freshwater to deserts, tropics to the subarctic. Angiosperms range in size from the microscopic duckweed to *Eucalyptus,* which exceeds 100 meters in height. Unlike gymnosperms, angiosperms completely enclose their ovules within sporophyte tissues, specifically within the *ovaries* of flowers. Fruits are mature ovaries that are often edible to attract organisms for the dispersal of seeds. They may utilize wind, gravity or water to disperse their seeds. *Vegetables* are edible parts of a plant that are not part of the reproductive organs.

The angiosperm flower consists of highly modified leaves and structures to facilitate sexual reproduction. The *receptacle* is a modified stem tip to which attach the flower parts. The outermost of these parts are the *sepals*, a whorl of modified leaves which protect the bud as a flower develops within. *Petals* are modified leaves interior to the sepals; their shape and coloration account for the attractiveness of many flowers to pollinators. Flowers may have only male or female reproductive organs or may have both. The ovules of an angiosperm are in the ovary of a flower, while the pollen grains are produced in the *anthers*. Glands located in the region of the ovary produce *nectar*, a nutrient gathered by pollinators as they travel among flowers. The diversity of flowers is related to the numerous ways they are pollinated. Color, size, shape, and scent all attract specific pollinators to disperse pollen. Coevolution with insects and other animals encourages dispersal of the pollen or fruit to locations suitable for germination and development.

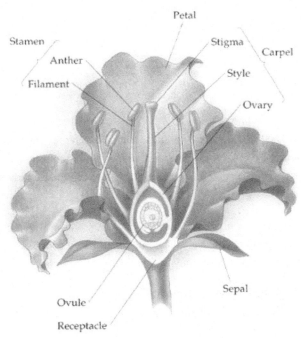

Structure and Organization of Plants

Plant Organs

Vascular plant anatomy can be divided into two systems, the root, and the shoot system. The *root system* consists of the primary root plus any side branches. It is generally equal in size to the aboveground shoot system and anchors the plant in the soil, providing support. Roots absorb water and minerals from the soil, while *root hairs*, heavily concentrated near the root tip, further increase the absorptive surface. Roots produce hormones which must then be distributed to the plant. The roots of *perennials*, plants with leaves and stems that die at the end of each growing season, remain in the ground so that they may regrow next season. *Annuals* completely die out at the end of the growing season and can only regrow from a seed.

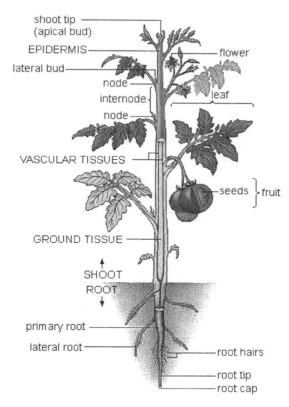

shoot tip
(apical bud)

EPIDERMIS

lateral bud

node

internode

node

flower

leaf

VASCULAR TISSUES

seeds } fruit

GROUND TISSUE

SHOOT

ROOT

primary root

lateral root

root hairs

root tip

root cap

A typical angiosperm plant body: shoot system (above ground) and root system (underground)

The *shoot system* consists of the stem, the branches, and the leaves. The *stem* forms the main axis of the plant, from which other plant organs branch out. Upright stems produce leaves and array them to be exposed to a maximum amount of sunlight. *Nodes* are the site at which a leaf attaches to the stem, and *internodes* are the regions of the stem between the nodes. Some modified stems may be underground but can be distinguished from the root system by their nodes and internodes. Like the roots, the stem has a vascular tissue to transport water, organic nutrients and inorganic minerals throughout the plant. Trees use a woody tissue to strengthen their stems and allow them to grow remarkably large. Stems often function in the storage of water and nutrients.

The *leaf* is the major organ of photosynthesis in most plants. Leaves receive water from roots by the stem. They are typically broad to maximize the surface area for CO_2 absorption and solar energy collection, and thin so that light can penetrate to the underside of the leaf. The *lamina* (blade) is the wide portion of a leaf where the majority of the photosynthetic tissue is located. A *petiole* is a stalk that attaches a leaf blade to the stem. The *leaf axil* is the upper acute angle between petiole and stem and is the location of where an *axillary bud* originates. Axillary buds may develop into additional leaves or flowers. Leaves are heavily modified for different functions, such as protection, structural support, and food storage.

Monocot vs. eudicot plants

The two main groups of angiosperms are the *Monocotyledons* (monocots) with 65,000 species and the *Eudicotyledons* (eudicots), formerly dicots, with 175,000 species.

Monocots produce one *cotyledon* at germination and often have flower parts in multiples of threes. Cotyledons are embryonic seed leaves which nourish the embryo before and throughout germination. Soon afterward they may fall off and be replaced by true, photosynthetic leaves or may mature with the plant. The cotyledon of monocot functions to transfer stored nutrients to the embryo.

Eudicots produce two cotyledons at germination and often have flower parts in multiples of four or five. These cotyledons are nutrients themselves and directly feed the embryo. Note that gymnosperm seedlings have cotyledons that are variable in number. The cotyledons of a gymnosperm are similar to monocots as they transfer nutrients to the embryo but do not feed it directly.

The distinction between monocots and eudicots represents an important evolutionary division. Some well-known monocots are palms, lilies, tulips, and grasses, including grains such as corn and wheat. The eudicots are a substantially larger group that include oaks, maples, dandelions, cotton, and tomatoes.

In addition to the difference in floral arrangement and the number of cotyledons per seed, monocots and eudicots have other distinguishing characteristics. Eudicots have vascular tissues which are arranged in a ring, while in monocots the vascular tissues spread throughout the stem. Nearly all monocots are *herbaceous* (relatively soft and pliable), while eudicots may be either herbaceous or *woody* (with rough bark). Monocots typically exhibit leaf veins which run parallel, while eudicots have branched, net-like venation. Their vascular systems differ in the stems; monocots have *vascular bundles* of xylem and phloem randomly spread throughout the stem, while eudicots have their bundles arranged in rings.

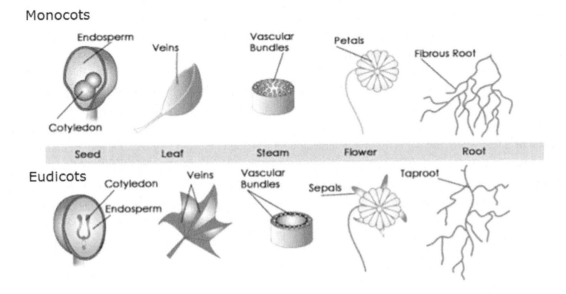

	Embryos	Leaf venation	Stems	Roots	Flowers
Monocots	One cotyledon	Veins usually parallel	Vascular bundles usually complexly arranged	Fibrous root system	Floral parts usually in multiples of three
Eudicots	Two cotyledons	Veins usually net-like	Vascular bundles usually arranged in a ring	Taproot usually present	Floral parts usually in multiples of four or five

All monocots exhibit a *fibrous root system* (adventitious root system), which consists of a mass of branching, slender roots originating from the base of the stem and holding the plant secure in the soil.

Conversely, eudicots begin life with a *taproot* (primary root system). A primary root (the taproot) grows straight down and is the dominant root of the plant. It is large, fleshy, and central to any lateral branches. Eudicots which retain a true taproot system throughout their lifetime include carrots, radishes, and beets. Most eudicots develop fibrous roots as they age.

Plant Tissues

Plant and animal cells have several key differences. Unlike animal cells, plant cells have cell walls with a thick, rigid framework of *cellulose*, a tough polysaccharide. Other polysaccharides, such as *lignin* for strength or *pectin* for elasticity, are in various concentrations depending on the cell type. The walls give the cell its shape, regulate the transport of water and other materials, and allow the plant to hold itself up against gravity. Plant cells must be connected by *plasmodesmata* (channels of cytoplasm) to allow the passage of materials through the otherwise impenetrable cell walls.

Plant cells contain *plastids* (organelles such as chloroplasts) not in animal cells. Vacuoles in plant cells are large and have a variety of functions, including carbohydrate, protein and pigment storage. The vacuoles help maintain *turgor pressure*. Plant cells must remain *turgid* (swollen) so that they can actively push against the cell wall and maintain its shape. In the absence of turgor pressure, the plant's cell membranes shrink from the cell walls, causing them to be *flaccid,* which compromises the structural integrity of the plant. A plant with inadequate turgor pressure wilts.

Plants experience *indeterminate growth*, as they do not have a pre-programmed body plan and continue to grow throughout their entire life. There are constants like leaf shape and branching patterns, but it is not possible to predict precisely where a new branch develops on a tree. Plants retain areas where rapidly dividing; undifferentiated stem cells remain through the life of the plant. Animals have stem cells as well, but in much smaller amounts and with more limited *potency* (differentiation capability).

Plant stem cells are in *meristematic* tissue, which rapidly divides throughout the plant's lifetime to produce completely undifferentiated cells for growth and repair. Meristem cells are generally small, six-sided, boxy structures with a large nucleus. Vacuoles may or may not be present. As the cells mature, the vacuoles grow into many different shapes and sizes, depending on the needs of the cell. The vacuole may fill 95% of the cell's total volume.

There are two regions of meristematic tissue in a plant: the apical meristem and lateral meristem. The *apical meristem* is located at or near the tips of roots and stems. As new cells form in the apical meristem, the plant elongates. Root apical meristem lengthens the roots, while shoot apical meristem allows the formation of new leaves and flowers, and lengthens the stem. This vertical growth is *primary growth*. Vertical stem growth enables the plant to reach towards the sunlight, while root lengthening is important to anchor the plant and probe for water and nutrients in deeper soil layers. While apical meristem is in all vascular plants, a special form, *intercalary meristem*, is only in monocots such as grasses. These plants grow lengthwise both from their tips via apical meristem, and from their nodes via intercalary meristem. This allows grasses to regrow even when their tips are cut.

Apical meristem gives rise to three *primary meristems*: protoderm, ground meristems, and procambium, which each produce one of the *primary tissues*. *Protoderm* is the outermost primary meristem which gives rise to epidermal tissue. *Ground meristem* is the inner meristem that generates *ground tissue*, as the inner bulk of the plant. Finally, the *procambium* develops into *primary vascular tissue*, the *primary xylem,* and *primary phloem*.

In some species, the procambium gives rise to the *lateral* (or secondary) *meristems,* which account for secondary growth and are located near the periphery of the plant. *Secondary growth* is generally horizontal in a direction (e.g., tree trunk in girth). This growth allows for extra xylem and phloem tissue production and provides stability for the plant to grow taller. Just as the primary meristems which produce primary tissues, the lateral meristems produce secondary tissues. Lateral meristems and secondary growth are in gymnosperms, woody eudicots, and the few woody monocots. These organisms still generally exhibit *apical dominance,* in which vertical or primary growth surpasses any lateral or secondary growth due to hormonal control that promotes the action of apical meristem and inhibits lateral meristem.

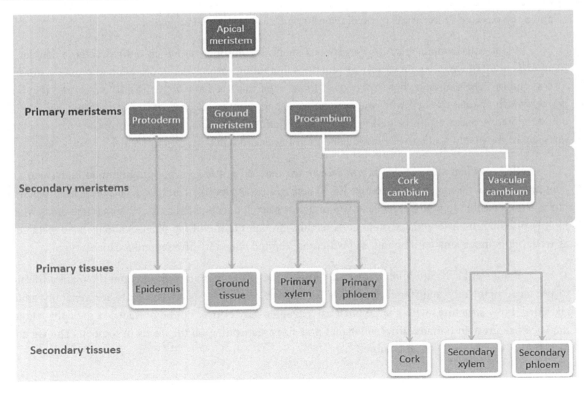

There are two types of lateral meristems: vascular cambium and cork cambium, both derived from the procambium. The *vascular cambium* is a thin, branching cylinder that, except for the tips where the apical meristems are located, runs the length of the roots and stems of most perennial plants and many herbaceous annuals. The vascular cambium is responsible for the production of *secondary xylem* and *secondary phloem*, which increase the girth of the plant.

The *cork cambium* is structurally similar to the vascular cambium, but it is only found in woody plants, where it produces the secondary tissue *cork*. As cork cells mature, they become encrusted with a waxy lipid of *suberin*, which renders them waterproof and inert, replacing the epidermis and producing outer bark. Cork protects a plant and makes it resistant to attack by fungi, bacteria, and animals. Both the vascular cambium and the cork cambium, if present, begin to produce cells and tissues only after the primary tissues produced by the apical meristems have begun to mature.

The primary tissues (epidermal tissue, ground tissue, and primary vascular tissue) are the *nonmeristematic tissues.* Nonmeristematic tissues are differentiated according to their intended function in the plant. Sometimes the tissues are composed of the same type of cells throughout, and sometimes they are mixed.

Epidermal tissue acts as an outer protective covering for the roots, leaves, and stems of non-woody plants. The epidermis is the outermost layer of closely-packed epidermal cells on all plant organs. In mature woody plants, the epidermis of the stem is replaced altogether by bark. The epidermis is in direct contact with the environment and therefore is subject to environmental conditions. Generally, it is one cell layer thick, but there are exceptions, as in the case of tropical plants where it may be several cells thick and act as a sponge. The cuticle, created by a fatty substance called *cutin* that is secreted by most epidermal cells, covers the walls of epidermal cells to prevent water loss and protect against pathogens. Epidermal cells may be modified to serve various functions. For example, certain epidermal cells in the roots are modified into root hairs to increase the surface area for water and nutrient absorption. Epidermal cells of stems and leaves produce hairs as *trichomes* that have a variety of functions, such as repelling insects and reducing water loss.

Ground tissues make up the bulk of the plant and consist of three specialized cell types: parenchyma, collenchyma, and sclerenchyma. *Parenchyma* has thin and somewhat flexible cells walls and is in all organs of a plant. Most parenchyma cells can differentiate into other cell types under special conditions, such as repair and replacement of organs after injury. In leaves, some parenchyma cells differentiate into *chlorenchyma cells,* which have chloroplasts and form the *mesophyll tissue,* the photosynthetic tissue of the plant. In storage organs, parenchyma cells have large vacuoles that store starch, fat, protein or water. Some parenchyma cells may store resins, tannins, and crystals which they secrete.

On the epidermis of leaves, each stoma is flanked by a pair of special parenchymal *guard cells*. Guard cells are elongated and attached at their ends forming a donut shape. When they receive a signal to open stomata, they intake potassium ions, driving water to enter and increase turgor pressure. The guard cells enlarge in width because they are restricted by cellulose microfibrils but can expand lengthwise. The elongation causes the cells to buckle outward and create an opening between them. Guard cells respond to changes in light, heat and CO_2 concentration, as well as internal signals of water scarcity. These factors are particularly important during plant development, as they affect the density of stomata of the adult plant. CO_2 concentration rises in the air near the plant when photosynthesis ceases. There is evidence that this activates a receptor in the plasma membrane of guard cells, which shuts off the proton pump to close the stomata. Blue wavelengths of sunlight trigger stomatal opening.

Guard cells store *flavin pigments* in their vacuoles, which absorb blue light. This pigment sets in motion a cytoplasmic response activating the proton pump that causes potassium ions to accumulate. However, plants kept in the dark open and close their stomata like clockwork, as if responding to the regular changes in sunlight over a 24-hour period. Some internal biological clock may be responsible for this phenomenon. The plant can actively control the guard cells with hormones. *Abscisic acid* (ABA) is a plant hormone with many uses, one of which is the initiation of stomata closure.

Collenchyma cells resemble parenchyma but have thicker primary cell walls, often with uneven thicknesses. They usually occur as bundles of cells just beneath the epidermis and are elongated. Their walls are pliable in addition to being strong, making them well suited to provide structural support in regions of the plant that are still growing.

The last category of ground tissue cells are the *sclerenchyma cells*, the strongest and least pliable. They provide support to mature regions of plants and have thick secondary cell walls embedded with tough *lignin*. Unlike other ground tissue cells, they are dead at functional maturity and cannot increase in length. Sclerenchyma cells may be long and fibrous, or more globular and irregularly shaped.

Primary vascular tissue includes primary xylem and phloem, which transport water, ions, minerals, and other nutrients, and provide support. Xylem and phloem extend from roots to leaves in organized bundles with supporting ground tissue. Secondary xylem and phloem serve the same function as primary xylem and phloem but are produced by the lateral meristem during secondary growth rather than by the procambium during primary growth. Normally, secondary vascular tissues are only in woody eudicots but may be in a few monocots and herbaceous eudicots.

Recall that xylem passively conducts water and mineral solutes (i.e., xylem sap), upward through a plant. The cells of xylem tissue have two cell walls, the primary and secondary. These secondary cell walls are reinforced with lignin and are often deposited unevenly in various patterns to stretch. The cell walls are all that is left in xylem cells, as the cytoplasm and plasma membrane has disappeared. Therefore, the xylem is dead at functional maturity. There are two types of xylem cells: vessel elements and tracheids. *Vessel elements* are hollow, non-living cells that are stacked end-to-end to reach from roots to leaves. They are connected vertically by *perforation plates* at their ends, allowing water and minerals to rise through a continuous pipeline. Each vessel element is connected horizontally to other vessel elements by *pits* in their side walls.

Tracheids are hollow and non-living but are longer and more slender than vessel elements. Columns of tracheids lie next to one another and reach throughout the length of the plant. They overlap at their tapered ends, which have pits but not perforation plates. Water rises through a tracheid until it reaches its end, at which point it moves laterally through a pit into an adjacent tracheid and then continues to rise until it reaches the next pit. Tracheids form not a continuous pipeline, but rather a "staircase" for sap transport.

Phloem transports sucrose and other organic compounds from the areas where they are produced or stored (i.e., the leaves or storage organs) to any region of the plant that requires nutrients. It includes sieve-tube cells and companion cells. *Sieve-tube cells* (sieve-tube members or sieve elements) are living cells arranged end to end and connected by plasmodesmata to form tubes. These porous connections between sieve tube cells (similar to the perforation plates between vessel elements) are *sieve plates.* Although they are living and have a cytoplasm actively involved in the conduction of food materials, sieve-tube cells do not have nuclei at maturity. Therefore, they depend on *companion cells* to sustain them. Companion cells are specialized parenchyma cells connected to sieve-tube cells by numerous plasmodesmata. They are smaller and more generalized than sieve-tube cells, with a nucleus that maintains the function of both cells. Some companion cells assist in transferring products from other plant cells to the sieve tube cells.

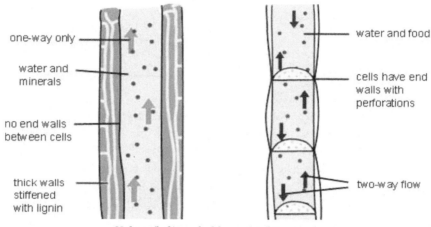

Xylem (left) and phloem (right) vessels

While most conduction in the vascular tissues is up and down, there is some side-to-side conduction via *vascular rays,* flat ribbons of parenchyma cells which intersect the vertical vascular system. In a sense, the rays are the "capillaries" of the plant, transferring materials between the major arteries and veins of the plant (xylem and phloem) and the outer tissues. In trees and other woody plants, rays radiate from the center of stems and roots and look like the spokes of a wheel in cross-section.

Organization of roots

Root and shoot tips are a site of prolific primary growth due to the apical meristem. Root and shoot tips are organized into zones of cells in various stages of differentiation. The *zone of cell division* contains apical meristem and adds cells to the tip. In the *zone of elongation*, cells become more specialized and undergo elongation, which pushes the root further down into the soil or the shoot upward towards the light. The *zone of maturation*, beyond the zone of elongation, is where cells are mature and fully differentiated. In roots, a *root cap* protects the delicate apical meristem; its cells constantly slough off and are replaced. The *quiescent center* is a population of cells in the apical meristem that are resistant to radiation and chemical damage, possibly as a reserve if the apical meristem becomes damaged.

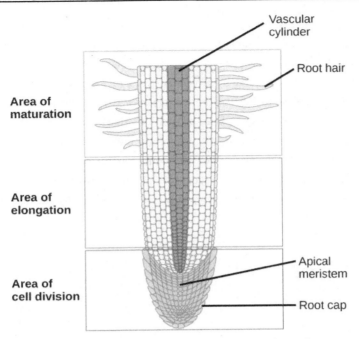

Root zone with root cap, apical meristem, and root hairs

The flexible roots found in non-woody plants (nearly all monocots and many eudicots and gymnosperms) are herbaceous roots. The tissues of an herbaceous eudicot root are comprised of several different cell types. In the center of the root is the *vascular cylinder,* called the *stele.* It contains the xylem and phloem along with some ground tissue. The *pericycle* is the outermost layer of the vascular cylinder, through which all materials pass on their way to the xylem and phloem. It is made of lateral meristems and is responsible for secondary growth in the form of root branching.

Surrounding the vascular cylinder is a single ring of cells as the *endodermis.* Its rectangular cells fit together and are sealed by the *Casparian strip*, an impermeable lignin and suberin layer which prevents water and minerals from entering the vascular cylinder by slipping between cells. Access to the vascular cylinder is by entering the endodermal cells. The endodermis marks the boundary between the vascular cylinder and the *root cortex*, a layer of large, thin-walled, irregularly shaped parenchyma cells. These cells contain starch granules, as the cortex functions in food storage. The cells of the cortex are loosely packed, allowing water and minerals to pass between them without entering the cells themselves. Finally, the outermost ring of the root is the epidermis, a single layer of thin-walled, rectangular cells that protect the root and extend root hairs for absorption.

Monocot roots look nearly identical to herbaceous eudicot (dicot) roots, but with an additional layer of the vascular cylinder in the center, the *pith.* Xylem and phloem are arranged around this pith.

Woody roots have many more layers than herbaceous ones. While the outer layers (epidermis, cortex, and endodermis) are the same, the area within the endodermis is much more complex. The endodermis surrounds a layer of cork, produced by the cork cambium in the next interior layer. Cork is what makes these roots tough and woody.

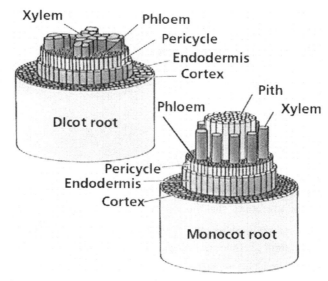

Herbaceous eudicot root and monocot root with an additional layer of pith vascular tissue

Moving from just inside the cork cambium to the center, the rest of the root contains the primary phloem, secondary phloem, vascular cambium, secondary xylem, and primary xylem. The vascular cambium continually divides to produce the secondary vascular tissues, which is why they are on either side of it.

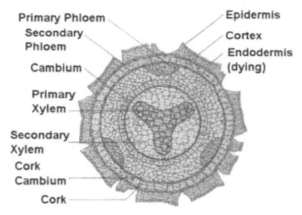

*Woody eudicot / gymnosperm root. A woody monocot root is identical
except for the pith within the primary xylem*

As previously discussed, plants may have a fibrous root system or a taproot system. However, the distinction is not entirely clear; a plant may have a taproot with lateral branches but have fibrous roots originating from the stem, or have a slender taproot with so many branches that it appears as a fibrous system. Fibrous roots do not need to originate from a stem, and can technically grow from any root.

Roots are heavily modified and do not exist in the underground root system. Many roots are *aerial* (above ground). *Prop roots,* for example, originate from the stem and reach to the soil or other support structures to anchor a plant. *Pneumatophores* project above the water to acquire oxygen. These roots are commonly seen in plants which live in swampy, aquatic environments, such as mangrove trees. *Haustoria* are another form of aerial root, which projects from parasitic plants and latches onto the host plant to invade vascular tissues from which they extract water and nutrients.

Highly specialized *propagative roots* allow many plants to reproduce asexually. These roots bud off from horizontal stems as *stolons* that "run" on or just under the soil, many feet away from the parent plant. The roots then develop into a clone of the parent.

Roots of many plant species develop mutualistic relationships with fungi and bacteria to maximize nutrient acquisition, as discussed later.

Organization of Stems

Some plants have adopted highly specialized stems that give them certain advantages. Stolons are stems that grow horizontally along the ground and give rise to new plants from their nodes. A *rhizome* is a horizontal, underground stem similar to a stolon, but while stolons are small, thin stems which project from the main stem, a rhizome is the main stem itself. Like stolons, some rhizomes have *stem tubers* which store nutrients and therefore allow the plant to survive winter, drought and other adverse conditions. Stem tubers serve as a means of asexual reproduction. An example of a stem tuber is a potato. Other plants, such as cacti, have modified their stems for water storage. Like prop roots, stems can be adapted to wrap around support structures. They are slim and provide an attachment as well as support, allowing the plant to climb upwards.

The stem tip, like the root tips, is a site of primary growth where cell division extends its length. Shoot apical meristem produces new leaves and primary meristems, increasing stem length. It is protected within a terminal bud of *leaf primordia* (immature leaves). *Bud scales* are scale-like coverings which protect terminal buds as they lie dormant in the winter.

Shoot apical meristem gives rise to the three primary meristems (protoderm, ground meristem, and procambium), just like root apical meristem.

Herbaceous stems have cross-sectional anatomy that features an epidermis with waxy cuticle surrounding ground tissues and vascular bundles. In monocot stems, vascular bundles are scattered throughout the stem; there is no well-defined cortex or pith. By contrast, the herbaceous stems of eudicots have vascular bundles arranged in a ring towards the outside of the stem, separating the outer cortex from the inner pith. Herbaceous eudicot stems resemble monocot roots.

A woody eudicot stem, like a woody root, is more complex than an herbaceous one. A mature woody stem, or *trunk,* has no epidermis or cortex, which have been replaced by bark and wood, respectively.

A woody plant in its early years undergoes only primary growth and is therefore herbaceous. However, it eventually enters a stage of secondary growth. During this time, the layer of vascular cambium between the primary phloem and primary xylem begins to produce new cells, resulting in a "sandwich" of primary phloem, secondary phloem, vascular cambium, secondary xylem, and secondary phloem. Secondary phloem is soft and does not build up, but the secondary xylem accumulates to form *wood.* The vascular cambium is dormant during the winter, so no wood is produced at this time.

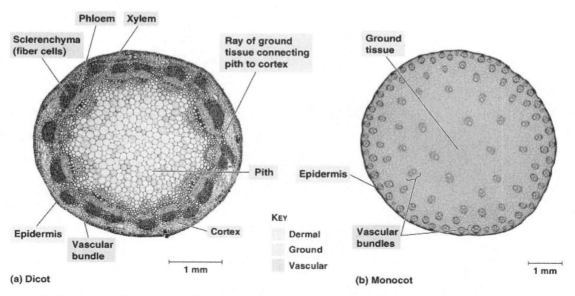

Herbaceous eudicot stem (left) with central pith and monocot stem (right) with ground tissue

Springwood is composed of wide xylem vessel elements with thin walls, necessary to conduct sufficient water and nutrients to sustain the abundant growth that occurs during spring. *Summerwood* forms when moisture is scarce; composed of a lower proportion of vessels, it contains thickly-walled tracheids and numerous fibers. An *annual ring* is one ring of spring wood followed by a ring of summer wood, which amounts to one year's growth.

Sapwood is the outer annual rings of wood where vascular transport occurs. *Heartwood* is made of the inner annual rings, found only in older trees. The vessels in the heartwood no longer function in transport; they have become plugged with resins and gums that inhibit the growth of bacteria and fungi. The pith inside the vascular cylinder is often replaced entirely by heartwood.

Meanwhile, the cork cambium just interior to the epidermis produces cork cells impregnated with suberin, which grow outward and eventually replace the epidermis. They die and form an impervious barrier to gas and water exchange, physical damage and pathogens. *Lenticels*, raised areas of loosely-packed cells, are the only areas that allow for gas exchange. The cork cambium sometimes produces a layer of living parenchyma cells. Together, the cork cells, cork cambium cells, and parenchyma cells are the *periderms*. After a few weeks, the cork cambium loses meristematic ability and expansion splits the original periderm. New cork cambium then forms deeper in the cortex of the stem. All tissues exterior to the vascular cambium are considered *bark*. From outside to inside, the bark includes the periderm, the primary phloem, and the secondary phloem.

It is advantageous to be woody when there is adequate rainfall, as woody plants can grow taller and larger as long as they can sustain their tissues. However, because they are long-lasting, woody plants need more defense mechanisms against attack by herbivores and parasites. They mature much more slowly and thus cannot reproduce as quickly. Before they have a chance to mature, they may be destroyed by attack or disease before they reproduce.

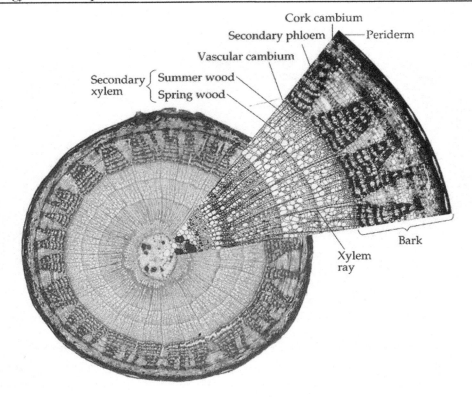

Organization of leaves

Leaves are the plant's photosynthetic organs and have a flattened blade and a petiole attaching the leaf blade to the stem. They can be categorized as either simple or compound leaves. *Compound leaves* are divided into smaller leaflets, and each leaflet may have its stalk, while *simple leaves* have a plain shape that is not deeply lobed or divided into smaller leaflets.

The *leaf veins* are vascular bundles within the leaves, which transport water and nutrients. The top and bottom sides of the leaves are covered by the epidermis and a waxy cuticle that helps keep the leaf from drying out. Along with stomata, the leaf may bear protective hairs or glands that secrete irritating substances. Mesophyll tissue is the inner body of a leaf, and the majority of photosynthesis occurs here. *Palisade mesophyll* is the main photosynthetic tissue and contains elongated chlorenchyma cells with many chloroplasts. Palisade is on the upper half of the leaf where light intensity is the greatest. *Spongy mesophyll* contains loosely packed parenchyma cells that increase the area for gas exchange and are in the lower half of the leaf.

As with roots and stems, leaves can be highly specialized to serve certain purposes. Plants adapt their leaves to their environment; shade plants have broad leaves to maximize sunlight collection, while desert plants have reduced leaves and sunken stomata to minimize water loss. These leaves may be large and fleshy, as in the case of succulents, to store water. Leaves may be protective; cactus spines are modified leaves. They can even be predatory organs, as with carnivorous plants that trap and digest insects within their leaves.

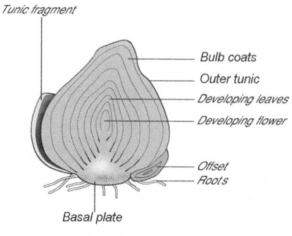

Bulb for food storage

Bulbs, such as onions, shallots, and tulips, have many layers of leaves surrounding a short stem. Like tubers, bulbs serve as food storage.

Nutrition and Transport in Plants

Plant nutrition and soil

Essential nutrients (i.e., carbon, hydrogen, oxygen, minerals) comprise 96% of a plant's dry weight. Carbon dioxide is the source of carbon for a plant, water is the source of hydrogen, and oxygen can come from either atmospheric oxygen, carbon dioxide or water. Minerals are extracted from the soil, often with the aid of fungi and bacteria. Essential nutrients fulfill the following criteria:

1. They have an identifiable nutritional role
2. No other element can substitute and fulfill the same role
3. A deficiency of the element causes the plant to die.

Elements are divided into macronutrients and micronutrients by their concentration in plant tissue. The *macronutrients* (nitrogen, potassium, and calcium) still make up less than 5% of the plant but are present in appreciable amounts. *Micronutrients* include many metals (zinc and copper) found in hundredths or even thousandths of a percentage. *Beneficial nutrients* improve the growth of a particular plant but are not strictly necessary. For example, sugar beets show better growth in the presence of sodium but can survive without it.

Soil formation begins with the weathering of rock by freezing, glacier flow, stream flow, and chemicals. Lichens and mosses are the first species to colonize barren rock. As they die, they leave decaying tissues which begin to accumulate and support other life. This decayed organic matter, *humus*, takes time to accumulate, but once present, its acidity leaches minerals from rocks and further enriches the soil. Depending on the parent material and weathering, a centimeter of soil may develop within 15 years.

The mature soil consists of soil particles, decaying organic matter, living organisms, air, and water. A mixture of 10 to 20% humus mixed with a top layer of soil particles is best for plants. Humus keeps the soil loose and crumbly, decreases runoff and aerates the soil. Soil particles are classified along a continuum, with clay less than 0.002 mm in diameter and the largest sand particles up to 2.0 mm in diameter. Particles that are larger than clay but smaller than sand are *silt*. The best soil includes particles of varying sizes which provide critical air spaces. Exclusively sandy soils lose water too readily, while clay holds water but packs too tightly. The nitrogen content of clay soil is low because clay particles are negatively charged and cannot retain negatively charged nitrate (NO_3^-). However, their charge does allow them to attract positively charged ions such as calcium (Ca^{2+}) and potassium (K^+). *Loam*, a mixture of roughly equal parts of sand, silt, and clay, retains water and nutrients while still allowing roots to take up oxygen in the air spaces.

Animals are essential for the soil. Burrowing mammals, insects and other organisms help turn over the soil. *Detrivores,* organisms which feed on decomposing plants and animals, break down this matter into absorbable nutrients. Fungi, protozoa, algae, and bacteria complete the decomposition process. Decomposition is particularly important because it converts the nitrogen in dead organisms into nitrate, a form which plants utilize.

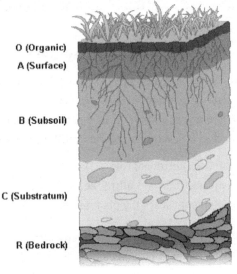

Soil horizons

Uptake of water and minerals

Water from the soil moves directly into the cytoplasm of root hair epidermal cells and is transported across the cortex and endodermis of a root. *Symplastic movement* is the movement of water and solutes through the continuous channel of cytoplasm that runs through plant cells, via plasmodesmata. Therefore, the *symplast* is the area within the plant's plasma membranes through which water and solutes can travel, i.e., the connected cytoplasms of many cells. *Apoplastic movement* of water and solutes through the regions outside of the symplast is collectively the *apoplast*. This includes cell walls and any extracellular spaces through which water may diffuse. It is more rapid than symplastic movement because it does not involve movement through plasma membranes and there is less resistance to the flow of water.

Water flows through both the apoplast and the symplast on its way to the xylem, though mostly through the apoplast. This water contains many minerals that the plant needs, but may have toxins and other undesirable substances. However, since the water diffuses freely through the apoplast, there is no way to prevent the passive transport of these toxins until the water reaches the endodermis. The Casparian strip rings the endodermal cells. The Casparian strip prevents the water from traveling further through the apoplast; it must pass through the plasma membrane and enter the symplast to continue to the pericycle.

Forcing all water and solutes to pass through the symplast of the endodermal cells provides a way for the plant to regulate transport into the xylem. The plant may control how water and any harmful molecules enter its vascular system. Once water passes under the Casparian Strip in the endodermal cells, it enters the apoplast again on its way to the xylem.

Minerals follow the same path of uptake as water, but many minerals are actively taken up by plant cells. Mineral nutrient concentration in roots may be many times greater than in surrounding soil. Active transport is required to uptake minerals against their concentration gradient. This may be done directly using membrane proteins that take ions into the cell.

They can enter indirectly through a *chemiosmotic mechanism*. Proton pumps on the cell membranes of the root cells use ATP to pump H^+ ions out of the cell and into the soil. This establishes an electrochemical gradient that causes potassium, calcium and other positively-charged ions to flow into cells, as H^+ displaces these cations in the soil and encourages them to diffuse into the roots. The H^+ ions associate with anions in the soil like PO_3^{4-} and bring them along when they re-enter the roots.

Along with the branching of roots and formation of root hairs to increase surface area, symbiotic relationships are important in nutrient acquisition. Plants cannot use atmospheric nitrogen because they lack enzymes to break the triple bond in N_2. *Rhizobia*, which are nitrogen-fixing bacteria living in *root nodules* of legumes, make nitrogen compounds available to the plants in exchange for organic compounds. Rhizobial bacteria reduce atmospheric nitrogen (N_2) to ammonium (NH_4^+), in *nitrogen fixation*. Other plants may not have nitrogen-fixing bacteria living in their roots but have a relationship with free-living nitrogen fixers.

The relationship between roots and fungi is crucial. The branching threads of the fungus, as *hyphae,* are robust nutrient absorbers due to their large surface area and ability to decompose decaying matter into inorganic nutrients. Hyphae penetrate the root tissues and deliver nutrients to the plant, which they absorb from the soil. The association between roots and hyphae is a *mycorrhiza*. A plant can uptake many more nutrients in conjunction with a fungus than it could on its own due to the increased absorption capabilities of the fungus. In return, the plant delivers carbohydrates and amino acids to the fungus. Mycorrhizae allow both plant and fungus to inhabit soils in which they may otherwise die.

Not all plants benefit those whom they gather nutrients from. Some plants are parasites, sending out haustoria to grow into the host plant and tap its xylem and phloem. In an interesting reverse of the mycorrhizal relationship, some plants parasitize fungi and steal organic compounds which they receive from other plants. Plants can even take their nutrients from animals, as with the Venus flytrap and the sundew, which are carnivorous and prey on insects.

Transport mechanisms in plants

Transportation in plants relies on the mechanical properties of water:

1) the uptake and release of water and solutes by individual cells, as when root cells absorb water and minerals from the soil

2) short-distance transport of substances from cell to cell (e.g., loading of sucrose from photosynthetic cells into the sieve tube cells of the phloem)

3) long-distance transport of sap within the xylem and phloem.

Both diffusion and water potential are important in the transport of water, minerals and organic molecules in plants. These two phenomena rely on the chemical properties of water: polarity and hydrogen bonding. *Water potential* is the potential energy of water and is a measure of the capacity to release (or take up) water. Water flows from a region of higher to a region of lower water potential. In cells, water potential includes both the *pressure potential*, which is the effect that pressure has on water potential and *osmotic potential*, the effect that solutes have on water potential. Water flows from an area of high to an area of low pressure, just as it flows from an area of low solute concentration (high water concentration) to an area of high solute concentration (low water concentration).

When the solute concentration is higher within a cell than outside of it, water enters and increases the pressure inside the cell. Plants' sturdy cell walls allow substantial water pressure to build up until osmotic potential is fully balanced and water ceases to enter. Turgor pressure is maintained by the pressure potential that increases due to this osmotic pressure. Cells must maintain a high solute concentration within the cytoplasm to encourage the movement of water into the cells. A cell in a *hypotonic solution* has a low solute concentration and high water concentration compared to the solution. Animal cells swell and undergo lysis if they are placed in a hypotonic environment, but plant cells *must* have hypotonic extracellular fluid to maintain the overall turgidity of the plant's body.

Water can rise upward through the xylem in two ways: pushing and pulling. These are forces, but they arise naturally rather than by active transport. The cells of the xylem are non-living and do not perform any processes to move water. Xylem movement is a passive process, created by natural properties.

At night, when transpiration is low, water entering root cells creates osmotic *root pressure*. This is a positive pressure that pushes the sap upward through the plant. It can cause water to be forced out of leaf veins, a phenomenon of *guttation*. Guttation is visible as water beads along the edges of leaves.

However, root pressure on its own is not usually sufficient to drive water movement. Tall trees generate almost no root pressure since the weight of the water pushing down on the xylem more than counteracts the positive pressure. For most plants, root pressure is of minimal importance compared to *transpiration pull,* which is primarily responsible for upward sap movement in the daytime. Transpiration pull is created by *transpiration*, the evaporation of water from leaves. This drives a negative pressure, or *tension,* which pulls the sap upward.

While the mechanisms behind upward movement through the xylem are not entirely understood, the most generally accepted theory is the *cohesion-tension model.* Under this model, the cohesive properties of water and the tension created by transpiration pull are the primary forces driving sap against gravity. *Cohesion* refers to water bonding together due to hydrogen bonds. The cohesion of water is strong enough to transmit the pulling force down to the roots. Another property of water implicit in the cohesion-tension model is *adhesion*, the ability of water to cling to other molecules, such as the walls of xylem vessels. Adhesion gives a water column extra strength to resist gravity and prevents it from slipping back down.

Transpiration pull, cohesion, adhesion and the structure of xylem vessels are all important in the carrying of water by the transpiration stream. The *transpiration stream* is the unbroken column of water that flows from the roots to the leaves and then evaporates into the air. The tension created by the transpiration of water coupled with the cohesive and adhesive forces is enough to support the stream against the forces of gravity, even in trees 100 meters tall. It must remain continuous to conduct water upwards properly.

Plants must constantly balance the requirements of the transpiration stream against the water loss incurred by the process of transpiration. Losing too much water through the leaves forces the plant to close its stomata, thereby inhibiting gas exchange and photosynthesis. Recall that guard cells regulate the opening and closing of stomata. They respond to four main abiotic factors that affect the rate of transpiration in a typical terrestrial plant:

1. *Light* – the rate of transpiration is much greater during daylight hours since photosynthesis occurs only when the light is available. To maximize photosynthesis, more stomata remain open and more water is transpired.

2. *Humidity* – water diffuses out of a leaf from a high concentration gradient inside the leaf to a lower concentration gradient in the air. A rise in humidity means there is a higher concentration of water vapor outside the leaf, resulting in a decrease in transpiration rate. As leaves transpire, the water vapor which they emit remains relatively close to the leaf. Humidity can increase greatly around the leaf and inhibit further transpiration.

3. *Temperature* – a rise in temperature excites water molecules and increases the transpiration rate. It can cause the enzymes of photosynthesis to work more efficiently, increasing the transpiration rate to sustain the increased rate of photosynthesis.

4. *Wind* – can offset the effect of humidity by removing the humidity from around the leaf, thereby increasing transpiration rate.

Phloem transport does not rely on transpiration pull, so it is only indirectly affected by changes in the four abiotic factors of transpiration. Phloem tissue transports sugars and other organic molecules from sources to sinks. During the growing season, leaves photosynthesize vigorously and produce large quantities of sugar. *Sources* are located in the leaves as well as storage organs, while *sinks* are common in the flowers, seeds, roots or any other part of the plant that requires nutrients.

Between the sinks and the sources are the sieve tubes, formed by sieve-tube cells arranged end to end. Although it seems counterintuitive, phloem transport requires energy even though it often moves in the direction of gravity. *Loading* is the movement of source sugars into the phloem, which goes against the concentration gradient. *Active translocation* describes this energy-requiring phloem transport.

The *pressure-flow model* outlines how a positive pressure potential drives the direction of phloem transport. The model states that sucrose is loaded into phloem via active transport due to an electrochemical gradient established by an H^+ pump. Phloem loading results in a high solute concentration at the source end. This creates high osmotic potential in the phloem, encouraging water to diffuse into the phloem. The buildup of pressure at the source creates a positive pressure potential which drives the sap to flow to a region of lower pressure – the sink.

At the sink, sap is unloaded down its concentration gradient, exiting the phloem and entering the tissue for which it is destined. Water is then recycled via transport in the xylem.

Contrary to popular belief, phloem sap does not necessarily move from the leaves to the roots. It merely flows from source to sink, meaning it can move in any direction along phloem. For example, the source may be at an underground tuber and the sink at a fruit; in this case, sap flows upwards against the direction of gravity. For this reason, the common mnemonic for phloem and xylem transport "*phlo low, xy high*," is misleading. While xylem transport does always move "high," phloem transport can move in any direction.

Control of Growth and Responses in Plants

Plant responses

One defining characteristic of life is an ability to respond to stimuli. Adaptive organisms respond to environmental stimuli because it leads to longevity and survival of the species. Plants lack the nerves and muscles of animals and therefore must respond by varying their growth patterns.

Tropisms are directional movement responses which occur due to external environmental stimuli. The stimulus comes from one direction and affects the direction of growth. Growth toward a stimulus is a *positive tropism*, while growth away from a stimulus is a *negative tropism*. Hormones respond to the stimulus by directing one side of the plant to elongate, resulting in a curving toward or away from a stimulus.

Phototropism is the growth of plants in response to light, commonly seen in shoots and roots. This response is controlled by a class of plant hormones called *auxins*. A shoot displays positive phototropism when auxins produced at the tip of the shoot are translocated to the side of the shoot tip that is receiving less light. Once auxins reach the shaded side of the plant, they initiate faster cell growth on that side. As the shaded side elongates faster than the sunny side, the plant shoot gradually bends towards the light. Some plants do this quickly, following the sun as it moves across the sky each day. Conversely, negative phototropism is seen in roots, where auxins cause them to grow away from light and into the soil. In this case, auxins initiate elongation on the upper surface of the roots, bending the roots downward.

Auxins similarly affect plant organs when it comes to gravity. *Gravitropism* is a response to the direction of gravity. Stems and leaves exhibit negative gravitropism when they grow upward in opposition to gravity. This can be seen in an upright plant placed on its side, which begins to project its leaves upward. Phototropism is observed, as the leaves re-orient themselves to best face the light. Positive gravitropism, on the contrary, aids phototropism in encouraging roots to grow downward. Root cap cells detect gravity using *statoliths*, which are starch grains within starch *amyloplasts* (storage organelles). Wherever the statoliths settle due to gravity, the root tip grows in that direction.

The hormone *ethylene* works with auxins in the *thigmotropism* response. This refers to a growth response due to contact with solid objects. A few minutes of contact can bring about a response that lasts for several days. The coiling of morning glory or pea tendrils around posts is a common example of positive thigmotropism. The cells in contact with the object are directed to grow less, while those on the opposite side elongate, causing the tendril to bend around the post. Combined with the effects of phototropism and gravitropism, the tendril spirals upward. Roots exhibit negative thigmotropism when they grow away from an obstacle in the soil, such as a rock.

In contrast to tropisms, *nastic movements* are responses independent of the direction of the stimulus. For example, *seismonastic movements* result from touching, shaking or thermal stimulation. The leaflets of *Mimosa pudica* exhibit a seismonastic movement when they curl up when touched. This response takes a second or two and is due to a loss of turgor pressure within cells, causing the petiole to droop. Another example is the Venus flytrap, which snaps down on an insect when it touches a trigger hair at the base of the trap.

Sleep movements are nastic responses to the daily changes in light level, as with the prayer plant that folds its leaves each night. This is an example of a *circadian rhythm*, a biological response which occurs on a 24-hour cycle. Circadian rhythms are driven by a *biological clock*, an internal mechanism that roughly maintains the rhythm even in the absence of stimuli. With the proper external stimuli (e.g., light from the sun), the biological clock synchronizes to exactly 24 hours. It detects sunlight using a red flavin pigment *cytochrome*, which absorbs the blue wavelength of light.

Plant hormones

For plants to respond to stimuli, the activities of plant tissues must be coordinated. Because plants do not have a nervous system, all communication is done through hormones. These chemical messengers are synthesized in one part of a plant after receiving a stimulus. They may travel long distances to another part of the organism via the phloem, or move from cell to cell via carriers. A response may be influenced by several hormones and may require a specific ratio of multiple hormones. Synthesized hormone imitators are *plant growth regulators* (PGRs) and can be used to manipulate plant growth.

Auxins are vitally important to processes which involve growth and development. For example, auxins direct fruit growth, prevent leaf *abscission* (falling) and *senescence* (aging or dormancy), and help initiate flowering. They are the driving force behind apical dominance, as they inhibit the growth of lateral buds while promoting the growth of the shoot tip. Most auxins are produced in the shoot apical meristem, so they must travel down the plant to their destination. They regulate growth by initiating cell elongation via mechanisms that are not well understood.

Cytokinins are a class of plant hormones that promote cell division rather than elongation. In some cases, they enhance the effects of auxins, as when they prevent the senescence of leaves. In other cases, they directly oppose auxins, promoting abscission and inducing lateral growth. The ratio of auxins to cytokinins determines which process wins out.

Gibberellins are another group of 70 chemically similar plant hormones. Like auxins, gibberellins promote growth by cell elongation and other mechanisms. The gibberellin GA_3 induces the release of sugars in seeds to feed the embryo and facilitate germination. It is the most common of the natural gibberellins.

Abscisic acid (ABA), despite its name, rarely plays a role in leaf abscission. Although ABA promotes abscission when applied by researchers, the hormone does not appear to have this function in the plant naturally. Rather, ABA competes with gibberellins, promoting dormancy instead of germination and growth. As with auxins and cytokinins, the ratio of ABA to gibberellins determines largely which hormone prevails. Reductions in ABA and increases in gibberellins break seed and bud dormancy, causing seeds to germinate and buds to produce leaves. High ABA concentrations keep a seed or bud dormant, stopping all growth. ABA initiates dormancy in other plant organs as well, usually in preparation for winter. It moves from the leaves to the vegetative buds in the fall, converting them to "winter" buds covered by thick, hardened scales. ABA responds to water deficiency by closing stomata.

Ethylene is a gaseous hormone that is responsible for positive feedback loops such as ripening, abscission, flowering, and senescence. At various stages of a plant's life cycle, ethylene may promote certain growth events or inhibit them. Ethylene gas ripens fruit by increasing the production of *cellulase*, an enzyme which hydrolyzes cellulose in plant walls and softens the fruit. As the fruit ripens, it releases further ethylene gas, which stimulates further ripening. The hormone easily diffuses out of cells and into the atmosphere, which is why a ripening apple can induce the ripening of a banana some distance away.

Ethylene stimulates cellulase production to initiate abscission of leaves or fruit, provided there are sufficiently low levels of auxin. Cellulase weakens the connections of the leaves or fruits to the stem and causes them to drop. Researchers cannot agree on why plants shed their leaves in the fall, but it may be for plants to conserve more energy for other activities during winter time, such as developing seeds in preparation for spring. Fruits abscess to become more available for animals to eat them and spread the seeds.

Photoperiodism

Many physiological changes in plants (e.g., seed germination, breaking of bud dormancy and onset of senescence) are related to seasonal changes in day length. *Photoperiodism* is a physiological response to relative lengths of daylight and darkness and is important for flowering since changes in day and night length signal the onset of spring.

Plants can be divided into three groups based on their photoperiodic flowering strategy. *Short-day plants* flower when the day is shorter than a critical length, or rather when the night is longer than a critical length. After the discovery of photoperiodism, researchers understood that it depends on the length of the night, not the length of the day. Short-day plants may more accurately be termed long-night plants. *Long-day plants* (short-night plants) have the opposite strategy, flowering when the day is long and thus the night is shorter than a critical length, often during the summer. *Day-neutral plants* are not affected by day length and may rely on different external or internal cues to flower.

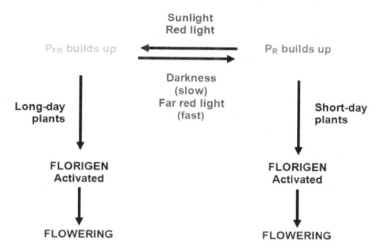

In addition to cytochrome for blue light, plants contain *phytochrome*, a blue-green leaf pigment that exists in two forms. P_r, or *phytochrome red*, absorbs red light (660 nm) and P_{fr}, or *phytochrome far-red*, absorbs far-red light (730 nm). P_r and P_{fr} can be converted interchangeably. During a 24-hour period, there is a shift in the ratio of pigments. Direct sunlight has more red than far-red light, so during the day, more P_r is converted into P_{fr}.

Shade and sunsets have farther-red than red light, and so more P_{fr} is converted to P_r at sunset and throughout the night. The amount of P_{fr} remaining as the night ends is most likely the way for plants to time the length of dark periods. A short night results in higher P_{fr} levels in the morning, while a long night depletes nearly all of it. P_{fr} promotes flowering in long-day (short-night) plants, which is why a short night usually induces flowering. Conversely, P_r promotes flowering in short-day (long-night) plants, and P_{fr} inhibits it, which is why they flower after a long night when P_r is high, and P_{fr} is low.

Reproduction in Plants

Reproductive strategies

Recall that plants have a two-generation life cycle with alternation of generations, which includes the sporophyte and gametophyte generation. The *gametophyte generation* is the dominant generation in nonvascular terrestrial plants (liverworts, hornworts, and mosses). The gametophyte represents the majority of the plant's body, while the sporophyte is a small, protruding structure that is reliant on the gametophyte. Once it grows from a spore, the gametophyte develops organs in which mitosis produces the gametes. The *archegonia* (singular, *archegonium*) produce eggs, while the *antheridia* (singular, *antheridium*) produce sperm. Like in animals, these sperms have flagella. The gametophyte must be in a damp environment so that its sperm can swim through water from the antheridium to an egg in an archegonium.

The sperm of one gametophyte may travel to the egg of an entirely different gametophyte, or the gametophyte may be self-fertilizing. In both examples, once the sperm reaches the archegonium, it fertilizes the egg to form a zygote. The zygote begins to divide via mitosis, developing into a small sporophyte growing on the larger gametophyte. The sporophyte is composed of a *foot*, which arises from the former archegonium and anchors it to the gametophyte, and a *capsule* at the end of a short stalk.

A sporangium is a spore-producing tissue that continually undergoes meiosis to form the haploid spores, which are stored in the capsule. The capsule splits to release spores, which travel on the wind or through the water some distance from the original plant. Once there, a *germ tube* emerges from the spore and begins to divide into the gametophyte. Rhizoids anchor the gametophyte to the soil and absorb waters and minerals. The gametophyte projects antheridia and archegonia, which produce sperm and eggs to begin the cycle again.

All three phyla of the nonvascular terrestrial plants roughly follow this life cycle plan, with small deviations. In hornworts (phylum *Anthocerotophyta*), the archegonia are recessed into the plant's body, while liverworts and mosses have long-necked, hollow archegonia down which the sperm must swim to reach the egg.

In contrast to the bryophytes, the dominant generation in most vascular plants is the sporophyte. However, note that many seedless vascular plants have gametophytes that are not dependent on the sporophyte. The life cycle of a fern begins with the production of spores by meiotic cell division within sporangia located in clusters on the underside of the fronds. Spores are then released and dispersed by the wind. Each spore germinates into a *prothallus*, which is the immature gametophyte. The prothallus develops antheridia and archegonia on its underside and begins generating gametes.

Fertilization occurs only if water is present to allow flagellated sperm to swim from antheridia to an archegonium. The resulting zygote begins its development inside an archegonium, but the embryo quickly outgrows the space. A sporophyte becomes visible when the first leaf grows above the prothallus and roots begin to develop below. Fronds are projected from the roots, eventually maturing into a large, leafy sporophyte.

While the gametophytes of the seedless vascular plants are small but still observable, those of the seed plants are tiny and entirely dependent on the sporophyte. Because seed plants are heterosporous, they have two different types of sporangia. Microsporangia produce microspores via meiosis. Each microspore divides mitotically into a microgametophyte. The gametophyte generates two sperm cells from its antheridia. Meanwhile, the ovules house megasporangia which produce the megaspores. Each megaspore divides into a megagametophyte, which has archegonium that produces the eggs.

The life cycle of a gymnosperm begins with the dominant sporophyte (the entire tree) and its sporangia borne in cones. Each scale-like sporophyll of a pollen cone, a *microsporophyll,* has two microsporangia. Microspores generated in the pollen cones develop into the male gametophyte, a small cluster of cells as a *pollen grain.*

The ovule (seed cones) have megasporophylls, with megasporangia generating megaspores. True to their name, these are much larger than the male microspores. One megaspore develops into a megagametophyte, which has 2 to 6 archegonia, each containing a single large egg.

Pollen grains make their way to the ovules of the seed cones and then develop a pollen tube that digests its way toward a megagametophyte, discharging two non-flagellated sperm. Fertilization takes place one year after pollination. The ovule matures and becomes the seed, composed of the embryo, reserve food, and the seed coat. The woody seed cone opens to release winged seeds in the fall of the second season. Some gymnosperms can produce fleshy *arils* which are edible by animals and dispersed similarly to angiosperm fruits.

The life cycle of an angiosperm is similar, only differing in the anatomy involved. Their sporangia and gametophytes are in flowers rather than cones. The flowers are reproductive structures which are produced by the shoot apical meristem.

Flowers may have petals, sepals, and both female and male reproductive structures, making them *complete flowers.* A complete flower is *bisexual* since it contains both male and female reproductive organs. Species with bisexual flowers are *hermaphroditic.*

However, a large number of angiosperms have *incomplete flowers,* which are missing one or more of these components. A flower which is only male is *staminate,* while one that is only female is *carpellate.* Plant species may be *dioecious,* in which there are designated male plants with staminate flowers and designated female plants with carpellate flowers. They may be *monoecious,* in which each plant has both staminate and carpellate flowers and therefore can be self-fertilizing. Note that this is different from a hermaphroditic species since the flowers are unisex, not bisexual.

Carpellate flowers are named for their *carpels,* the female reproductive structures. A carpel is a modified sporophyll located at the center of a flower. The *pistil* is the female reproductive structure; it may refer to a single carpel or several fused carpels. The vase-like carpel is generally composed of three parts. Within the enlarged base of a carpel is the *ovary.* A slender stalk of the *style* extends from the ovary. At its tip is the *stigma,* an enlarged sticky knob which receives pollen grains.

An ovary contains ovules, making the carpel analogous to the seed cone of a gymnosperm. Each ovule is covered by parenchymal cells except for one small opening, the *micropyle*. One parenchyma cell enlarges to become a *megasporocyte* (sporocytes being the cells which make up sporangia), which generates four haploid megaspores. Three of the megaspores are nonfunctional, but one divides mitotically into a seven-celled megagametophyte as an *embryo sac*. Five of the cells are support cells, two of which are *synergids* and three of which are *antipodal*. The other two cells are the egg cell and binucleate cell, with two nuclei as *polar nuclei*.

Staminate flowers are named for their *stamens,* which are the male reproductive structures. The *filament* of the stamen is a slender stalk that supports a saclike container at the top: the *anther*. The stamen is analogous to the pollen cone of a gymnosperm; it has microsporangia within the anther that generate microspores, which develops into the microgametes that produce pollen grains. An anther has four pollen sacs, each containing many *microsporocytes*. Microsporocytes undergo meiotic cell division to produce four haploid microspores; each divides into two haploid cells enclosed in a sculptured wall. This is a pollen grain, containing a *tube cell* and a *generative cell*. The larger tube cell eventually forms the pollen tube, while each generative cell divides mitotically to form two sperm. Once both events have occurred, the pollen grain has become the mature male gametophyte.

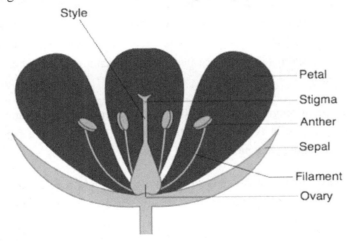

The structure of a dicotyledonous animal-pollinated flower

Pollination occurs when pollen is transferred from an anther to a stigma. It begins the breakdown of walls around the pollen sacs in the anther. Wind or animals carry pollen to the stigma of a flower, where it fully matures into a pollen grain. *Self-pollination* is the transfer of pollen from the anther to a stigma on the same plant, while *cross-pollination* refers to the transfer of pollen from the anther to the stigma of another plant. Plants often have mechanisms that promote cross-pollination, such as the carpel only maturing after anthers have released their pollen. Cross-pollination is advantageous because it increases genetic diversity.

Once a pollen grain reaches a stigma, it must travel down the style into the ovary. This is accomplished by the tube cell, which grows into a pollen tube that extends through the stigma and style to the micropyle opening of an ovule. The pollen tube injects its two sperm cells into the embryo sac so that fertilization can occur.

Unlike many other organisms, angiosperms utilize *double fertilization*, with two fertilization events. One of the sperm cells in the pollen grain fertilizes an egg to form a diploid zygote, while the other unites with the two polar nuclei to form a triploid cell as the *endosperm*. The entire ovule develops into a seed with three main components: zygote divides mitotically to form the embryo, endosperm nucleus divides into a multicellular tissue that nourishes the embryo, and the walls of the ovule develop into a protective seed coat or *integument*.

This strategy helps angiosperms conserve energy by only investing in endosperm production if an egg is fertilized. Without double fertilization, they would have to produce nutritive tissues in anticipation of fertilization. As with menstruating mammals, this tissue goes to waste if the egg is not fertilized. Additionally, double fertilization is an advantage because it speeds the development of the endosperm, with three nuclei rather than one.

Note that gymnosperms do not undergo double fertilization. One of the sperm cells fertilizes the egg, while the other recedes. The embryo is fed by nutrients passed along by its cotyledons from the tissue of the mother cone.

Pollen falls on stigma → pollen germinates → pollen tube grows → two sperm enter ovule → egg is fertilized and becomes embryo → polar bodies are fertilized and become endosperm → the entire ovule becomes a seed, and the ovary becomes a fruit.

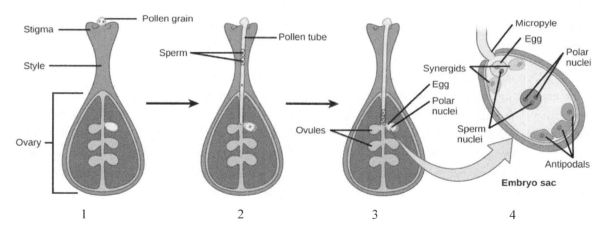

1

The pollen grain adheres to the stigma, which contains two cells: a generative cell and a tube cell.

2

The pollen tube cell grows into the style. The generative cell travels inside the pollen tube. It divides to form two sperm.

3

The pollen tube penetrates an opening in the ovule called a micropyle.

4

One of the sperm fertilizes the egg to form the diploid zygote. The other sperm fertilizes two polar nuclei to form the triploid endosperm, which will become a food source for the growing embryo.

Seed development

In the days after double fertilization, the single-celled zygote of an angiosperm divides asymmetrically to form the proembryo and the suspensor. The *proembryo* is an undifferentiated ball of cells which the *suspensor*, analogous to an umbilical cord, anchors to the endosperm. At this stage, the embryo is globular. It begins to have a defined symmetry as cells near the suspensor differentiate into root cells, and those at the opposite end become shoot cells to establish the *root-shoot axis.*

After differentiation into the proembryo and the suspensor, one or two cotyledons develop, depending on whether the plant is a monocot or eudicot. The cotyledon of a monocot embryo rarely stores food. Instead, it absorbs food molecules from the endosperm and passes them to the embryo. Conversely, the eudicot embryo has two cotyledons which usually store nutrients for the embryo. The cotyledons are noticeable in a eudicot embryo and may fold over. Whether monocot or eudicot, the embryo has a heart-shape when the cotyledons appear but then grows to a torpedo shape. With elongation, the root and shoot apical meristems become distinguishable at the ends of the plant. The embryo continues to differentiate into three regions. The *epicotyl* is the embryonic shoot tip above the cotyledons; it contributes to shoot and leaf development. Below the cotyledons is the embryonic stem, the *hypocotyl*. At the bottom is the *radicle*, the embryonic root.

Seed germination requires certain conditions to be fulfilled; otherwise, the seed remains dormant, undergoing sufficient respiration and metabolism to keep the seed alive but at a low level. Seeds retain their viability for varying times: maple seeds last a week, while lotus seeds are viable for hundreds of years.

Dormancy allows the seed to survive adverse conditions and ensure germination begins during optimum conditions. One such condition is sufficient oxygen so that the plant can undergo aerobic cell respiration to provide energy until the first leaves emerge. Once the leaves emerge, photosynthesis can then provide the energy needed for growth if there is sufficient light. Appropriate temperatures are needed to allow enzymes to act. If it is too hot or too cold, the enzyme activity is too low for germination. Seeds do not germinate if there is not sufficient water, needed to rehydrate the cells of the seed and activate certain enzymes which initiate its metabolism. Water causes the seed to swell and burst out of its seed coat, enabling the seedling to emerge.

Once germination of a eudicot or gymnosperm embryo begins, the radicle is the first to emerge and burrow down into the soil; this becomes the taproot of the plant. The epicotyl then begins to grow upward through the soil. At this point, germination may proceed differently depending on the species. In some instances, the cotyledons travel with the epicotyl aboveground, while in others they remain underground. If the cotyledons are above ground, young leaves sprout from the tip of the epicotyl (now a rudimentary shoot, the *plumule*); the cotyledons may die or persist on the plant and become photosynthetic. If below ground, the cotyledons may die or convert into storage organs.

In monocots, the plumule and radicle are enclosed in protective sheaths as the *coleoptile* and the *coleorhiza*, respectively. The radicle bursts from the coleorhiza upon germination, while the plumule escapes the coleoptile when it reaches the surface of the soil, upon which the first true leaves burst through their casing.

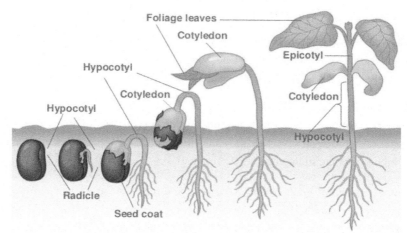

Seed germination from radicle (the embryonic root of the plant) through hypocotyl, cotyledon and flowering leaves

Fruit types and seed dispersal

Fruit is a mature ovary that encloses seeds to protect them and facilitate seed dispersal. They develop as the ovary wall ripens and thickens into a protective and nutritious covering as the *pericarp*. Other flower parts may become incorporated into the fruit. Often, they, rather than the ovary, are the bulk of the fruit and is an *accessory fruit.* There are two types of fruit – simple fruits and compound fruits.

Simple fruits develop from a single carpel, made of one ovary or several fused together. Legumes and cereal grains are examples of simple *dry fruits*; such fruits are often mistaken for seeds because they have a dry pericarp which tightly adheres to the seed within. Simple *fleshy fruits* can be identified by their fleshy pericarp (e.g., tomatoes, plums and peaches).

A *compound fruit* develops from several individual ovaries. If the ovaries are all present in a single flower, it is an *aggregate fruit* (e.g., a blackberry). Several flowers may ripen together to form *multiple fruits* (e.g., a pineapple). An aggregate fruit develops from many ovaries on a single flower, while multiple fruits develop from many flowers on a single plant. In both cases, the fruit is made of many different fruits which are fused.

The goal of a fruit is to facilitate seed dispersal. Dispersal moves seeds some distance from the original plant, expanding the species' range, incorporating greater genetic diversity and reducing intraspecies competition. Fruits accomplish this in a variety of ways, using other organisms or the plant's agency. For example, plants such as clover, bur, and cocklebur have hooks and spines, which attach to the fur of animals. Other plants have lightweight hairs, plumes or wings that disperse in the wind. Note that these fruits do not resemble the familiar apples or bananas; they are completely inedible to humans and many other animals. Many fruits are dispersed by water or simply by gravity, as they fall to the ground. Some plants can forcefully eject their seeds from the plant rather than relying on gravity alone.

Plants utilize other organisms, especially birds and mammals, by making their fruits enticing. Edible fruits are eaten by birds and mammals and defecated some distance away. In some cases, they may be buried by rodents like squirrels.

Asexual reproduction in plants

In addition to sexual reproduction, many plants are capable of asexual reproduction which can confer significant advantages. Many algae and bryophytes use asexual reproduction by *fragmentation* when they detach *gemmae cups,* single cells or masses of cells that develop into new individuals.

The vascular plant version of fragmentation is *vegetative propagation,* which utilizes the meristematic tissue of a parent plant to produce clonal offspring. For example, the eyes on a potato plant tuber undergo vegetative propagation when they bud to produce new plants. Stolons and rhizomes are another way in which plants facilitate vegetative propagation.

This strategy can be of great benefit because it is rapid, less energy-consuming, and less reliant on pollinators than sexual reproduction. It allows the species to expand its range and ensure survival by creating a *clonal colony.* A plant on its own may die in harsh conditions and be unable to regrow, meaning its genes fail to propagate. However, a colony of identical clones greatly increases the chance that at least one plant survives and continues to propagate the colony's genome.

Unfortunately, vegetative propagation can have adverse side effects because it rapidly multiplies the effects of the disease. Plants are vulnerable to disease because their cells are connected via plasmodesmata, creating a superhighway for pathogens throughout the entire organism. Vegetative propagation enhances the likelihood that the parent plant spreads any pathogens to the entire clonal colony, leading to a mass die-off.

Some plants have developed a workaround to this downside, *apomixis,* a form of asexual reproduction that involves seeds. Rather than propagating itself by sprouting clones from its body, the plant disperses seeds as it would in sexual reproduction. However, the seeds are not formed from a combination of sperm and eggs but are generated only from the mother. Normally, once the mother generates a megaspore within an ovule, it undergoes meiosis to generate four haploid spores. During apomixis, however, the megaspore divides by mitosis instead to form two diploid spores. These spores develop into gametophytes which are capable of autonomously generating an embryo without fertilization from a father plant. Asexual reproduction by apomixis lowers the risk of disease in vegetative propagation but has the disadvantage of decreased genetic diversity and a slow evolution.

Apomixis is of great interest to agricultural scientists who wish to engineer sexually reproductive crops to reproduce via apomixis instead, making crop production more efficient. Most interferences in plant reproduction have historically focused on breeding or vegetative propagation. Humans have been generating plants for thousands of years by planting cuttings from the parent plant. Modern day advances in tissue culture and manipulation of plant hormones have resulted in new methods for producing commercially valuable plants.

Today, hybridization has become highly sophisticated. Science is capable of genetic engineering, in which genes are altered to create *transgenic plants* with traits such as insect, pathogen, herbicide, drought or frost resistance. Scientists are currently attempting to introduce C_4 photosynthesis and improve the efficiency of water and CO_2 uptake in crop plants, to improve agriculture. They wish to induce plants to produce chemicals with industrial applications. The field of genetic engineering has medical applications as well: researchers have designed plants which produce certain antibodies that are promising treatments for cancer and other diseases. As molecular biology and genetics become ever more advanced, the possibilities continue to grow.

Chapter 9

Photosynthesis

- **Photosynthetic Organisms**

- **Photosynthetic Reactions**

- **Types of Photosynthesis**

Photosynthetic Organisms

Photosynthesis is the conversion of solar energy into chemical energy, which is then used to assemble the organic molecules which fuel an organism's metabolic activities. Photosynthesis includes the *light-dependent reactions*, when solar energy is captured, and the *Calvin cycle* (light-independent or "dark" reactions) when carbohydrates are synthesized. Only plants, algae, and certain bacteria are capable of carrying out photosynthesis, but all organisms perform cellular respiration (i.e., the breakdown of nutrients into energy). In eukaryotes, photosynthesis occurs in chloroplasts, while cellular respiration occurs in mitochondria. Photosynthesis is considered a "backbone process" as it is crucial for life on Earth.

Photosynthesis is a complex metabolic pathway, but can be expressed as:

light energy + carbon dioxide + water → carbohydrate + oxygen + water

light energy + 6 CO_2 + 12 H_2O → $C_6H_{12}O_6$ + 6 O_2 + 6 H_2O

Cellular respiration breaks down carbohydrates (e.g., glucose) produced from photosynthesis into energy (ATP) to fuel metabolic activities:

carbohydrate + oxygen → carbon dioxide + water + energy

$C_6H_{12}O_6$ + 6 O_2 → 6 CO_2 + 6 H_2O + ~34 ATP

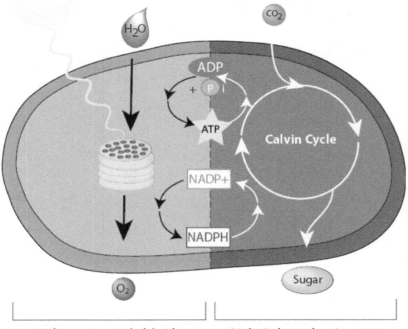

Light reaction in thylakoid   Light-independent in stroma

Light-dependent and light-independent reactions during photosynthesis

Photosynthesis as the transformation of solar energy

Solar radiation is composed of a range of wavelengths, most of which are filtered out as they pass through the atmosphere. 42% of the total light from the sun that hits the Earth's atmosphere reaches the surface. The ozone layer filters the high-energy wavelengths (e.g., gamma rays, x-rays, and UV radiation). Many of the low-energy wavelengths (e.g., radio waves, microwaves and some infrared radiation) are also filtered. Visible light is the majority of solar radiation reaching Earth's surface.

Einstein named *photons* as the discrete particles of light. Today, a photon is considered both a particle and a wave, as it exhibits properties of both. The energy of photons varies with their wavelength. Longer wavelength light has lower energy, while shorter wavelength light has higher energy. *Excitation* is when a molecule absorbs light energy, and the energy levels of its electrons are elevated. *Fluorescence* is when the energy is emitted immediately upon absorption. If it is emitted after a delay, then the effect is *phosphorescence*. Light energy may be converted into many forms; a property exploited by photosynthesis.

Primary producers (autotrophs) are organisms that transform inorganic elements of their surroundings into organic compounds. Nearly all primary producers are *photosynthetic autotrophs* that use light energy to create carbohydrates. Other producers are *chemosynthetic autotrophs* that use inorganic chemical reactions to create organic compounds. Except for the rare life based on chemosynthetic autotrophs, all food chains rely on the photosynthesizing organisms. The organic molecules created by photosynthesis fuel both the primary producers and all organisms above them in the food chain.

The rate of photosynthesis by an organism can be determined by measuring oxygen production and carbon dioxide uptake, or indirectly by an increase in biomass. For example, the oxygen bubbles released by aquatic plants during photosynthesis can be collected and measured to determine oxygen production. Measuring the uptake of carbon dioxide is more difficult and is usually done indirectly. When plants absorb carbon dioxide from water, the pH of the water rises, so measuring pH levels indicate carbon dioxide uptake.

Abiotic factors of photosynthesis

There are four abiotic (nonliving) factors necessary for photosynthesis: carbon dioxide (CO_2), water (H_2O), light and temperature. These elements influence the efficiency of photosynthesis. Usually, one of these is the *limiting factor* in a plant at a given time.

Earth's atmosphere contains approximately 78% nitrogen and 21% oxygen, with the remaining 1% being a mixture of gases such as carbon dioxide. CO_2 in the atmosphere reaches photosynthetic tissues via *stomata* (openings) on the underside of plant leaves. Carbon dioxide then dissolves in a thin film of water that covers the outside of leaf cells and diffuses through the cell walls to reach the chloroplasts. The rate of photosynthesis increases with CO_2 concentration but eventually levels off at high concentrations.

Another abiotic factor, water, may or may not be plentiful at the location of an individual plant. In hot, dry climates, plants often close their stomata to conserve water, although this reduces the CO_2 supply to the chloroplasts. Less than 1% of the water that is absorbed by plants is used in photosynthesis;

the remainder is either *transpired* (evaporated from leaves) or incorporated into cell components. The water utilized in photosynthesis is the source of the O_2 gas byproduct.

Light is a crucial component of photosynthesis. Light wavelengths within the visible light spectrum, which ranges from the red light at 780 nm to violet light at 390 nm, are the only forms of solar radiation useful for photosynthesis.

Mnemonic ROY G BIV (red, orange, yellow, green, blue, indigo, violet) describes the order of colors according to decreasing wavelength (or increasing energy). Wavelength and energy are inversely proportional. In the visible spectrum, red has the longest wavelength but the lowest energy.

Of the visible light that reaches a leaf, approximately 80% is absorbed. Light intensity may vary widely depending on the time of day, temperature, season, altitude, latitude and other atmospheric conditions. While photosynthetic pigments utilize the entire spectrum of visible light, the rate of photosynthesis varies across different wavelengths. Violet-blue light (400 to 525 nm) and orange-red light (625 to 700 nm) are the wavelengths most often absorbed for photosynthesis. Light intensity is a limiting factor because if there is no sunlight, then *photolysis* (splitting of water by photons) cannot occur during the light-dependent reactions. This results in a shortage of ATP and NADPH, products of the light-dependent reactions necessary for the Calvin cycle. At low and medium light intensity, the rate of photosynthesis is directly proportional to the light intensity. However, high-intensity light is not necessarily beneficial for plants, as it can decrease photosynthetic efficiency.

Temperature, like the other abiotic factors, has an optimum range. At low temperatures, the enzymes involved in photosynthesis work slowly and therefore less efficiently. As temperature increases, the rate of photosynthesis increases steeply until the optimum temperature is reached. If temperature increases beyond this point, then the rate of photosynthesis begins to decrease rapidly.

Plants as photosynthesizers

Plant photosynthesis requires the intake of CO_2 and H_2O. Absorption of water is primarily handled by roots which move the water up through vascular tissue in the stem until it reaches the leaves. Within the leaves are specialized *mesophyll cells*, where photosynthesis takes place. The exchange of O_2 and CO_2, as well as some water, occurs through the stomata. These pores are opened and closed by special *guard cells*. The density of stomata is dependent upon ecological conditions like humidity and CO_2 concentration.

The plant's leaves, composed of the *lamina* (blade) and the *petiole* (stalk), are its primary photosynthetic organs and typically have a large surface area to maximize light harvesting. A *simple leaf* has one lamina, while a *compound leaf* has many distinct laminae. Leaves can be highly variable in shape and are usually thin so that light can penetrate to cells on the underside. The underside of leaves is often covered with hairs as *trichomes*, which serve to catch water, reduce airflow and produce wax. The waxy cuticle covers the outer epidermis of the leaf to prevent water loss but has the tradeoff of limiting gas exchange.

The upper and lower epidermises of the leaf feature the stomata and serve a protective function, while loosely arranged *spongy mesophyll* tissue creates air spaces. *Palisade mesophyll* is more tightly packed and contains the highest concentration of chloroplasts. Chloroplasts usually remain near the cell wall, since this arrangement guarantees optimal use of light.

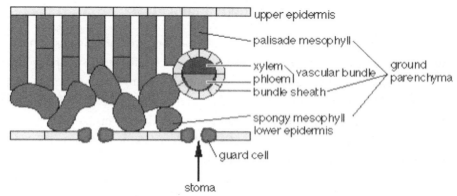

Cross section of a typical (C4) plant's leaf

Once CO_2 and H_2O have entered a mesophyll cell, they diffuse into the chloroplasts, the site of both the light-dependent reactions and the Calvin cycle. A chloroplast is a double-membraned organelle, a *plastid,* each of which has its ribosomes and identical copies of a double-stranded, circular DNA molecule unique to the cell's DNA.

The membrane structure of a chloroplast includes an outer plasma membrane, an intermembrane space, and an inner plasma membrane. Within the inner membrane, a fluid-filled space of the *stroma* serves as the site of the Calvin cycle. *Thylakoids* are flattened disc-like sacs, which are often organized into stacks, *grana,* also reside in the stroma. The light-dependent reactions occur in the thylakoids. These structures have a large surface area for light absorption and an inner *lumen* where protons accumulate. The accumulation of protons in the thylakoid lumen creates an electrochemical gradient used in photosynthesis. Chlorophyll and other pigments involved in the absorption of solar energy are embedded in the thylakoid membranes.

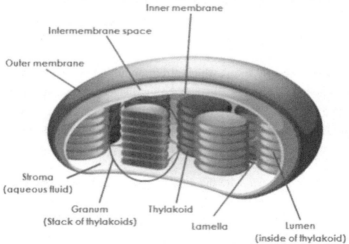

Structure of chloroplast as a double-membered structure showing stroma as the site of the Calvin cycle

Photosynthetic Pigments

Every pigment molecule has a distinctive *absorption spectrum* for light absorption. A graph of the percent of light absorbed at each wavelength is used to visualize this spectrum. Photosynthetic activity under different wavelengths of light can be plotted on an *action spectrum* graph. Action spectrums, which resemble absorption spectrums, indicate that chlorophyll is the main pigment involved in photosynthesis. It is the most abundant photosynthetic pigment in plants, which is why the rate of photosynthesis is highest at the low wavelengths of visible light (400 to 525 nm). These wavelengths, which comprise violet and blue light, are the most readily absorbed by chlorophyll. Chlorophyll absorbs an appreciable portion of red and orange light, which is why the significant photosynthetic activity is found there.

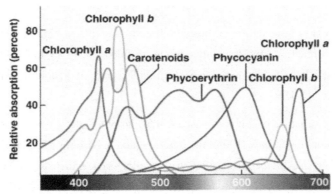

Relative absorption of visible light wavelengths (nm) by plant pigments

Very little light is absorbed by chlorophyll at wavelengths of yellow and green light (525 to 625 nm) and most is reflected, giving plants their green color. Photosynthetic organisms that are not green utilize alternate photosynthetic pigments. For example, red algae use *phycobilins,* which absorb blue, yellow and green light and reflect red light.

Chlorophyll exists in several forms which differ slightly in molecular structure. All chlorophyll molecules have a long lipid tail that anchors them in the lipid layers of thylakoid membranes, a *porphyrin ring* of alternating double and single bonds, and a single atom of magnesium in the center. Chlorophyll is analogous to *heme,* the iron-containing pigment in hemoglobin protein within the red blood cells which carries O_2.

Chlorophyll a and *chlorophyll b* are the most common forms, with chlorophyll *a* generally being three times as abundant as chlorophyll *b*. Chlorophyll *a,* a bluish-green pigment, has the formula $C_{55}H_{72}MgN_4O_5$ while chlorophyll *b*, which is yellow-green, is $C_{55}H_{70}MgN_4O_6$. The main role of chlorophyll *b* is to broaden the spectrum of light available for photosynthesis. It absorbs light energy and transfers the energy to a chlorophyll *a* molecule.

Green plants contain other pigments that contribute to photosynthesis. *Carotenoids* are yellow-orange pigments that absorb light in the violet, blue and green ranges of the spectrum. When chlorophyll breaks down in the autumn, plant leaves turn yellow, orange and red as carotenoids become visible. Even though carotenoids are present in small amounts, they allow a low rate of photosynthesis to occur at wavelengths of light that chlorophyll cannot absorb. *Accessory pigments* (antenna pigments) are the light-absorbing pigments that aid chlorophyll *a*, including chlorophyll *b, c, d,* and carotenoids.

Light-harvesting complexes (antenna complexes) contain 250 to 400 antenna pigment molecules and other compounds. They surround the *reaction center,* a pair of special chlorophyll *a* molecules. Together, the antenna complexes and reaction center make up a *photosystem.* In the chloroplasts of green plants, two photosystems, *Photosystem I* (PS I) and *Photosystem II* (PS II), operate together to initiate the first phase of photosynthesis. There are countless numbers of these two photosystems throughout the grana (granum, singular) of the chloroplasts.

Photosynthetic Reactions

Light-dependent reactions require solar energy as energy-capturing reactions. The primary function of the light-dependent reactions is to trap solar energy and store it as chemical energy as ATP or NADPH. These reactions begin with light striking the chlorophyll molecules. Subsequent reactions result in the conversion of some light energy to chemical energy.

When solar energy excites electrons in the antenna complex, it is *photoactivated.* The energized electrons then pass from one pigment molecule to the next until they reach the special chlorophyll *a* pair at the reaction center. The reaction center then passes on the excited electron to the first in a chain of electron carriers called the *electron transport chain* (ETC). These excited electrons in the chlorophyll are unstable and re-emit absorbed energy as they travel from molecule to molecule along the electron transport chain. The result is the production of ATP and NADPH. The light-reactions comprise all events from initial photoactivation of a photosystem to ATP and NADPH synthesis. This process is constant during daylight hours as the plant produces ATP and NADPH for the Calvin cycle.

Two photosystems play critical roles in the light-dependent reactions of photosynthesis. Photosystem I (PS I) is best excited by light at about 700 nm, while Photosystem II (PS II) cannot use photons of wavelengths longer than 680 nm. Each photosystem has an antenna complex composed of chlorophyll, carotenoids, accessory proteins and cofactors such as magnesium and calcium.

Photosynthesis then proceeds to the Calvin cycle reactions in the stroma, where the NADPH and ATP created by the light-dependent reactions is used to reduce CO_2 into carbohydrates. Oxidation-reduction reactions are throughout the processes of photosynthesis and respiration. Oxidation results in the net loss of an electron or electrons, while reduction results in the net gain of an electron or electrons. The mnemonic device "OIL RIG" stands for Oxidation Is Loss and Reduction Is Gain.

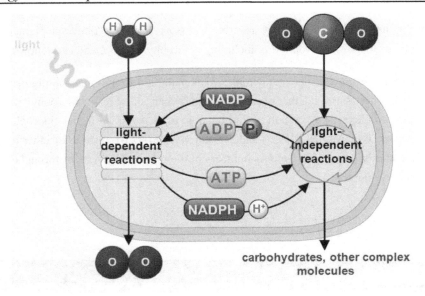

Examples of oxidation-reduction reactions in photosynthesis:

- Overall, photosynthesis is a redox reaction: water is oxidized to oxygen and CO2 is reduced to sugar and water

- During photolysis, water is oxidized by the energy of photons into oxygen, protons, and electrons

- During the electron transport chain, each molecule in the chain is reduced (gain an electron) as it receives an electron and is then oxidized (loses an electron) as it passes the electron to the next

- At the end of the electron transport chain, $NADP^+$ is reduced to NADPH

- During the Calvin cycle, CO_2 is reduced into sugar and water

Light-Dependent Reactions

There are two sets of reactions in the thylakoid membrane: the *noncyclic electron pathway* and the *cyclic electron pathway*. Both are light-dependent reactions. Each pathway produces ATP, but only the noncyclic electron pathway produces NADPH. *Photophosphorylation* is ATP production via photosynthesis, so the two pathways are the *cyclic* and *noncyclic photophosphorylation*.

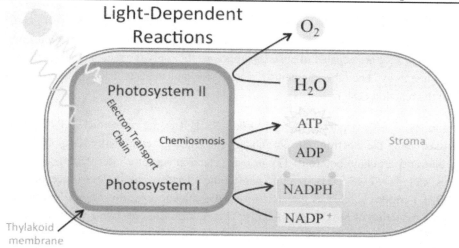

Light-dependent reactions within the chloroplast

Noncyclic Electron Pathway

The noncyclic electron pathway is a light-dependent reaction in thylakoid membranes and requires the participation of both PS I and PS II. Despite their names, the pathway begins at PS II, which contains chlorophyll *a*, a carotenoid called *beta-carotene*, several electron-carrier proteins and numerous cofactors. The reaction center of PS II is a special pair of chlorophyll *a* molecules known as *P680*. The noncyclic electron pathway begins when a photon of light strikes accessory pigments surrounding P680 near the inner surface of a thylakoid membrane. The light energy excites the electrons of these pigments, which transfer their energy to the reaction center to excite electrons in P680. This is an unstable reaction, and thus most of the energy is lost to heat. Up to four photons at a time may strike PS II, but P680 accepts one electron at a time.

Once excited, the electron leaves the chlorophyll and passes from one molecule to the next along the electron transport chain. The first molecule in this chain, the primary electron acceptor, is *pheophytin*. Pheophytin accepts the excited electron and transfers it to the next acceptor, the *plastoquinone complex*. Plastoquinone then delivers the electron to the *cytochrome complex*, which is passed to *plastocyanin*, and finally PS I.

The electrons are passed through a chain of oxidation-reduction reactions. After each molecule in the electron transport chain passes an electron to the next, it is reoxidized, readying it to accept a new electron from the molecule behind it in the chain. However, for P680 to continually send electrons down the chain, it must be resupplied. This is accomplished by the splitting of water by photons (i.e., photolysis). Photolysis occurs in the *oxygen-evolving complex* (OEC), which contains manganese and an enzyme complex. Photolysis produces 4 electrons, 4 protons, and O_2. The 4 electrons are transferred to P680; the protons remain in the thylakoid lumen. The O_2 is released into the atmosphere as a waste product of this process.

As electrons are traveling along the electron transport chain, protons (H^+ ions) are pumped across the thylakoid membrane into the thylakoid lumen, creating a proton gradient. The protons can travel back across the membrane, down their concentration gradient, but to do so, they must pass through ATP synthase, generating ATP. The synthesis of ATP via the noncyclic electron pathway is noncyclic photophosphorylation.

The activity occurring at PS I resembles that of PS II, with chlorophyll *a,* beta-carotene, electron carriers, and cofactors. The PS I reaction center is *P700.* It is a special pair of chlorophyll *a* molecules. Like PS II, pigments in PS I are capable of absorbing photons, but only the reaction center molecules can truly utilize the light energy. The other pigments are accessory molecules which help harvest light energy and transmit it to the reaction center.

Just as in PS II, after a photon of light strikes PS I, it excites electrons in P700. The excited electrons are passed along the electron transport chain. The electrons continually resupply P700 shuttled along the electron transport chain before it. The primary electron acceptor after P700 is a series of iron-sulfur molecules called *4Fe-4S.* 4Fe-4S then passes the electrons to the next acceptor molecule, *ferredoxin* (NADP). Ferredoxin releases the electrons to the enzyme *ferredoxin-NADP reductase* (FNR), which catalyzes the reduction of $NADP^+$ to NADPH. This process requires two electrons and one proton ($NADP^+ + 2e^- + H^+ \rightarrow NADPH$). NADPH is integral in providing hydrogen ions to the second series of major photosynthetic reactions: the Calvin cycle or *carbon-fixing reactions.*

The flow of electrons in the noncyclic pathway is linear:

Photon strikes antenna complex of PS II $\rightarrow$ 4 electrons are released $\rightarrow$ P680 $\rightarrow$ Pheophytin $\rightarrow$ Plastoquinone $\rightarrow$ Cytochrome complex $\rightarrow$ Plastocyanin $\rightarrow$ P700 $\rightarrow$ 4Fe-4S $\rightarrow$ Ferredoxin $\rightarrow$ 2 NADPH

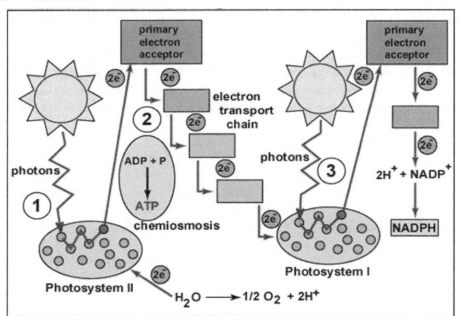

Light-dependent reaction with photosystem II and photosystem I as reactive centers to absorb electrons

Cyclic Electron Pathway

Sometimes an organism has the reductive power (NADPH) that it needs to synthesize new carbon structures, but still needs ATP to power other activities in the chloroplast. If the light intensity is not a limiting factor, there usually is a shortage of $NADP^+$ as NADPH accumulates within the stroma. $NADP^+$ is needed for the normal flow of electrons in the thylakoid membranes, as it is the final electron acceptor. If $NADP^+$ is not available, then the normal flow of electrons is inhibited. This promotes the *cyclic electron pathway*, which bypasses NADPH production in favor of ATP synthesis only.

The cyclic electron pathway begins when P700 receives an electron from plastocyanin and passes it to ferredoxin. However, because $NADP^+$ is not available in this case, ferredoxin does not pass electrons onward to $NADP^+$. Rather, it returns the electrons to the cytochrome complex. This causes the cytochrome complex to pump protons across the thylakoid membrane and create a proton gradient to drive ATP synthesis. The electrons are then returned to P700 via plastocyanin, and the process repeats.

The process is cyclic because the electrons return to PS I rather than move on to $NADP^+$; in this case PS I produces the only ATP. In the cyclic electron pathway, PS I receives an electron not from PS II but from itself. The electron must be recycled continuously. The role of PS I as ATP producer may seem counterintuitive, since in the noncyclic electron pathway it is responsible for NADPH production, while PS II produces ATP. However, PS I is excellent at transferring an electron, as it is a powerful reducing agent (electron donor) and passes an electron to ferredoxin rather than $NADP^+$ to produce ATP when NADPH levels are high.

Chemiosmosis

Both the noncyclic and cyclic electron pathways produce ATP by generating a proton gradient across the thylakoid membrane. The movement of protons down their concentration gradient across the membrane is *chemiosmosis*. Chemiosmosis drives ATP synthesis.

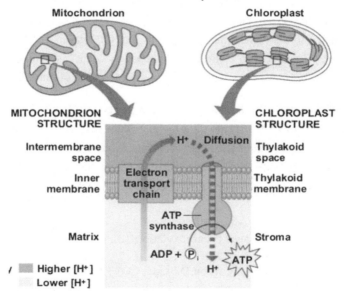

Chemiosmotic gradient comparison for cellular respiration (mitochondria) and photosynthesis (chloroplast)

Cyclic photophosphorylation is driven by the events along the cyclic pathway between PS I and cytochrome. The proton-motive force produced as electrons travel from P700 to ferredoxin to cytochrome causes cytochromes to pump protons into the lumen. The thylakoid lumen acts as a reservoir for H^+ ions. The concentration gradient created by the accumulation of H+ in the lumen causes the H^+ to move back across the thylakoid membrane to the stroma, down the concentration gradient. From chemiosmosis, ATP synthase uses the energy released from the movement of hydrogen ions down their concentration gradient to synthesize ATP from ADP and inorganic phosphate.

Noncyclic photophosphorylation is a more complex process. Hydrogen ions enter the thylakoid lumen at two points along the noncyclic electron pathway. The first occurs during photolysis when 2 H_2O is split into O_2, 2 electrons, and 2 H^+. The O_2 is a waste product and is released into the atmosphere. The electrons are donated to P680, which resides in the transmembrane PS II complex. Meanwhile, the H^+ are released into the thylakoid lumen.

The electrons travel from P680 to pheophytin to plastoquinone and then reach the cytochrome complex. The electron transfer from plastoquinone to cytochrome, via an intermediate carrier molecule, creates energy which pumps more H^+ from the stroma into the thylakoid lumen. As in cyclic photophosphorylation, the protons move down their concentration gradient and drive ATP synthase. The hydrogen ions return to the stroma to synthesize NADPH from $NADP^+$ and 2 electrons from PS I.

The Calvin Cycle Reactions

Both ATP and NADPH are important products of light-dependent reactions, and both are used in the synthesis of carbohydrates from atmospheric CO_2. The reactions that accomplish this have historically been known as light-independent (or dark) reactions. Despite their names, these reactions do not typically occur at nighttime. In fact, they indirectly require light, because they rely on light-dependent reactions which only take place during daylight. For this reason, the term "dark reactions" is no longer used, and some scientists even claim that "light-independent" is a misnomer. In recent years, the term *carbon-fixing reactions* have emerged as a more apt description of the Calvin cycle.

These reactions take place in the stroma of chloroplasts and occur if the end products of the light-dependent reactions are available. Depending on the plant involved, the carbon-fixing reactions may progress in different ways. Commonly, CO_2 from the atmosphere is combined with a 5-carbon sugar of *ribulose-1,5-bisphosphate* (RuBP). The CO_2 and RuBP are converted via several steps into a 6-carbon sugar such as glucose. Some of the sugars are further combined into polysaccharides for storage within the plant.

The Calvin cycle has three stages: (1) carboxylation, (2) reduction, (3) regeneration.

1) Carboxylation or Fixation of Carbon Dioxide: CO_2 + RuBP $\Rightarrow$ PGA

   Carboxylation is the attachment of CO_2 to 5-carbon RuBP to form an unstable 6-carbon intermediate. The enzyme *ribulose-1,5-bisphosphate carboxylase/oxygenase* (RuBisCO) catalyzes this reaction. RuBisCO comprises 20–50% of the protein content of chloroplasts. Its ubiquity most likely makes it the most common protein in the world. As soon as the 6-carbon intermediate is formed, it splits to form two molecules of the 3-carbon compound *3-Phosphoglyceric acid* (3-PG, 3-PGA, or simply PGA). 3-Phosphoglyceric acid is *glycerate 3-phosphate*.

2) Reduction of 3-Phosphoglyceric acid: PGA + ATP + NADPH $\Rightarrow$ G3P + ADP + Pi + NADP+

Using ATP produced by light-dependent reactions, each PGA is phosphorylated to form an intermediate. The intermediate is then reduced by NADPH, the other product of light-dependent reactions, to form *glyceraldehyde 3-phosphate*. This is a 3-carbon compound, just like PGA. Glyceraldehyde 3-phosphate (G3P, GP, GA3P, or GAP) is *triose phosphate* (TP) and *3-phosphoglyceraldehyde* (PGAL). Along with each G3P, the reduction produces ADP, inorganic phosphate (Pi) and NADP$^+$.

G3P can be converted into many useful molecules, such as glucose. Glucose can be combined with fructose to form sucrose, an important plant carbohydrate. Glucose is the starting point for the synthesis of polysaccharides (i.e., starch and cellulose).

3) Regeneration of RuBP: G3P + ATP $\Rightarrow$ RuBP + ADP + Pi

However, only one G3P can be used for conversion into glucose. The remaining G3P is used to regenerate RuBP, which is essential for carbon fixation to continue. These G3P are met with a carbon acceptor and undergo a series of reactions, requiring energy from ATP, which convert them into RuBP. At this point, the cycle begins again.

While this description followed the fate of one original RuBP molecule, in reality, the process uses six RuBP and six CO_2 for each cycle.

Carboxylation converts six 5-carbon RuBP and six CO_2 into six 6-carbon unstable intermediates. The six intermediates then break down into twelve 3-carbon PGA.

$$6 \text{ RuBP} + 6 \text{ CO}_2 \rightarrow 12 \text{ PGA}$$

During reduction, the 12 PGA plus 12 ATP and 12 NADPH are converted into 12 3-carbon G3P, 12 ADP, 12 Pi, and 12 NADP$^+$. Two of the G3P create glucose phosphate.

$$12 \text{ PGA} + 12 \text{ ATP} + 12 \text{ NADPH} \rightarrow 12 \text{ G3P} + 12 \text{ ADP} + 12 \text{ Pi} + 12 \text{ NADP}^+$$

Finally, during regeneration the other 10 G3P, using 6 ATP, are remade into 6 RuBP, producing 6 ADP and 4 Pi as a result.

$$10 \text{ G3P} + 6 \text{ ATP} \rightarrow 6 \text{ RuBP} + 6 \text{ ADP} + 4 \text{ Pi}$$

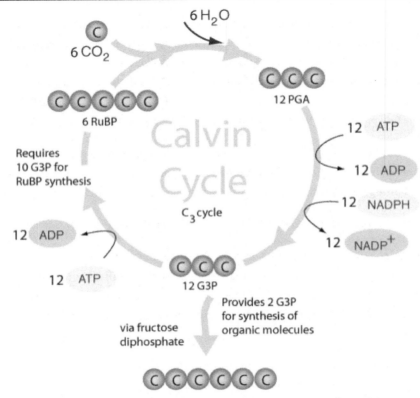

Calvin cycle (light-independent) synthesizes sugars from CO_2

Types of Photosynthesis

Not all plants perform the Calvin cycle. The terms C_3, C_4, and *CAM plant* describe the different ways in which plants perform photosynthesis. C_3 and C_4 refer to the carbon length of the first photosynthetic carbohydrate product. C_3 plants produce a 3-carbon molecule, 3-phosphoglyceric acid (PGA), and C_4 plants produce a 4-carbon molecule, *oxaloacetate* (OAA). Like C_4 plants, CAM plants produce oxaloacetate but undergo photosynthesis differently.

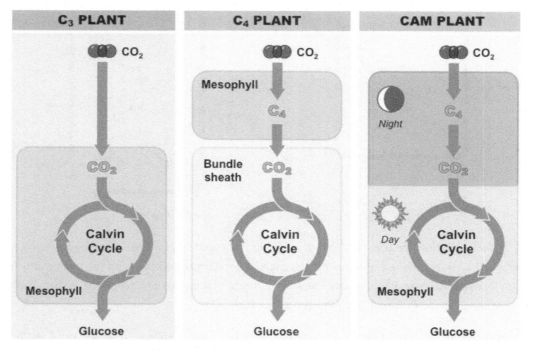

Comparison of C_3, C_4 and CAM plants for the synthesis of glucose

C3 Photosynthesis

More than 90% of all angiosperms are C_3 plants. In C_3 plants, the Calvin cycle proceeds as described, in which CO_2 is fixed directly into the 3-carbon molecule PGA.

In hot weather, stomata close to save water, but this has the disadvantage of decreasing CO_2 concentration in leaves while increasing O_2 concentration. High O_2 concentration is a disadvantage because RuBisCO can use O_2 in the Calvin cycle rather than CO_2. Using O_2 in this way is *photorespiration* because it involves intake of O_2 and production of CO_2, like cellular respiration. During photorespiration, RuBP reacts with O_2 to create CO_2, in contrast to photosynthesis, when RuBP reacts with CO_2 to form carbohydrates. The former is an oxygenation reaction, while the latter is a carboxylation reaction. RuBisCO performs oxygenation rather than carboxylation around 25% of the time, and this may be promoted by low CO_2, high O_2 or high temperature.

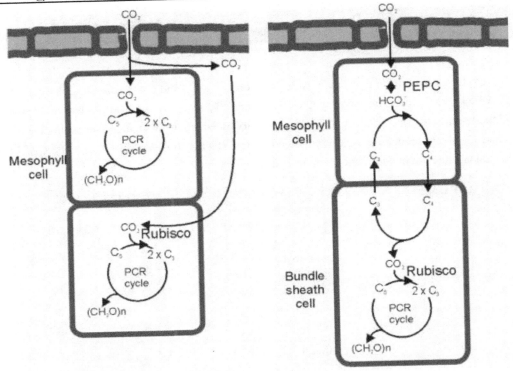

Comparison of C₃ (left) and C₄ plants for the fixation of CO₂

Photorespiration results in the production of one PGA and one *phosphoglycolate*. Phosphoglycolate is of little use in the plant and inhibits normal carboxylation, so the plant must spend energy to convert the phosphoglycolate back to a useful molecule and reclaim the two carbons. Peroxisomes break down the products of this process.

Photorespiration is a wasteful and inefficient process that results in a net energy loss for the plant. Plants must spend up to 40% of their energy stored in sugars to deal with the inevitable damage created by the oxygenation reaction.

The oxygenation reaction developed as an adaptation to Earth's early atmosphere. When photosynthesis first evolved billions of years ago, the concentration of atmospheric O_2 was low compared to today, making photorespiration rare. Photorespiration grew to be a great inconvenience to plants. Plants have since evolved ways to reduce the damage caused by O_2 in the Calvin cycle.

Humans have attempted to circumvent the issues created by the oxygenation reaction. Molecular genetics has been used to modify the properties of RuBisCO to eliminate the oxygenation reaction while retaining the carboxylation reaction. Results so far indicate that the two reactions cannot be separated, because modifications in RuBisCO that reduce oxygenase activity reduce carboxylase activity. Nature developed a system to avoid photorespiration.

C4 Photosynthesis

C_4 photosynthesis is in less than 5% of plants, mostly in hot, dry climates. It is especially well represented in the grasses. Corn and sugar cane, two of the most important crops, are C4 plants. Because C4 plants avoid photorespiration, their net photosynthetic rate may be 2 to 3 times that of a C3 plant. However, in moist or cold environments, C3 plants are more efficient. C_4 photosynthesis has evolved from C3 plants independently many times in the evolutionary timeline. All of the enzymes that were co-opted for C_4 photosynthesis were already present in the C3 plants from which they evolved. Researchers are currently attempting to artificially convert certain C3 plants into C4 plants to maximize crop yields and allow agriculture in more hostile environments.

While C3 plants have chloroplasts in their mesophyll cells, C4 plants have additional chloroplasts in special *bundle sheath cells*. These cells surround the veins of the leaves; an arrangement of *Kranz anatomy*. In the leaves of a C4 plant, mesophyll cells are arranged concentrically around the bundle sheath cells. The light-independent reactions are split between both cell types, with preliminary carbon fixation occurring in the mesophyll cells and the Calvin cycle occurring in the bundle sheath cells. Rather than RuBisCO, the mesophyll cells contain an alternate *phosphoenolpyruvate* (PEP) enzyme. RuBisCO is in the bundle sheath cells.

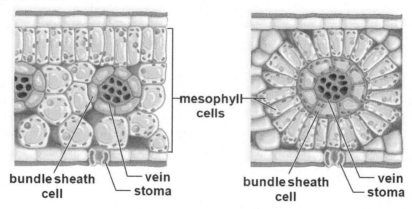

The comparison of the anatomy of C_3 (left) and C_4 plants.
C_4 photosynthesis uses carbon concentration to improve the efficiency of photosynthetic carbon fixation. The leaves of most C_4 plants have a Kranz leaf anatomy consisting of bundle sheath and mesophyll cells.

When CO_2 enters the leaves of a C4 plant, it is first absorbed by mesophyll cells, as in a C3 plant. However, rather than being fixed by RuBisCO into PGA, the CO_2 is combined with PEP to form 4-carbon *oxaloacetate* (OAA). This reaction is catalyzed by the enzyme *PEP carboxylase* (PEP-C or PEPCase). PEPCase has no affinity for O_2, unlike RuBisCO, and has a higher affinity for CO_2 at high temperatures than does RuBisCO. PEPCase assimilates carbon more efficiently while avoiding photorespiration.

Oxaloacetate is then usually converted to *malate,* a reduced form of oxaloacetate, and pumped into the bundle sheath cells. Some plant species reduce oxaloacetate to aspartate rather than malate, and many do not alter oxaloacetate before shuttling it to the bundle sheath cells.

Once in the bundle sheath cells, the malate is converted back to oxaloacetate and then decarboxylated (CO_2 is removed) to form 3-carbon *pyruvate*. The CO_2 remains in the bundle sheath cells to begin the Calvin cycle, while the pyruvate is returned to the mesophyll cells to be regenerated into PEP using ATP and its efficiency is lost in cooler temperatures.

The initial fixed carbon form, whether oxaloacetate, malate or aspartate, does not substitute for any of the carbon compounds of the Calvin cycle, such as RuBP or PGA. These acids merely serve as sources of CO_2 for the conventional Calvin cycle in the bundle sheath cells. CO_2 is fixed into RuBP as normal and then proceed throughout the cycle. In this regard, C_4 plants fix carbon twice: once in the mesophyll and once in the bundle sheath.

CAM Photosynthesis

CAM (Crassulacean acid metabolism) *photosynthesis* is another alternative to the C3 strategy, found almost exclusively in plants from dry environments. Some CAM plants include cacti, stonecrops (family Crassulaceae, for which the strategy is named), orchids, bromeliads, and succulents. Like C4 plants, CAM plants use PEPcase to fix CO_2 into oxaloacetate and separate this step from the Calvin cycle. However, while C4 plants separate them spatially, with the former occurring in mesophyll cells and the latter in bundle sheath cells, CAM plants separate them temporally. In this case, both steps proceed in the mesophyll cells, but initial CO_2 takes place at night, while the Calvin cycle is during the day.

In a CAM plant, stomata are opened only at night, when CO_2 is taken up into the plant and incorporated into the mesophyll. They proceed to use PEPCase to react CO_2 and PEP to form oxaloacetate, which they then convert into malate and store as the malic acid in large vacuoles in the mesophyll cells. During the day, malic acid is returned to the chloroplasts for conversion into oxaloacetate, then decarboxylated into pyruvate and CO_2. The CO_2 is then introduced into the Calvin cycle, while the pyruvate is used to regenerate PEP.

No CO_2 intake occurs during the daytime, as stomata are closed to avoid transpiration. The CAM strategy was evolved mainly to avoid transpiration rather than to reduce the effects of photorespiration, although a benefit as well. The major advantage of a CAM strategy over a C4 strategy is the ability to conserve water by closing stomata during the day. Photosynthesis in a CAM plant is minimal due to the limited amount of CO_2 fixed at night, but this does allow them to live under stressful conditions.

Please, leave your Customer Review on Amazon

Chapter 10

Ecosystems, Biosphere and Conservation Biology

- Ecosystems, Energy Flow and Nutrient Cycles

- The Biosphere and Biomes

- Conservation Biology

Ecosystems, Energy Flow and Nutrient Cycles

The nature of ecosystems

An *ecosystem* includes a biotic community as well as its abiotic environment. The biotic community is organized by which organisms fuel their metabolic activities. *Autotrophs* are organisms which create their organic compounds using energy from the sun or inorganic compounds. They can be divided into photoautotrophs and chemoautotrophs. *Photoautotrophs* (e.g., algae, plants, and cyanobacteria) use photosynthesis to convert solar energy into organic compounds. *Chemoautotrophs* are bacteria which oxidize inorganic compounds such as ammonia, nitrite, and sulfide to generate organic compounds. They are rare and are typically in caves, hydrothermal ocean vents and other environments lacking light. Autotrophs are the basis of the ecosystem and feed heterotrophs.

Heterotrophs must obtain their nutrients by consuming other organisms. Herbivores feed on autotrophs and are typically prey animals for carnivores and omnivores. *Detritivores* are organisms which break down dead organic matter, recycle energy and nutrients throughout the ecosystem. The smallest detritivores, such as fungi and bacteria, are *decomposers.* Many decomposers are *saprotrophs,* which digest organic matter externally by secreting enzymes into the surrounding environment and then absorbing the broken-down products.

Energy flow

Ecosystems are characterized by nutrient production, movement, consumption, and recycling. Nutrients are forms of chemical energy which flow throughout the system in predictable ways, governed by the laws of thermodynamics. The first law of thermodynamics states that energy can neither be created nor destroyed, while the second law states that when energy is transformed into a different form, some energy is lost to the environment.

Energy flow is assembled into the *food chain* of a community. *Trophic levels* are the food chains organized into levels which describe how an organism feeds. Arrows in the food chain show the direction of energy flow. The greater the number of energy pathways in a food web, the more stable is the community. At the base of every food chain are the autotrophs as *primary producers.* They are responsible for *primary production*, the creation of organic compounds using energy from the sun or inorganic compounds. The chemical energy created by primary production nourishes both the producers and other organisms in the food chain. It can be quantified by *biomass* or the amount of organic material in a given area. Biomass may be either living or dead; the only requirement is that it contains usable energy.

Herbivores that consume primary producers are *primary consumers.* Those organisms which consume the primary consumers are *secondary consumers,* and so on. Typically, food chains do not exceed the level of a tertiary consumer. At the top of the food chain are *apex predators,* organisms which have no natural predators.

As an example, a primary producer in a food chain may be a carrot plant. Its direct predator, the eastern cottontail rabbit, is a primary consumer. The red fox is a secondary consumer, and the golden eagle is the tertiary consumer and apex predator. However, golden eagles often prey on rabbits in addition to foxes. Here, the basic linear food chain fails to capture the complex interactions in an ecosystem. In reality, an ecosystem involves many food chains interconnected into a *food web.*

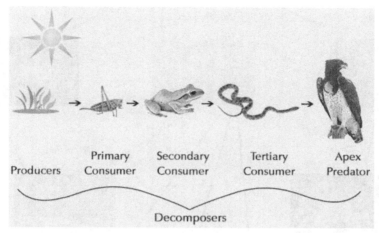

Example of a food chain from producers to apex predator

Detritivores are difficult to place on a food web and are often conceptualized as being separate. A food web which does not include detritivores is a *grazing food web;* it places primary producers at the basal level. A *detrital food web* places detritivores at the bottom.

The food web begins when producers receive energy from the sun through photosynthesis. *Gross primary productivity* is the total amount of energy they generate via photosynthesis. However, much of this energy must be used by the producers themselves to fuel their metabolism. The *net primary productivity* passed to heterotrophs is the unused energy.

Due to the second law of thermodynamics, energy is always lost at each trophic level, because all organisms lose heat through cellular respiration. Furthermore, some organic matter remains undigested at each trophic level and is lost as waste. Detritivores help recycle this energy.

Ecological efficiency describes the proportion of energy at each trophic level that is transferred to the next. On average, about 90% of an organism's energy is used for metabolism, and 10% is passed to the next trophic level. Therefore, 1,000 kilograms of plant biomass supports 100 kilograms of primary consumers, which converts to 10 kilograms of secondary consumers and finally 1 kilogram of tertiary consumers. Note that massive amounts of biomass are needed at the lower levels to support the apex predators at the top. This rapid loss of energy is why food chains rarely have more than four links. However, energy reclamation by decomposers greatly improves the efficiency of the overall ecosystem.

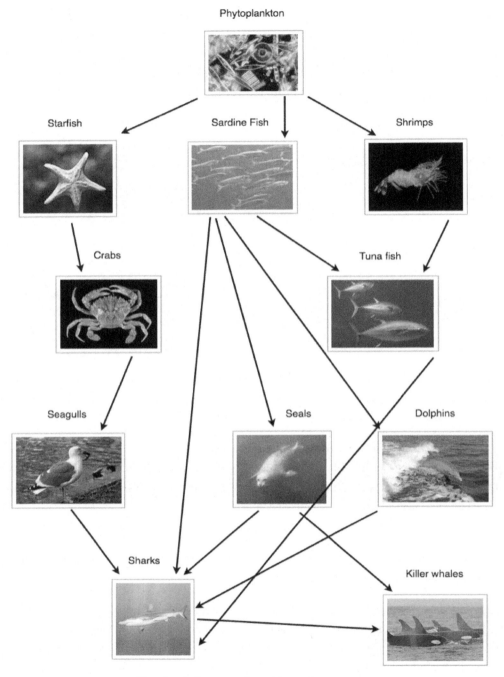

Food web from producer through consumers

Energy and biomass are greatest at the primary producer level and lowest at the apex predator level. The result is that apex predator populations are the least stable and the most heavily impacted by population fluctuations at lower trophic levels. *Ecological pyramids* represent the trophic levels in a food web by their energy content, biomass or number of species. Naturally, the base of the pyramid is the producer trophic level, with consumer trophic levels rising upwards.

An *ecological pyramid of numbers* is based on the number of organisms at each trophic level, while a *pyramid of energy* is based on how much energy each level generates. A *pyramid of biomass* is based on the biomass at each trophic level.

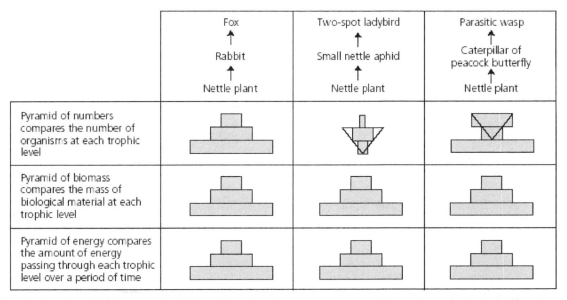

Ecological pyramids; the ecological pyramid of numbers, the pyramid of energy and pyramid of biomass

Most pyramids resemble a typical upright triangle. However, at particular points in time, the pyramid may have areas of inversion. For example, this occurs if an herbivore population feeds on a fast-growing producer, keeping its biomass low in comparison.

Biogeochemical cycles

Nutrients such as water, carbon, nitrogen and phosphorous are in limited supply in any ecosystem. They must be continually recycled through the biotic and abiotic components of an ecosystem via *biogeochemical cycles.* Reservoirs and exchange pools are abiotic portions of the biogeochemical cycle. A *reservoir* is an area that stores a particular nutrient for long periods of time, perhaps hundreds, thousands, or even millions of years. Some nutrients in the reservoir are made accessible to organisms by an *exchange pool,* which is temporary storage. Nutrients then flow through living organisms (i.e., the biotic community).

The water cycle

The *hydrologic cycle* describes the movement of water throughout Earth's crust, atmosphere, water bodies, and living organisms. At a basic level, the hydrologic cycle involves evaporation of water from freshwater and saltwater bodies, which eventually condenses and falls as precipitation. Some water flows below Earth's surface and becomes *groundwater,* contained in *aquifers.* The depth at which an aquifer is completely saturated with water is the *water table.* Some groundwater seeps back to the surface and forms freshwater bodies, or runs off into the ocean. Groundwater is an example of a biogeochemical reservoir, which makes up about 20% of the world's freshwater. Freshwater makes up about 3% of the world's supply of water and is considered a renewable resource. Freshwater can, however, become locally unavailable when consumption exceeds supply or when it becomes polluted.

Freshwater bodies and the atmosphere are exchanged pools that store water temporarily and make it available to organisms. Ice in the polar regions and water in the deep oceans are reservoirs, like groundwater.

The carbon cycle

The *carbon cycle* involves the exchange of carbon between organisms and their environment. All terrestrial organisms and marine mammals exchange carbon dioxide (CO_2) directly with the atmosphere. Aquatic organisms that do not breathe air do this exchange indirectly by inhaling or exhaling dissolved carbon dioxide as bicarbonate (HCO_3^-).

Both terrestrial and aquatic autotrophs uptake of carbon and convert it into organic compounds, which then cycle through the food web. As heterotrophs respire, they return carbon dioxide to the air and water. Photosynthesis and respiration occur at relatively equal rates, keeping the cycle in balance.

Organisms return carbon as organic compounds, in the form of waste or ultimately from their decomposing bodies, which can be recycled by detritivores. Some organic compounds are not decomposed but rather become preserved as coal, oil and natural gas fossils. The global reservoirs of the carbon cycle are these fossil fuels, along with bicarbonate that remains deep in the oceans or becomes trapped as limestone.

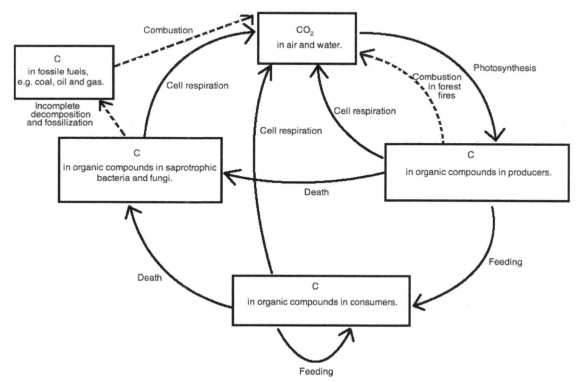

Carbon cycle with the exchange of carbon among organisms and the atmosphere

Carbon dioxide is one of three primary *greenhouse gases* that contribute to increasing the atmospheric temperature. The burning of fossil fuels and forests causes additional carbon dioxide to enter the atmosphere. Solar energy reaches the Earth and warms the planet, but a significant amount should be reflected into space, primarily in the form of infrared radiation. Greenhouse gases absorb some of this reflected infrared radiation and trap it in the atmosphere, causing a rise in atmospheric temperatures. This additional CO_2 interferes with the normal exchange between photosynthesis and respiration and alters the carbon cycle. A source of carbon pollution is oil spills, which contribute millions of tons of oil to the ocean each year.

The Nitrogen Cycle

The cycle of nitrogen throughout its various forms is the *nitrogen cycle.* Nitrogen gas (N_2) makes up 78% of the atmosphere and is vital for plant and animal growth. Organisms use nitrogen compounds to assemble amino acids, nitrogenous bases and other nucleotides such as ATP and $NADP^+$.

Plants rely on relationships with certain bacteria to convert nitrogen into forms which they can use, a process of *nitrogen fixation.* Plants can absorb nitrogen as either ammonium (NH_4^+) or nitrate (NO_3^{-1}). Nitrogen-fixing bacteria may live within plant roots, or freely in the soil and water. They reduce nitrogen gas into ammonium which can be taken up by the plant. Ammonium can be created from the urea in animal wastes.

Bacteria can further convert ammonium into nitrate (NO_3^{-1}) by *nitrification.* This is a two-step process requiring certain bacteria to convert ammonium to nitrite (NO_2^-), and other bacteria to convert nitrite to nitrate. Nitrogen can be converted to nitrate when lightning or cosmic radiation provides energy for a reaction between atmospheric nitrogen and oxygen. *Denitrification* refers to the conversion of nitrate back to unusable nitrous oxide (N_2O) and nitrogen gas, thus completing the nitrogen cycle. This is performed by another class of bacteria.

The nitrogen cycle has been significantly altered by human production of fertilizer, which adds massive amounts of ammonium to the soil. Runoff from nitrogen-rich fields results in over-enrichment, or *eutrophication,* of lakes and oceans. Eutrophication causes large algal blooms which overtake the environment and kill off other organisms.

The burning of plants and fossil fuels adds nitrogen oxide to the atmosphere, contributing to air pollution. These emissions combine with atmospheric water vapor to form acids which precipitate and acidify the soil, *acid deposition.* Nitrogen oxides react with hydrocarbons in the atmosphere to form *photochemical smog,* which contains dangerous compounds that cause respiratory distress. These air pollutants can accumulate near the ground due to *thermal inversions,* in which warm air traps cold air just above the Earth's surface.

The Phosphorus Cycle

In the *phosphorus cycle,* plants take up phosphate ions (PO_4^{3-} and HPO_4^{2-}) made available in the soil, primarily from the weathering of rocks. However, while most phosphorus is in sediments, some phosphorus which runs off into water bodies is incorporated into organic compounds by algae.

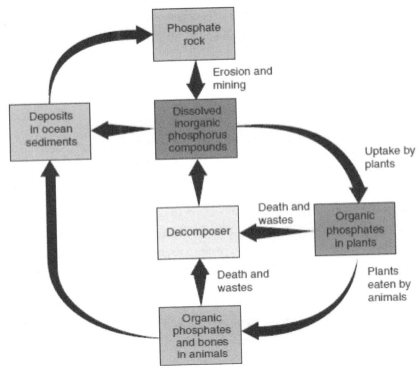

Phosphorus cycle is taken up by plants and primarily from soil

Animals eat these producers and incorporate some of the phosphates into phospholipids, ATP, and nucleotides. The decay of organisms and decomposition of animal wastes eventually makes phosphate ions available again. Phosphate incorporated into teeth, bones, and shells, however, does not decay for long periods of time. Phosphate is the limiting nutrient in most ecosystems, and any available phosphate is rapidly taken up by organisms.

Like the other biogeochemical cycles, humans have disrupted the phosphorus cycle. This is primarily due to mining of phosphate ores and runoff from livestock wastes and fertilized fields. Human and animal sewage is a contributor to phosphate buildup in water bodies. This pollutes the water, leading to eutrophication.

Climate

Climate refers to the prevailing weather conditions in a region over time. It is primarily dictated by temperature and rainfall and is the result of complex interactions between different elements, primarily solar radiation due to the tilt of the Earth, topography and the presence (or absence) of a nearby body of water. Water and air circulate through the oceans and atmosphere and create powerful climatic effects which drive weather conditions.

The tropics are the hottest region on Earth because the sun's rays strike the equator directly. Since the Earth is a sphere, sunlight strikes the higher and lower latitudes at an angle, lessening the effect of the light. Earth is tilted on its axis, causing one pole to receive more sunlight than the other depending on Earth's position in its journey around the sun. These changes in sunlight through the year cause seasons.

Ocean water is warmest at the equator and coldest at the poles due to the distribution of the sun's rays. As air takes on the temperature of the water below, warm air moves from the equator toward colder latitudes, taking the heat with it. Moisture that evaporates into the air is carried with this heat energy. This movement of air, caused by the oceans, creates winds. The cooling and warming of air in the atmosphere drives *circulation cells,* closed circuits of wind circulation. The *Hadley cell* forms as warm, moist air rise at the equator and flows northward, high above Earth's surface. This causes hot, humid conditions in the tropics. At about 30° N, the air cools and sinks, absorbing moisture and causing desert conditions at this latitude. After it descends, the air circles back south to the equator, traveling close to the Earth's surface.

Meanwhile to the north, warm air rises at about 60° N. It flows northward as it cools and eventually descends at the pole (90° N), causing dry, frigid conditions. The air then moves back south to 60° N to continue this cycle, deemed the *Polar cell.*

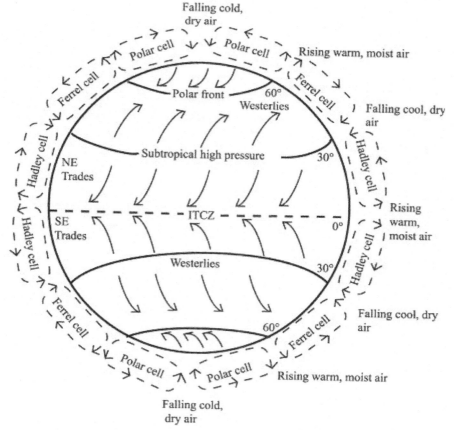

Atmospheric circulation with Hadley, Ferrel, and Polar cells

The air of the Hadley cell sinking at 30° N and the air of the Polar cell rising at 60° N drives the circulation of an intermediate cell, the *Ferrel cell*. In the Ferrel cell, sinking air at 30° N flows north across the Earth's surface to about 60° N and then rises as it warms. It moves back south to 30° N and cools, restarting the cycle.

Hadley, Ferrel, and Polar cells are in the same sequence south of the equator.

Due to the Earth's rotation, air does not move directly north to south but is deflected to the east or west. This is the *Coriolis Effect*. It influences both circulation cells and the *wind belts* which encircle the globe. Six wind belts are on Earth: *trade winds* on either side of the equator, *westerlies* in the temperate regions between 30° and 60° N and S, and weak *polar easterlies* between 60° and 90° N and S.

Topography refers to the physical features of land, which can have a tremendous effect on the climate. Mountains, for example, can cause *rain shadows* in certain areas. Air blowing over a mountain is forced to rise, cooling the land. The windward side, therefore, receives moist air, producing more rainfall. By the time air descends on the leeward side of the mountain range, it is dry. As an example, rainfall on the windward side of the Hawaiian Islands measures over 750 cm but 50 cm in the rain shadow.

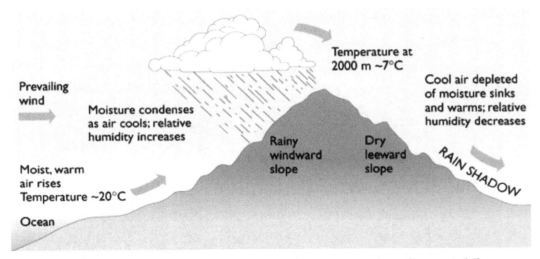

Mountains cause rain shadows with one side experiencing significant rainfall

Oceans help regulate land temperatures since they retain heat and remain cool longer than landmasses. During the day, land heats up, and warm air rises, allowing cool sea breezes to blow inland and replace the rising air. At night, the land cools and the cold air sinks, blowing out to sea. Water's excellent capacity for heat regulation (water has a high specific heat and can absorb energy) generally keeps coastal areas cooler than the inland and minimizes the temperature difference between day and night.

The Biosphere and Biomes

The Earth can be divided into layers, each with its distinct environmental conditions. The *geosphere* is the solid portion of Earth's surface that includes the cryosphere and lithosphere. The *cryosphere* contains all frozen water on Earth, most are at the poles. The *lithosphere* is the rocky surface of the Earth that extends down about 100 kilometers. It comprises the crust as well as the upper mantle. The hydrosphere covers three-quarters of the lith*osphere*, the zone of liquid water primarily contained in oceans. The hydrosphere supports a vast diversity of life and helps regulate global temperatures by absorbing and slowly releasing large amounts of heat.

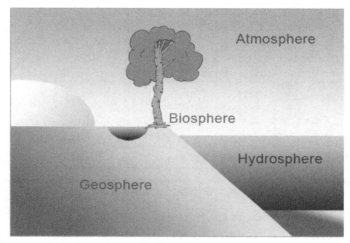

The intersection of geosphere, hydrosphere, atmosphere, and biosphere

The *atmosphere* is the gaseous layer wrapped around the Earth, which regulate temperatures by insulating the Earth from the frigid temperatures of outer space. It is held in place by the planet's gravity and is most concentrated near the Earth's surface. The major gases within the atmosphere are nitrogen and oxygen. Nitrogen makes up more than 75% of the atmosphere and is essential for the growth of plants and animals. Oxygen, about 20% of the atmosphere, is necessary for cellular respiration (ATP production) and makes up the protective ozone barrier which absorbs potentially damaging solar radiation. Other gases in the atmosphere include carbon dioxide, water vapor and noble gases such as argon and neon.

Interacting with the geosphere, the atmosphere, and the hydrosphere is the *biosphere*, a thin layer that comprises all biomass on Earth. In a sense, the biosphere is the global ecosystem. It is further divided into large *biomes*, characterized by a certain climate which supports a unique community of plants and animals. Biomes change with latitude as well as elevation. For example, tundra yields to coniferous forests and then to deciduous forests as one moves from the poles towards the equator. This same sequence can be seen as one moves from a mountain peak to ground elevation.

Terrestrial Biomes

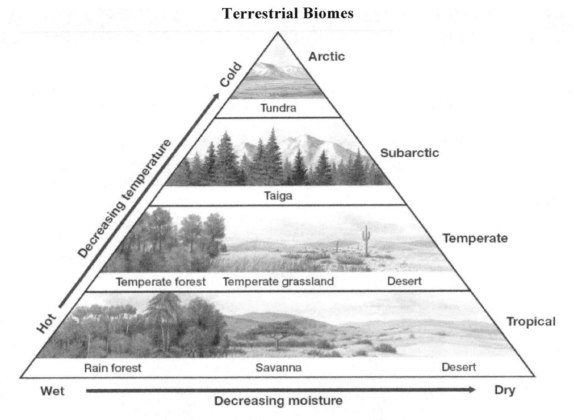

Biomes changes can be related to latitude and elevation

Tundra

The tundra is near the north and south poles, wherever there is no exposure to rock or ice-covered seas. It covers about 20% of the Earth's surface, including Greenland, Scandinavia, Siberia, northern Canada and the coasts of Antarctica and its surrounding islands. A form of tundra, *alpine tundra*, is at the tops of all high mountains, even at the equator.

Tundra is characterized by a cold, dry climate that receives less than 20 cm of annual rainfall. If not for the water provided by melting snow, the tundra would be defined as a desert. Except for alpine tundra, most tundra has a perpetually frozen layer of soil of *permafrost*. Tundra is a difficult environment for plants, due to high winds, a short growing season and hard permafrost. Trees cannot grow in the tundra, but short woody shrubs can survive, as they flower and seed quickly during the short period of available sunlight.

In the summer, the ground of the tundra is covered with bogs, marshes, and streams, with abundant grasses and mosses. Tundra has low biodiversity, and only a few animals are adapted to live in the tundra year-round. During the summer, the tundra supports a wider variety of insects and migratory birds and mammals, such as shorebirds, wolves and reindeer.

Coniferous forests

Coniferous forests have a milder climate than the tundra, as well as higher annual precipitation. They are primarily made of conifers such as spruce, fir, hemlock, and pine trees. These evergreens are well adapted to a snowy, dry climate and have thick, protective leaves or needles. Mountainous coniferous forests are *montane coniferous forests.*

Coniferous forests are divided into two or more biomes, depending on convention. The coniferous forest biome just below the tundra is a *taiga* or *boreal forest.* It extends across northern Europe, Asia, and North America, and makes up nearly 30% of the world's forest cover. It is the largest land biome on Earth and is exceeded in coldness and dryness only by the tundra. The soil does not contain permafrost but is relatively thin and nutrient-poor. It is acidic and covered by lichens, mosses, and fallen needles. The taiga supports greater animal diversity than the tundra, with hundreds of bird species and many year-round, large mammals such as bears, wolves, moose, elk, and bison. It supports many small mammals, such as beavers, hares, and squirrels. Unlike the tundra, cold-blooded reptiles and amphibians can survive in the taiga.

At the latitudes or elevations below the taiga are *temperate coniferous forests.* They have milder summers and winters than the taiga, along with higher precipitation. Temperate coniferous forests can be throughout Europe, Asia, North America, and South America. They include the evergreens of the taiga as well as cedars, redwoods, juniper, and some deciduous trees. The understory is larger and more diverse, with a variety of shrubs and herbaceous plants.

Temperate deciduous forests

Temperate deciduous forests are at latitudes lower than the coniferous forests, primarily in eastern North America, eastern Asia and much of Europe. The climate in these areas is moderate with mild winters and appreciable rainfall, about 75 to 150 cm per year. They have four well-defined seasons, with a growing season that ranges between 140 and 300 days. Deciduous trees are characterized by their leaves, which they shed in the fall and regrow in the spring.

Deciduous forests support a wide variety of life, including countless birds, mammals, amphibians and reptiles. The forest is well-stratified, with large, mature trees shading saplings, shrubs, herbaceous plants, lichens, and mosses. This well-developed understory is possible because of the broad leaves of the deciduous trees, which allow sunlight to penetrate the canopy. Some major tree species in a typical deciduous forest are maples, oaks, and elms.

Tropical forests

Tropical forests are at or near the equator and can be subdivided into several categories, depending on the biome classification system in use. *Tropical rainforests* are within 30° north and south of the equator, in South America, Central America, Africa, India, Southeast Asia, and Oceania. They are characterized by a warm climate and abundant rainfall, from 190 to 1,000 cm per year. Tropical rainforests are warm and humid year-round, with little change between the seasons. This climate sustains the highest diversity of any land biome with a staggering amount of insect life. Colorful birds and amphibians are abundant, as are snakes and lizards.

Many primates and other mammals can be in the rainforest. The largest carnivores are big cats such as jaguars and leopards. The forest is highly stratified, with a tall canopy of evergreens and dense understory. The forest floor has rich soil but sustains few plants due to the heavy shade. A special type of tropical rainforest in high mountains is a *montane rain forest* or *cloud forest* and has a remarkably cold climate compared to lowland tropical rain forests.

Other tropical forests are drier and may exhibit a mix of deciduous and evergreen trees. They are north and south of the equatorial rain forest belt, in Africa, India, Southeast Asia, and South America. One of the largest of these biomes is the *tropical seasonal forest*, which is warm year-round but has long dry seasons. The tropical seasonal forest is less diverse than the tropical rainforest but still has a wide variety of fauna. This biome often gives way to grasslands.

Shrublands

Shrubland, called *scrubland,* is dominated by short, woody shrubs with thick evergreen leaves. These shrubs must be highly resistant to forest fires and drought since shrubland is characterized by hot, dry summers and mild winters. The seeds of many plants in the shrubland even require heat from fires to induce germination. Shrublands found throughout California and along the coasts of South America, Western Australia, and the Mediterranean are *chaparral.* The chaparral is highly adapted to the climate, with extensive root systems and large leaves that retain water. In California, the chaparral is composed of oaks, manzanitas, sages and other short, thorny shrubs.

Drier shrubland found mixed with interior desert regions is *xeric shrubland.* The American West and other desert regions in Asia, South America and Africa have large expanses of xeric shrubland.

Grasslands

Grasslands are relatively arid but still receives greater than 25 cm of annual rainfall. Grasslands are in areas too wet for desert and too dry for forests, and once covered 40% of Earth's surface. They are now significantly diminished since their rich soil is ideal for agriculture. Grasslands are highly adapted to droughts, flooding, fires and grazing from the many herbivores in these regions.

The two types of grassland are temperate grasslands and tropical grasslands. *Temperate grasslands* are characterized by a mild climate and relatively low and predictable diversity. They are throughout North America, South America, Eurasia, and South Africa, where they are *prairies, pampas, steppes,* and *veldts,* respectively. Many temperate grassland animals are large grazing mammals like bison and antelope. Other grassland prey animals are mostly birds and rodents. They are preyed on by coyotes, foxes, lynxes, wolves, snakes and predatory birds. Temperate grasslands can be further divided into the *tall-grass* and *short-grass* regions. Tall grasslands can support trees, are more humid and mild, and are in the lowlands. Short grasslands are drier and cannot support trees, and may be in cold highlands, (e.g., steppes of Russia and Ukraine). Many short grasslands are mixed with deserts and shrublands.

The tropical grasslands are *savannas*, which contain some trees but are mostly open. The most well-known savannas are in Africa, although they can be in South America and Australia. They have a relatively cool dry season followed by a hot, rainy one. Savannas support higher biodiversity than temperate grasslands and have the largest variety of herbivores on Earth. Antelopes, zebras, wildebeests, water buffalo, elephants and giraffes make their homes in savannas along with large carnivores like lions, cheetahs, hyenas, and leopards. Insect life in savannas is plentiful and varied.

Deserts

Deserts are at latitudes about 30° north and south of the equator, where dry air descends from the Hadley cell. They have an annual rainfall of less than 25 cm and subsequently a lack of cloud cover. The absence of clouds makes the days hot and the nights cold. Due to their inhospitable conditions, deserts have some of the lowest biodiversity on Earth. Their plants are highly adapted to heat and drought, but some deserts such as the Sahara are nearly devoid of vegetation. Most desert animals are small insects, reptiles, birds, and rodents since a large size poses a problem for heat regulation. However, large birds, camels, kangaroos, and coyotes can be in various deserts.

The most well-known deserts are in low, interior regions, but deserts can be on coasts or at high altitudes, where they are *cold deserts*. Some scientists classify the tundra as a *polar desert* since it supports little life and is dry.

Aquatic Biomes

Aquatic biomes make up the majority of Earth's biosphere. They can be roughly classified as either freshwater or saltwater (i.e., marine).

Marshes, swamps, and bogs are all aquatic biomes called *wetlands* that can be on coastlands across the globe. Wetlands may be freshwater, marine or a mix, called *brackish.* They have incredibly high diversity, with countless amphibians, reptiles, and birds. *Hydrophytes,* plants which are adapted to live in water, are in abundance.

Estuaries are where a freshwater river merges with the ocean. This may be a bay, a lagoon, an inlet, a sound or other partially enclosed aquatic body. They are intertwined with wetlands. Estuaries are brackish, but they are still considered to be marine biomes because of their importance to the seas. Over half of all marine fish are believed to have been born or raised in estuaries. Estuaries are provided with nutrients from rivers, ocean tides, and decayed vegetation, making them rich environments that can support a wide variety of aquatic flora and fauna. They are a unique biome that has been greatly threatened by habitat destruction and pollution, leading to the collapse of many important ecosystems.

The oceans comprise the other marine biome. The movement of water in the oceans is largely influenced by temperature, the friction of surface winds, salinity differences and the Coriolis effect. These factors all combine to create ocean currents. Since ocean currents are bounded by land, they move in a circular path, counterclockwise in the Northern Hemisphere and clockwise in the Southern Hemisphere. These currents are vital to regulating ocean temperatures and the global climate. Currents induce *upwelling,* the circulation of cold nutrient-rich waters to the surface. Ocean regions which experience upwelling are typically the most diverse and productive because they fuel the activities of *plankton,* a large group of free-swimming microorganisms. *Phytoplankton* is microscopic photosynthetic algae, while *zooplankton* is animals that feed on other plankton. Plankton is the basis of aquatic food chains because they are a food source for many other organisms. More importantly, phytoplankton generates the majority of primary productivity in oceans; other contributors include macroalgae, cyanobacteria, and hydrophytes.

Plankton can be in the freshwater biomes, which include lakes, ponds, rivers, and streams. They contain salt in less than 1% concentration. Streams and rivers are connected to other ecosystems and can exhibit remarkable spatial heterogeneity. The *headwaters* of the river are cool, clear and well oxygenated. As the river travels towards the sea or a lake, nutrient content and species diversity increase. The waters at the mouth of the river are murkiest due to sediment accumulation, which affects light penetrance. There may be lower diversity, lower oxygen, and lower productivity as a result. Plant life in rivers and streams must be able to anchor tightly to the riverbed. In calmer waters, bottom-dwellers can stay in place. In fast-moving waters, nearly all animals are fish that traverse great distances.

Ponds and lakes lack currents and have a more stable, permanent life. Aquatic plants, algae, insects, mollusks, crustaceans, amphibians, and fishes may be in lakes. Birds and reptiles such as turtles, snakes or crocodiles may prey on these organisms. Many lakes and ponds have limited diversity due to their isolation from other regions.

Lakes

Lakes can be divided into four zones defined by their depth and distance from the shore. The *littoral zone* includes the shallow areas closest to shore, where the water is warmest and completely penetrated by light. Plants root themselves to the lakebed in this region and support animals such as mollusks, crustaceans, insects, amphibians and small fishes. Many larval organisms are reared in nurseries in the littoral zone.

The *limnetic zone* is the sunlit area in the open waters of the lake and is home to many plankton and fish. Most photosynthesis occurs here.

The *profundal zone* is below the limnetic zone in deeper waters, home to larger fish, turtles, and snakes. Many birds dive to the profundal zone to capture prey. There is little photosynthesis because sunlight cannot sufficiently penetrate the deeper waters.

The *benthic zone* is at the bottom of a pond or lake and is made of soft sediment that receives little if any, sunlight. It is inhabited by organisms that can tolerate low oxygen levels, including worms, mollusks, and crustaceans. These filter feeders thrive on the debris which falls from the higher zones.

In the temperate latitudes, deep lakes are stratified depending on the season. In summer, the surface waters are warm due to the heat of the sun, while the depths are cold. These layers are separated by the *thermocline,* a layer of abrupt temperature change. In winter, this is reversed, with the depths remaining temperate due to insulation by surface ice and the surface being cold since it is closest to the frigid outside temperatures. In fall and spring, changing temperatures cause mixing that returns the lake to a uniform temperature without a significant thermocline. Lake animals are adapted to these seasonal changes and may migrate to different depths for more favorable conditions

Lakes can be classified along a continuum from *oligotrophic* (nutrient poor) to *eutrophic* (nutrient rich). Oligotrophic lakes are not able to sustain much plant life and have low levels of primary productivity; consequently, they are dominated by fish. They have clear waters and are often in cold, alpine regions. By contrast, eutrophic lakes have high productivity and can support an abundance of plants that outcompete fish. Excessive eutrophication results in massive algal blooms which deplete oxygen as they decompose. This creates hypoxic conditions that kill off animals. An influx of nutrients can change an oligotrophic lake into a eutrophic lake. Lakes which are *mesotrophic* have moderate nutrient levels and can sustain both plant and animal life.

Oceans

Oceans are bordered by continents, each of which is situated on a *continental shelf,* the continental crust which extends under the water for some distance from the coast. After several hundred meters it begins to drop off rapidly, forming a steep, downward *continental slope.* A continental rise follows the slope; contrary to the name, this region is still sloping downward, albeit less steeply than the slope. Finally, the rise levels off into the vast *abyssal plain,* which is the bottom of the majority of the world's oceans. Large ridges and deep trenches mark it.

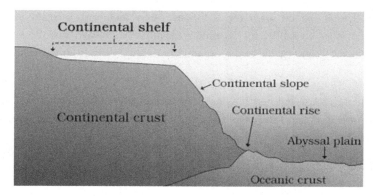

Oceans are characterized by the continental shelf, slope, and rise

Ocean zones are much larger and more complex than lake zones. Rather than the entire littoral zone, scientists typically refer to the *intertidal zone,* the area of the littoral zone that is periodically covered and uncovered by water. The changing of sea levels due to tides presents a challenge to intertidal organisms. Organisms must anchor themselves to rocks, hide in crevices or burrow in the sediment to avoid desiccation and being swept away.

Past the intertidal zone, the waters begin to deepen, but the continental shelf continues below; this region is the *neritic zone.* The neritic zone is highly productive because sunlight and nutrients are relatively abundant. It is well-oxygenated and has stable conditions which make it suitable for the majority of ocean life. Organisms from microscopic plankton to large fish are in the neritic zone, as well as vast coral reefs. It is the *coastal zone* because it marks the boundary between the shore and the open ocean.

Once the continental shelf drops off into a slope, this marks the open seas of the *oceanic zone.* It is vast and sustains a variety of sea life, though relatively little compared to the neritic zone. The largest sea animals, such as whales, sharks, and giant fishes, are in the oceanic zone. Smaller animals and plankton can be in the surface waters.

All waters of the ocean that are neither close to the bottom nor the shore are the *pelagic zone*. Pelagic waters are throughout the neritic and oceanic zones and are vertically stratified based on light level. The *photic zone* extends from the surface to about 1000 meters deep and is where photosynthesis occurs due to phytoplankton activity. Fish, jellies, dolphins, and seaweed live in the photic zone. Apex predators in the upper waters of the photic zone include sharks, mackerels and tunas. The bottom of this zone is poorly lit or *disphotic*; consequently, it is dominated by predators with excellent photoreceptors and other adaptations to the low light. Prey organisms often have translucent or red coloring, which is both well-disguised in dark waters.

Below the photic zone is the *aphotic zone,* where no photosynthesis takes place because it is mostly, if not entirely, dark. It extends to the abyssal plain 6,000 meters below and often even deeper into massive ocean trenches. Familiar squids and sperm whales can be in this zone, but most inhabitants are strange and poorly understood. Many animals at these depths let out occasional flashes of light to communicate with others or attract prey. Animals in this area are carnivores, filter feeders or scavengers which feed on dead organisms falling from above.

The *benthic zone* is the seabed, which may be completely exposed to the air in the intertidal zone or thousands of meters below the surface at the abyssal plain of the oceanic zone. Benthic organisms are adapted to living on or in the sediment, often anchoring themselves to the underlying substrate. The sediment may be sandy, rocky, muddy or silty. In many areas, the seabed is covered in coral. Benthic organisms rely on organic and inorganic nutrients falling from the water column above, as *marine snow.*

In the intertidal, littoral and neritic zones, the benthic zone receives sunlight and can support a varied food web with seaweed and filter feeders at the first trophic level. Starfish, crustaceans, mollusks and some bottom-dwelling fish occupy the upper trophic levels. In the oceanic zone, the benthic zone receives little to no sunlight, and therefore no photosynthetic organisms can be found. Rather, filter feeders and scavengers feed on marine snow. Starfish, crustaceans, mollusks and bottom-dwelling fish are common on the continental slope and rise. On the pitch black abyssal plain, there is a huge diversity of microbes but little macrofauna. Only a few select filter feeders and scavengers are found. These include sponges, worms, sea lilies, and other invertebrates. This is a region of extreme cold and intense pressure, except where hydrothermal vents expel superheated, sulfurous water. These vents support chemosynthetic bacteria, tube worms, and clams.

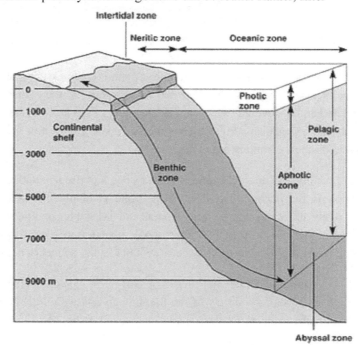

Ocean with zones at respective depths

Conservation Biology

Conservation biology applies ecology and social sciences to the study of preserving Earth's biodiversity. It stresses that biodiversity is vitally important for the health of humans and all other organisms, and strives to prevent extinction due to habitat destruction, pollution, invasive species, and overexploitation. Most conservation biology emphasizes practical applications, but there is still appreciable effort focused on the intrinsic value of nature to human cultures.

Biodiversity

The term *biodiversity* describes some concepts. *Species diversity* is the variety of species living in a given area. It is believed that there could be anywhere from 5 to 100 million species on Earth, and a fraction of this possible diversity has been cataloged. *Genetic diversity* describes the variety of alleles within the gene pool. It is the ultimate measure of biodiversity but is challenging to assess. Genetic diversity maintains the fitness and adaptability of a population; threats to genetic variation pose the risk of extinction. *Landscape diversity* refers to the variety of interacting ecosystems in a region. Homogenous landscapes can support only a limited diversity of species, similar to isolated landscapes.

Biodiversity is not evenly distributed, but rather is highest at the tropics and lowest at the poles. It forms pockets of *biodiversity hotspots*, regions with unusually high concentrations of species. Biodiversity hotspots are characterized by their *endemic species,* organisms which are unique to that area alone. These regions make up a fraction of Earth's area but are estimated to contain 60% of all species; examples include the Mediterranean, the Amazon rainforest and Madagascar. Biodiversity frontiers such as the deep sea likely have more species than was formerly suspected.

Value of biodiversity

The benefits of biodiversity to humans are *ecological services,* which may be economic, scientific, aesthetic or ethical. Diversity strengthens an ecosystem and makes it healthier, increasing the efficiency of these services. For example, a greater diversity of crop plants increases their yield, while the greater diversity of fish keeps fisheries more stable and productive.

Economic benefits of diversity include goods such as food, oxygen, clothing, medicine, fuel, consumer products, and habitats. The rainforest contains up to $150 billion of possible prescription drugs, in addition to the many already discovered. The importance of natural products in the manufacturing of material goods cannot be overstated. Plastics, rubber, and lumber are just a small fraction of the materials derived from the environment. Fossil fuels, such as coal, oil and natural gas, are all provided by dead organisms.

Nature provides economically advantageous services, such as pest control, pollination, soil maintenance, and the biogeochemical cycles. Unless these cycles can be someday approximated by technology, humans are reliant on the environment to purify water, break down pollutants, trap carbon dioxide, provide oxygen and recycle organic compounds into edible food.

Some ecosystem services are lifesaving. River and wetland ecosystems are natural buffers against flooding, storm surges, and other extreme weather events. Trees and other plants hold the soil and prevent erosion that clogs reservoirs, creates landslides, chokes coastal ecosystems and ruins fisheries. Trees are important because they are the largest providers of oxygen and removers of atmospheric carbon dioxide. Deforestation worsens the issue of global warming by diminishing the carbon dioxide sink.

Tourism of natural environments, or "ecotourism," is a multibillion-dollar industry; many cities and even countries base an entire economy on ecotourism. If these natural resources are threatened, it will cause debilitating economic collapses.

Even if the technology were able to replace the economic services of the environment, humans would be deeply impacted by the aesthetic loss and ethical conflict. Erosion and extinctions threaten relationships with plants, animals and natural habitats, and thus human culture as a whole.

Causes of extinction

The greatest threats to biodiversity are those that cause species extinctions. Research shows that the most powerful catalysts of extinction are habitat loss, introduced or invasive species, overexploitation of flora and fauna, and pollution.

Habitat loss may be caused by destruction, fragmentation or degradation. The most common form is direct human destruction. Examples include clearing forests for agriculture, draining wetlands, diverting rivers and building dams, mining, trawling the ocean floor with fishing nets and urbanization. Unfortunately, biodiversity hotspots are often the most threatened by habitat loss. A classic example is the tropical rainforests, which are being cleared at an astonishing rate. Other biomes that are particularly threatened are grasslands, deciduous forests, wetlands, estuaries, and the oceans. Habitat fragmentation is usually the result of urban sprawl and infrastructure, which can divide an area into fragments that are too small to support a species. Fragmentation hinders migration, which is crucial for the maintenance of diversity. Habitat degradation is an indirect form of habitat loss that is caused by pollution, climate change, and exotic species. For example, an invasive herbivore may weaken a grass population that is used as a habitat by native species.

Exotic species are often introduced accidentally or deliberately into new ecosystems. Human circumvention of natural barriers has resulted in the transport of many alien species to new environments. Ecosystems evolve with their native organisms in balance, and the introduction of a new species often disrupts the food chain and leads to extinction.

Overexploitation occurs when harvesting from a wild population becomes unsustainable. Animals may be overexploited for food, hides, furs, ivory, pets or sport. Nearly a third of fisheries have been dramatically overexploited, and tigers and elephants have been hunted nearly to extinction. Plants are typically overexploited for food, building materials, medicine, and agriculture. Often, overexploitation hinders an ecosystem's ability to recover even decades after it ceases.

The introduction of any harmful element to an ecosystem is pollution. For example, air pollution is mostly caused by exhaust from factories and motor vehicles. It results in the accumulation of greenhouse gases, acid deposition, disease, and other ill effects. Soil and water pollution are typically the result of agricultural, industrial and sewage runoff, which may contain pesticides, heavy metals, oil, fertilizers, animal wastes, and other toxic chemicals. The littering of trash is a form of physical pollution that affects the land and water.

However, pollution may take the more abstract form of light and noise pollution. Over-illumination from cities disorients animals and impacts their circadian rhythms, migration patterns, and reproduction. Noise pollution interferes with natural communication between animals, particularly those who use sonar. *Genetic pollution* is a recent phenomenon which is caused by human interference with the genetic diversity of a population. For example, breeding or engineering hybrids, homogenizing the gene pool or introducing new species— all affect the natural population. It can destabilize the population, cripple its fitness and even decimate it.

The causes of extinction are worsened by climate change. Global warming melts ice caps, increasing sea level and destroying coastal habitats. It warms the seas and increases pH, threatening many aquatic species which are adapted to certain conditions. The coral reefs, in particular, are already suffering from temperature and acidity shock. Global warming promotes the growth of pests and pathogens, increasing rates of disease. Climate change alters global weather patterns, causing extreme weather events which destroy habitats and kill wild populations. Regions with a suitable climate for species shift rapidly, probably faster than organisms can migrate or adapt, allowing exotic species to outcompete native species in the changing environment.

Conservation techniques

Rather than attempt to conserve every ecosystem on Earth, conservation biologists focus their efforts on biodiversity hotspots to maximize effectiveness. Hotspots may be detected in a variety of ways. Widespread field studies provide accurate counts of species diversity but are labor-intensive. An easier but sometimes misleading method is to look for *indicator species*, which highlight certain conditions of an ecosystem that may otherwise go undetected. For example, the presence of the spotted owl indicates a stable, old growth forest, while bleaching of corals indicates acidic waters. Indicator species are often keystone species, which can be useful targets for conservation since they prevent the extinction of several other species.

Endemic, rare and endangered species merit the primary attention of conservation efforts. Many of these organisms become *flagship species,* adored by humans for their looks or symbolism. Flagship species are almost always mammals, such as polar bears and giant pandas, and can be a powerful tool for mobilizing conservation. Unfortunately, many vitally important invertebrate and plant species are disregarded in favor of flagship species.

Umbrella species are spread over a range of habitats; they are often a flagship, keystone or indicator species. Because the umbrella species has a wide range and is easily observable, targeting conservation of the umbrella species undoubtedly assures conservation of other species. For example, protecting the spotted owl requires the protection of its habitat, old-growth forests in North America. These forests are home to hundreds of other species which benefit from the spotted owl's conservation.

Conservation efforts require cost-benefit analysis, statistical models and detailed proposals to achieve success. For example, *population viability analysis* helps determine how much habitat a species needs to survive. This can guide conservationists to balance the costs of habitat preservation with the benefits for the species. Governments take on conservation efforts if they are efficient and effective. One of the best ways to compel governments is to describe the practical, monetary advantages of taking action. It is helpful to provide clear guidelines for what actions should take place. For example, the *IUCN Red List* describes the conservation status of thousands of organisms, with categories such as "least concern," "vulnerable," "endangered," and "extinct." Hunting and fishing laws, the creation of nature preserves, sustainable building practices and restriction of development are conservation efforts which governments can enforce.

Hunting, poaching, and fishing may be banned outright for certain species, or somewhat restricted. Sustainable hunting and fishing practices aim to prevent the overexploitation of biodiversity.

Conservation aimed at invasive or introduced species often attempts to cull the populations of the problem species. This can be costly, difficult and often requires an ongoing effort. Introducing a *new* exotic species which preys on the problematic one can be effective if it is done carefully and scientists can be confident it will not backfire.

As habitat loss is the main cause of extinctions, habitat preservation is the foremost aim of conservation biology. One of the most difficult but important ways to protect biodiversity is to restore a degraded habitat to its former health. Habitat restoration falls under the category of *restoration ecology*, which studies strategies to restore ecosystems.

Planting vegetation may reduce erosion, for example, or controlled burns can clear species which are outcompeting all others. The ultimate goal of habitat restoration is to return the habitat to its natural state and ensure it can maintain this state without human intervention. This often takes the form of nature preserves, where species and their environment are protected from human interference. Wildlife corridors are a more recent technique used to connect habitats which have been fragmented by land development. Preserves and corridors are often selected using *gap analysis,* which involves overlaying land-use maps with species maps to highlight areas where biodiversity is high but unprotected or fragmented. Of course, it is always better to be proactive rather than reactive. Low impact development, reduction of logging and mining and pollution control are ways habitat loss can be prevented.

Pollution control is a facet of conservation that has received great attention and even international agreement in the *Kyoto Protocol.* This treaty was drafted to target climate change, the most important frontier of conservation at this time, primarily in the reduction of greenhouse gases and other pollutants. Pollution can be prevented or reversed by enacting an eco-friendly industry, better waste management, cleanup, recycling and new technologies which reduce the need for old, harmful substances.

The most effective conservation is *in situ,* efforts occurring in the ecosystem in question. *Ex situ* conservation occurs outside the original ecosystem and may include relocating species to a new environment or breeding them in zoos and then re-releasing them into the wild. *Gene banks* are a new focus of ex situ conservation, where scientists maintain samples of plant seeds, cuttings, and animal gametes. The idea is that these organisms may later be reintroduced in case the species become endangered or extinct.

Chapter 11

Population and Community Ecology

- **Ecology of Populations**

- **Community Ecology**

Ecology of Populations

Scope of ecology

Ecology is the study of the distribution and abundance of organisms and their interactions with one another and the environment. Humans have been studying the natural world for hundreds of years, but the term ecology was not coined until the 19th century by the German zoologist Ernst Haeckel. The field was greatly advanced by widespread acceptance of the theory of evolution. Evolution allowed scientists to fully understand how ecological pressures such as natural selection to shape the environment. Since then, ecology has become modernized, with rigorous, comprehensive studies and sophisticated statistical models. Modern ecology has applications to a wide range of fields, including conservation, agriculture, social science, and many others.

Ecology is hierarchical, as it can be studied at many levels, from cellular to the *biosphere,* the entire region of Earth in which all living organisms reside. Most ecologists begin their study at the organismal level. An organism's *habitat* includes both its physical and biological surroundings, including nearby organisms. Organisms of the same species often live in groups of *populations,* which occupy roughly the same region. The *community* includes the populations of all species in a given locale. For example, a freshwater lake is a community that may include algae, plants, fish and microorganism populations. All living organisms and their interactions are *biotic factors.*

An *ecosystem* includes all biotic factors as well as the physical environment. In the lake example, the ecosystem would include the populations described above as well as water salinity, temperature, pH, density, light level, soil composition, and other factors. These are *abiotic factors.* Together, all ecosystems with similar abiotic factors make a *biome,* which is usually defined by its climate. For example, the tropical rainforest is a biome that includes all hot, humid ecosystems supporting a high diversity of life. Collectively, biomes comprise the biosphere.

Characteristics of populations

Population size and growth models

Population ecology is the study of the growth, abundance, and distribution of populations. The size of the population is denoted as *N*, the total number of individuals. Population size about living space is *population density* or the number of individuals per given unit of area. The way in which density is patterned over the population's range is *population dispersal.* Populations may be spread uniformly, randomly or be clumped. Ecologists often study the changes in population distribution across space or over time.

Generally, a population is densest near the center of its range and sparsest at the edge. This edge is the *zone of physiological stress* because it has suboptimal conditions for the species in question. Physiological stressors may include extreme temperatures, inadequate water or pollution. Beyond this is the *zone of intolerance,* where no individuals of the species can survive. The species' theoretical range is determined by the range of physiological stressors which it can tolerate. However, this range may be restricted by other species due to *biological stressors,* such as competition and predation.

Stressors are *limiting factors*, conditions which act to limit growth or abundance of a population. Limiting factors may be *density-independent,* which are independent of population density (e.g., light availability and precipitation). Limiting factors can be *density-dependent,* whereby the effect becomes more severe as population density increases. Density-dependent factors include competition, disease, parasites, and food scarcity. They typically fluctuate and drive a *population cycle,* a cyclic change in the population size.

The population size (N) over time can be predicted by *natality* (birth rate) and *mortality* (death rate). Together, natality and mortality can be used to calculate the *intrinsic rate of natural increase* (r).

$$r = \frac{(\text{birth rate} - \text{death rate})}{N}$$

However, population increase is usually subject to many factors. For example, population ecologists must consider the *immigration* of individuals out or the *emigration* of other individuals into the population.

Population growth typically exhibits one of two patterns. The first is *discrete growth* (discrete breeding or discrete reproduction), in which organisms breed all at once at a particular time of year. They may breed once in their lifetime, as *semelparous,* or they may reproduce year after year, as *iteroparous.* The former strategy produces *discrete generations,* in which the adult generation reproduces and soon dies, leaving behind the next generation. This results in a population that has one generation at any given time. Iteroparity produces *overlapping generations*, in which an elderly generation is living at the same time as a reproductive generation and a sexually immature generation. At any time, at least two generations can be observed in the population.

Continuous growth is the other pattern of population growth, which occurs when organisms reproduce continuously without regard for a certain breeding season. Populations which exhibit continuous growth are always iteroparous and have overlapping generations. Most organisms do not fit neatly into one of these two patterns and instead exhibit a combination of the two. For example, plants may reproduce sexually at a certain time each year but reproduce asexually at any time.

Exponential growth often occurs in populations which are iteroparous and have overlapping generations. It is represented by a J-shaped, *exponential growth curve.* The first phase of the curve is the *lag phase.* At this time, growth is slow because the population is small. At a certain critical size, the population enters the *exponential growth phase* when growth accelerates rapidly.

A population that is undergoing the maximum population growth possible, with no hindrance from limiting factors, is fulfilling its *biotic potential.* Biotic potential is calculated by taking into account the number of offspring produced by each reproductive event (*clutch size*), the frequency and a total number of reproductive events, offspring survival rate and the age at which an individual reaches sexual maturity. Populations can only reach their biotic potential when they have ample space, resources, and absence of predation.

However, these factors are often in limited supply, and the environment can only support so many organisms, the *carrying capacity* (K). When populations approach carrying capacity and deplete resources, they encounter *environmental resistance* and growth begins to slow. An S-shaped *logistic growth curve* represents growth under environmental resistance. The first portion of the curve is exponential, with the lag phase and exponential phase described earlier. However, the population eventually reaches a *transitional phase* or *deceleration phase*, when the birth rate begins to fall below the death rate due to resource competition, predation, disease, and other density-dependent factors. At this point, growth begins to slow. At carrying capacity, the population enters a *stable equilibrium phase* with minimal growth. During stable equilibrium, natality and mortality are roughly equal.

$$\text{The logistic growth curve:} \quad \frac{\Delta N}{\Delta t} = rN \left(\frac{K-N}{K} \right)$$

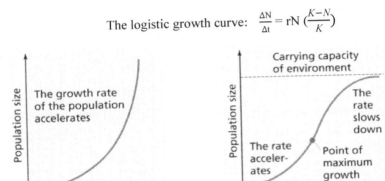

Exponential (unrestricted) growth (left) and logistic (restricted) growth (right)

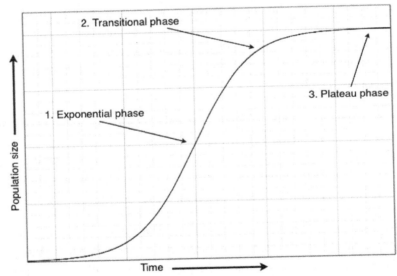

S-shaped logistic growth curve of population growth with environmental resistance

Age distribution

A population with overlapping generations typically exhibits *pre-reproductive, reproductive* and *post-reproductive* generations. The abundance of individuals of each gender and in each age group of a population is represented in an *age structure diagram.* Horizontal bars represent the number of individuals in each age group.

A pyramid shape indicates an expanding population with high birth rate and exponential growth. The pre-reproductive generation is largest because offspring are rapidly reproduced, while the reproductive generation is intermediate. The post-reproductive generation is smallest as the elderly die. A bell shape represents a relatively stable population, in which the pre-reproductive and reproductive generations are roughly equal, and the post-reproductive generation is smaller by a narrow margin. Finally, an urn-shaped diagram is indicative of a declining population. The post-reproductive generation is largest because of few new individuals. Individuals from the reproductive generation enter the post-reproductive generation and continually die, while the pre-reproductive generation is too small to sustain growth.

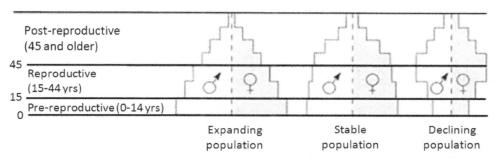

Age structure diagrams: pyramid (left), bell-shaped (center), urn-shaped (right).

Carrying capacity

Carrying capacity is determined by the available water, space, food, light, and other factors. It is density-dependent, becoming more restricted as growth increases. Stable populations do not attempt to maximize biotic potential but rather remain just under carrying capacity. However, overshooting carrying capacity can be a valuable strategy as well, provided the population can introduce new individuals before it crashes.

Carrying capacity is an important regulator of population size and a powerful driver of evolution. Populations respond to carrying capacity by expanding their range or evolving adaptations, which relieve them of some of the restrictions of carrying capacity. No organism can indefinitely evade carrying capacity, and if it is overshot, high mortality results.

Mortality Patterns

In demography, predictions can ascertain the probability that an individual will die before their upcoming birthday based upon their age. The tool used to make these predictions is a *life table* (*mortality table* or *actuarial table*). This information ultimately signifies the survivorship of certain age-based populations.

In actuarial science, two varieties of life tables are used. A *period table* is used to calculate mortality rates during a set period for a specified population. A *cohort life table* (generation life table) is used to represent a certain population's overall mortality rates.

Life tables track cohorts over their lifetime. A *cohort* is a group of individuals born at the same time that are aging together. The life table shows how members of the cohort die at various ages. *Survivorship* refers to how many individuals remain alive at a given point in time.

Region	1990	1995	2000	2005	2010	2012	MDG target 2015	Decline (percent) 1990–2012	Annual rate of reduction (percent) 1990–2012	Annual rate of reduction (percent) 1990–2000	Annual rate of reduction (percent) 2000–2012
Developed regions	15	11	10	8	7	6	5	57	3.8	3.9	3.8
Developing regions	99	93	83	69	57	53	33	47	2.9	1.8	3.8
Northern Africa	73	57	43	31	24	22	24	69	5.4	5.3	5.5
Sub-Saharan Africa	177	170	155	130	106	98	59	45	2.7	1.4	3.8
Latin America and the Caribbean	54	43	32	25	23	19	18	65	4.7	5.1	4.4
Caucasus and Central Asia	73	73	62	49	39	36	24	50	3.2	1.6	4.5
Eastern Asia	53	46	37	24	16	14	18	74	6.1	3.7	8.0
Excluding China	27	33	31	20	17	15	9	45	2.7	−1.2	5.9
Southern Asia	126	109	92	76	63	58	42	54	3.5	3.1	3.9
Excluding India	125	109	93	78	66	61	42	51	3.3	3.0	3.5
South-eastern Asia	71	58	48	38	33	30	24	57	3.9	3.9	3.8
Western Asia	65	54	42	34	26	25	22	62	4.4	4.4	4.5
Oceania	74	70	67	64	58	55	25	26	1.4	1.0	1.7
World	90	85	75	63	52	48	30	47	2.9	1.7	3.8

This life table, developed by the United Nations, indicates the mortality levels and trends for children under five years old worldwide.

The three general types of survivorship curves are:

Type I survivorship curve shows a long curve with a relatively short drop-off near the end in which most individuals survive until they die of old age (e.g., the human population).

Type II survivorship curve, which is negative and linear, shows individuals dying at a constant rate over their lifespan (e.g., some birds and lizards).

Type III survivorship curve is when most individuals in the population die at an early age. Those that do survive, however, tend to live for a relatively long time. It is essentially the opposite of a Type I curve and is seen in many invertebrates, plants, and fish.

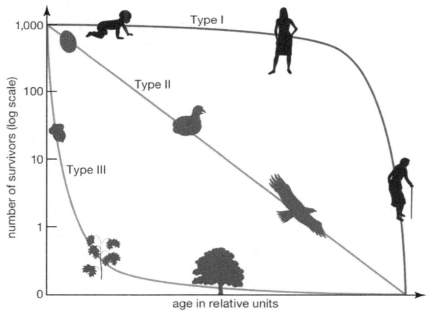

Survivorship curves comparing type I, type II and type III organisms

Regulation of population size

As previously discussed, populations are regulated by some density-dependent and independent factors. These may be *intrinsic* (i.e., within the population) or *extrinsic*. Intrinsic factors include the behavior and anatomy of the species, while extrinsic factors include climate, habitat and other organisms.

The exponential and logistic growth curve models are a straightforward view of population growth that only takes into account classic limiting factors. However, real populations may be affected by hundreds of complex and interrelated factors. These are usually intrinsic factors such as social behavior, which influence how competition proceeds and individuals immigrate or emigrate. For example, populations may actively recruit new members or exclude them due to territoriality.

Straightforward models have difficulty predicting randomness. Populations may deviate from the expectation because of chance or unpredictable events, like a natural disaster.

Life history patterns

Ecologists have historically divided organisms into two major life history patterns: *r-selection* and *K-selection*. However, most species are not strictly r-strategists or K-strategists; it is more common for a species to exhibit characteristics of both. They may be able to shift strategy in response to environmental factors.

r-Selection

The r-selected species attempt to maximize their rate of natural increase. They typically overshoot carrying capacity, causing the population to crash suddenly. Therefore, their population growth may show severe fluctuations. Typically, *opportunistic species*, r-strategists rapidly seize the opportunity to proliferate. They are often the first to colonize a new habitat and do well in unstable environments subject to density-independent factors. This is because they reproduce quickly and reach sexual maturity at a young age, maximizing the opportunity for reproduction before death. They may get the chance to reproduce once, making semelparity (single reproductive event) a common characteristic of r-strategists. The trade-off is that they must produce many offspring at one time because they cannot protect them from infant mortality. To survive, r-selected organisms have a short lifespan and must quickly adapt to new environments.

K-Selection

K-selected species attempt to maintain their rate of natural increase. They exist near the carrying capacity at a state of equilibrium, making them *equilibrium species*. Unlike r-strategists, K-strategists are specializers uniquely suited to their environment. This makes them successful but vulnerable to disturbances. They typically enter a new habitat after r-strategists have colonized it. To maintain their existence, K-selected species are typically large and invest considerable energy in caring for their offspring, of which they often produce one at a time. K-strategists reach sexual maturity slowly, but they can live for a long time. They can reproduce several times throughout their lifespan, making them *iteroparous* (multiple reproductive cycles during their lifetime).

Human population growth

The human population is in the exponential phase of a J-shaped growth curve. The global population is currently increasing by about 80 million people per year. This tremendous growth has been fueled by technological advances, an increase in food supply, reductions in disease and habitat expansion. Current estimates predict that the world population will level off sometime in this century between 8 and 10 billion. Some claim this number is a vast underestimate and that the human population may reach 14 billion by the end of the 21st century.

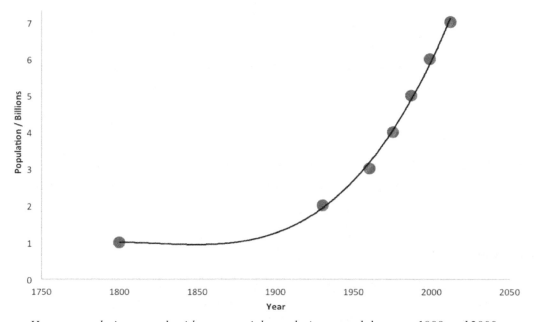

Human population growth with exponential population growth between 1800 and 2000

In the mid-20th century, the developed countries had a significant decline in mortality rate, soon followed by a decline in birth rate. Currently, the growth rate in first world nations is at about 0.1%. Their age structures are relatively stable, with some countries even exhibiting a declining population.

Nearly all population growth in the coming years will be seen in the less developed countries, especially in Africa, Asia, and Latin America. Even so, the growth rate in these countries is lower than it was at its peak of 2.5% in the 1960s. The way to level off population growth is an aggressive campaign of family planning, birth control, and encouragement to produce fewer children. However, due to cultural attitudes and a high infant mortality rate in underdeveloped world nations, it is difficult to convince people to have fewer children or delay childbearing until later years.

The growing populations of the less developed countries and the high consumption of the more developed countries put stress on the environment. Currently, the *ecological footprint* of those in the United States and other developed nations is unsustainable. The ecological footprint is determined by the amount of land which is required to sustain an individual's lifestyle, including the area in which he or she lives, the farmland required to produce food, the factories required to produce material goods and the distances across which these products must travel to reach the individual. An average American family, regarding consumption and waste production, equates to the lifestyle habits of thirty people in India.

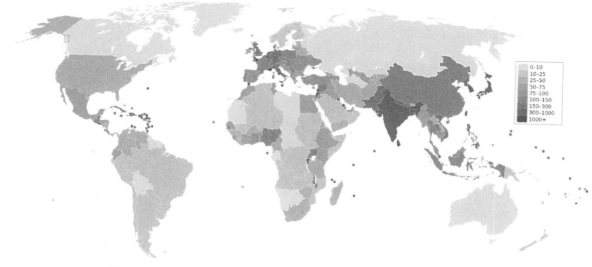

This world map indicates the average population density per km²

Developed countries account for one-fourth of the world population but provide 90% of the hazardous waste production. Intense resource consumption affects the cycling of chemicals and contributes to pollution and the extinction of species. Without drastic reductions in the collective ecological footprint, humans will overshoot carrying capacity and possibly experience catastrophic disease, famine, and other density-dependent factors.

Community Ecology

Community ecology is the study of the composition, diversity, interactions, and relationships between populations and how these characteristics change over time.

The concept of the community

A community includes all populations within a given environment that interact with one another. They may be large, as in an entire forest, or small, as in a bacterial community in the gut of an animal. Because of this variability in scale, it can be difficult to delineate the boundaries of a community.

Studies show that the composition of a community is constantly in flux due to natural selection, migration, environmental changes, and random chance. Many communities fluctuate regularly with the seasons; for example, the tundra is inhospitable in winter but supports plant and animal life in the summer. Some communities may be transient, as with an animal corpse, which supports a community until decay is complete.

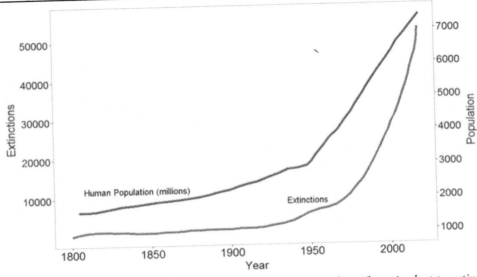

The correlation between human population levels and the number of species lost to extinction

Various models were proposed in the 20th century to outline the concept of a community. The American ecologist Frederic Clements described a *holistic theory,* in which the community acts together as a "superorganism." By contrast, his contemporary Henry Gleason proposed the *individualistic theory,* which states that each population can inhabit a community due to its unique adaptations. Under this model, the range of a species is due to its tolerance to physiological stressors and is independent of the distribution of other species. Finally, Stephen Hubbell's *neutral theory* proposes that at a basic level, all species have the same ability to adapt, grow and disperse. Their differences are negligible, and the variation seen in different communities is essentially due to randomness. Biodiversity is a product of chance and not of inherent characteristics of the species. Likely, none of these models are exactly correct. It seems that communities are affected by biotic and abiotic factors, as well as by random chance.

Structure of the community

Species composition, or *species richness,* is the number of different species within a community, without any further information as to their relative abundances. Relative abundance is *species evenness* or the relative size of each population in the community; for example, a forest with 40 yellow poplars and 40 American elms has more species evenness than a forest with 70 poplars and 10 elms. Together, evenness and richness describe the *species diversity* of a community. The most diverse communities have a high level of both richness and evenness.

The structure of a community depends on abiotic factors along with complex interactions between populations. *Niche theory* is essential to understanding how organisms and populations fit into the community. The *ecological niche* is the role an organism plays in its community, including its habitat, resources consumed and interactions with other organisms. Two deer species may occupy the same geographic range but inhabit different niches because they feed on different species of plants.

The *fundamental niche* is the place which an organism occupies in the absence of competition from other species. However, organisms are usually limited to a *realized niche,* the more restricted niche which allows it to avoid overlap with other organisms. For example, one barnacle species may be able to live on rocks that are exposed to the full range of tides, making their fundamental niche the entire tidal area. However, a second barnacle species may be capable of out-competing the first, but can only live in the lower tidal region where desiccation is minimal. This forces the first species into a realized niche that includes only the higher tidal region. The two species typically have an area of niche overlap at the middle region where intense competition takes place.

Organisms have developed different strategies to maximize success under the constraints of niche theory. *Generalists,* such as humans, can occupy a variety of niches. They can quickly adapt to a new niche if the current one is threatened. *Specialists* become highly adapted to a single niche.

Interactions between populations can be complex and include competition for resources, predation, coevolution, and symbiotic relationships.

Competition between populations

Interspecific competition occurs when different species utilize a resource that is in limited supply. The *competitive exclusion principle* states that no two species can occupy the same niche at the same time. Over time, either one population replaces the other, or the two species evolve to occupy different niches.

Resource partitioning occurs when two species occupy the same habitat but pursue slightly different resources or secure their resources in different ways, minimizing competition and maximizing success. This is commonly observed in birds, which may inhabit the same tree but spend time in different tree zones to avoid competition. Another example is the mixed flocks of swallows, swifts, and martins, which fly together and eat the same insects but have different nesting sites. Studies show that when bird species are separated, they have intermediate beak sizes, but when forced together they begin to diversify and develop beaks of many different sizes. This is *character displacement*, or a *niche shift*, and is a powerful form of evolution. Natural selection favors adaptations which minimize competition.

Predator-prey interactions

Predation is a community interaction which occurs when one organism feeds on another. In a broad sense, this includes not only carnivores killing prey, but filter feeders which strain microorganisms from the water, parasites that feed on a host and herbivores which eat plants. Herbivores that are *grazers* (grass eaters) and *browsers* (leaf eaters) are examples of animals that eat part of their prey, weakening it in the process. This is similar to most parasitic relationships.

The predator-prey relationships of the food web may fluctuate over time. Predator-prey population densities show regular peaks and valleys with the predator population often lagging slightly behind the prey. This may occur because the predator can overconsume the prey, which causes the prey population to decline, followed by the predator population. It may be the opposite, in which the prey is unable to keep pace with the prey population, causing the prey to overshoot carrying capacity and then crash. In many cases, several scenarios affect predator-prey cycling. For example, the snowshoe hare and the Canadian lynx exhibit cycling, which is affected by both predation as well as the hare's food supply.

The grouse population cycles, perhaps because the lynx switches to grouse when the hare populations decline, evidence that predators and prey do not normally exist as simple two-species systems.

Coevolution

Predator-prey interactions drive *coevolution,* in which one species evolves in response to the adaptation of another. Prey have evolved a variety of predator defenses.

Crypsis is a common adaptation which allows an organism to escape detection by predators. It may include camouflage, hiding behavior, or *mimicry* (imitation of another species). *Batesian mimicry* is when one defenseless species (the mimic) imitates another species (the model), which has successful antipredator defenses. For example, the harmless corn snake exhibits similar coloration to the venomous copperhead snake. *Mullerian mimicry* occurs when several different species, all with some defense mechanism, have coevolved to have similar coloration serving as a universal warning. This warning coloration is *aposematic coloration.* Aside from toxicity, prey may have other passive defenses that make the prey difficult, unpalatable or even fatal to eat. This may include large body size, sharp spines, tough skin or shells and a foul odor.

Defenses against predation may be active, such as fleeing from, frightening or fighting back against a predator. Active defenses are more costly in energy but can be quickly adapted to a particular situation and may be effective.

Coevolution need not be competitive, as many beneficial relationships result from coevolution. Pollination, for example, is the result of millions of years of evolution between flowers and pollinators. Flowers have evolved energy-rich nectar along with coloration and pheromones which attract the pollinators. The pollinators have evolved anatomy which is well adapted for nectar consumption and pollen transfer. This coevolution is an example of a symbiotic relationship.

Symbiotic relationships

Symbiosis is an intimate, often permanent, association between members of two populations that may (or may not) be beneficial and may be *obligatory*, where one or both organisms cannot survive without the other.

Parasitism

Parasitism involves an organism that derives nourishment from the host. This relationship benefits the parasite at the expense of the host and allows the parasite to live with minimal expenditure of energy. Parasites may be *ectoparasites*, which cling to the exterior of the host using special appendages, or *endoparasites* that live within the host. Parasites occur in all kingdoms of life, and even viruses are classified as parasitic. Many parasites have several hosts. The primary host is the main source of nutrition, while the secondary host may serve as a vector to transport the parasite to other hosts. *Parasitoids* are organisms which invariably kill their host, unlike parasites, which usually permit the host to live.

Commensalism

In *commensalism*, one species benefits and the other is neither benefitted nor harmed. It is difficult to determine true commensalism because classification must determine that one organism is not affected. For example, barnacles on a whale could be considered commensalism, but because they slow the whale, this may be parasitism. However, some relationships which may be reasonably defined as commensalism include insects that land on other animals for transport and crustaceans that inhabit the discarded shells of other organisms. Some relationships are so loose that it is difficult to know if they exhibit true commensalism. Many ecologists argue that any relationship between the two species has some subtle degree of parasitism or commensalism.

Mutualism

Mutualism is a cooperative relationship which is essentially the opposite of competition. A mutualistic relationship benefits both individuals provided the relationship is balanced. The imbalance is commonly seen in mutualistic relationships, whereby one individual violates the terms of the relationship to gain an unfair advantage. For example, *mycorrhizae* of plant roots and fungi are common mutualists, in which the fungus provides inorganic minerals to the plant in exchange for organic nutrients. However, the fungus may continue to receive nutrients but refuse to supply the plant. When this happens, the mutualistic relationship becomes parasitic.

Plant-fungus mutualism is often not obligatory, because both organisms can survive without their relationship. By contrast, the relationship between ant colonies and the bullhorn acacia tree is necessary for the acacia's survival. The ants are protected within the tree and receive nutrients, while the ants defend against herbivores and other plants that might block light to the tree. If the ants are killed, the tree is quickly overrun and dies.

Cleaning symbiosis is another common mutualism; crustaceans, fish, and birds clean ectoparasites from other animals, improving the "client's" health while providing the cleaner with food. However, cleaners may feed on the client's tissues, turning this into a case of parasitism.

Mutualistic relationships may involve the exchange of resources (e.g., mycorrhizae), a service for a resource (e.g., cleaning symbiosis, pollination), or the exchange of services (e.g., sea anemone and clownfish protecting one another from their respective predators).

Community development

All communities experience *ecological succession* as they change over time. This process can be observed over years or vast periods of geologic time. *Primary succession* occurs after an event creates or exposes a substrate that has never previously supported life. This may be bare rock left behind after a glacier moves, or a lava flow after volcanic activity. *Pioneer species* are the organisms first to colonize a newly exposed habitat. They are opportunistic, r-selected species that can tolerate harsh conditions. In primary succession, moss and lichens are the pioneer species, which die and leave behind organic matter that accumulates into the soil. After the soil is built, grasses begin to grow, followed by shrubs and finally trees. Herbivores enter the community, followed by their predators. As succession progresses, many K-selected species outcompete the pioneers.

If an existing community is sufficiently disturbed, all life will be destroyed, and the habitat reverts to a barren state. Common events are forest fires or abandonment of farmland to cause this scenario. In this case, the habitat undergoes *secondary succession.* The soil is already present and does not need to be built. Unlike primary succession, secondary succession is strongly influenced by the previous conditions of the habitat.

Clements popularized the idea that species diversity and total biomass increase until a final equilibrium, or *climax community,* is reached. This is the stable state of the community, which remains relatively unchanged until a catastrophic event destroys it. Stability of communities is seen in three ways: persistence through time, resistance to change and recovery after a disturbance.

However, most modern ecologists recognize that true climax communities are rare. Realistically, disturbances are so frequent that most communities are succession and never truly reach a climax community. Furthermore, the "steady" state of a community is dynamic. Many ecologists refer to the steady state as the *mature community* or the *old growth community.* The transitional states are known as a *seral community* or *sere.*

Clements' ideas are summarized under the *climatic climax theory*, where each region has one possible climax community, determined by the climate alone. Other factors are negligible compared to the climate. However, many ecologists subscribe to the *polyclimax theory*, which states that a region has multiple possible climax communities depending on the climate plus other environmental conditions such as topography and the nature of the disturbance. By contrast, the *climax pattern theory* proposes that there are multiple possible climax communities, but that they are influenced by both the environment *and* the species present. This theory places more emphasizes on how organisms respond to the changing environment and the way this shapes how succession proceeds. Finally, many ecologists believe that succession is cyclical, with a habitat passing through several alternating climax communities over time.

Three main models describe succession. The *facilitation model* applies to species which alter the environment to make it more hospitable to the next species. Soil building is an important example. The *inhibition model* describes species that attempt to hold on to their place in the community and make it more difficult for new species to succeed them. It assumes that each successional stage has a dominant species, which must be outcompeted or destroyed by a disturbance to allow the next species to establish itself. Finally, the *tolerance model* describes species that neither help nor hurt succeeding species. This model assumes that the climax community is made of species that can co-exist with one another.

All three models have been observed in nature, but it is difficult to predict which model a particular environment will experience.

Community biodiversity

The *intermediate disturbance hypothesis* states that a moderate level of disturbance yields the highest community diversity. Frequent disturbances may cause extinction, while few disturbances allow one species to grow dominant and outcompete others. Occasional disturbances, however, periodically inhibit dominant species and dampen competition to allow new species to enter, maximizing biodiversity. Disturbances may alter the physical environment to favor different species than before. Many communities rely on disturbances to keep them healthy, which is why forest management personnel set controlled forest fires.

Communities that experience widespread, frequent disturbances are dominated by r-strategists, while K-strategists, which outcompete others dominate undisturbed communities. Competition must be at moderate levels to maintain the most diverse communities. *Keystone predators* are essential species which regulate competition by controlling the population of species that would otherwise overrun a community. The starfish *Pisaster* is a classic example of a keystone predator that keeps the mussel *Mytilus* from outcompeting other invertebrates and algae for space.

Biodiversity is best served by intermediate migration, i.e., emigration that is sufficient to offset mortality and immigration. Isolated communities have difficulty sustaining migration and may experience extinctions. *Insular biogeography* studies how isolation affects community structure. "Island communities" may be on islands or in an isolated region, such as a lake or patch of forest surrounded by cropland. Islands which are most distant from other communities have the lowest rates of migration and usually exhibit low diversity. They are more likely to experience extinction since new individuals cannot easily emigrate and replenish the population. Insular biogeography proposes that larger islands are more capable of supporting high diversity, especially because they can sustain keystone predators.

The *spatial heterogeneity model* is a concept in biogeography that explains why heterogeneous habitats can support higher diversity. The best habitats are patchy, or heterogeneous, creating a wider variety of niches. Heterogeneity may be seen in the topography, soil, and climate. Forests usually have vertical heterogeneity, or *stratification,* which creates drastically different habitats at different heights of the forest. Spatial heterogeneity is self-reinforcing since greater heterogeneity leads to greater biodiversity, and diversity in itself is a form of heterogeneity.

Global biodiversity has been on the decline for some time due to the effects of humans. Pollution, habitat destruction and hunting cause extinctions, which have a domino effect that can lead to the collapse of entire communities. Invasive species are rampant, which is a grave threat to diversity. Invasive species often have no natural predators in the environment they are introduced to, allowing them to out-compete other species. Many communities have become overrun by invasive species in a matter of decades and are now barren. The problem has been compounded by misguided efforts to introduce an even more invasive species to cull the original invader.

In many cases, the new invaders take the place of the old one. However, some efforts have successfully suppressed invasive species. Ecologists continue to research methods to minimize the effects of human disturbances and interventions, and preserve biodiversity.

Notes

Please, leave your Customer Review on Amazon

Chapter 12

Endocrine System

- **Hormones and their Sources**

- **Mechanisms of Hormone Action**

Hormones and their Sources

The function of the endocrine system:
specific chemical control at cell, tissue and organ levels

The *endocrine system* synthesizes and secretes hormones into the bloodstream ("*endo*" means "within"). The target cell of a particular hormone has a specific receptor that allows it to respond to the stimulus. The *exocrine* system secretions travel through ducts to external environments. Several glands in the body create different hormones of the endocrine system, such as the sudoriferous (sweat), sebaceous (oil), mucous, digestion and mammary glands. A few glands, such as the pancreas, have both endocrine and exocrine functions.

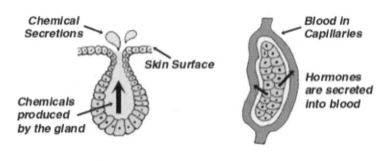

Exocrine Gland   Endocrine Gland

Definitions of endocrine gland and hormones

Hormones are molecules created by glands that affect distant organs through the circulatory system. Hormones regulate metabolism and other functions of the cells. Even small amounts of the hormone may have a widespread effect. Hormones can be divided into two categories based on what they stimulate: *direct hormones* (or *non-tropic hormones*) are involved in directly stimulating target hormones, and *tropic hormones* stimulate other endocrine glands.

Endocrine glands are ductless glands in the endocrine system that secrete hormones into the bloodstream which affect specific target cells. A single gland may produce several different hormones. Multiple glands may produce the same hormone. The principal human endocrine glands include the hypothalamus, the pineal, and the pituitary glands, all located in the brain; the thyroid and parathyroid glands (located in the neck); the ovaries (located in the abdomen); the testes (in the scrotum); and the thymus (in the thorax).

The chemical demands of an organism vary by daily function and life cycle. When a particular demand is recognized, genes that encode for a particular hormone are transcribed, resulting in the secretion of the hormone to the bloodstream and eventually resulting in an effect on the target cells. In general, hormones instruct cells in tissues to devote their resources to the production of proteins. For example, human adolescents require the production of gonadotropic hormones, which target the sex organs during puberty. Adolescent changes include the ability to produce functional gametes for reproduction, as well as growth, the development of secondary sex characteristics.

Posterior Pituitary Hormones

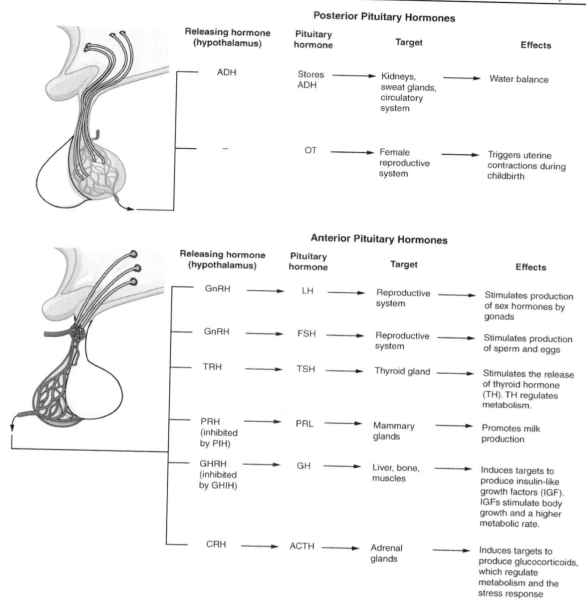

Releasing hormone (hypothalamus)	Pituitary hormone	Target	Effects
ADH	Stores ADH	Kidneys, sweat glands, circulatory system	Water balance
–	OT	Female reproductive system	Triggers uterine contractions during childbirth

Anterior Pituitary Hormones

Releasing hormone (hypothalamus)	Pituitary hormone	Target	Effects
GnRH	LH	Reproductive system	Stimulates production of sex hormones by gonads
GnRH	FSH	Reproductive system	Stimulates production of sperm and eggs
TRH	TSH	Thyroid gland	Stimulates the release of thyroid hormone (TH). TH regulates metabolism.
PRH (inhibited by PIH)	PRL	Mammary glands	Promotes milk production
GHRH (inhibited by GHIH)	GH	Liver, bone, muscles	Induces targets to produce insulin-like growth factors (IGF). IGFs stimulate body growth and a higher metabolic rate.
CRH	ACTH	Adrenal glands	Induces targets to produce glucocorticoids, which regulate metabolism and the stress response

Releasing hormones from the hypothalamus and the effects on the posterior and anterior pituitary

The releasing hormone from the hypothalamus finds its target (posterior or anterior pituitary cell). It binds to the receptor, where a chemical messenger instructs the cell to release the desired hormone. Finally, the secreted hormone targets a specific tissue (e.g., thyroid gland) and instruct the tissue to produce a particular substance.

Major endocrine glands: names, locations, and products

Humoral glands directly respond to blood levels of ions and nutrients, such as the parathyroid hormone's response to low blood calcium levels. Neural glands release hormones when stimulated by a nerve impulse or an action potential. These glands are involved primarily with the fight or flight response. Hormonal glands release hormones when stimulated by other hormones from another gland.

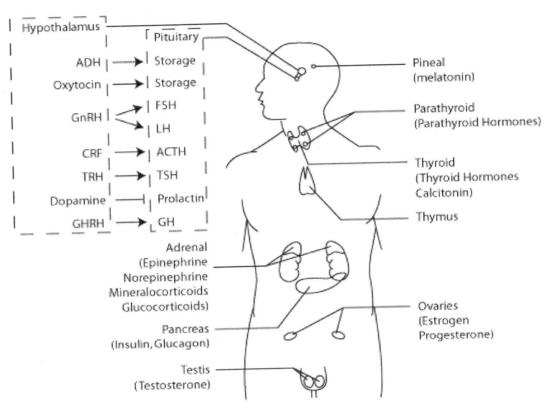

Endocrine glands and associated hormones

The *hypothalamus* monitors the external and internal conditions of the body. It regulates the internal environment through the autonomic nervous system, affecting heartbeat, temperature and water balance. Additionally, the hypothalamus contains neurosecretory cells that link it to the pituitary gland, where it controls glandular secretions. The hypothalamus secretes *ADH* (vasopressin), *oxytocin* and *GnRH* (gonadotropin-releasing hormone).

The *pituitary gland* situated at the back of the brain is responsible for the creation of many hormones related to growth. The pituitary gland is connected to the hypothalamus by a stalk-like structure and therefore can store hormones produced by the hypothalamus. It is comprised of two parts: the posterior pituitary and the anterior pituitary.

The *posterior pituitary* does not synthesize hormones; rather, it stores vasopressin and oxytocin produced by the hypothalamus and secretes these hormones into the bloodstream. The posterior pituitary contains neurosecretory cells that originate in the hypothalamus and respond to neurotransmitters.

The *anterior pituitary* (adenohypophysis) functions in the regulation of hormone production from other glands. The hypothalamus regulates the stimulation of the anterior pituitary. The hypothalamus controls the release of anterior pituitary hormones through a *portal system* consisting of two capillary systems connected by a vein.

The anterior pituitary produces seven types of hormones: (1) thyroid stimulating hormone (TSH), (2) adrenocorticotropic hormone (ACTH), (3) gonadotropic follicle stimulating hormone (FSH), (4) gonadotropic luteinizing hormone (LH), (5) prolactin, (6) melanocyte stimulating hormone (skin color change in fishes, reptiles and amphibians) and (7) growth hormone (GH), somatotrophic hormone.

The regulation of the pituitary gland is via a negative feedback mechanism and by the secretion of releasing and inhibiting hormones. The hypothalamus produces hypothalamic-releasing and hypothalamic-inhibiting hormones which pass to the anterior pituitary by the portal system. *Releasing hormones* that originate in the hypothalamus target cells in the anterior and posterior pituitary to stimulate the production and secretion of a particular hormone. Conversely, *inhibiting hormones* released from the hypothalamus target cells in the anterior pituitary to inhibit the production and secretion of a particular hormone. In general, each hormone from the anterior pituitary has a specific releasing and inhibiting hormone from the hypothalamus that controls its release.

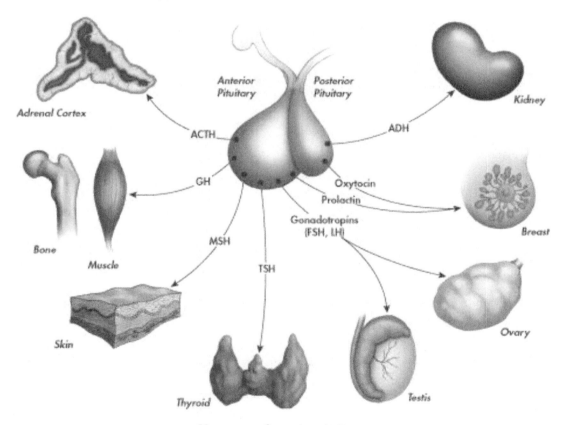

Hormones of anterior pituitary

Tropic effects:	FSH, LH, TSH, ACTH, ADH
Non-tropic effects:	Prolactin, oxytocin, MSH
Non-tropic and tropic effect:	GH

Tropic hormones have other endocrine glands as targets.

The *thyroid gland* is located in the neck, attached to the ventral surface of the trachea just below the larynx. The thyroid can be stimulated by the anterior pituitary gland, which secretes *thyroid-stimulating hormone* (TSH). Thyroid hormones increase metabolism and require iodine to function adequately. The two hormones, thyroxine (T_4) and triiodothyronine (T_3), are produced by follicles of the thyroid gland. These hormones possess four and three iodine atoms, respectively. Iodine is necessary for growth and neurological development in children and increases in basal metabolic rates in the body. Iodine, actively transported into the thyroid, may reach concentrations 25 times greater than in the bloodstream. Iodine deficiency causes the enlargement of the thyroid (goiter). Goiter is easily prevented by supplementing with fortified salt, which contains iodine. The thyroid gland also produces *calcitonin*, which decreases the calcium levels in the blood.

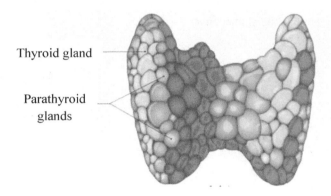

Thyroid gland

Parathyroid glands

Posterior view of thyroid showing parathyroid glands

Hyperthyroidism (Graves' disease) occurs when the thyroid gland is enlarged or overactive. The eyes protrude because of edema in the eye socket tissue as exophthalmic goiter. Additional symptoms include an increased metabolic rate and sweating. Removal or destruction of some thyroid tissue by surgery or radiation often cures it.

Hypothyroidism is attributed to decreased secretion of the thyroid gland. It lowers the heart rate and respiratory rate. Other disorders of the thyroid are *achondroplasia* (dwarfism) and *progeria* (premature aging). *Cretinism*, a condition of severely stunted physical and mental growth, occurs in individuals suffering from hypothyroidism since birth. Thyroid treatment can help but must be begun in the first two months of life to prevent developmental delay.

Four *parathyroid glands* are embedded in the posterior surface of the thyroid gland in a conformation resembling four peas. Parathyroid glands produce *parathyroid hormone* (PTH). Low calcium levels stimulate the release of parathyroid hormone, which increases calcium levels by stimulating osteoclasts to release calcium from the bone.

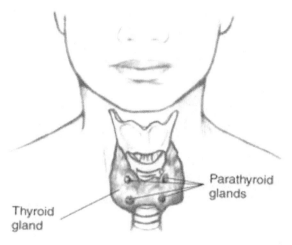

Anterior view of the thyroid gland. Parathyroid glands are on the posterior surface

The two *adrenal glands* are on top of each kidney. Each gland consists of two sections: an outer *adrenal cortex* and an inner *adrenal medulla*. The hormones released from the adrenal cortex provide a sustained response to stress. The adrenal cortex secretes *glucocorticoids* and *mineralocorticoids*, which are both steroid hormones. It secretes a small amount of male and female sex hormones in each sex. The adrenal medulla releases *epinephrine* and *norepinephrine*, hormones that provide an immediate response to stress (as opposed to the sustained response provided by the hormones from the adrenal cortex).

The hypothalamus exerts control over both adrenal glands. Nerve impulses travel via the brain stem through the spinal cord to sympathetic nerve fibers to the medulla. The hypothalamus uses ACTH-releasing hormone to control the anterior pituitary's secretion of ACTH, a hormone that stimulates the adrenal glands to release a variety of hormones including *dehydroepiandrosterone* (DHEA) and *cortisol*. Adrenal hormones increase during times of physical and emotional stress.

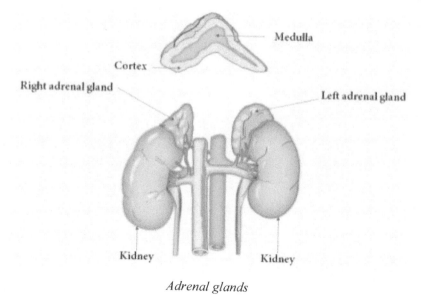

Adrenal glands

The *pancreas* lies transverse in the abdomen between the kidneys and near the duodenum of the small intestine. The pancreas is composed of both exocrine and endocrine tissue. Exocrine tissue produces and secretes digestive juices into the small intestine via the ducts. *Pancreatic islets of Langerhans* are endocrine tissues that produce *insulin* and *glucagon*.

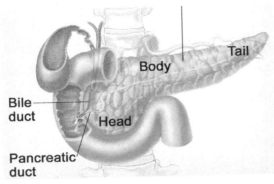

Pancreas

The islets of Langerhans contain two cell types, alpha cells, and beta cells. *Alpha cells* of the islet secrete glucagon. The release of glucagon is catabolic and occurs when the energy charge is low, working to raise blood glucose levels. The term energy refers to a measure of the energy status of biological cells and is related to ATP, ADP and AMP concentrations. *Beta cells* secrete insulin, which is secreted anabolically and is released when the energy charge is high, where the beta cells work to lower the blood glucose levels. Insulin stimulates the liver and other body cells to absorb glucose. All body cells utilize glucose; therefore, its level is subject to homeostasis and must be tightly regulated.

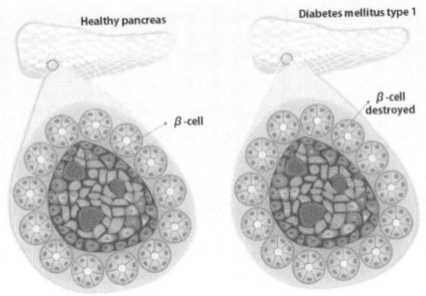

Pancreatic Islets of Langerhans; beta cells secrete insulin in non-diabetic patients

Male *testes* are located in the scrotum and function as the human gonads, producing *androgens*, including the male sex hormone *testosterone*. Testosterone stimulates the development of male secondary sex characteristics, such as large vocal cords, increased muscle mass and facial hair.

Female sex hormones include estrogens and progesterone, which are secreted by the *ovaries* (the female gonads). *Estrogen* secreted at puberty stimulates the maturation of ovaries and other sexual organs. *Progesterone* is a steroid hormone that is important for the menstrual cycle, in pregnancy and embryogenesis of humans and other species.

The *pineal gland*, located near the center of the brain, produces *melatonin*, primarily at night. Melatonin helps establish *circadian rhythms*, which are the basis for a 24-hour physiological cycle. The pineal gland may be involved in human sexual development; children with a damaged pineal gland due to a brain tumor tend to experience puberty earlier.

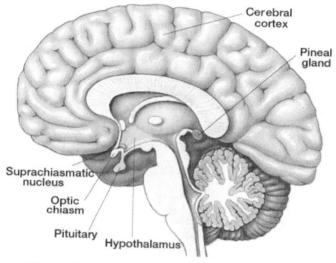

The pituitary gland, hypothalamus and pineal gland

The *thymus* is a lobular gland that lies just beneath the sternum in the upper thoracic cavity. Thymus hormones (*thymo-, thymic*) stimulate the development of T cells and are involved in the immune response. The thymus reaches its largest size and is most active during childhood; with age, it shrinks and becomes fatty.

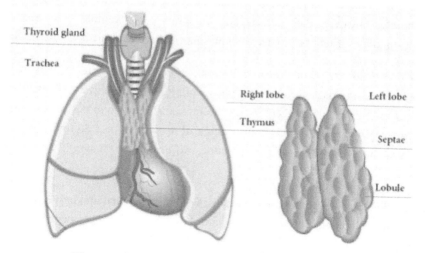

Thymus gland stimulates the development of T cells

Some lymphocytes that originate in the bone marrow pass through the thymus and change into T lymphocytes. The thymus produces and secretes *thymosins* which aid in the differentiation of T cells and may stimulate other immune cells.

Major types of hormones

There are three major classes of hormones: peptide hormones, steroid hormones, and amino acid-derived hormones.

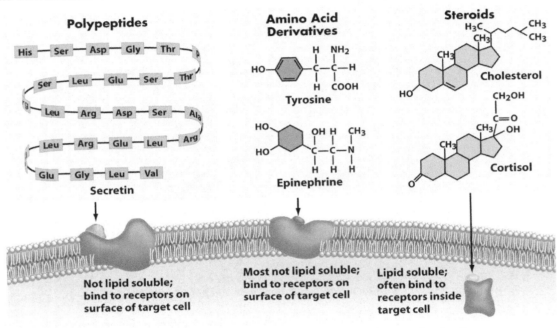

Three classes of hormones and schematic of their mode of action

Peptide hormones are made from amino acid as peptides or proteins. Their specific receptors are on cell surfaces. Their mechanism of action utilizes the secondary messenger system. They affect their target organs rapidly; however, their effects are temporary. An example of a peptide hormone is calcitonin, which stimulates osteoblast to decrease plasma calcium levels as bone density increases.

Steroid hormones are made from cholesterol. Their receptors are located in the cytosol (or nucleus) of a cell. Steroids pass through the phospholipid bilayer of the cell membrane. Once the hormone binds to an intracellular receptor, the hormone-receptor migrates to the nucleus, and the complex binds to a DNA target to affect transcription of specific genes. Steroid hormones affect their target organs slowly, but their effects are long-lasting. An example of a steroid hormone is the growth hormone.

Amino acid-derived hormones are a third class of hormones that are derived from the amino acid tyrosine. Amino acid-derived hormones are catecholamines (e.g., epinephrine, norepinephrine, dopamine). Epinephrine is released from the adrenal medulla of the adrenal glands as the "fight-or-flight" response. Amino acid-derived hormones are generally small molecules.

Comparison of peptide, steroid and amino acid-derived hormones

	Peptide hormones	Steroid hormones	Amino acid derivatives	
			Catecholamines	Thyroid hormones
Synthesis and storage	Made in advance; stored in secretory vesicles	Synthesized on demand from precursors	Made in advance; stored in secretory vesicles	Made in advance; precursor stored in secretory vesicles
Release from the parent cell	Exocytosis	Simple diffusion	Exocytosis	Simple diffusion
Half-life	Short	Long	Short	Long
Transport in blood	Dissolved in plasma	Bound to carrier proteins	Dissolved in plasma	Bound to carrier proteins
Location of receptor	Cell membrane	Cytoplasm or nucleus; some have membrane receptors also	Cell membrane	Nucleus
Response to receptor-ligand binding	Activation of second messenger systems; may activate genes	Activation of genes for transcription and translation; may have non-genomic actions	Activation of second messenger systems	Activation of genes for transcription and translation
General Target Response	Modification of existing proteins and induction of new protein synthesis	Introduction of new protein synthesis	Modification of existing proteins	Induction of new protein synthesis
Examples	Insulin, parathyroid hormone	Estrogen, androgens, cortisol	Epinephrine, norepinephrine, dopamine	Thyroxine (T_4)

Peptide hormones

The majority of hormones are peptide hormones. They are initially synthesized together as larger *preprohormones*, which are then cleaved to inactive *prohormones* in the lumen of the rough ER. The prohormone is then cleaved to form the active hormone in the Golgi apparatus. In the Golgi, they may be modified with carbohydrates.

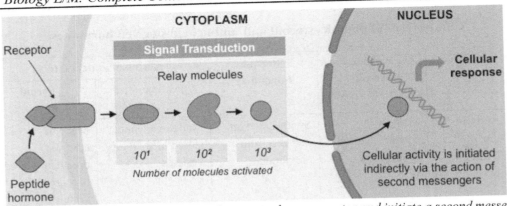

Peptide hormones bind as a ligand to a plasma membrane receptor and initiate a second messenger cascade to amplify the cellular response. The hormone may initiate phosphorylation or dephosphorylation of molecules in the cytoplasm as the mechanism for cellular response

Insulin is a peptide hormone secreted by the beta (β) cells of the islets of Langerhans, in the pancreas. Its secretion is increased during the absorptive state and decreased during the post-absorptive state. Insulin targets are muscle, adipose, and liver tissues. The main roles of insulin are to 1) stimulate the movement of glucose from extracellular fluid into the cells by facilitated diffusion, 2) stimulate glycogen synthesis and 3) inhibit glycogen catabolism. Insulin is an anabolic building hormone and inhibits protein degradation. It promotes cell division and differentiation because it is required for the production of *insulin-like growth factors* (IGF-I), which is important for the regulation of normal physiology. Insulin is controlled by an increase in plasma glucose or amino acid concentration. *Glucose-dependent insulinotropic peptide* (GIP) is a hormone secreted in the GI tract to stimulate insulin secretion. Parasympathetic nerve fibers stimulate insulin secretion.

Hypoglycemia refers to low plasma glucose levels that result from an excess of insulin (beta cells) or a deficiency of glucagon (alpha cells). *Glucagon* is a peptide hormone secreted by alpha (α) cells of the pancreas. Its target is the liver tissue, and its actions are antagonistic to insulin. Glucagon increases glycogen breakdown and gluconeogenesis to increase the plasma concentration of glucose during the post-absorptive state or when plasma glucose is low (hypoglycemia). Sympathetic nerves stimulate glucagon secretion.

Atrial natriuretic hormone (ANH) is a peptide hormone secreted by cardiac (heart) cells when the atria of the heart are stretched due to increased blood volume. ANH inhibits the secretion of renin by the kidneys, inhibits the secretion of *aldosterone* from the adrenal cortex and decreases sodium reabsorption. When sodium is excreted along with water, the blood volume, and pressure decrease.

Antidiuretic hormone (ADH) is a peptide hormone produced in the hypothalamus. Antidiuretic hormone (ADH) increases reabsorption of water by increasing the permeability of the nephron's collecting ducts, which results in water reabsorption and increased blood volume and pressure. ADH is released when there is low water content (high osmolarity) in the blood. Caffeine blocks ADH, tricking the brain that the body is over-hydrated. This causes an increased volume of fluid output, which is why coffee drinkers may urinate more often.

Calcitonin is a peptide hormone produced by the thyroid gland. Calcitonin lowers calcium levels in the blood by inhibiting the release of calcium from bone. Calcitonin increases deposits in the bone by reducing the activity and number of *osteoclasts* (i.e., bone cells that resorbs bone tissue). Calcium is necessary for blood clotting. If blood calcium is lowered, the release of calcitonin is inhibited.

The *parathyroid hormone* (PTH), a peptide hormone produced by the parathyroid gland, stimulates the absorption of Ca^{2+} by activating vitamin D, retaining Ca^{2+}, excreting phosphates by the kidneys and demineralizing bone by promoting the activity of osteoclasts. PTH is antagonistic to calcitonin. When the blood calcium level reaches the right level, the parathyroid glands inhibit the synthesis of PTH. If PTH is not produced in response to low blood Ca^{2+}, the body goes into tetany. In *tetany*, the body shakes uncontrollably from involuntary, continuous muscle spasms. Ca^{2+} is vital to proper nerve conduction and muscle contraction.

Gonadotropins are peptide hormones produced in the adenohypophysis. The two principal gonadotropins, *follicle-stimulating hormone* (FSH) and *luteinizing hormone* (LH), act on the gonads (ovaries and testes) to secrete sex hormones. FSH stimulates the maturation of ovarian follicles in females, and in males, it acts on the Sertoli cells of the testes to stimulate sperm production (spermatogenesis). LH triggers ovulation and the formation of the corpus luteum in females. In males, it stimulates the interstitial cells of testes to produce testosterone. *Gonadotropin-releasing hormones* (GnRH) is released from the hypothalamus to stimulate the production of the gonadotropins.

Oxytocin is a peptide hormone produced in the hypothalamus which stimulates uterine muscle contraction in response to uterine wall nerve impulses, markedly during childbirth. It stimulates the release of milk from mammary glands for nursing.

Prolactin, a peptide hormone produced in the adenohypophysis, is secreted after childbirth; it causes the mammary glands to produce milk and plays a role in carbohydrate and fat metabolism. The neurotransmitter dopamine inhibits prolactin.

Gastrin is a peptide hormone important during digestion in the stomach. It stimulates the secretion of HCl. *Secretin* is in the small intestine and is activated when acidic food enters from the stomach. It neutralizes the acidity of chyme (partly-digested food) by the secretion of alkaline bicarbonate. *Cholecystokinin* (CCK), found in the small intestine, causes contractions of the gall-bladder and the release of bile, which is involved in the digestion of fats.

Growth hormone (GH), or somatotropic hormone, is a peptide hormone produced in the adenohypophysis that promotes skeletal and muscular growth. GH acts to stimulate the transport of amino acids into cells and to increase the activity of ribosomes. GH promotes fat metabolism rather than glucose metabolism. When there are low plasma levels of GH, *growth hormone releasing hormone* (GHRH) is released from the hypothalamus, and GHRH stimulates GH production.

Gigantism is a condition that occurs from oversecretion of growth hormone during childhood and results in a person who is significantly taller and bigger than a standard human. *Acromegaly* is a condition that occurs when too much growth hormone is produced during adulthood and results in the disproportioned growth of certain areas of the body (the parts that still respond to growth hormone).

Adrenocorticotropic hormone (ACTH) is a peptide hormone that stimulates the adrenal cortex to release glucocorticoids and cortisol, which are steroid hormones involved in the regulation of the metabolism of glucose.

Steroid hormones

Steroid hormones are produced by the adrenal cortex and the gonads and are specifically synthesized from cholesterol in the smooth ER. They are lipids and can freely diffuse across membranes, but require protein transport molecules to dissolve in the blood. All steroid hormones have the same complex of four carbon rings but have different side chains. Steroid hormones include the glucocorticoids (e.g., cortisol) and mineralocorticoids (e.g., aldosterone) of the adrenal cortex. They include the gonadal hormones: estrogen, progesterone, and testosterone. The placenta produces estrogen and progesterone.

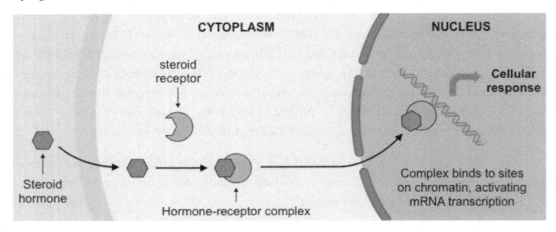

Steroid hormones pass the plasma membrane and bind to an intracellular receptor which then migrates as a complex into the nucleus to affect gene transcription

Glucocorticoids are stress hormones that help regulate blood glucose levels. They raise blood glucose levels by stimulating gluconeogenesis in the liver and affect fat and protein metabolism. Like glucocorticoids, *cortisol* increases energy by raising blood glucose levels. Cortisol inhibits the immune system and exerts antigrowth effects by stimulating protein catabolism. Cortisol and corticosterone affect the metabolism of glucose and other organic nutrients. Cortisol counteracts the inflammatory response and helps to medicate arthritis and bursitis. When the body is under stress, corticotropin-releasing hormone (CRH) from the hypothalamus is released, which then triggers ACTH from the adenohypophysis to be released.

Aldosterone, a type of mineralocorticoid produced in the adrenal cortex, participates in mineral balance by regulating Na^+, K^+ and H^+ ions by the kidney. Its primary role is helping the kidneys move sodium and water from urine into the bloodstream, thus regulating the body's sodium and potassium balance. Additionally, it acts to reduce the sodium sensitivity of taste buds, and it acts on sweat glands to reduce sodium loss during perspiration. Aldosterone is stimulated by a decrease in blood volume and acts by increasing Na^+ reabsorption.

Hyposecretion (i.e., diminished secretion) of glucocorticoids and mineralocorticoids results in *Addison's disease*. When ACTH is in excess, it can lead to the buildup of melanin. The lack of cortisol results in low glucose levels. This can lead to severe fatigue, perpetuated and worsened by stress. The lack of aldosterone lowers blood sodium levels; a person then has low blood pressure and dehydration. Left untreated, Addison's disease can be fatal.

Hypersecretion (i.e., excessive secretion) of corticosteroids result in *Cushing's syndrome*. Excess cortisol leads to changes in carbohydrate and protein metabolism and causes a tendency toward diabetes mellitus. As a result, muscular protein decreases and subcutaneous fat collects, creating an obese torso, while arms and legs remain normally proportioned. A "puffy" appearance typically characterizes sufferers of this syndrome.

Gonadocorticoids are another group of steroid hormones produced in the adrenal cortex. They are responsible for the onset of puberty and the onset of the female libido.

Adrenal *androgens*, responsible for the development of male sex characteristics, are produced in the adrenal cortex and regulated by ACTH. The most potent androgen, testosterone, is produced primarily in the testes. The adrenal androgens are weak steroids and some function as precursors to testosterone (such as DHEA, which is androstenolone). In addition to influencing male development, androgens play some role in female puberty and the adult female. The renin-angiotensin-aldosterone system controls mineralocorticoid secretion. Under low blood volume and sodium levels, the kidneys secrete renin, which cleaves angiotensin I, promoting a cascade that increases blood pressure.

Testosterone is the major androgen secreted by the testes. Testosterone is largely responsible for the male sex drive. Anabolic steroids are synthetic variants of testosterone that can be used to treat some hormone problems in men. Testosterone affects the sweat glands and the expression of the baldness gene, among many other characteristics and functions of the body.

Estrogen, a female sex hormone secreted by the ovaries, is necessary for oocyte development. It is responsible for the development of female secondary sex characteristics including the formation of a layer of fat beneath the skin and a larger pelvic girdle. Estrogen and progesterone are required for breast development and the regulation of the uterine cycle.

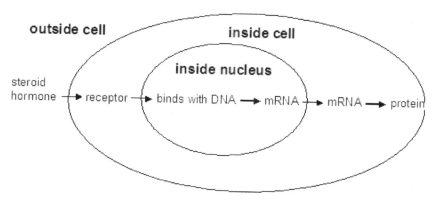

Steroid hormone pathway mechanism

Amino acid-derived hormones

Amino acid-derived hormones are usually derivatives of the amino acid tyrosine. Enzymes form *tyrosine derivative*s in the cytosol or on the rough ER. The adrenal medulla secretes the amino acid-derived hormones *epinephrine* (adrenaline) and *norepinephrine* (noradrenaline), which stimulate "fight or flight" responses. They are considered to be stress hormones because both hormones bring about body changes corresponding to an emergency. The fight-or-flight response initiates the glycogen to glucose conversion, vasoconstriction to the internal organs and skin, vasodilation to the skeletal muscles and increased heartbeat. This causes the blood glucose level and the oxygen content to rise and metabolic rate to increase. Also, the bronchioles dilate and breathing rate increases. Blood vessels to the digestive tract and skin constrict, while blood vessels to the skeletal muscles dilate and the cardiac muscle contracts more forcefully, causing the heart rate to increase.

The adrenal medulla secretes more epinephrine than norepinephrine. Both hormones are catecholamines. *Catecholamines* solubilize in water, dissolve in blood, bind to receptors on the target tissue and mainly act via the secondary messenger system.

Norepinephrine is formed from the amino acid tyrosine, and *epinephrine* is formed from norepinephrine. Although the effects of these two catecholamines are generally similar, there are differences. Norepinephrine constricts almost all of the body's blood vessels, while epinephrine constricts only the minute blood vessels and dilates the larger blood vessels of the liver and skeletal muscle. Epinephrine increases metabolic rate by the calorigenic effect and stimulates the catabolism of glycogen and triacylglycerols. It increases energy by raising glucose and oxygen content in the blood. It is released when the autonomic nervous system must respond to stress.

Other types of amino acid-derived hormones include the two iodine-containing hormones *thyroxine* (T_4) and *triiodothyronine* (T_3), which are secreted by the follicles of the thyroid glands. T_4 is secreted in larger amounts but is mostly converted to T_3, the more active form. *Thyroid hormones* (TH) are lipid soluble, require a protein carrier in blood and bind to receptors in the nucleus. TH regulate oxygen consumption, growth and brain development. Thyroxine is responsible for controlling the body's metabolic rate and therefore is responsible for the amount of energy consumed and the volume of proteins produced. Thyroxine is released when there is a low concentration of it in the blood. *Hyperresponsiveness* is the hypersecretion of thyroid hormones and can lead to hyperresponsiveness to epinephrine and an increase in heart rate.

Although tyrosine is the primary amino acid from which this class of hormones is derived, other amino acids are hormone precursors. For example, the amino acid tryptophan is the precursor to melatonin. *Melatonin*, produced in the pineal gland, targets the body as a whole. It aids in maintaining balanced circadian rhythms (daily biological cycles), such as sleep patterns. *Glutamic acid* is an amino acid and the precursor to *histamine*, a hormone that is part of the body's natural allergic response. Histamine plays an important role in the immune system.

Neuroendocrinology:
the relationship between neurons and hormonal systems

Secretion of some hormones is under the control of the nervous system. This was first recognized in the brain's role (specifically the hypothalamus) in stimulating the pituitary gland to release hormones. The nervous system controls hormone release based on the body's current state of being. For example, control through the hypothalamus is seen when blood glucose levels are higher under times of stress. The endocrine system is significantly slower than the nervous system, and hormones can modulate the nervous system. For example, low estrogen levels during menses tend to lead to mood disruptions that are characteristic of premenstrual syndrome (PMS).

Neurons secrete *hypophysiotropic hormone*s in response to action potentials. Each of these hormones is named after the anterior pituitary hormone that it controls.

Endorphins, technically a neuro-hormone, inhibit the perception of pain.

Mechanisms of Hormone Action

Cellular mechanisms of hormone action

Lipid soluble hormones cross the plasma membrane and directly activate genes.

Water soluble hormones are unable to cross the plasma membrane. Instead, they bind to membrane receptors on the outside of cells. Secondary messengers (e.g., tyrosine kinase or G-coupled proteins) then relay the signal inside the cell. An example of a secondary messenger is cyclic AMP (cAMP). In a *cAMP pathway*, an amino acid hormone binds to a membrane receptor. A G-protein is then activated alongside adenylate cyclase, and cAMP is produced (discussed elsewhere).

In the *phospholipid pathway,* an amino acid hormone binds to a membrane receptor, in which a G-protein is activated. *Phospholipase C* is activated, and the membrane phospholipids split into two secondary messengers: *diacylglycerol* (DAG) and *inositol triphosphate* (IP$_3$). DAG triggers a protein kinase cascade and IP$_3$ releases Ca^{2+} from the ER.

Hyposecretion occurs when a gland is secreting too little hormone because it is unable to function normally; the disorder is *primary hyposecretion*. Possible causes include a genetic lack of an enzyme, dietary deficiency of a precursor or an infection.

Primary hypersecretion occurs when a gland secretes too much hormone. *Secondary hypersecretion* is the excessive stimulation of a gland by its tropic hormone.

Hyporesponsiveness occurs when target cells do not respond to a hormone due to a deficiency of receptors, a defect in the signal transduction mechanism or a deficiency of an enzyme that catalyzes the activation of the hormone. In diabetes mellitus, the target cells of the hormone insulin are hyporesponsive.

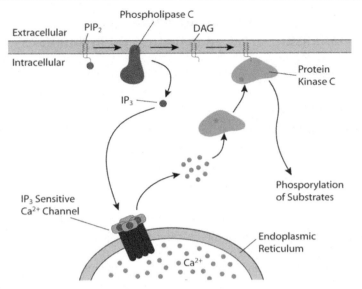

Phospholipid pathway: Cleavage to IP₃ and DAG to release calcium for the ER

Transport of hormones through blood

Hormones travel long distances via the blood and lymph. *Unbound hormones* (peptide hormones) are water-soluble hormones that dissolve in blood plasma, while *bound hormones* (steroid hormones) circulate in blood while bound to plasma proteins (e.g., serum carrier or transport proteins). These transport proteins are produced by the liver and bind to hormones within the serum. Free hormones diffuse across capillary walls to encounter their target cells.

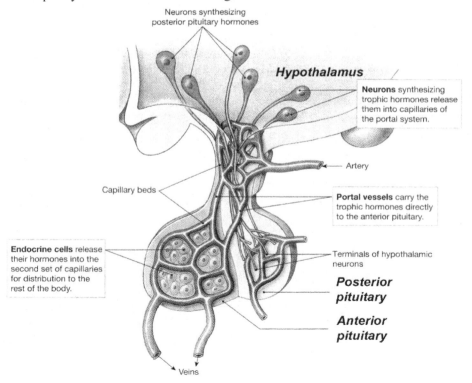

Copyright © 2019 Sterling Test Prep. Any duplication (copies, uploads, PDFs) is illegal.

Hypothalamic-hypophyseal portal circulation permits the transportation of neuro-hormones by the neuroendocrine cells of the hypothalamus directly to the cells of the pituitary. Usually, these hormones are excluded from the general circulation. In this system, capillaries from the hypothalamus go through the plexus of veins around the pituitary stalk and then finally into the anterior pituitary gland.

Hepatic portal circulation is a type of circulation that allows for the transportation of hormones from the Islets of Langerhans in the pancreas. The hormones of the pancreas include insulin and glucagon. Various nutrients absorbed from the intestine and transported into the liver use this pathway. This pathway occurs before general circulation occurs. In this system, capillaries originating from the gastrointestinal tract and the spleen merge to form the portal vein. The vein then enters the liver and divides to form portal capillaries.

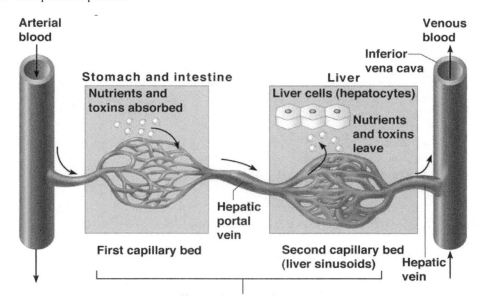

Hepatic portal circulation

Specificity of hormones at the target tissue

Hormones can reach all tissues via the blood, but only cells that have specific receptors act as target cells for a particular hormone. Specificity depends on the unique receptors on the target cells and the lack of receptors on non-target cells for the hormone. Hormones bind to cells with receptor sites for that specific hormone. Peptide hormone receptors are on the surface of the cell, while steroid hormones bind a receptor in the cytoplasm (or nucleus). Cells can either *upregulate* (increase production) or *downregulate* (decreased production) the receptors they express. A low concentration of a hormone may be compensated for by an increase in the number of receptors while a high concentration of a hormone can lead to a decrease in the number of receptors.

Receptors for peptide hormones and catecholamines are present on the extracellular surface of the plasma membrane, while those for steroids are mainly in the cytoplasm. Hormones that bind to surface receptors can influence ion channels, enzyme activity, G proteins, and secondary messengers.

Genes can be activated or inhibited, resulting in a change in the synthesis rate of proteins encoded for by these genes.

Some hormones can reduce the number of receptors available for a second hormone, resulting in decreased effectiveness of the second hormone. The hormone that blocks the second hormone is the *antagonist,* and this process is *antagonism.* A hormone can induce an increase in the number of receptors for a second hormone, increasing the latter's effectiveness, a process of *permissiveness.*

There are specific mechanisms of cell-to-cell signaling pathways that allow hormones to bind to various target cells. In *endocrine signaling*, hormones are distributed into the blood and bind to long-distance target cells.

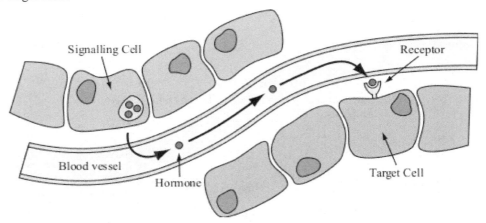

Hormones are released from a gland, travel in the blood and affect a target cell that has the receptor

Autocrine signals are local chemicals that are released by cells that bind to receptors on the same cell. *Paracrine* signals act through short distances on other cells. Both of these signals usually involve messengers with short lifespans since a short distance needs to be covered. Neurotransmitter secretions are a type of paracrine signaling.

Some signals are local since they affect neighboring cells and do not travel through the bloodstream. *Prostaglandins* are potent chemical signals produced within cells from arachidonate, a fatty acid. For example, prostaglandins cause the uterus to contract and are involved in menstrual discomfort. The effect of aspirin on reducing fever and controlling pain is due to its effect on prostaglandins. *Growth factors* promote cell division and mitosis.

*Pheromone*s are chemical signals that act at a distance between individual organisms. Ants produce a pheromone trail for other members of the ant colony to find food. Female silkworm moths release pheromones to lure male moths from miles away. Axillary secretions from the armpits of men and women are thought to have some effect on the opposite sex. Some women may prefer the axillary odor of men with a different plasma membrane protein. Also, women may synchronize their menstrual cycles with other women living nearby.

Integration with the nervous system by feedback control

The effects of hormones are controlled by negative feedback and by antagonistic hormone action.

Negative feedback is where the product of a process decreases the rate of that process. *Long-loop negative feedback* occurs when the last hormone in a chain of control can exert negative feedback on the hypophysis-pituitary system. If an anterior pituitary hormone exerts a negative feedback effect on the hypothalamus, it is *short-loop negative feedback*. This mechanism is seen in pituitary hormones that do not influence other endocrine glands. The pancreas produces insulin when blood glucose rises, causing the liver to store glucose. When glucose is stored, the blood glucose level goes down, and the pancreas stops insulin production.

Antagonistic actions of hormones are important factors in regulation. For example, the effect of insulin is offset by the production of glucagon by the pancreas. The thyroid hormones lower blood calcium levels, while the parathyroid hormones raise blood calcium levels.

The concentration of a hormone in the plasma depends on its rate of secretion and rate of removal. The kidneys can excrete hormones or they are metabolized by their target cells. Plasma concentrations of ions (or nutrients) may control the secretion of a hormone, and the hormone may control the concentration of its regulators through negative feedback.

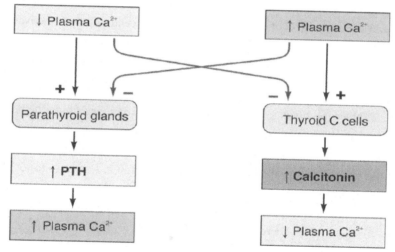

Positive and negative feedback loops to regulate plasma Ca²⁺ levels

When nerve cells in the hypothalamus determine that blood is too concentrated, ADH is released, and the kidneys respond by reabsorbing water. As the blood becomes more dilute, ADH is no longer released, an example of negative feedback. Oxytocin is made in the hypothalamus and stored in the posterior pituitary; it is involved in *positive feedback* as it increases intensity. Such positive feedback does not maintain homeostasis.

During the normal control of hormones, *humoral glands* directly respond to chemical levels in the blood (e.g., parathyroid responds to low blood calcium). *Neural glands* release hormones when stimulated by nerves (fight-or-flight response). *Hormonal glands* release hormones when stimulated by other hormones (tropic hormones).

Regulation by second messengers

The secondary messenger system utilizes non-steroid hormones. In this system, the hormone binds to a membrane receptor but does not enter the cell. As soon as the hormone binds to a membrane receptor, it initiates some reactions that activate an enzyme. The binding of the hormone receptor causes the adenylate cyclase, located in the membrane, to be activated. This activation converts ATP to cAMP, and cAMP activates protein kinases. These protein kinases activate various enzymes, stimulate cellular secretions and open ion channels.

Peptide hormones cannot enter a cell on their own, so they use vesicles to cross the membrane. Typically, they act on the surface receptors via secondary messengers. The peptide hormones bind to a receptor protein on the plasma membrane, a process of receptor-mediated endocytosis. Epinephrine is an example of a peptide hormone that binds to a receptor protein. A relay system then leads to the conversion of ATP to cAMP. Peptide hormones are considered first messengers, cAMP, and calcium second messengers. The second messenger may set an enzyme cascade in motion. Activated enzymes can be used repeatedly, resulting in a thousand-fold response. These hormones may serve as neurotransmitters.

The nucleotide cAMP is produced from ATP with the assistance from the enzyme adenylate cyclase, and numerous hormones manipulate the amount of cAMP within the cell. In turn, cAMP affects many cell processes. A common effect of an increase in cAMP concentration is the activation of the cAMP-dependent protein, protein kinase A. Protein kinase A is commonly in a catalytically-inactive state; however, it becomes active when it binds to cAMP. When protein kinase A is activated, it phosphorylates many different enzymes. These enzymes are either activated or suppressed by the phosphorylation process.

For example, glucagon binds to its receptors on the plasma membrane. The bound receptor can then interact with the adenylate cyclase through the use of various G proteins. The active adenylate cyclase starts converting ATP to cAMP, which then increases the cell's cAMP content. The increased level of cAMP in the cytosol allows for increased binding of protein kinase A to cAMP, thus making protein kinase A catalytically active. The active protein kinases proceed to phosphorylate various enzymes within the cell, which changes the enzymes' conformation and controls their catalytic behavior. Finally, the levels of the cAMP decrease due to destruction by cAMP-phosphodiesterase and the inactivation of adenylate cyclase.

Along with affecting the activity of pre-existing components of the cell, an increase in cAMP can have major effects on the transcription of a gene by phosphorylating transcription factors that regulate gene expression.

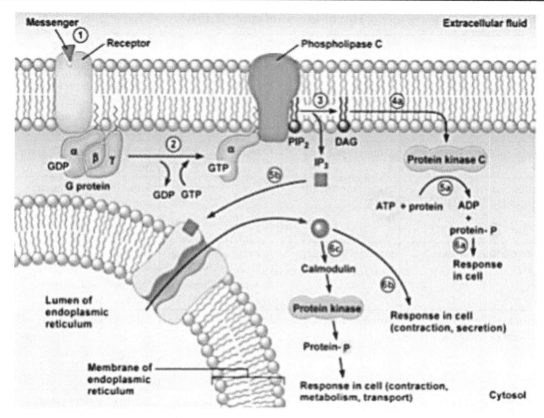

Peptide hormone using G protein secondary messenger to phosphorylate cytosolic enzymes

Tyrosine kinase is another secondary messenger molecule. Protein kinases are often the receptors for protein hormones. The kinase activity of these receptors results in the phosphorylation of various tyrosine residues located on other proteins. Insulin is a prevalent example of a hormone whose receptor is tyrosine kinase.

For the activation of tyrosine kinase, a hormone binds to the exposed receptor on the cell surface. This results in a conformational change, which activates kinase domains in the cytoplasmic regions of the receptor. Usually, the receptor can phosphorylate itself as a part of the kinase activation process. As with cAMP, the activation of the tyrosine kinase cascade causes rapid changes by phosphorylating a variety of intracellular targets, usually enzymes, which become active or inactive after phosphorylation.

In other cases, the binding of a hormone to a surface receptor initiates a tyrosine kinase cascade, even though the receptor might not be a tyrosine kinase. The growth hormone receptor is an example. The interaction of the growth hormone and its receptor leads to the activation of cytoplasmic tyrosine kinases. The results are similar to receptor kinases.

Hormone Summary Table

Hormone	Secreted by:	Target(s)	Effect(s) at target site
Growth hormone (GH)	Anterior pituitary	Bone, muscle, fat	Growth of tissues
Thyroid stimulating hormone (TSH)	Anterior pituitary	Thyroid	Stimulates the release of T_3/T_4 hormones from the thyroid which increase the basal metabolic rate
Prolactin (PRL)	Anterior pituitary	Mammary glands	Production of milk in the breasts
Adrenocorticotropic hormone (ACTH)	Anterior pituitary	Adrenal cortex	Stimulates the adrenal cortex to secrete stress hormones called glucocorticoids
Luteinizing hormone (LH)	Anterior pituitary	In males: interstitial cells in testes; In females: mature ovarian follicle	Males: testosterone secretion Females: ovulation; estrogen secretion
Follicle stimulating hormone (FSH)	Anterior pituitary	Males: seminiferous tubules of testes; Females: ovarian follicle	Males: sperm production Females: follicle growth during menstruation; ovum maturation
Triiodothyronine (T_3) & thyroxine (T_4)	Thyroid	All cells	Regulates metabolism
Aldosterone	Adrenal cortex	Kidney tubules	Increases Na^+ reabsorption and K^+ secretion at the distal convoluted tubule and the collecting duct; net increase in salts in the plasma, increasing osmotic potential and subsequently, blood pressure
Cortisol	Adrenal cortex	All cells	Stress hormone that increases gluconeogenesis in the liver thus increasing blood glucose levels; stimulates fat breakdown
Estrogen	Ovarian follicle	Gonads (Ovaries)	Stimulates female sex organs; causes LH to surge during menstruation
Progesterone	Corpus luteum	Uterine endometrium	Preparation for implantation (thickens lining), growth and maintenance of uterus
Testosterone	Seminiferous tubules	Gonads (Testes)	Stimulates formation of secondary sex characteristics and closing of epiphyseal plates

Hormone Summary Table (continued)

Hormone	Secreted by:	Target(s)	Effect(s) at target site
Anti-diuretic hormone (ADH)	Posterior pituitary	Distal convoluted tubule (DCT)	Causes collecting duct of the kidney to become highly permeable to water; concentrating the urine
Oxytocin (OT)	Posterior pituitary	Uterine smooth muscle	Contraction during childbirth; milk secretion during nursing
Parathyroid hormone (PTH)	Parathyroid	Kidney tubules and osteoclasts	Reabsorption of Ca^{2+} into blood, bone resorption (increases blood Ca^{2+})
Calcitonin	Thyroid	Kidney tubules and osteoblasts	Secretion of Ca^{2+} into urine, bone formation (decreases blood Ca^{2+})
Insulin	β Islets	All cells, liver and skeletal muscle	Pushes glucose into cells from blood, glycogen formation (decreases blood glucose)
Glucagon	α Islets	Liver and skeletal muscle	Stimulates gluconeogenesis, the breakdown of glycogen (increase in blood glucose)
Epinephrine	Adrenal medulla	Cardiac muscle, arteriole, and bronchiole smooth muscle	Raises heart rate, constricts blood vessels, dilates pupils and suppresses the immune system
Norepinephrine	Adrenal medulla	Cardiac muscle, arteriole, and bronchiole smooth muscle	Raises heart rate causing glucose to be released as energy and blood to flow to the muscles
Melatonin	Pineal gland	Limbic system	Emotions/behavior; circadian rhythm

Notes

Chapter 13

Nervous System

- **Neuron or Nerve Cell**

- **Organization of Vertebrate Nervous System**

- **Major Functions**

- **Sympathetic and Parasympathetic Nervous Systems**

- **Reflexes**

- **Integration with Endocrine System: Feedback Control**

Neuron or Nerve Cell

Neurons (nerve cells) generate electric signals that pass from one end of the cell to the other. Neurons release chemical messengers, *neurotransmitters*, to communicate with other cells. A neuron conducting signals toward a synapse (the junction where a neuron communicates with a target cell) is a *presynaptic neuron*, while a neuron conducting signals away from a synapse is a *postsynaptic neuron*. Neurotrophic factors are proteins that guide the development of neurons.

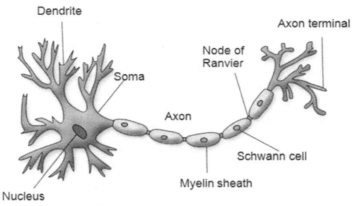

Neurons contain dendrites to receive signals that propagate along the axon toward the axon terminal

Neurons outside the central nervous system (CNS) can repair themselves, but neurons within the CNS cannot. In the autonomic nervous system, the transmission of an impulse involves the interaction between at least two different neurons. The first neuron is the *preganglionic neuron*, and the second is the *postganglionic neuron*. Neurons vary in size and shape, but all have three parts: cell body, dendrites, and axon.

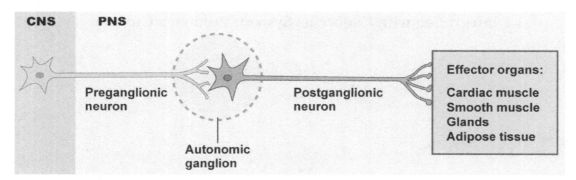

A motor neuron relays signals from the brain or spinal cord to a muscle or gland

Cell body: site of nucleus and organelles

Because nerve cells are specialized for signal transmission and pathway-formation, the form of neurons reflects their function. The *cell body* contributes much to a neuron's density. The cell body houses the neuron's organelles, such as the nucleus and mitochondria. The cell body has a well-developed, rough endoplasmic reticulum and a Golgi apparatus for synthesis and modification of important proteins for proper neuronal function. This part of the neuron is similar in form to most somatic cells, except for the dendrites.

Dendrites: structure and function

Dendrites are the branched receptive areas of a neuron that extend from the cell body. They receive information and conduct impulses toward the cell body. The branching of dendrites increases the surface area for reception. The number of dendrites depends on the function of the neuron. For example, *unipolar neurons* have one dendrite.

Axon: structure and function

The *axon* is one key feature that differentiates neurons from other cells. The axon is crucial in the neuron's impulse generation and is responsible for carrying outgoing messages from the cell. A long axon is a *nerve fiber*. A nerve fiber is a single axon, while a nerve is a bundle of axons bound together by connective tissue. This axon can originate from the CNS and extend to the body's extremities. This effectively provides a pathway for messages from the CNS to the periphery. Axons conduct impulses away from the cell body to stimulate or inhibit a neuron, muscle or gland.

Axon terminals, secretory regions of the nerve, are located at the end of the axon, away from the neuron's cell body. Other names for the axon terminal are the synaptic knob or synaptic bouton. The neuron's axon terminal is the site of signal transmission toward the receptor of another cell.

Glial cells and neuroglia

Nervous tissue is made of neurons and neuroglia. *Neuroglia* support and nourish the neurons. *Glial cells* are nervous tissue support cells capable of cellular division. Oligodendroglia and Schwann cells are glial cells that support neurons physically and metabolically. *Astroglia* regulates the composition of the extracellular fluid in CNS. *Microglia* perform immune functions. Other glial cells include ependymal cells (use cilia to circulate cerebrospinal fluid), satellite cells (support ganglia) and astrocytes (provide physical support to neurons of CNS; maintain mineral and nutrient balance).

Myelin sheath, Schwann cells, oligodendrocytes for insulation of axon

The *myelin sheath* is a phospholipid layer that surrounds a neuron's axon. The myelin sheath insulates the axon, increasing the conductivity of the electrical messages sent through a nerve cell. Myelin is a good insulator because it is fatty and does not contain channels. By preventing leakage of charge, myelin increases the speed of propagation, enabling axons to be thinner. It is formed by the membranes of highly-specialized, tightly-spiraled *neuroglia cells*. These neuroglia cells are either Schwann cells or oligodendrocytes.

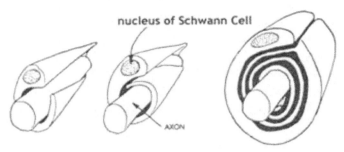

Schwann cell sheath is a phospholipid coat growing around a nerve axon

In the peripheral nervous system (PNS), the *Schwann cells* produce myelin for nerve cells. Many of these specialized cells wrap myelin around the neuron's axon, providing an insulating sheath that prevents the loss of signal transmission. *Oligodendrocytes* are the central nervous system analog of Schwann cells. They make myelin sheaths (insulation) around CNS axons. Insulation occurs at intervals, punctuated by opening, which exposes the plasma membrane of the axon, causing an action potential to jump along nodes of Ranvier.

Only vertebrates have myelinated axons. Myelinated axons appear as white matter, while neuronal cell bodies appear as gray matter.

Many neurodegenerative autoimmune diseases result from a lack of myelin sheath. For instance, in multiple sclerosis, the lack of insulation from a myelin sheath slows or leaks the conductivity of signals across neural pathways, severely decreasing the efficiency of the patient's nervous system.

Nodes of Ranvier: role in the propagation of nerve impulse along an axon

The spaces between adjacent sections of myelin where the plasma membrane of the axon is exposed to extracellular fluid are *nodes of Ranvier* (neurofibrillary nodes). The myelin sheath prevents the flow of ions between intracellular and extracellular compartments. Therefore, action potentials occur at the non-insulated nodes of Ranvier; *saltatory conduction* is this jump of action potentials from one node to another.

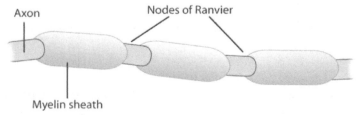

Nodes of Ranvier for saltatory conduction occurring at exposed sections of plasma membrane

Synapse as the site of impulse propagation between cells

Synapse is the space between the axon bulb and the dendritic receptor of the next neuron. A synapse is a junction between two neurons that permits a neuron to pass an electrical (or chemical) signal to another cell. A synapse consists of a *presynaptic membrane*, a *synaptic cleft*, and the *postsynaptic membrane*. In a synapse, the electrical activity in the presynaptic neuron influences the electrical activity in the postsynaptic neuron. The influence can be either excitatory (positive response) or inhibitory (negative response).

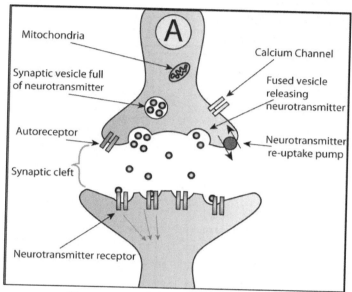

The synaptic cleft between two neurons

Between the presynaptic membrane and the postsynaptic membrane, there is a narrow fluid-filled space of the synaptic cleft. Neurotransmitters are released into synaptic vesicles from the presynaptic neuron's axon terminal and into the synaptic cleft. The vesicles migrate the synaptic cleft and travel toward the postsynaptic neuron, binding to postsynaptic receptors.

In *convergence*, many presynaptic neurons affect a single postsynaptic neuron. This allows information from many sources to influence the activity of one cell. In *divergence*, a single presynaptic nerve cell affects many postsynaptic nerve cells, allowing one information source to affect multiple pathways.

The nervous system uses several types of synapses to create complex pathways for relaying information. *Axodendritic synapses* exist between the axon terminal of one presynaptic neuron and one dendrite of the postsynaptic neuron. An *axosomatic synapse* resides between the axon terminal of the presynaptic neuron and the cell body of the postsynaptic neuron. *Axoaxonic synapses*, while rare, can exist between the presynaptic and postsynaptic axon terminals.

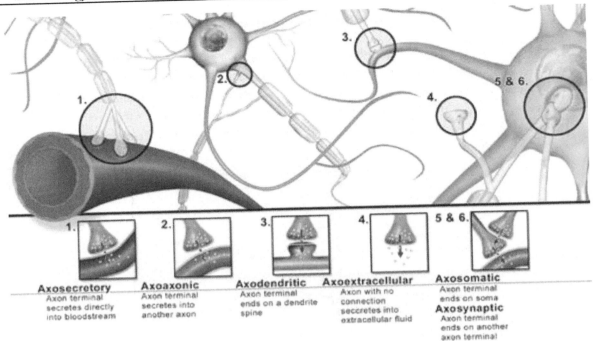

Synapse types

At *electric synapses*, the presynaptic and postsynaptic cells are joined by *gap junctions*, allowing action potentials to flow directly across the junction. Due to the short distance and the direct, physical link between the two neurons, these junctions provide an incredibly fast transmission of signals. Since chemical synapses are usually fast enough for signal transmission, electrical synapses are relatively rare.

At a *chemical synapse*, the axon of the presynaptic neuron ends in swelling as the axon terminal. An extracellular space of the synaptic cleft separates the presynaptic and postsynaptic neurons, preventing a direct propagation of current. Chemical synapses are unidirectional; a signal can only be transmitted from presynaptic to the postsynaptic neuron.

Synaptic activity: transmitter molecules, synaptic knobs, fatigue and propagation between cells without resistance loss

An action potential travels along the axon and reaches the end of the presynaptic neuron. The depolarization of the presynaptic membrane results in the opening of voltage-gated calcium channels. Calcium ions flow into the presynaptic neuron, causing vesicles with neurotransmitters inside the neuron to fuse with the plasma membrane.

The Ca^{2+} ions induce reactions that allow the vesicles holding neurotransmitters to fuse with the plasma membrane and liberate their contents into the synaptic cleft by exocytosis. These synaptic vesicles store neurotransmitters that diffuse across the synapse towards the postsynaptic membrane. When an action potential arrives at the presynaptic axon bulb, synaptic vesicles merge with the presynaptic membrane. When vesicles merge with the neuron's plasma membrane, neurotransmitters are discharged into the synaptic cleft. Neurotransmitter molecules diffuse across the synaptic cleft to the postsynaptic membrane where they bind with specific receptors.

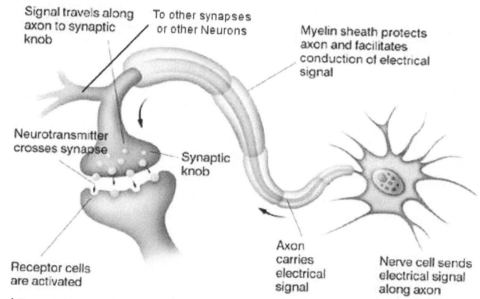

Signal travels along axon to synaptic knob

To other synapses or other Neurons

Myelin sheath protects axon and facilitates conduction of electrical signal

Neurotransmitter crosses synapse

Synaptic knob

Axon carries electrical signal

Nerve cell sends electrical signal along axon

Receptor cells are activated

Dendrites receive stimuli, propagate impulse with diminution along the axon and release neurotransmitter into the synaptic cleft toward the postsynaptic neuron

The neurotransmitters are then released into the synaptic cleft via exocytosis. These neurotransmitters then diffuse via Brownian motion (a type of irregular motion) and bind within the synaptic cleft to specific receptors located on the postsynaptic plasma membrane. The receptors are ligand-gated ion channels, which open and let sodium and other positively charged ions into the postsynaptic neuron. As these positively charged ions enter the postsynaptic neuron, they cause the neuron's membrane to depolarize, which results in an action potential, which moves down the postsynaptic neuron. The calcium ions are pumped back into the synaptic cleft from inside the presynaptic neuron. Finally, the neurotransmitters are degraded and recycled by enzymes in the synaptic cleft.

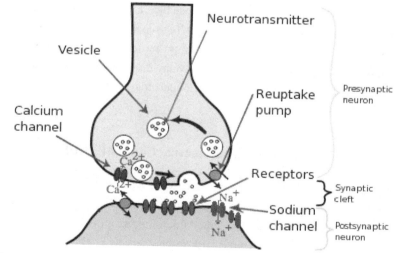

Neurotransmitter

Vesicle

Reuptake pump

Presynaptic neuron

Calcium channel

Receptors

Synaptic cleft

Ca^{2+}

Na^+

Sodium channel

Postsynaptic neuron

Na^+

Synaptic transmission with Ca⁺ causing the release of the neurotransmitter from the presynaptic membrane into the cleft

Neurotransmitters are chemicals that cross the synapse between two neurons, or between the neuron and a muscle or gland. After an action potential has traveled down the axon, it induces the release of neurotransmitters from the presynaptic neuron's axon terminal into the synapse. The axon terminal, or synaptic knob, contains vesicles of neurotransmitters waiting to be exocytosed. An action potential reaching the synaptic knob causes an influx of calcium, which signals the vesicles to fuse with cell membranes (exocytosis) to release the neurotransmitters into the synaptic cleft.

Once the postsynaptic membrane receives the neurotransmitter, the chemicals bind to a receptor (usually on the dendrite) and open ion channels. This causes a change in the membrane potential of the postsynaptic neuron. If this *graded potential* stimulus is large enough, it triggers an *all-or-none response*, inducing the propagation of the signal down the axon of the postsynaptic neuron. Enzymes quickly degrade these neurotransmitters or taken up by the presynaptic terminal so that they do not persistently stimulate the postsynaptic neuron.

At least 25 different neurotransmitters have been identified. *Acetylcholine* (ACh) and *norepinephrine* (NE) are two common neurotransmitters. *Cholinergic fibers* release ACh. In some synapses, the postsynaptic membrane contains enzymes that rapidly inactivate the neurotransmitter. For example, acetylcholinesterase degrades acetylcholine. Once a neurotransmitter is released into a synaptic cleft, it initiates a response and is then removed from the cleft. *Biogenic amines* are neurotransmitters containing an amino group such as *catecholamines* (e.g., dopamine, norepinephrine, epinephrine and serotonin). Nerve fibers that release epinephrine and norepinephrine are adrenergic and noradrenergic fibers, respectively. Amino acid neurotransmitters are the most prevalent neurotransmitters in the CNS. These include glutamate, aspartate, GABA (gamma-aminobutyric acid) and glycine.

Neuropeptides are composed of two or more amino acids. Neurons that release neuropeptides are peptidergic (e.g., beta-endorphin, dynorphin and enkephalin groups). Nitric oxide, ATP and adenine act as neurotransmitters. Many neurons of the PNS end at neuroeffector junctions on muscle and gland cells. Neurotransmitters released by these efferent neurons then activate the target cell. In other synapses, the presynaptic membrane reabsorbs neurotransmitters for repackaging in synaptic vesicles or for the molecular breakdown. The short existence of neurotransmitters in a synapse prevents continuous stimulation (or inhibition) of postsynaptic membranes. Many drugs that affect the nervous system act by interfering with (or potentiating) the action potentials of neurotransmitters.

Resting potential and electrochemical gradient

Luigi Galvani discovered in 1786 that an electric current stimulates a nerve. An impulse is too slow to be caused simply by an electric current in an axon. Julius Bernstein proposed that the impulse is the movement of unequally distributed ions on either side of an axon-membrane, the plasma membrane of the axon. The 1963 Nobel Prize went to the British researchers A. L. Hodgkin and A. F. Huxley, who confirmed this theory.

Hodgkin, Huxley and other researchers inserted a tiny electrode into the giant axon of a squid. The electrode was attached to a voltmeter and an oscilloscope to trace the change in voltage. The voltmeter measured the difference in the electrical potential between the inside and the outside of the membrane. The oscilloscope indicated any changes in polarity. Since the plasma membrane is more permeable to potassium ions than to sodium ions, there are always more positive ions outside the cell; this accounts for some polarity. The large, negatively charged proteins in the cytoplasm (e.g., Cl−) contribute to the resting potential of − 70 mV.

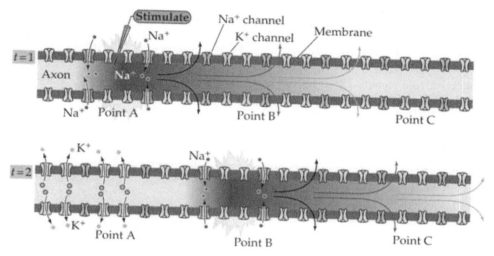

Movement of ions along a synapse changes the voltage across the membrane

When an axon is not conducting an impulse, an oscilloscope records a membrane potential of − 70 mV, indicating that the inside of the neuron is more negative than the outside. *Resting potential* is the electrical potential across the plasma membrane of a cell's axon that is not conducting an impulse. This polarization is due to the difference in electrical charge on either side of the axon membrane.

The magnitude of the voltage potential is determined by the differences in specific ion concentrations between the intracellular and extracellular fluids, as well as the differences in membrane permeability of different ions as a function of the number of open ion channels for these ions. Na⁺ and K⁺ are the most important ions in generating the resting membrane potential. At rest in a living cell, Na⁺ is greater outside the cell, while K⁺ is greater inside the cell. The ions flow about the *electrochemical gradient*, which is the combined electrical and chemical concentration differences on each side of a membrane. This difference is attributed to the net flow of charge across the nerve cell's membrane.

The *sodium-potassium pump* moves three Na⁺ ions to the outside of the membrane for every 2 K⁺ ions it moves into the cell along with Cl⁻ ions inside the cell and creates a net negative charge inside the cell. Additionally, the plasma membrane is more permeable to K⁺ ions as K⁺ moves out of the cell more easily than Na⁺ moves into the cell, accentuating the relatively negative resting potential inside the neuron. In general, K⁺ moves out of the cell and Na⁺ moves into the cell down their concentration gradients. However, the intracellular concentration of these two ions are kept constant by an active transport system that pumps Na⁺ out of the cell and K⁺ into it.

Action potential: threshold, all-or-none, sodium-potassium pump

Action potentials are large, rapid alterations in membrane potential. It is the reversal and restoration of the electrical potential across the plasma membrane of a cell as an electrical impulse passes (i.e., depolarization and repolarization). Membranes capable of producing action potentials are excitable membranes (e.g., membranes of nerve and muscle cells).

When the membrane becomes depolarized, sodium channels open and positive sodium ions rush inside. During *depolarization*, the ion concentration is opposite from resting potential. Sodium ions dominate the inside, while potassium ions dominate the outside. In response, the membrane potential moves in the positive direction. The membrane potential goes from –70 mV at resting potential to +30 mV in the depolarization phase.

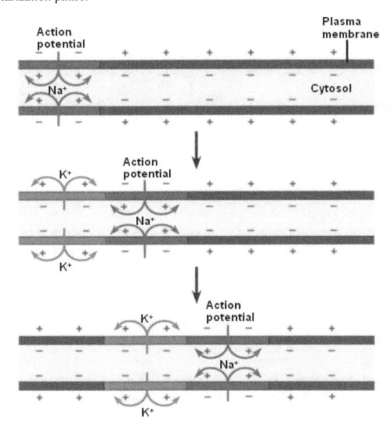

During *repolarization*, potassium channels open and sodium channels close. The positive potassium ions rush outside, and the membrane potential drops down. Now, sodium ion concentration is higher on the inside, while potassium ion concentration is higher on the outside. This is the opposite of the resting state. Thus, the membrane potential returns to its resting value, and the potential returns to – 70 mV again in the repolarization phase.

A hyperpolarization is an event in axon potential propagation where potassium channels do not close fast enough. Consequently, the membrane potential briefly drops below the normal resting potential, to around –80 mV.

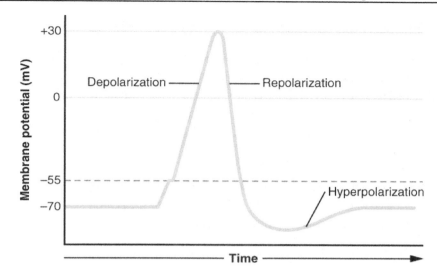

Action potential and voltage changes; depolarization occurs when Na⁺ rushes in, and K⁺ exits the cell; hyperpolarization restores the resting potential with the Na⁺ / K⁺ pump

After an action potential is propagated along the axon of a neuron, there is a period when a second stimulus does not produce a second action potential. This is the *absolute refractory period* and occurs because once the voltage-gated Na^+ channels close, the membrane needs to repolarize before the channels can open again. The sodium-potassium pump works to re-establish the original resting state (i.e., potassium inside and sodium outside). It maintains this unequal distribution of Na^+ and K^+ ions. The sodium-potassium (Na^+ / K^+) pump is an active transport system that moves Na^+ ions out and K^+ ions into the axon. The pump is always working because the membrane is permeable to these ions and they diffuse toward the lesser concentration.

Until the resting potential of -70 mV is restored, the neuron cannot generate another action potential. Following the absolute refractory period, a second action potential can be produced if the stimulus strength is greater than usual. This is the *relative refractory period* and results from hyperpolarization. The relative refractory period begins after hyperpolarization and lasts until the resting potential is re-established. The refractory period prevents an action potential from reversing direction, even though theoretically ions are rushing in and diffusing in both directions.

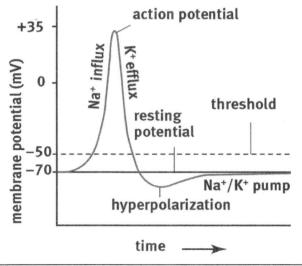

In local anesthetics, Na$^+$ ion channels are blocked, and pain signaling is absent.

Since a neuron is a long cell, it gets depolarized part-by-part and not all at once. The area of the membrane that gets depolarized has a difference in potential with the adjacent area of the membrane that is still at resting potential, thereby causing a local current. This current then depolarizes the adjacent resting membrane, and an action potential continues onward. Since depolarization of an area is followed by a refractory period, the action potential moves unidirectionally.

Similar to water flow through a pipe, the velocity of an action potential across an axon is positively correlated with fiber diameter; a larger fiber offers less resistance, and thus, greater diameter allows for less resistance to the flow of ions. Also, myelination of the neuron's axon increases efficiency by preventing ions from escaping, referred to as "charge leakage."

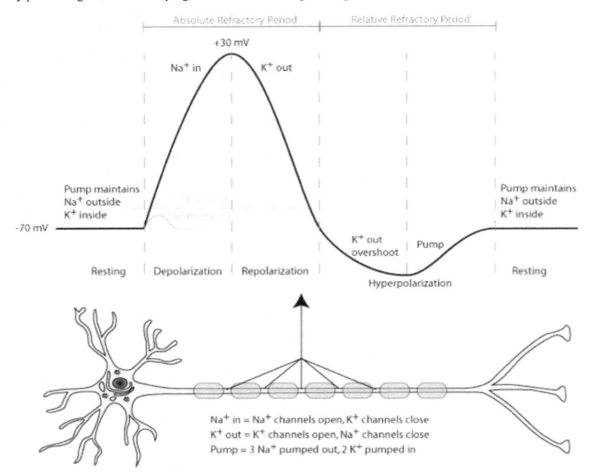

The direction of impulse through a neuron is dendrite → cell body → axon.

In summary, an action potential is all-or-none. This means that if neurotransmitters cause the postsynaptic cell to reach a certain threshold potential, the action potential induced is just as large as the presynaptic action potential. Propagation between cells involves no resistance loss because the postsynaptic action potential is as large as the presynaptic potential.

Excitatory and inhibitory nerve fibers and summation

Graded potentials are changes in membrane potential confined to a small region of the plasma membrane. The magnitude of these potentials is related to the magnitude of the initiating stimulus. When a stimulus (graded potential) depolarizes above a threshold value, an action potential (AP) occurs, initiating a signal. If one graded potential barely makes the threshold value and another overshoots it a lot, both cause the same action potential. A graded potential is an all-or-none response. From –70 mV up to the threshold of ~ –55 mV (or –70 downward), the graded potential cannot travel, but it can potentially (if it surpasses threshold) open the voltage-gated channels. If the threshold is exceeded, the action potential travels down the axon by opening other voltage-gated channels. The other gated types cannot spread unless they trigger this AP. Since the AP is an all-or-none response, the strength of a neural signal is based on other factors (frequency of AP firing or how many nerve cells contribute APs, etc.).

Most synapse interactions are either excitatory or inhibitory. Whether the response is excitatory or inhibitory depends on the type of neurotransmitter or receptor. *Excitatory neurotransmitters* use gated ion channels and are fast acting. *Inhibitory neurotransmitters* affect the metabolism of the postsynaptic cells and are slower. *Neuromodulators* modify the postsynaptic cell's response to neurotransmitters by changing the presynaptic cell's synthesis, or by releasing or metabolizing the neurotransmitter. Neurotransmitters may be taken back into the nerve terminal (active transport), be degraded by synaptic cleft enzymes (recycled back to presynaptic neuron), or diffuse out of the synapse.

Excitatory chemical synapses occur when the activated receptor on the postsynaptic membrane opens Na^+ channels. Na^+ ions move into the cell, resulting in depolarization. This potential change in the postsynaptic neuron is an *excitatory postsynaptic potential* (EPSP). EPSPs are graded potentials. *Inhibitory chemical synapses* occur when the activated receptor on the postsynaptic membrane opens Cl^- channels. Cl^- ions move into the cell, resulting in hyperpolarization. The potential change in the postsynaptic neuron is an *inhibitory postsynaptic potential* (IPSP). Like EPSP, IPSP are graded potentials.

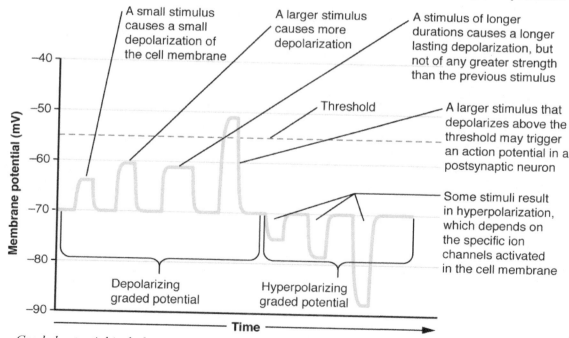

Graded potential includes temporal and spatial summation. Stimuli can be excitatory or inhibitory

Integration is the summing up of excitatory and inhibitory signals. If a neuron receives many excitatory signals from one synapse (consecutive neuron firing), the axon probably transmits a nerve impulse. If both excitatory and inhibitory signals are received, the summing may prohibit the axon from firing.

In most neurons, one EPSP is not enough to cross the threshold in the postsynaptic neuron. Only the combined effects of many excitatory synapses can exceed the threshold and initiate an action potential. *Temporal summation* is when the number of EPSP arriving at different times creates a depolarization. *Spatial summation* is when the number of EPSPs arriving at different locations creates a depolarization. IPSP also show similar summations, but the effect is a hyperpolarization and the inhibition of an action potential.

Interneurons

Interneurons (association neurons) are typically located within the structures that make up the central nervous system (i.e., the spinal cord and brain). They account for 99% of all neurons in the human body. Interneurons are multipolar, consisting of many dendrites used for receiving information and a single axon to send the collected information toward the synapse.

Interneurons form complex brain pathways throughout the central nervous system, transmit signals to the periphery via motor neurons and act as integrators to evaluate impulses for the appropriate response. Generally, the term interneuron is used to refer to small neurons that only connect to other nearby neurons (as opposed to *projection neurons* that can connect over long distances). Interneuron pathways play important roles in human survival and advancement, accounting for phenomena such as memory and language. Interneurons are usually inhibitory, although excitatory interneurons do exist.

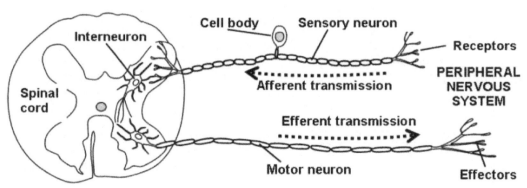

CNS and PNS structures with the direction of propagation shown for afferent sensory neurons (receptors → CNS) and efferent motor neurons (CNS → effectors)

Sensory and Effector neurons

Afferent neurons (sensory neurons) send impulses from the PNS towards the CNS. Afferent neurons are unipolar, as a single dendrite collects information and transmits it through one axon. A sensory receptor at a dendrite of an afferent neuron conveys signals from tissues and organs to the brain and spine.

Receptors are specialized endings of afferent neurons or separate cells that affect ends of afferent neurons. They collect information about the external and internal environment in various energy forms. This stimulus energy is first transformed into a graded potential (receptor potential). *Stimulus transduction* is the process by which a stimulus is transformed into an electrical response. The initial depolarization in afferent neurons is achieved by either a *receptor potential* (in receptors) or by a spontaneous change in the neuron membrane potential as a *pacemaker potential*.

Efferent neurons (motor neurons) carry signals away from the CNS to cells of muscles or glands in the peripheral system. In total, 43 main nerves are branching off the CNS to the peripheral nervous system. Efferent neurons are structurally multipolar and stimulate *effectors*, which are target cells that elicit a response. For example, neurons may stimulate effector cells in the stomach to secrete gastrin.

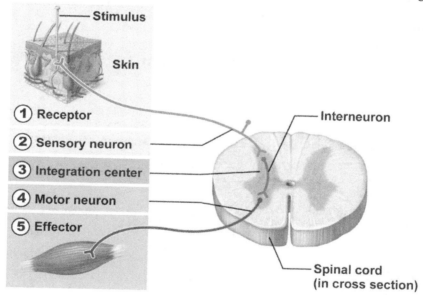

A stimulus is processed through a sensory neuron to the spinal column.
An interneuron communicates the information to a motor neuron for a response at the effector

Organization of Vertebrate Nervous System

The three specific functions of the nervous system are:

- receiving sensory input

- transferring and interpreting impulses

- generating motor output to muscles and glands

High-level control and integration of body systems

The nervous system is a highly organized arrangement of neural pathways that extends to nearly every part of the body. *Neurons* are cells specialized to quickly transmit electrical impulses by forming pathways toward or away from the brain. The nervous system is organized so that the brain can use these neural pathways to interpret stimuli from the environment and subsequently direct the appropriate response to the body. These responses involve other bodily systems (e.g., the endocrine, muscular and cardiovascular systems).

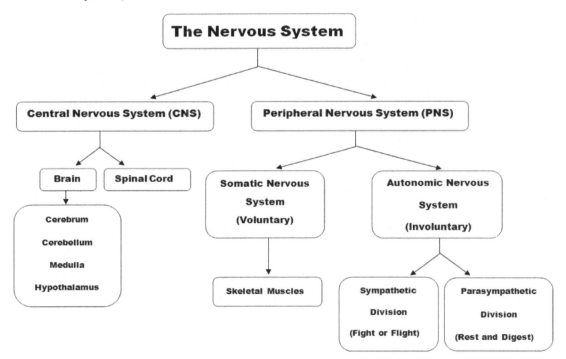

The nervous system divides into the central and peripheral nervous system

The central nervous system (CNS) consists of the brain and the spinal cord, while the peripheral nervous system (PNS) consists of all other nerves and ganglia (collections of cell bodies). The PNS contains the *somatic nervous system*, which includes the pathways of voluntary control over our skeletal muscles, and the involuntary *autonomic nervous system*, which includes the *sympathetic* (fight or flight) and *parasympathetic* (rest and digest) branches.

The PNS sends signals from sensory neurons toward the CNS. The CNS then sends signals to muscles and organs via effector neurons to cause certain actions.

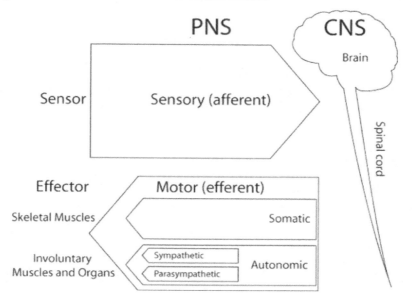

Major Functions

Adaptive capability to external influences

The nervous system's highly adaptive capability helps the brain to interpret external influences efficiently. The brain's interpretation of the environment begins at the sensory receptor. The brain then communicates the corresponding response using other bodily systems.

A *stimulus* is an energy that activates a receptor. Stimulus energy is first transformed into graded or receptor potentials. S*timulus transduction* is the process by which a stimulus is transformed into an electrical response. Each receptor is specific to a certain type of stimulus, which is its *adequate stimulus*. Furthermore, receptors can respond to a specified intensity range of that type of stimuli. However, the nervous system has adapted so that receptors can still be activated even when stimuli are not specified for a particular receptor. This adaptation allows the nervous system to communicate with the rest of the body in case a receptor is not responding to the environment appropriately. This is an example of *nonspecific stimuli* and occurs when the nonspecific stimulus is of high intensity.

The *threshold potential* is the potential at which a membrane is depolarized to generate an action potential. A stimulus that is strong enough to depolarize the membrane is the threshold stimulus. A stimulus greater than threshold magnitude elicits an action potential of the same amplitude. This reaction occurs because once the threshold is reached, membrane events are no longer dependent upon the strength of the stimulus. Action potentials, therefore, occur maximally or do not occur at all, generating an all-or-none response. This is why a single action potential cannot convey information about the magnitude of the stimulus that initiated it.

Efferent control

Efferent neurons (motor neurons) stimulate *effectors* or target cells that elicit a certain response. Effectors include muscles, sweat glands or cells in the stomach that secrete gastrin. Motor neurons have many dendrites and a single axon. They conduct impulses from the central nervous system (CNS) to muscle fibers or glands.

The efferent system is divided into a somatic and an autonomic system.

The *somatic nervous system* innervates skeletal muscles. It consists of myelinated axons without any synapses. The activity of these neurons leads to excitation (contraction) of skeletal muscles; therefore, they are motor neurons. Motor neurons are never inhibitory. The somatic fibers are responsible for the voluntary movement.

The *autonomic nervous system* innervates smooth and cardiac muscles and consists of sets of two neurons connecting the CNS and effector cells. The synapse between these two neurons in the autonomic ganglion. The nerve fibers between the CNS and the ganglion are pre-ganglionic fibers. The post-ganglionic fibers are between the ganglion and the effector cells.

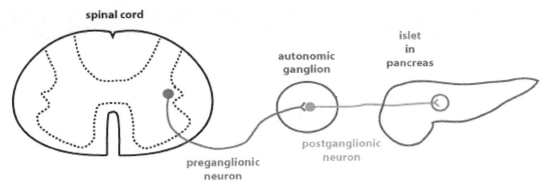

An autonomic nervous system with information traveling from spinal cord to effector cell via pre-ganglionic and post-ganglionic fibers.

Within the autonomic nervous system, all impulses are involuntary. Examples of these involuntary impulses include respiratory system control and heart rate. The autonomic nervous system can further be divided into the *sympathetic* (fight or flight) and *parasympathetic* (rest and digest) systems. Both systems use negative feedback mechanisms, whereby the presence of high levels of a compound inhibits its production. For example, the alternating release of the antagonistic hormones insulin and glucagon is essential for the homeostatic regulation of blood sugar.

The ganglia (collection of cell bodies) of sympathetic neurons are located close to the spinal cord and are arranged to act as a single unit. The parasympathetic ganglia neurons are located close to the organs and are arranged so that the parts can act independently. The sympathetic system is involved in responses to stress. Many organs and glands receive dual innervation from both sympathetic and parasympathetic fibers. The two systems generally have opposite effects and work together to regulate a response.

Sensory input

Afferent neurons have sensory receptors at their ends and convey signals from tissues and organs into the CNS.

The *somatic system* has two main pathways in opposite directions. One pathway uses *nerves* (bundles of axons) to carry sensory information from the peripheral skeletal muscles back to the CNS. The other pathway works in the opposite direction, allowing humans to use their muscles consciously by sending impulses from the CNS to skeletal muscles.

The receptor is the site of stimulation. Once the receptor receives the stimuli, the sensory neuron carries the impulse to the *integration center*. The integration center connects sensory neurons to motor neurons via synapses inside the CNS. Integration can be monosynaptic or polysynaptic. There are no *interneurons* (neurons that serve as a link between sensory and motor neurons) involved in *monosynaptic* integration. Monosynaptic integration involves a direct synapse from the sensory to a motor neuron. *Polysynaptic* integration requires at least one interneuron. Once the signal has reached the motor neuron, it is then carried toward the effector, the site of response to the stimulus.

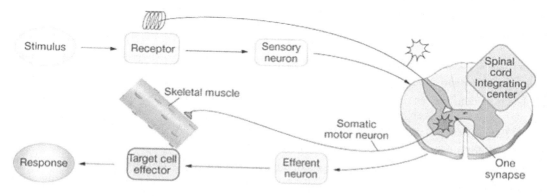

(b) **Polysynaptic reflexes** have two or more synapses.

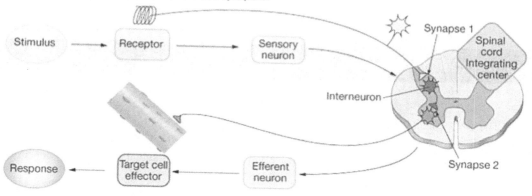

A monosynaptic reflex has a single synapse between the afferent and efferent neurons.
A polysynaptic reflex has two (or more) synapses (i.e., interneurons)
between the afferent and efferent neurons.

Integrative and cognitive abilities

The *limbic system* is a complex network of tracts in the brain. It incorporates multiple brain areas, such as medial portions of cerebral lobes, subcortical nuclei, and the diencephalon. The two major structures of the limbic system are the *hippocampus* and the *amygdala.*

The limbic system's coordination of multiple areas of the brain allows us to interpret stimuli from the environment and act accordingly. The *hippocampus* delivers sensory input to the prefrontal area of the brain. People are aware of past experiences because the hippocampus stores this information in designated association areas. The *amygdala* is the area of the brain that associates emotions with thoughts or experiences. After the amygdala and the hippocampus process sensory information, the highly developed prefrontal cortex allows for reasoning, preventing humans from acting purely upon basic emotion. A general interpretation area receives information from the sensory association areas, allowing to quickly integrate signals and send them to the prefrontal area for immediate response. The prefrontal area in the frontal lobe receives input from other association areas, and reasons and plans.

Memory is the brain's ability to retain and recall information. *Learning* takes place when a person retains and utilizes memories. The prefrontal area in the frontal lobe is active in *short-term memory* (e.g., telephone numbers). *Long-term memory* is a mix of semantic memory (numbers, words) and episodic memory (persons, events).

Skill memory is the ability to perform motor activities, commonly referred to as "muscle memory." The *hippocampus* serves as an intermediary between processed memories and the prefrontal cortex. The amygdala is responsible for fear conditioning and associates danger with certain sensory stimuli.

Long-term potentiation (LTP), the strengthening of specific neural pathways in the hippocampus following associated learning, is essential for memory storage. Some excited postsynaptic cells may die due to excessive amounts of glutamate neurotransmitter. The death of too many neurons in the hippocampus may be the underlying cause of Alzheimer's disease, a disease that gradually causes a person's memory to deteriorate.

Sympathetic and Parasympathetic Nervous Systems

The PNS consists of the sensory and motor branch. It lies outside the CNS and contains the cranial and spinal nerves. The PNS transmits signals to and from the CNS using sensory and motor neurons. 12 pairs of cranial nerves connect to the brain and 31 pairs of spinal nerves that connect to the spinal cord.

Cranial nerves mostly connect to the head, neck and facial regions, and *spinal nerves* lie on either side of the spinal cord. Nerves are made from many parallel nerve fibers that contain axons and myelin sheaths. S*pinal roots* are the paired spinal nerves that leave the spinal cord by two short branches. The *dorsal root* (or sensory root) contains fibers of sensory neurons conducting nerve impulses to the spinal cord. The *ventral root* contains the axons of motor neurons that conduct nerve impulses away from the spinal cord.

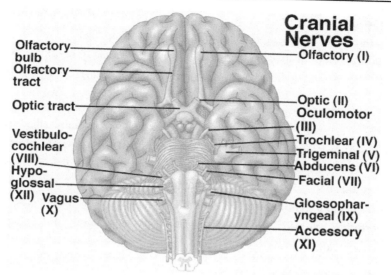

Twelve pairs of cranial nerves

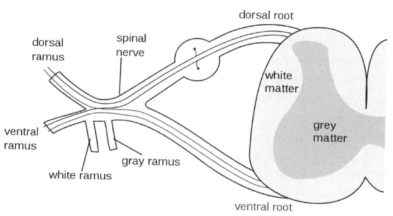

Dorsal root carries impulses to the spinal cord. Ventral root carries impulse away from the spinal cord

Spinal nerves are mixed nerves that conduct impulses to and from the spinal cord. Spinal nerves contain sensory and motor fibers, with each type of fiber serving its region. *Sensory nerves* only contain sensory nerve fibers. *Motor nerves* only contain motor nerve fibers. The cell bodies of neurons are in the CNS or ganglia.

The PNS is divided into the somatic and autonomic systems. The *somatic system* is responsible for the voluntary movement of skeletal muscles.

The *autonomic system* controls the involuntary movement of the cardiac muscle, smooth muscle, and glands. There are two divisions: the sympathetic and parasympathetic systems. Both systems function in an involuntary manner, innervate all internal organs and utilize two neurons and one ganglion for each impulse. The first neuron has a cell body within the CNS and a preganglionic fiber. The second neuron has a cell body within the ganglion and a postganglionic fiber. Breathing rate and blood pressure are regulated by reflex actions to maintain homeostasis.

The *sympathetic system* is responsible for fight or flight responses, generally raising blood pressure and heart rate. Most preganglionic fibers of the sympathetic system arise from the *thoracic-lumbar* (middle) portion of the spinal cord and almost immediately terminate in ganglia that lie near the spinal cord. The preganglionic fiber is short, while the postganglionic fiber that contacts an organ is long.

The sympathetic system is important during emergency situations (the "fight or flight" response). To defend itself or flee, the body activates the sympathetic system to accelerate the heart rate and dilate the bronchi. These responses require a supply of glucose and oxygen. To divert energy from less necessary digestive functions, the sympathetic system inhibits digestion. It increases blood pressure, dilates the pupils to allow more light into the eye, and breaks down glycogen to release glucose into the blood. The neurotransmitter released by the postganglionic axon is mainly norepinephrine, which is similar to epinephrine (adrenaline).

The *parasympathetic system* is responsible for rest and digest responses, lowering heart rate and promoting non-emergency functions (e.g., digestion, relaxation, and sexual arousal).

The parasympathetic system consists of a few cranial nerves and nerves exiting the spinal cord, including the vagus nerve (innervates the heart and branches to the pharynx, larynx and some internal organs) and fibers that arise from the sacral region of the spinal cord. Since the parasympathetic system contains efferent nerves from both the cranial and the sacral regions, the efferent nerves are *craniosacral* (as opposed to the thoracic-lumbar sympathetic nerves). The parasympathetic system is a "housekeeper system," meaning that it promotes internal responses resulting in a relaxed state. The parasympathetic system causes the eye pupil to constrict, promotes digestion and slows the heartbeat. The neurotransmitter released by the parasympathetic system is the neurotransmitter acetylcholine.

Reflexes

Reflexes involve the spinal cord and do not require the participation of the brain. There are two advantages to this. First, the brain is constantly working, and any additional tasks for the brain would result in precious resources being diverted. Second, reflexes are designed to be fast responses. For instance, if someone puts his or her hand on a hot stove, they probably would not analyze the pain before removing their hand from the heat. As a result of bypassing the brain, the reaction time to these stimuli is much faster.

The feedback loop, reflex arc, effects on flexor and extensor muscles

Feedback loops occur when the outputs of a system are fed back into the system as inputs, leading to a change within the system. The two types are positive and negative feedback loops.

Positive feedbacks are mechanisms that encourage the continuation of a cycle. An example of positive feedback is uterine contractions when one contraction leads to oxytocin release and subsequently more contractions. Another example is blood clotting when platelets are activated at the wound site and then work to attract more platelet activation and clumping. Positive feedbacks are not common because they disrupt homeostasis.

Negative feedbacks are mechanisms that counteract the continuation of a cycle. For example, a drop in blood pressure causes the release of ADH, which increases blood pressure. Conversely, an increase in blood pressure causes a drop in ADH. Another example is in the Golgi tendon reflex, a sudden contraction of the quads (extensor muscles) causes negative feedback that relaxes the quads and contracts the hamstrings (flexor muscles).

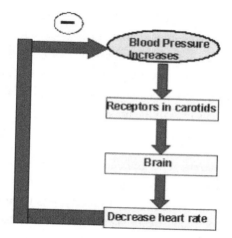

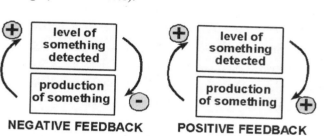

NEGATIVE FEEDBACK   **POSITIVE FEEDBACK**

Reflex arcs are a type of negative feedback. They are rapid, involuntary responses to stimuli involving two or three neurons, but the brain does not integrate the sensory and motor activities. Instead, the reflex arcs synapse in the spinal cord. Even though the reflex arc bypasses the brain, the brain is aware that the response took place. One example of a reflex arc is immediate withdrawal from a painful stimulus. Another example is the knee-jerk reaction. Tapping the knee tendon causes a sudden stretching of the muscle, and the reflex leads to contraction of that muscle, creating the knee-jerk (negative feedback).

The mechanism of a reflex arc occurs in the following sequence:

1. Sensory receptors generate an impulse in a sensory neuron that moves along sensory axons toward the spinal cord.

2. Sensory axons enter the cord dorsally and pass signals to interneurons.

3. Interneurons pass the signals to motor neurons.

4. Impulses travel along motor axons to an effector.

5. The effector causes an effect, such as a muscle contracting, to cause withdrawal from a pain stimulus.

Reflex response occurs because the sensory neuron stimulates interneurons. Some impulses extend to the cerebrum. When this occurs, a person becomes conscious of the stimulus and the reaction.

Reflexes often affect both flexor and extensor muscles. During the knee-jerk reflex, the extensors of the quads are contracted while the flexors of the hamstrings are relaxed.

Roles of the spinal cord and supraspinal circuits

The CNS, located in the midline of the body, consists of the brain and the spinal cord. It integrates sensory information and controls the body. The CNS is extremely important because it controls the biological processes of our body and all conscious thought. Because of the critical importance of the spinal cord and brain, these organs are safely encased within bones. The *cranium* protects the brain, and the *spine* protects the spinal cord. Both are wrapped in three connective tissue coverings as *meninges*. The spaces between the meninges are filled with *cerebrospinal fluid* to nourish and protect the CNS from injury. The cerebrospinal fluid is within the central canal of the spinal cord. It is and produced within the *ventricles* of the brain.

The meninges in the CNS have three layers: the outermost *dura mater*, the middle *arachnoid*, and the inner *pia mater*. The space between the pia and the arachnoid, the *subarachnoid space*, is filled with *cerebrospinal fluid* (CSF), acting as a shock absorber for neural tissue. Because the brain is unable to store glycogen adequately, it is highly dependent on a continuous supply of glucose and oxygen from the bloodstream. The exchange of substances between blood and extracellular fluid in CNS is highly restricted via a complex group of blood-brain barrier mechanisms. The CSF and the extracellular fluid in the brain are in diffusion equilibrium with each other but are separated from the blood.

A group of nerve fibers traveling together in the CNS is a *pathway* (or tract). A band of nerve fibers that connects the left and right halves is a *commissure*.

Information in the CNS passes along two types of pathways:

(1) Long neural pathways, in which neurons with long axons carry information directly between the brain and spinal cord or between different regions of the brain. There is no diminution in the transmitted information.

(2) Multineuronal (or multisynaptic) pathways are comprised of many neurons or synapses. New information can be integrated into the transmitted information.

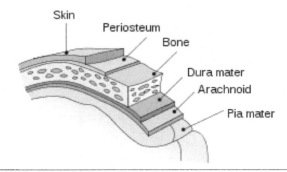

Cell bodies of neurons having similar function cluster together as ganglia in the peripheral nervous system and nuclei in the CNS.

The spinal cord has two main functions. It provides the means of communication between the brain, the spinal nerves and the synapse (or synapses if it is polysynaptic) for the reflex arc. Sensory information enters through the *dorsal horn* in the spinal cord and exits through the *ventral horn* as motor information towards the periphery.

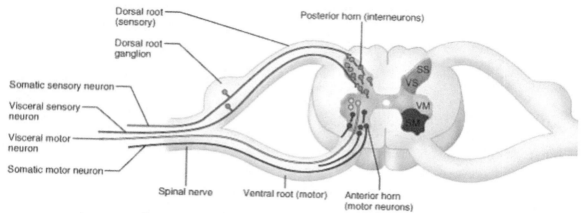

Sensory neuron (dorsal root) and a motor neuron (ventral root)

The spinal cord and brain are composed of gray and white matter. These areas of matter house either unmyelinated or myelinated neurons, signifying whether the axons of the neuron have been covered with myelin, a phospholipid substance that allows for a faster propagation of axon potentials in neurons.

Gray matter is composed of interneurons, cell bodies, dendrites, and glial cells. Unmyelinated cell bodies and short fibers give gray matter its color. In a cross-section, the gray area looks like a butterfly (or the letter H). It contains portions of sensory neurons, motor neurons and the short interneurons that connect them. Deep in the brain, multiple nuclei make up the *basal ganglia* gray matter. Common malfunctions in this area lead to conditions such as Huntington's disease and Parkinson's disease.

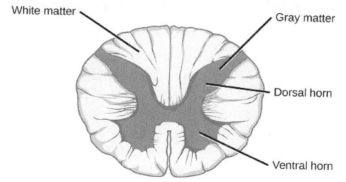

Comparison of gray (unmyelinated) matter and white (myelinated) matter

In *white matter*, the long myelinated fibers of interneurons run together in tracts, giving the white matter its color. Tracts conduct impulses between the brain and the spinal nerves; ascending tracts are dorsal, and *descending tracts* from the brain are ventral. *Ascending tracts* from the lower brain centers relay sensory information up to the primary somatosensory area. Descending tracts from the primary motor area communicate with the lower brain centers. Near the brain, tracts cross over from one side of the body to the other; therefore, the left side of the brain controls the right side of the body.

Afferent fibers enter from the peripheral system on the dorsal side of the spinal cord via the dorsal roots (containing the dorsal root ganglia). *Efferent fibers* leave the spinal cord on the ventral side via the ventral roots. The two roots combine to form a spinal nerve on each side of the spinal cord.

There are 31 pairs of spinal nerves, designated by 4 levels of exit: cervical (8), thoracic (12), lumbar (5), sacral (5), and coccygeal (1).

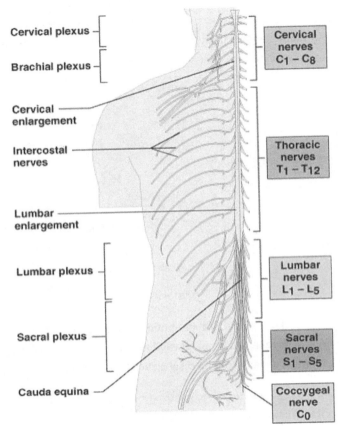

Integration with Endocrine System: Feedback Control

The nervous system and endocrine system interact in many ways to influence various aspects of human behavior. The *hypothalamus*, the brain structure that regulates basic needs as well as emotional and stress responses, is the link between the nervous system and the endocrine system because it controls the release of hormones by regulating the pituitary gland.

One example of an interaction between the nervous system and endocrine system is the release of the hormones *epinephrine* (adrenaline) and *norepinephrine* from the adrenal medulla, triggered by the sympathetic "fight or flight" system. These hormones are produced at the ends of synaptic nerve fibers. The two hormones are extremely similar in both structure and function, and they play an important role in stress response. Epinephrine is primarily released from the adrenal medulla and can act on receptors while in the bloodstream, whereas norepinephrine is primarily released from the synaptic nerve fibers and binds with receptors near the synaptic cleft.

Chapter 14

Respiratory System

- **General Structure and Function**

- **Structure of Lungs and Alveoli**

- **Breathing Mechanisms**

- **Thermoregulation: Evaporation and Panting**

- **Alveolar Gas Exchange**

- **pH Control**

- **Regulation by Neural Control**

General Structure and Function

The human *respiratory system* includes the structures that conduct air to and from the lungs. Its purpose is to supply the body's tissues with oxygen. The pathway of oxygen in the respiratory system involves cooperative transportation and diffusion within the circulatory system. Oxygen is taken in from the environment by the lungs. Here, the circulatory system's red blood cells meet the lung's alveoli, and oxygen diffuses into the cells. Red blood cells then carry the oxygenated blood towards the body's tissues. Carbon dioxide follows a reverse path, traveling in deoxygenated blood from the tissues and back to the lungs, where it is expelled through the nose (or mouth) into the external environment.

Gas exchange and thermoregulation

Cellular respiration involves the breakdown of organic molecules (e.g., glucose) to produce ATP. A sufficient supply of oxygen is required for aerobic respiration of the Krebs cycle and the electron transport chain to efficiently convert potential energy within food into the energy of ATP. Carbon dioxide is generated by aerobic cellular respiration and must be removed from the cell. There is an exchange of gases, where carbon dioxide leaves the cell and oxygen enters. Animals have organ systems that are involved in facilitating respiration and regulating the transport of gases between the environment and the body's cells.

Breathing involves inspiration (bringing air into the lungs) and expiration (moving air out of the lungs). In general, gas exchange is the process of exchanging one gas for another. External respiration involves gas exchange with the external environment at a particular respiratory surface. Internal respiration is more complex in animals and involves gas exchange between blood and tissue fluid. The process occurs in the alveoli (small air sacs) of the lungs. Oxygen diffuses into the capillaries surrounding the alveoli. Carbon dioxide diffuses out of the capillaries and into the alveoli.

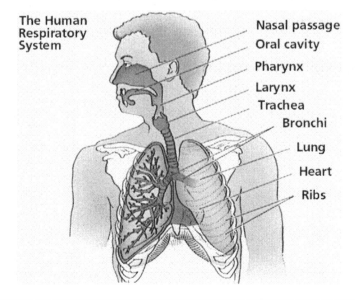

The lungs are located deep within the thoracic cavity so that they are protected from desiccation. The pathways in the respiratory system connect the lungs to the outside environment. Human lungs have at least 50 times the skin's total surface area.

First, air enters the body through either the nose (or mouth). It is warmed, filtered and passed through the nasal (or oral) cavity. Air then passes through the *pharynx* into the upper part of the trachea that contains the *larynx*. The larynx is held open by cartilage that forms the *laryngeal prominence* (Adam's apple). The *vocal cords* are two bands of tissue that extend across the opening of the larynx. As air passes across them, these tissues vibrate, creating sounds. The walls of the *trachea* are reinforced with C-shaped rings of cartilage. As food is swallowed, the larynx rises and the *glottis* is closed by a flap of elastic, cartilaginous tissue, as the *epiglottis*. A backward movement of the soft palate covers the entrance to the nasal passages; this movement then directs food downward. Airways beyond the larynx are divided: the *conducting zone* and *respiratory zone*.

The trachea divides into two *bronchi*. After passing the larynx, the air moves into the bronchi that carry air to the lungs. Bronchi are reinforced to prevent their collapse and are lined with ciliated epithelium and mucus-producing cells. The C-shaped rings of cartilage diminish as bronchi branch within the lungs.

The *conducting zone* has no gas exchange and consists of the tracheal tube which branches into two bronchi before each enters a lung and branches further. Walls of the trachea and the bronchi contain cartilage for support. *Terminal bronchioles* are the first branches without cartilage.

Within the lungs, each bronchus branches into numerous smaller tubes as *bronchioles* that conduct air to *alveoli*, which are grape-like sac clusters. The *respiratory zone* is where gas exchange occurs and consists of the respiratory bronchioles with alveoli attached.

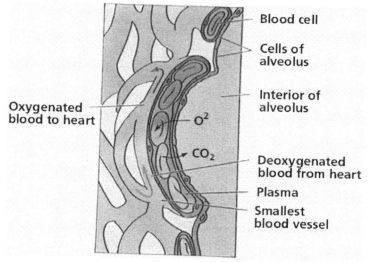

Blood cell

Cells of alveolus

Interior of alveolus

Oxygenated blood to heart

O^2

CO_2

Deoxygenated blood from heart

Plasma

Smallest blood vessel

The respiratory bronchioles with capillaries and alveoli for gas exchange.
Oxygen moves from the alveoli to the red blood cell, while carbon dioxide moves
from the red blood cell (or dissolved in the plasma) to the alveoli for expiration by the lungs

Protection against disease and particulate matter

The respiratory system has several mechanisms to protect against disease and particulate matter. The nasal cavity contains visible hairs as *vibrissae*. Nostril hairs filter out coarse particles. The *paranasal sinuses* secrete mucus, which makes its way through the mucous glands into the *lamina propria* and finally into the *respiratory mucosa,* where the mucous can trap foreign particles and pathogens.

The epithelial cells of the respiratory mucosa secrete *defensins*, which are natural antibiotics that aid in defense against invading microbes. *Serous glands* supply the *lamina propria* with water fluid that contains lysozyme (antibacterial enzyme). Macrophages, which engulf pathogens, are also present.

The respiratory mucosa is a part of the nasal cavity's ciliated epithelial lining. Cilia on the mucous lining of the respiratory tract sweep pathogens and particles out of the body by creating a slight current.

The nasal cavity wall contains three projections: superior, medial and inferior turbinate (or conchae). These projections increase the surface area of the nasal mucosa. As air passes, the heavier non-gaseous components tend to be deflected into the mucosal surfaces.

The entire respiratory tract has a warm, wet, mucous membrane lining exposed to environmental air. The condition of the airways and pressure difference between the lungs and atmosphere are important factors in the flow of air in and out of the lungs. Many diseases can affect the condition of the airways.

Strep throat is a severe infection caused by the bacteria *Streptococcus pyogenes*, which results in a high fever, increased difficulty swallowing and a high risk of systemic infection.

Sinusitis is an infection of the sinuses; 1–3% of upper respiratory infections are accompanied by sinusitis.

Asthma narrows the airways by causing allergy-induced spasms of surrounding muscles or by clogging the airways with mucus.

Tonsillitis is when the tonsils and adenoids of the pharynx become inflamed as the first line of defense.

Laryngitis is an infection of the larynx, causing hoarseness and an inability to speak. However, persistent hoarseness without any upper respiratory infection is one of the warning signs of laryngeal cancer.

Lung cancer, formerly more common in men, now surpasses breast cancer as a cause of death in women. It is often due to smoking. Lung cancer develops in the lung tissue in steps. First, a thickening and callusing of the cells lining the bronchi begins. Also, cilia are lost, so it becomes impossible to prevent dust and dirt from settling in the lungs. Next, cells with atypical nuclei appear in the callused lining.

Bronchitis is an inflammatory response that reduces airflow and is caused by long-term exposure to irritants such as cigarette smoke, air pollutants or allergens. *Acute bronchitis* is an infection of the primary and secondary bronchi and is usually preceded by a viral upper respiratory infection. *Chronic Obstructive Pulmonary Disease* (COPD) is chronic bronchitis or the production of excessive mucus in bronchi that obstructs the airways.

Pneumonia is usually caused by a bacterial (or viral) lung infection. The bronchi and alveoli fill with fluid. Pneumonia can be localized in specific lobules. AIDS patients are subject to a rare form of pneumonia caused by the protozoan *Pneumocystis carinii.*

Cystic fibrosis is a life-threatening genetic disorder that causes excessive mucus production that clogs the lungs, digestive system and other organs.

Pulmonary tuberculosis (TB) is caused by the bacterium *Mycobacterium tuberculosis* (tubercle bacillus). A TB skin test is a highly diluted extract of the bacilli injected into the patient's skin. If a person has been exposed, the immune response causes an area of inflammation. Bacilli that invade lung tissue are isolated by the lung tissue in tubercles (tiny capsules). If the person is highly resistant, the isolated bacteria die. If the person is not resistant, the bacteria can spread. A chest X-ray can detect active tubercles. Appropriate drug therapy can ensure localization and the eventual destruction of live bacteria. The resurgence of pulmonary tuberculosis has accompanied the increased incidence of AIDS and is often seen in lower socioeconomic demographics. The new strains of tuberculosis are often resistant to standard antibiotics.

Hypoxia is the deficiency of oxygen at the tissue level. There are four types: *hypoxic hypoxia,* which is characterized by reduced arterial pO_2; *anemic hypoxia* is when the total oxygen content of the blood is reduced due to an inadequate amount of erythrocytes and hemoglobin; *ischemic hypoxia* is when blood flow to the tissues is low; and *histotoxic hypoxia* is when tissues are unable to utilize oxygen due to interference from a toxic agent.

Emphysema is a chronic and incurable disorder involving distended and damaged alveoli. It is a disease characterized by increased airway resistance, decreased surface area for ventilation due to alveolar fusion, and ventilation-perfusion inequalities (i.e., the difference in the amount of air reaching the alveoli and the amount of blood sent to the lungs). The lungs often balloon due to trapped air and ineffective alveoli. Emphysema is often preceded by *chronic bronchitis.* The elastic recoil of the lungs is reduced, and the airways are narrowed, making expiration difficult. Since the surface area for gas exchange is reduced, insufficient O_2 reaches the heart and brain. This triggers the heart to work furiously to force more blood through the lungs, which may lead to a heart condition. Lack of oxygen to the brain makes the patient feel depressed, sluggish and irritable. Exercise, drug therapy and supplemental oxygen may relieve the symptoms and slow the progress of the condition.

Structure of Lungs and Alveoli

The *lungs* are two large, lobed organs in the chest. The *thorax* is a closed compartment, bound at the neck by muscles and separated from the abdomen by the *diaphragm*, a sheet of skeletal muscle. The wall of the thorax is composed of ribs, the *sternum* (breastbone) and intercostal muscles between ribs.

Lungs are ingrowths of the body wall and connect to the outside by a series of tubes and small openings. The *pleura* (a thin sheet of cells) separates the inside of the chest cavity from the outer surface of the lungs. The *pleural sac* (closed sac) surrounds each lung. The pleural surface coating the lung is the *visceral pleura* and is attached to the lung by connective tissue. The *parietal pleura* (outer layer) is attached to the thoracic wall and diaphragm. A thin layer of intrapleural fluid separates the two layers of the pleura. Changes in the hydrostatic pressure of the intrapleural fluid (the intrapleural pressure P_{ip}, the intra-thoracic pressure) cause the lungs and thoracic wall to move together during breathing.

Lung breathing evolved about 400 million years ago. Lungs are not solely in the domain of vertebrates. Some terrestrial snails have gas exchange structures similar to those in frogs. Each lung is composed of alveoli, the air sacs where gas exchange with blood takes place. Lungs have a non-respiratory function and release many biologically active substances into the blood and remove them. They trap and dissolve small blood clots.

Alveoli are open-ended, hollow sacs as continuous lumens of airways. Their inner walls are lined with a single layer of flat epithelial cells called *type I alveolar cells*, interspersed by thicker, specialized *type II alveolar cells*. Alveolar walls contain capillaries and a small interstitial space with

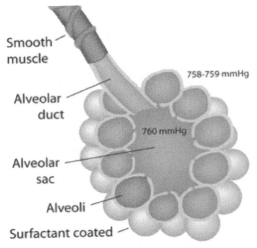

Gas exchange of oxygen and carbon dioxide occurs at the alveoli in the lungs

interstitial fluid and connective tissue. Cells in the alveolar walls secrete a fluid that keeps the inner surface of the alveoli moist, allowing gases to dissolve. *Surfactant*, a natural detergent, prevents the sides of the alveoli from sticking together. Blood within an alveolar capillary wall is separated from air within an alveolus by a thin barrier of 0.2 μm thick. Pores in the walls permit the flow of air. The extensive surface area and thin barrier permit rapid exchange of large quantities of oxygen and carbon dioxide by diffusion.

Breathing Mechanisms

Tidal volume is the amount of air entering the lungs during a single, normal inspiration, or leaving the lungs in a single expiration. The maximal amount of air that can be increased above this value during the deepest inspiration is the *inspiratory reserve volume*. After the expiration of a resting tidal volume, the volume of air remaining in lungs is the *functional residual capacity*. The additional volume of air, after the expiration of the resting tidal volume, is the *expiratory reserve volume*. The expiratory reserve volume can be expired via active contraction of expiratory muscles. The *residual volume* is the air remaining in the lungs after maximal expiration. The *vital capacity* is the maximal volume of air that can be expired after a maximal inspiration. *Minute ventilation* is the volume of gas inhaled (or exhaled) from a person's lungs per minute.

Minute ventilation (ml/min) = Tidal volume (ml/breath) × Respiration rate (breaths/min)

Anatomic dead space is space within the airways that does not permit gas exchange with blood. *Alveolar ventilation* is the total volume of air entering the alveoli per minute.

Ventilation (ml/min) = [Tidal volume (ml/breath) – anatomic dead space (ml/min)] × respiratory rate (breaths/min)

Since a fixed volume of each tidal volume goes to dead space, increased depth of breathing is more effective in elevating alveolar ventilation than increased breathing rate. *Alveolar dead space* is the volume of inspired air that is not used for gas exchange from reaching the alveoli with a lack of blood supply. *Physiologic dead space* is the sum of the anatomic and alveolar dead space.

Diaphragm, rib cage, and differential pressure

The *diaphragm* is a muscle that pulls down when contracting, which increases chest volume, decreases air pressure and stimulates inspiration.

The *rib cage* expands outward during inspiration. Intercostal muscles help this expansion. At rest, the rib cage maintains lung volume, prevents the lungs from collapsing and forms a cage around the lungs for protection.

The volume of the lungs is dependent on two factors. *Trans-pulmonary pressure* is the difference in pressure between the inside and outside of the lungs, and *lung compliance* is the lung's ability to stretch.

Muscles used in respiration are attached to the chest wall. When they contract (or relax), they change the chest dimensions, which changes trans-pulmonary pressure, the lung volume and alveolar pressure, causing air to flow in (or out) of the lungs.

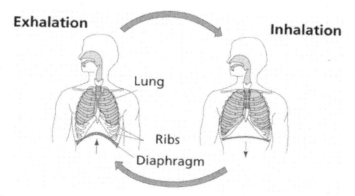

Diaphragm contraction during inspiration and relaxation during expiration

When a person inhales, muscles in the chest wall contract, lifting the ribs and pulling them outward. The diaphragm contracts and moves downward, enlarging the chest cavity. The reduced air pressure in the lungs causes air to enter the lungs. Exhaling reverses these steps.

Intrapulmonary pressure is the atmospheric pressure because the lungs are open to the atmosphere and have the same pressure as the outside environment. If the intrapleural pressure is less than the atmospheric pressure, it sucks on the lungs and prevents them from collapsing. During inspiration, intrapleural pressure decreases even further and causes the lungs to expand.

Differential pressure is the difference between intrapulmonary (inside lung) pressure and intrapleural (outside lung) pressure.

Similar to all other mammals, humans breathe through a *negative pressure* mechanism. The negative pressure mechanism is akin to sucking. During inhalation, lowering the diaphragm and raising the ribs forms negative pressure by increasing the volume of the thoracic cavity. The air, which is under greater outside pressure, flows into the lung. Increases in the CO_2 and H^+ concentrations in the blood are the primary stimuli that increase the breathing rate.

Resiliency and surface tension effects

The lung is elastic; due to surface tension, it recoils when relaxed after inspiration. If not for the rib cage, the lung would collapse even further.

$$\text{Trans-pulmonary pressure} = P_{alv} - P_{ip}$$

where P_{alv} is alveolar pressure and P_{ip} is intrapleural pressure.

P_{alv} is zero and equal to the atmospheric pressure. P_{ip} is negative, or less than the atmospheric pressure, because the elastic recoil of the lung inwards and the elastic recoil of the chest wall outwards increases the volume of the intrapleural space between them and decreases the pressure within.

Therefore, the trans-pulmonary pressure is greater than zero, which creates an expanding force equal to the elastic recoil force of the lung (keeping it from collapsing). The volume of the lungs is kept stable with air inside the lungs. By a similar phenomenon, the pressure difference across the chest ($P_{atm} - P_{ip}$), which is directed inward, keeps the elastic chest wall from excessively moving outward.

Resistance is determined mainly by radius. Trans-pulmonary pressure exerts a distending force, which keeps the airways from collapsing. This makes the airways larger during expiration and smaller during inspiration.

Lung compliance (LC) is a measure of elasticity (or the magnitude of change) in lung volume (ΔLV) that can be produced by a given change in transpulmonary pressure.

$$LC = \Delta LV / \Delta(P_{alv} - P_{ip})$$

When lung compliance is low, P_{ip} must be made lower to achieve lung expansion. This requires more vigorous contractions of the diaphragm and intercostal muscles.

Since the surface of alveolar cells is moist, the surface tension between water molecules resists stretching of lung tissue. Type II alveolar cells secrete a pulmonary surfactant that decreases surface tension and increases lung compliance. Surfactants produced in the alveoli decrease surface tension and help the alveoli to stay open. Respiratory distress syndrome in newborns is a result of low lung compliance.

Thermoregulation: Evaporation and Panting

One of the functions of the nose and the nasal cavity is to warm and moisten the inspired air, along with retrieving moisture and heat from the expired air.

The high water content of the mucus film acts to humidify inhaled air. A bed of capillaries and thin-walled veins lie under the nasal epithelium. The capillaries and thin veins are involved in warming the incoming air. If the air is cold, the *plexus* (a network of veins and capillaries) work together and fill with blood. The heat-intensifying process is then further amplified.

Heat and moisture can be exchanged from expired air. The conchae have been cooled by incoming inspired cold air. As warm air leaves, it precipitates moisture on the conchae, which acts to extract heat from the humid air flowing over them.

The surfaces involved in respiratory evaporation must be kept moist, as water is constantly evaporating from them. On a cold day, individuals can see water vapor in their breath, some of which is water that has evaporated from their lungs. Water is a product of cellular respiration and it, along with carbon dioxide, are expired during breathing.

Thermal panting is caused by a buildup of body heat, resulting from an increase in the environmental temperature or from additional activity and increased metabolic rates. Appropriate chemical receptors perceive the oxygen and carbon dioxide levels.

When the oxygen levels are too low (or carbon dioxide levels are too high), the respiratory system responds by increasing the overall rate of respiration. The overall panting rate increases at high altitudes (due to low pO_2 levels).

Alveolar Gas Exchange

Diffusion and differential partial pressure

Even though alveoli are small, their abundance in the lungs results in a large surface area for gas exchange. The walls of the alveoli have a single layer of thin cells, which creates a short diffusion distance for the gases.

Diffusion is the passive movement of materials from a higher to a lower concentration. A short diffusion distance allows for rapid gas exchange.

Ventilation is the mechanism of breathing. It is the process of bringing fresh air into the alveoli and removing stale air. Ventilation maintains the concentration gradient of carbon dioxide and oxygen between the alveoli and capillaries of the blood. Partial pressures measure the differences between oxygen and carbon dioxide concentrations. The greater the difference in partial pressure, the greater the rate of diffusion.

The body must expel carbon dioxide, the product of cell respiration. The body needs to take in oxygen for cell respiration to make ATP. There must be a low concentration of carbon dioxide in the alveoli so that carbon dioxide can diffuse out of the blood in the capillaries (where it exists in a high concentration) and into the alveoli. Also, there must be a high concentration of oxygen in the alveoli compared to a low concentration of oxygen in the capillaries, so that oxygen can diffuse into the blood of the capillaries from the alveoli.

The O_2 in the alveoli flows down its partial pressure gradient from the alveoli into the pulmonary capillaries. Here, O_2 binds to hemoglobin for transport. Meanwhile, CO_2 flows down its partial pressure gradient from the capillaries into the alveoli for expiration.

Blood entering pulmonary capillaries is the venous blood of systemic circulation. This blood has a high pCO_2 and a low pO_2. Differences in partial pressures of oxygen and carbon dioxide on the two sides of the alveolar-capillary membrane result in net diffusion of oxygen from the alveoli to the blood and carbon dioxide from the blood to the alveoli. With this diffusion, capillary blood pO_2 rises, and its pCO_2 falls, and net diffusion of these gases ceases when partial pressures in the capillaries become equal to those in the alveoli.

Diffusion improves with vascularization (i.e., more blood vessels); O_2 delivery to cells is promoted by erythrocytes (red blood cells) with the O_2-binding protein hemoglobin.

Diffuse interstitial fibrosis is a disease where the alveolar walls thicken with connective tissue and reduce gas exchange. Ventilation-perfusion inequality can result from ventilated alveoli with no blood supply or from blood flow through the alveoli with no ventilation, resulting in reduced gas exchange.

In steady state, the volume of oxygen consumed by body cells per unit time is equal to the volume of oxygen added to the blood in the lungs, and the volume of carbon dioxide produced by cells is identical to the rate at which it is expired. *Respiratory quotient* (RQ) is the ratio of CO_2 produced / O_2 consumed and depends on the type of nutrients used for energy.

Ventilation is the exchange of air between the atmosphere and the alveoli. Air moves by bulk flow from high pressure to low-pressure regions. The flow rate (F) can be found with:

$$F = (P_{atm} - P_{alv}) / R ,$$

where P_{atm} is the atmospheric pressure, P_{alv} is the alveolar pressure, and R is airflow resistance.

During ventilation, air is moved into the lungs by changing alveolar pressure through changes in lung dimensions.

Alveolar pO_2 (partial pressure of O_2) is lower than atmospheric pO_2 because oxygen in the alveolar air keeps entering the pulmonary capillaries. Alveolar pCO_2 is higher than atmospheric pCO_2 because carbon dioxide enters alveoli from pulmonary capillaries. pO_2 is positively correlated with the rate of alveolar ventilation and inversely correlated with the rate of oxygen consumption. pCO_2 is inversely correlated with the rate of alveolar ventilation and positively correlated with the rate of oxygen consumption.

Hypoventilation is an increase in the ratio of carbon dioxide production to alveolar ventilation, while *hyperventilation* is a decrease in this ratio.

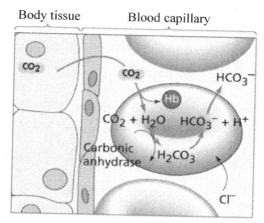

Erythrocyte (red blood cell) and gas exchange of O_2 and CO_2

In the alveoli capillaries, bicarbonate (HCO_3^-) combines with an H^+ ion (a proton) to form carbonic acid (H_2CO_3), which dissociates into carbon dioxide and water. The carbon dioxide then diffuses into the alveoli and out of the body with the next exhalation.

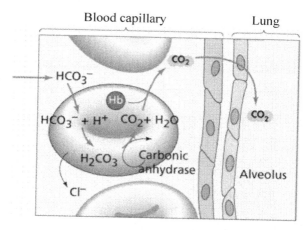

$$CO_2 \, (g) + H_2O \, (l) \leftrightarrow H_2CO_3 \, (aq) \leftrightarrow HCO_3^- + H^+ \, (aq)$$

Henry's Law

The mechanism of the gas exchange follows Henry's law, which states that there is an equilibrium concentration of oxygen that should be dissolved in the blood. When blood reaches the lungs, it has less than the equilibrium concentration of oxygen, because the body has used the oxygen.

Therefore, oxygen diffuses into the blood. The CO_2 in the blood that reaches the lungs is higher than the equilibrium concentration because the body releases CO2. Therefore, CO_2 diffuses out of the blood.

pH Control

A change in H+ concentration stimulates ventilation (e.g., due to lactic acid), input from mechanoreceptors in joints and muscles, increase in body temperature and increase in plasma epinephrine. Lactic acid in exercising muscles can cause metabolic acidosis or metabolic alkalosis (too much or too little acid, respectively), changing H^+ concentration and stimulating peripheral chemoreceptors.

Respiratory regulation refers to the variability in pH due to pCO_2 changes from alterations in ventilation. These changes in ventilation can occur rapidly and have a significant effect on pH. Carbon dioxide is lipid-soluble molecules that can freely transverse cell membranes at a rapid pace. Therefore, changes in pCO_2 result in rapid changes in [H+] in all body fluid compartments.

Two concepts portray the connection between alveolar ventilation and pH through pCO_2.

The following equation shows the relationship between alveolar ventilation (V_A) and pCO_2. Changes in alveolar ventilation are inversely related to changes in arterial pCO_2 and directly proportional to the total body CO_2 production (V_{CO_2}). The equation involves a constant of 0.863:

$$pCO_2 = 0.863 \times (VCO_2 / V_A)$$

The following is the Henderson-Hasselbalch equation, which states that the changes in pCO_2 cause changes in pH.

$$pH = pK_a + \log (HCO_3) / (0.03 \times pCO_2)$$

If a person is hyperventilating, arterial H^+ concentration rises due to increased pCO_2, which is respiratory acidosis.

On the contrary, hyperventilation lowers [H^+], which is respiratory alkalosis. Deoxy-hemoglobin has a higher affinity for H^+ ions than oxy-hemoglobin and binds most of the H^+ produced. In the lungs, when deoxy-hemoglobin is converted to oxy-hemoglobin, H^+ ions are released.

Hemoglobin (Hb) combines with H^+ ions as reduced hemoglobin (HHb). HHb is vital in maintaining normal blood pH. As blood enters the pulmonary capillaries, most of the CO_2 is in plasma as HCO_3^-.

$$HbO_2 \rightarrow Hb + O_2$$

Acidic pH levels and warmer temperatures promote this dissociation.

Regulation by Neural Control

The *vagus nerve* exercises influence over the respiratory system, heart, and viscera. This is an involuntary control, so breathing cannot be consciously stopped for long periods of time. Thus, the body breathes when it is able and needs to.

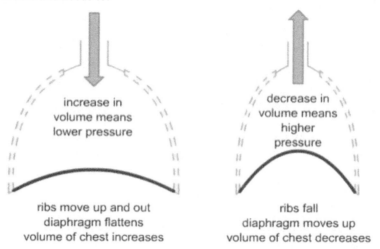

Ventilation with inspiration where diaphragm contracts (air in, on the left)
and expiration where diaphragm relaxes (air out)

The diaphragm and intercostal muscles are skeletal muscles. Therefore, breathing depends on the cyclical excitation of these muscles. Control of this neural activity resides with *medullary inspiratory neurons* in the medulla oblongata. Peripheral chemoreceptors (carotid bodies) and aortic bodies that are in close contact with arterial blood are stimulated by a steep decrease in arterial pO_2 and an increase in H^+ concentration. The medullary inspiratory neurons receive inputs from apneustic and pneumotaxic centers in the pons. Negative feedback from pulmonary stretch receptors controls respiration, the *Hering-Breuer reflex*. This reflex is triggered to prevent over-inflammation of the lung.

Protective reflexes (e.g., coughing and sneezing) protect the respiratory system from irritants. Receptors for sneezing are located in the nose or pharynx, while those for coughing are located in the larynx, trachea, and bronchi. These reflexes are characterized by a deep inspiration followed by a violent expiration.

Voluntary control of breathing is accomplished by the descending pathways from the cerebral cortex. It cannot be maintained when involuntary stimuli are high.

Tachypnea (rapid breathing) is a reflex effect from the J receptors, which are located in the lungs. It is stimulated by an increase in interstitial lung pressure due to the occlusion of a pulmonary vessel as a pulmonary embolus.

CO₂ sensitivity

Carbon dioxide concentration in metabolically active cells is much greater than in capillaries. Thus, CO_2 diffuses from the cells into the capillaries. About 7% of the CO_2 directly dissolves in the plasma. Another 23% of the CO_2 binds to the amino groups in hemoglobin to form *carb-amino-hemoglobin*. The remaining 70% is transported in the blood as a bicarbonate ion (HCO_3^-). CO_2 combines with water, forming *carbonic acid* (H_2CO_3), which dissociates to H^+ and bicarbonate ions (HCO_3^-):

$$CO_2 + H_2O \leftrightarrow H_2CO_3 \leftrightarrow H^+ + HCO_3^-$$

Carbonic anhydrase is an enzyme in red blood cells that speeds the reaction to remove carbon dioxide from the blood. Hence, the diffusion of even more carbon dioxide from the cells into the capillaries occurs for expiration out of the body.

The release of H^+ ions could drastically lower blood pH; however, the hydrogen ions are absorbed by the globin portions of hemoglobin, and the HCO_3^- diffuses out of the red blood cells and into the plasma.

The remaining CO_2 diffuses out of the blood across the walls of the pulmonary capillaries and into the alveoli. Any decrease in plasma CO_2 concentration causes the reverse reaction, also catalyzed by carbonic anhydrase for the release of $CO_2 + H_2O$ from the body:

$$H^+ + HCO_3^- \leftrightarrow H_2CO_3 \leftrightarrow CO_2 + H_2O$$

Chapter 15

Circulatory System

- **Circulatory System Overview**

- **Arterial and Venous Systems: Arteries, Arterioles, Venules, and Veins**

- **Oxygen and Carbon Dioxide Transport by Blood**

Circulatory System Overview

Functions: circulation of oxygen, nutrients, hormones, ions, and fluids; removal of metabolic waste

The circulatory system is responsible for *thermoregulation* (regulates body temperature), transport, fluid balance, and immune system function. It transports oxygen, carbon dioxide, nutrients, waste products, and hormones. The concentrations of the various components in the blood direct fluid retention or excretion, maintaining appropriate blood volume and composition. They affect how materials move out of or into the bloodstream, and to or from surrounding tissues.

Gases are taken in or released via the process of respiration, and nutrients are absorbed from the small intestine and distributed to tissues via the bloodstream. Blood filtration occurs at the kidneys and liver, where waste products and poisons are metabolized and removed. Steroid and peptide hormones are circulated throughout the body via the bloodstream, allowing for cellular communication. Also, the circulatory system protects the body from disease and injury by facilitating the movement of immune system cells.

Role in thermoregulation

The circulatory system maintains core body temperature by redirecting how blood flows to the skin and extremities. When temperatures drop, the body initiates *vasoconstriction* of the blood vessels to the skin, reducing blood flow near the surface and thus limiting heat loss to the air. Conversely, high temperatures cause *vasodilation* of these blood vessels, increasing blood flow to the skin and promoting heat loss to the surroundings.

Four-chambered heart: structure and function

The heart is a cone-shaped, muscular organ about the size of a fist located between the lungs and directly behind the sternum. It circulates blood throughout the body by a rhythmic pumping action. The chambers of the heart are lined with epithelial tissue as *endocardium,* while the outermost tissue is the *epicardium.* The *myocardium* (cardiac muscle) is the thick middle layer as the bulk of the heart. The cardiac muscle is entirely under the involuntary control of the autonomic nervous system and hormones. Unlike skeletal muscle, it has branching bands of muscle fibers connected by *intercalated disks.* These disks allow electrical signals to travel through the cardiac muscle cells and contract them as a single unit. Each cardiac muscle cell has a single nucleus, unlike the multi-nucleated skeletal muscle cells. The heart is surrounded by a double-layered sac of the *pericardium,* cushioned by the *pericardial fluid.*

Bands of connective tissue separate the chambers of the heart. The *septum* is an internal wall that divides the heart into its right and left halves. Each half of the heart contains an upper, thin-walled *atrium* and a lower, thick-walled *ventricle.* The four chambers function together as a double-sided pump. The atria are thinner and weaker than the muscular ventricles, but hold the same volume of blood. Each atrium receives blood from the veins (blood flowing towards the heart) of the cardiovascular system and delivers it to its respective ventricle through a valve. The ventricles then pump the blood to the rest of the body through arteries (blood flowing away from the heart), also regulated by a valve. The valves between atria and ventricles are the *atrioventricular valves,* and the valves between the ventricles and arteries are the *semilunar valves* for their half-moon shape when closed. Heart valves are supported by *chordae tendineae* (strong fibrous tendons) attached to muscular projections of ventricular walls to prevent valves from inverting.

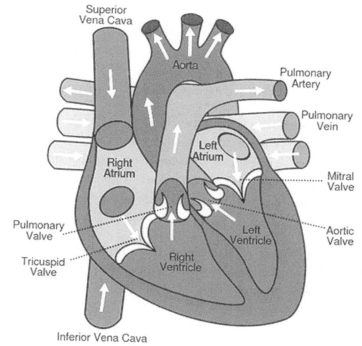

The heart has four chambers, associated blood vessels, valves and the route of blood through the heart is indicated with arrows

Deoxygenated blood returning from the body enters the *right atrium* through two major veins: the *superior vena cava* and the *inferior vena cava*. The right atrium then sends blood through the right atrioventricular valve (tricuspid valve) to the *right ventricle*. The *tricuspid valve* has three flaps, which snap shut when the right ventricle contracts, so that blood does not flow back into the right atrium. The right ventricle then ejects the blood through the *pulmonary semilunar valve* into the *pulmonary arteries.* These are the only arteries that contain deoxygenated blood, as all other arteries carry oxygen-rich blood. The pulmonary semilunar valve keeps blood from back-flowing into the right ventricle once it is in the pulmonary arteries. The pulmonary arteries deliver the deoxygenated blood to the lungs.

Oxygenated blood returns from the lungs through *pulmonary veins* and is delivered to the *left atrium*. The left atrium pumps this newly oxygenated blood through the left atrioventricular valve (*bicuspid* or *mitral valve*) to the *left ventricle*. From the left ventricle, blood is ejected into the *aorta* (body's largest artery) through the *aortic semilunar valve.* The left ventricle is the strongest and thickest chamber of the heart since it must pump blood throughout the body. This blood delivers oxygen to the body tissues and then returns deoxygenated blood to the right atrium to restart the cycle.

The right atrium and ventricle are separated from the left atrium and ventricle so that oxygenated blood does not mix with deoxygenated blood. Furthermore, blood is pumped out of the heart through one set of vessels and returns to the heart via another set. This structure ensures appropriate blood pressure and gas concentrations in different areas of the circulatory system.

Endothelial cells

The interior of the heart and blood vessels are lined with *endothelium,* a thin layer of *endothelial cells.* This semi-permeable layer controls the passage of cells and molecules in and out of the bloodstream. Endothelial cells help prevent blood clots and plaque formation and facilitate an inflammation response to invasive agents and damaged cells.

Basal muscle tone is a muscle's passive, baseline resistance to stretching, which varies throughout the body. Endothelial cells can rapidly change a blood vessel's smooth muscle tone from its basal level. Hormones and nerve signals regulate this vasoconstriction and vasodilation. Endothelial cells control blood pressure by producing nitric oxide (a potent vasodilator) and *Endothelial-1* (a potent vasoconstrictor). The controlled production of nitric oxide is essential to maintaining basal tone and therefore balanced blood pressure.

Endothelial cells are involved in *angiogenesis*, the formation of new capillaries. This process is vital to wound healing and the formation of collateral vessels, which provide alternate routes for blood flow in the event of a circulatory system blockage.

Systolic and diastolic pressure

Blood pressure is the pressure blood exerts on the walls of the blood vessels. It is usually measured in the arteries since this is where blood pressure is strongest. *Systolic pressure* is the pressure measured as the ventricles contract, and blood is forcefully ejected from the heart. *Diastolic pressure* is the blood pressure when blood is not being pumped since the ventricles are relaxing. Systolic pressure is the maximum blood pressure and is higher than the diastolic blood pressure. Reference blood pressure is 120 (systolic) over 80 (diastolic).

Both measurements of blood pressure represent a stage of a single heartbeat. *Systole* is the phase of contraction and blood ejection. When the term systole is used alone, it refers to the ventricles only. Note that throughout ventricular systole, the atria are relaxed (i.e., in *diastole*).

Diastole and systole are a part of the *cardiac cycle*, known as a *heartbeat* and lasts about 0.85 seconds. Systole makes up 0.30 seconds of this time, and diastole makes up 0.55 seconds. The heart beats about 70 times a minute in an average adult human. It undergoes over 3 billion contraction cycles during a normal lifetime.

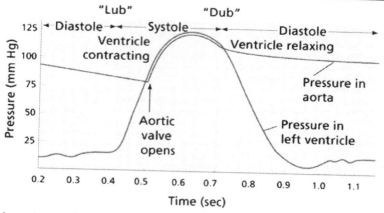

When the heart beats, the lub-dub sound is heard as the valves of the heart close. The "lub" results from the vibrations of the heart when the atrioventricular valves close, while the "dub" is heard as the semilunar valves close. These sounds mark divisions in the cardiac cycle.

At the beginning of systole, atrioventricular valves close and cause the lub sound. Semilunar valves have already been closed for some time. The ventricles contract while all valves are still closed, building up the pressure in the ventricles before the blood is ejected. This is the *isovolumetric ventricular contraction*, as the volume of blood in the heart remains the same (*iso-*). At 0.10 seconds into systole, semilunar valves open as the ventricles continue to contract, expelling blood out of the right ventricle into the pulmonary artery, and out of the left ventricle into the aorta. This is a *ventricular ejection*. The volume of blood ejected from each ventricle is the *stroke volume* (SV) while the amount of blood remaining after ejection is the *end-systolic volume* (ESV).

At the end of ejection, semilunar valves close and cause the dub sound. This marks the beginning of diastole when all chambers rest for 0.40 seconds. The first part of diastole is *isovolumetric ventricular relaxation*, as the semilunar valves close and the ventricles relax. In the next phase, *passive ventricular filling*, the atrioventricular valves open and allow blood from the atria to passively fill the ventricles. There is still no contraction taking place in either the atria or ventricles. In the last 0.15 seconds of ventricular diastole, atrial systole occurs; the atria contract and actively fill the ventricles. This period is *active ventricular filling*. The amount of blood in the ventricle at the end of diastole is the *end-diastolic volume* (EDV).

The stroke volume can ultimately be calculated by subtracting the end-systolic volume from the end-diastolic volume (SV = EDV – ESV). At this point, the atrioventricular valves close and the cycle repeats, as the ventricles re-enter systole.

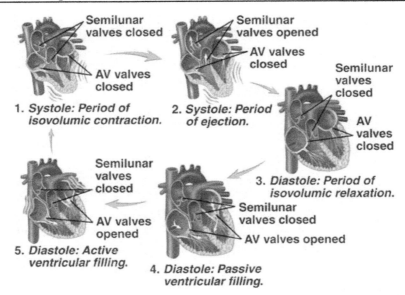

1. *Systole: Period of isovolumic contraction.* — Semilunar valves closed, AV valves closed

2. *Systole: Period of ejection.* — Semilunar valves opened, AV valves closed

3. *Diastole: Period of isovolumic relaxation.* — Semilunar valves closed, AV valves closed

4. *Diastole: Passive ventricular filling.* — Semilunar valves closed, AV valves opened

5. *Diastole: Active ventricular filling.* — Semilunar valves closed, AV valves opened

Values of the heart open and close during rhythmic contractions of the atria and ventricles

The strength of contraction is not constant but rather increases in response to a greater degree of stretch, up to an *optimum length* (OL). A higher volume of blood entering the ventricles (i.e., a greater EDV) stretches the ventricular muscles and stimulates a more forceful contraction. This empties the ventricles more completely and results in a lower ESV. This phenomenon, whereby a greater EDV produces stronger contraction, is the *Frank-Starling mechanism*. The overall effect of a higher EDV and a lower ESV is a higher stroke volume.

Heart issues can usually be diagnosed by observing irregularities in the cardiac cycle. During heart failure, the heart does not pump an adequate volume of blood. This may be due to diastolic dysfunction, in which the wall of a ventricle has reduced compliance and reduced ability to fill adequately, resulting in reduced EDV and therefore a reduced SV. Heart failure may be caused by systolic dysfunction, which results from myocardial damage that impairs cardiac contractility and therefore decreases SV. Adaptive reflexes to counter the reduced SV strive to maintain blood pressure, by retaining fluid to increase blood volume or constricting vessels. Unfortunately, excess fluid retention can impair respiration, and vasoconstriction makes it more difficult for the heart to pump. If the heart does not quickly recover from heart failure, it becomes even weaker.

A *pulse* can be measured in arteries far from the heart since it is a wave effect that passes down the walls of arterial blood vessels. The pulse occurs when the aorta expands and then immediately recoils following ventricular systole. Since there is one arterial pulse per ventricular systole, the arterial pulse rate can be used to determine the heart rate. For adults, a normal pulse rate (resting heart rate) is between 60 and 100 beats per minute (bpm); children have high resting pulse rates.

Generally, a lower heart rate, *bradycardia*, implies a more efficient heart function and better cardiovascular fitness (e.g., a well-conditioned athlete may have a pulse rate closer to 40 bpm). Bradycardia refers to a low heart rate (less than 60 bpm), while *tachycardia* refers to a heart rate greater than 100 bpm.

Pulmonary and systemic circulation

The human cardiovascular system has two major circulation pathways. *Systemic circulation* includes the loop between the heart and nearly all arteries, capillaries and veins of the body. *Pulmonary circulation* includes the path from the heart to the pulmonary arteries, pulmonary capillaries, pulmonary veins, and finally back to the heart. This is the only part of the circulatory system in which the arteries transport deoxygenated blood, and the veins transport oxygenated the blood. Blood pressure is lower in the pulmonary circulation since the blood vessels are shorter and provide less *vascular resistance* to blood flow.

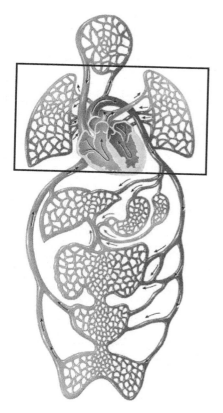

Pulmonary (enlarged) between heart and lungs and systemic circulation between heart and body

Pulmonary circulation begins when deoxygenated blood from systemic circulation enters the right atrium and then drains to the right ventricle through the right atrioventricular (tricuspid) valve. The blood is then ejected through the pulmonary semilunar valve into the right and left pulmonary arteries. The pulmonary arteries branch many times to form hundreds of thousands of *pulmonary capillaries* in the lungs. These capillaries surround *alveoli,* the small sacs in which gas exchange takes place. Carbon dioxide and water from the blood diffuse through the alveoli and are exhaled, while oxygen is inhaled and diffuses from the alveoli into the blood.

The oxygenated blood is returned to the left atrium of the heart via the pulmonary vein. Once oxygenated, blood returns to the heart, delivered through the left atrioventricular (bicuspid or mitral) valve into the left ventricle. It is pumped out of the left ventricle into the aorta via the aortic semilunar valve.

Systemic circulation begins as oxygenated blood travels through the body. The aorta is the body's largest and thickest artery. It arches and branches into the major arteries of the upper body before descending through the diaphragm, where it then branches further into arteries that supply the lower parts of the body. Arteries diverge into capillaries around organs and tissues to exchange gases, nutrients, and wastes.

The capillaries then merge into venules and veins as the blood is returning toward the heart. Veins of the lower body coalesce into the inferior vena cava, while those of the upper body coalesce into the superior vena cava. These two vessels empty into the right atrium of the heart so that the blood may then enter pulmonary circulation and be re-oxygenated. When oxygen levels are low, systemic capillaries which feed tissues in need of oxygen undergo vasodilation, so that more blood may be delivered to where it is needed most. By contrast, pulmonary capillaries, which feed low-oxygen alveoli, undergo vasoconstriction so that blood can be diverted to alveoli where gas exchange (uptake of O_2) is more efficient.

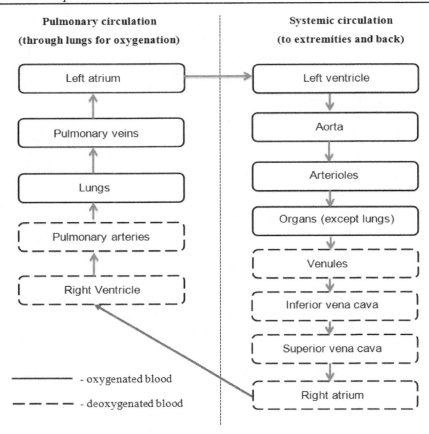

Coronary circulation supplies blood to the heart itself. *Coronary arteries* originate from the aorta and feed into capillaries that cover the heart and nourish myocardial cells. Deoxygenated blood from the coronary capillaries drains into the *cardiac veins,* which merge directly into the right atrium. Blocked flow in coronary arteries can result in chest pain and heart attacks.

In a fetus, circulation is more complicated, since the fetal lungs are immersed in the fluid and cannot perform gas exchange (O_2 from inspiration and CO_2 released from expiration). Oxygenated, nutrient-rich blood from the maternal side of the placenta is carried to the fetus via the *umbilical vein.* Most of this blood passes first through the liver before entering the inferior vena cava, but some bypass the liver via the *ductus venosus.* The oxygenated blood ends up in the inferior vena cava and then enters the right atrium. From the right atrium, most blood flows directly into the left atrium via the *foramen ovale,* which allows blood to bypass the fetal pulmonary circulation. The foramen ovale in the fetus is in the septum of the heart, which must close up before or shortly after birth to avoid heart problems. From the left atrium, the blood enters the left ventricle and is ejected into the aorta, where it circulates throughout the body. Once fully deoxygenated, the blood flows back into the placenta via the two *umbilical arteries* to be replenished.

Fetal blood that did not pass through the foramen ovale enters the right ventricle and is pumped into the pulmonary artery. This is mixed blood but is mostly deoxygenated. The *ductus arteriosus* is a vessel branching from the pulmonary artery and joins the aorta, shunting most of the deoxygenated blood into the aorta. Therefore, mixed blood travels from the aorta throughout the fetus' body. Only some of the blood in the pulmonary artery of the fetus travels throughout the pulmonary circulation and is returned via the pulmonary veins.

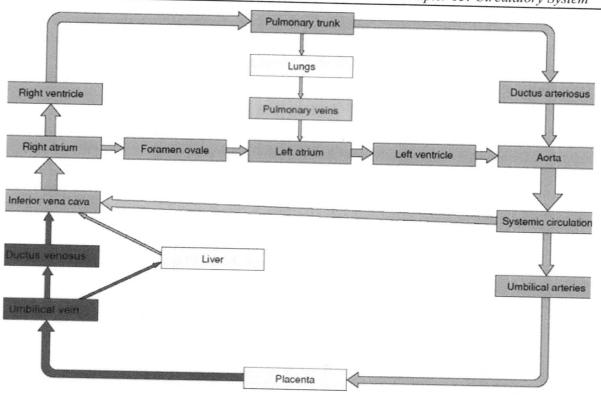

*Fetal circulation for the pathway of blood before lungs function for gas exchange.
Note the foramen ovale and ductus arteriosus*

Arterial and Venous Systems: Arteries, Arterioles, Venules and Veins

Structural and functional differences

The circulatory system divides into the arterial, capillary and venous portions. The arterial system carries blood away from the heart. Systemic arteries carry oxygenated blood from the heart to the body, while pulmonary arteries carry deoxygenated blood from the heart to the lungs.

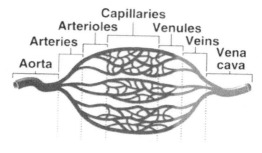

*Blood is received from the arteries and gas exchange occurs at the capillaries
before the blood returns to the heart through the veins*

Arteries are large, thick and elastic, with multi-layered walls of connective tissue, smooth muscle, and endothelium. The outermost layer is made of longitudinal collagen and elastic fibers to avoid leaks and bulges (an aneurysm). Within the connective tissue layer is a layer of circular smooth muscle and elastic fibers, which maintain muscle tone and contract to force blood through the vessels. Finally, an inner layer of endothelium lines the narrow lumen.

The strong, elastic arterial walls can maintain high blood pressure in the arterial system and aid in the pumping of blood throughout the body. During systole, contraction of ventricles ejects blood into arteries, distending the arterial walls. During diastole, the walls recoil elastically and force blood through. There is always some blood in the arteries to keep them semi-inflated, which is why diastolic blood pressure is not zero (averages 80 mmHg). Three major types of arteries: elastic arteries, muscular arteries, and coronary arteries.

The *elastic arteries* include the aorta and its major branches. They are the largest, thickest and most elastic of the arteries, providing an elastic pipe for blood directly out of the heart. Elastic arteries are not able to undergo vasoconstriction.

Muscular arteries (distributing arteries) distribute blood to specific organs. They contain a large amount of smooth muscle to direct blood flow and are involved in some vasoconstriction.

The *coronary arteries* directly supply and nourish the heart. They originate from the base of the aorta just above the aortic semilunar valve and branch across the heart's outer surface, the epicardium.

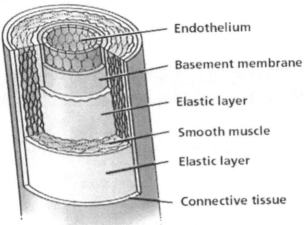

- Endothelium
- Basement membrane
- Elastic layer
- Smooth muscle
- Elastic layer
- Connective tissue

Cross section of distributing arteries

Arterioles are smaller vessels which branch from the arteries. They have walls of endothelium, smooth muscle, and connective tissue, but these layers are thinner. Arterioles are active in vasoconstriction and vasodilation, allowing the body to redirect and control blood flow and pressure.

Small arterioles branch into a network of *capillary beds*. The capillary beds surround the organs and tissues of the body and connect the arterial and venous systems. Capillaries are microscopic blood vessels which contain no muscle or connective tissue, a single layer of endothelial cells across which gases, nutrients, enzymes, hormones, and wastes diffuse.

After blood from the systemic arterial system exchanges nutrients and wastes with organs and tissues, deoxygenated blood leaves the capillaries and flows into the systemic venous system. The vessels of the venous system are thinner, more porous and less elastic. They contain an outer layer of fibrous connective tissue, a middle layer of smooth muscle and elastic tissue and an inner layer of endothelium. Vasoconstriction and dilation occur in the venous system.

Venules, which are thin and porous, are analogous to arterioles. They gather blood from capillary beds and merge into the larger veins. The veins have thicker and more resistant walls but are still much weaker compared to the arteries. Their thin walls and wide lumens make venous blood pressure lower than arterial blood pressure.

Unlike arteries, veins collapse when they contain little blood and are compressed by neighboring skeletal muscles. The squeezing action of skeletal muscles helps the low-pressure, slow-moving blood to travel to the heart. Veins depend on the action of the diaphragm and their smooth muscle walls to deliver blood to the heart. Larger veins have valves that prevent backflow of blood when moving against gravity.

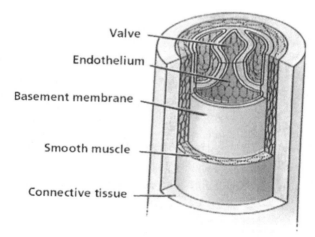

Cross section of veins. Note the presence of a valve and the absence of a thick layer of smooth muscle, as seen in arteries

Arteries are the thickest and strongest of the blood vessels, followed by the veins, arterioles, venules and finally the capillaries. The sympathetic nervous system innervates the walls of arteries, arterioles, veins, and large venules.

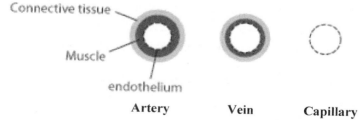

Artery   **Vein**   **Capillary**

Cross-sectional comparison of the three main vessels of the circulatory system

Not all blood received from the capillaries is delivered to the heart. Blood from the gastrointestinal capillary beds, for example, is first delivered to the liver to process and filter nutrients via the *hepatic portal vein*. The hepatic vein then leaves the liver and enters the inferior vena cava. All blood below the diaphragm is returned through the inferior vena cava, while all blood from the upper body is returned through the superior vena cava.

Pressure and flow characteristics

The pressure differential between the high-pressure arteries and low-pressure veins is what facilitates blood flow throughout the circulatory system.

At a basic level, the flow can be calculated by the following equation:

$$F = \Delta P / R$$

where ΔP is the difference in pressure between two points, and R is *resistance to flow*.

The resistance to flow is the opposing force which blood must counteract to move through the circulatory system. *Total peripheral resistance* (TPR) is the total amount of resistance throughout the circulatory system. TPR is a function of blood vessel diameter, blood vessel length, and viscosity.

Cardiac output (CO) is a measure of the rate of blood flow from the heart. It equals the stroke volume per minute (SV):

$$CO = HR \text{ (heart rate)} \times SV$$

CO is increased by a large increase in heart rate, resulting from increased activity in the sinoatrial (SA) node, and a small increase in SV, which is caused by an increased ventricular contractility mediated by sympathetic activity. The Frank-Starling mechanism accounts for an increase in end-diastolic volume.

Blood velocity describes the speed at which blood is moving through a particular vessel. Since blood volume flow rate (cardiac output) is approximately constant, blood velocity depends on the total cross-sectional area. *Bernoulli's principle* states that velocity is inversely proportional to cross-sectional area and can be calculated by dividing cardiac output by the cross-sectional area of the vessel.

Blood pressure is highest in arteries and decreases as it travels to arterioles, capillaries, venules, and veins. It is dependent on cardiac output, resistance to flow, total blood volume, vessel elasticity, and other factors. In most cases, systemic arterial blood pressure is measured, but systemic venous blood pressure is of interest. Additionally, the pressure in the pulmonary circulation can be measured.

The difference between systolic and diastolic pressure is *pulse pressure* (PP), while the average of the two is *mean arterial pressure* (MAP). The MAP is calculated by multiplying cardiac output by *total peripheral resistance* (TPR).

$$MAP = CO \times TPR$$

Blood pressure is highest in the arteries for many reasons; arteries are closest to the forceful ejections of the heart, have a small total cross-sectional area and high resistance. Arterioles have the greatest resistance to flow since they are the most capable of vasoconstriction, but the aorta has the highest pressure since it is closest to the heart.

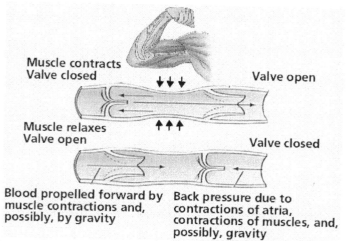

Muscle contracts
Valve closed   Valve open

Muscle relaxes
Valve open   Valve closed

Blood propelled forward by muscle contractions and, possibly, by gravity

Back pressure due to contractions of atria, contractions of muscles, and, possibly, gravity

The veins have the lowest blood pressure because the force of cardiac output diminishes over time, and a larger cross-sectional area to provide less resistance. Veins require the *skeletal muscle pump* (skeletal muscle contraction), movement of the *respiratory pump* (diaphragm), and valve action to maintain flow. Total blood volume is an important determinant of venous pressure. At any given time, most of the blood is in veins. Walls of veins are less elastic; they can stretch to accommodate large volumes of blood without recoiling (i.e., resisting flow).

Consequently, veins have low blood pressure, and vasoconstriction is often necessary to increase pressure in the veins and drive blood towards the heart. *Varicose veins* are an abnormal distention that develops when the veins' valves become weak and ineffective. Commonly, they are observed in the back of the leg and result in pooling of blood under the pressure of gravity.

Lastly, capillaries have slow, and even blood flow due to their high total cross-sectional area. Even though capillaries are the narrowest vessels, the total cross-sectional area of them is higher than any other vessels.

There is a decrease in the effective systemic circulating blood volume during a transition from horizontal to vertical body positioning. In a horizontal position, all blood vessels are at the same level, and almost pressure is due to cardiac output. In a vertical position, there is additional pressure at every point below the heart, equal to the weight of the blood column above it. This results in distension of blood vessels due to pooling of blood (e.g., varicose veins) and increased capillary filtration in lower parts of the body. This effect of gravity to pool blood can be offset by the contraction of skeletal muscles in the legs. People can become faint when standing for extended periods because of the pooling of blood in the lower extremities and less blood flow to the brain.

While the systemic circulation is considered a high-pressure system, the pulmonary circulation is low pressure. This is why the right ventricle is weaker than the left ventricle since it does not need to work as hard. The pulmonary circulation has shorter blood vessels and does not need as much force to work against gravity. The elasticity and lack of smooth muscle in pulmonary vessels further decrease resistance.

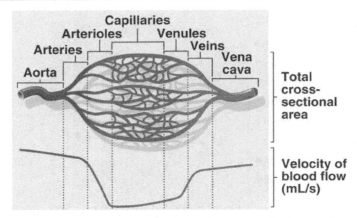

The relationship between total cross-sectional area and velocity of blood flow

Human blood pressure is usually measured at the brachial artery of the upper arm, which reflects systemic arterial pressure due to the contraction and relaxation of the right ventricle. A *sphygmomanometer* is an instrument which uses a pressure cuff to measure blood pressure. Healthy young adults should have a systolic pressure of 120 mmHg and a diastolic pressure of 80 mmHg (millimeters of mercury).

Hypotension is low blood pressure due to low blood volume, excessive vasodilation, anemia or heart conditions. If prolonged, hypotension can lead to *arrhythmia* (irregular heartbeat), dizziness and fainting. *Shock* is severe hypotension that may cause tissue or organ damage due to reduced blood flow. *Hypovolemic shock*, in which blood volume is reduced, may be caused by severe external bleeding (hemorrhage), dehydration, diarrhea or vomiting. Blood volume refers to the dissolved substances in the blood and not the fluid itself; therefore, a person can have a healthy 5 liters of blood but experience hypovolemia due to a lack of electrolytes. *Low-resistance shock* is a consequence of excessive vasodilation, which may occur due to endocrine or nervous system malfunction, weakened blood vessels and various drugs. *Cardiogenic shock* is when heart conditions reduce cardiac output to dangerous levels. Severe cardiogenic shock may be due to a heart attack and cardiac arrest.

When the body detects hypotension, it activates the sympathetic nervous system, which immediately increases stroke volume, heart rate, and total peripheral resistance to raise mean arterial pressure. Interstitial fluid enters the bloodstream due to reduced capillary pressure. In the long term, fluid ingestion and kidney excretion are altered, and erythropoiesis is stimulated to replace blood volume. Usually, the body can offset hypotension rapidly, but if it is severe and long-lasting, it is fatal.

Hypertension occurs when there is increased arterial pressure, generally due to an increased total peripheral resistance resulting from the reduced arteriolar radius. *Renal hypertension* results from increased secretion of *renin,* which generates the vasoconstrictor *angiotensin II*. Prolonged hypertension results in an increase in muscle mass of the left ventricle since it pumps against increased arterial pressure. This could decrease contractility and ultimately lead to heart failure.

Hypertension may be caused by stress, obesity, high salt intake or smoking, along with a genetic predisposition. Two genes are involved in hypertension for some individuals; they work together to produce angiotensin. Persons with this form of hypertension may one day be cured by gene therapy. Hypertension is often not detected until a stroke, or heart attack occurs but can be monitored by blood pressure. Pressure higher than 160 / 95 mmHg in women or   130 / 95 mmHg in men is indicative of hypertension.

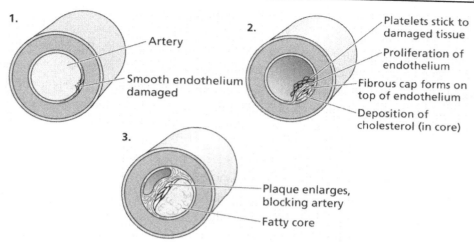

Progressive degeneration of the arterial lumen due to atherosclerosis (arteriosclerosis) and the associated obstruction in the blood vessel due to cholesterol plaque

Hypertension is seen in individuals with *atherosclerosis*. This condition occurs when soft masses of fatty materials, mostly cholesterol, accumulate beneath the inner linings of arteries. As this *plaque* accumulates, it protrudes into a vessel and interferes with blood flow. The thickened wall reduces blood flow and releases vasoconstrictors, further exacerbating the problem. Platelets may detect the plaque as a vascular irregularity and adhere to it, forming a clot. A *thrombus* is a clot that remains stationary. An *embolus* is a clot that dislodges into the bloodstream, which can be deadly if it reaches the heart or brain. Atherosclerosis can develop in early adulthood, but the symptoms may not appear until age 50 or older. In some families, atherosclerosis is inherited as *familial hypercholesterolemia*.

Atherosclerosis of a coronary artery may cause occasional chest pain, *angina pectoris*, which flares during periods of stress or physical exertion. Nitroglycerin and related drugs are used to dilate the blood vessels and relieve pain.

Hypertension and atherosclerosis are major contributors to heart disease, the leading cause of death in the US. Angina indicates oxygen demands are greater than the capacity to deliver it and is a warning sign for heart disease. If atherosclerosis and clotting in the coronary arteries create insufficient blood flow (*ischemia*), a portion of the heart muscle may die due to a lack of oxygen. This leads to *myocardial infarction* (MI), a *heart attack*. Damaged myocardial cells may create abnormal impulses which cause *ventricular fibrillation*, uncoordinated ventricular contractions. If the ventricular fibrillation severely impairs the heart's ability to deliver blood to the systemic circulation, it results in *cardiac arrest*.

A *stroke* is insufficient blood flow to the brain, which may occur if an embolus blocks a small cranial arteriole burst or; it can result in paralysis, severe neurological impairment, and even death. A stroke is a *cardiovascular accident* (CVA). Warning symptoms that foretell stroke include numbness in the hands or face, difficulty speaking, and blindness in one eye. If a cerebral artery is partially blocked, it is a temporary and less serious impairment as a *transient ischemic attack*.

Capillary beds

Capillaries in the human body span approximately 60,000 miles in total, permeating every tissue in the body to exchange nutrients, gases, and metabolic byproducts. They are most highly concentrated around organs such as the liver and intestines, which undergo high levels of metabolic activity. Blood velocity decreases as blood flows through capillaries because the total cross-sectional area of capillaries is relatively large.

Narrow, water-filled spaces often separate the endothelial cells of a capillary as *intercellular clefts*. There are three types of capillaries: continuous capillary, fenestrated capillary and sinusoidal capillary. *Continuous capillaries* contain no pores on their endothelial cells, but many have clefts at their cell boundaries. They exchange materials through their clefts or via endocytosis and exocytosis. Continuous capillaries are predominantly in skin and muscles, as well as in the cranium, where their tight junctions seal the blood-brain barrier.

Fenestrated capillaries contain small pores that are large enough for molecules to leak through, but not blood cells. They are predominantly in the small intestine (facilitate nutrient absorption), in the endocrine glands (allow passage of hormones) and in the kidneys (filter blood).

Sinusoidal capillaries contain pores that are large enough for blood cells to pass through. They are predominantly in lymphoid tissues, the liver and bone marrow, where they facilitate lymphocyte travel and blood cell modifications.

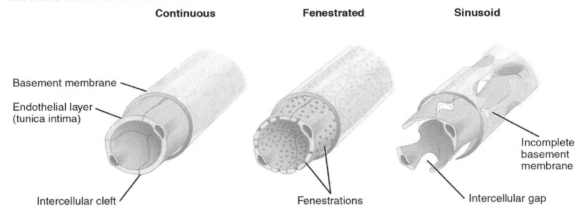

Capillary beds have two kinds of vessels: *true capillaries* (involved in the exchange between cells and blood) and the *metarterioles* (allow some blood to bypass the capillary bed). Metarterioles directly connect the arteriole and venules on opposite sides of the bed. Capillaries branch from the arterial side of the metarteriole; where they connect is a muscular sphincter. Contraction of the sphincters shuts off blood flow to the bed and causes blood to pass directly through to the venule. During times of exercise, metarterioles divert blood from the capillary beds of the skin and digestive organs so that it can be directed to the muscles.

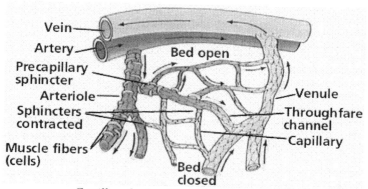

Capillary beds with metarterioles shown

Autoregulation permits constant blood flow in capillary beds because arterioles reflexively stretch and constrict to counteract changes in pressure.

Angiogenesis of capillaries is continuous to respond to growth, injury and changing metabolic activity. This process is activated by growth factors which direct partial digestion of an existing capillary, so that it may split in two or "sprout" a new capillary. Angiogenesis is particularly active in wound repair, muscle development, fat deposition, and tumor formation since these processes require an increased blood supply to the new tissues.

Mechanisms of gas and solute exchange

Capillaries are so narrow that only single red blood cells can pass. Some capillaries have small pores between the cells of the capillary wall, allowing white blood cells and other substances to flow in and out of capillaries by *paracellular* transport.

There are three basic mechanisms by which substances move across capillary walls to enter or leave the interstitial fluid: diffusion, vesicle transport, and bulk flow.

Diffusion is the passive movement of substances through the plasma membrane. Small molecules such as glucose and oxygen leave the capillaries and diffuse into the tissues, while carbon dioxide and small wastes leave the tissues and enter the capillaries.

Vesicle transport allows for the passage of materials via endocytosis and exocytosis. Larger, hydrophobic molecules typically travel through capillary walls by vesicle transport.

Bulk flow is a result of *hydrostatic pressure* and causes fluid to move from capillaries to the tissue fluids. It is higher in sinusoidal and fenestrated capillaries and lowers in continuous capillaries. It is opposed by *osmotic pressure*, which moves fluid from tissues to the capillaries due to differences in protein concentration. Both forces are described in the *Starling equation*, which explains how fluid and dissolved solutes either leave the capillaries (filtrate) or enter the capillaries (reabsorb).

The *net filtration pressure* (NFP) can be calculated using four forces, the *Starling Forces*. P_c is the hydrostatic capillary pressure, which favors fluid movement out of the capillary. P_{IF} is the interstitial fluid hydrostatic pressure, which favors fluid movement into the capillary. π_P is the osmotic pressure due to plasma protein (e.g., albumin) concentration, favoring fluid movement into the capillary. Finally, π_{IF} is the osmotic pressure due to interstitial fluid protein concentration, favoring fluid movement out of the capillary. The net hydrostatic pressure is the difference between P_c and P_{IF}, while the net osmotic pressure is the difference between π_P and π_{IF}. Together, the difference is the net filtration pressure.

A positive NFP indicates movement out of the capillary (filtration), while a negative NFP indicates movement into the capillary (reabsorption). Net filtration pressure equation:

$$NFP = Pc - PIF - \pi P + \pi IF$$

At the arterial end of a capillary, hydrostatic pressure is higher than osmotic pressure, so water leaves the capillary and enters the tissues. Midway along a capillary, there is no net movement of water. At this point, oxygen and nutrients diffuse into the tissue fluid, while carbon dioxide and other metabolic wastes diffuse into the capillaries. At the venous end of a capillary, osmotic pressure is higher than hydrostatic pressure, so water is reabsorbed in the bloodstream.

Mechanism of heat exchange

The hypothalamus is responsible for monitoring the temperature of blood, which is about 37 °C. If there are significant fluctuations from this set point, the hypothalamus sends nerve signals to the blood vessels to restore proper body temperature. When external conditions are hot, vasodilation allows blood to flow near the skin's surface so that heat is lost through convection and radiation. The opposite process occurs when external conditions are cold. Blood is kept away from the skin's surface through the mechanism of vasoconstriction. Blood flow is most reduced in the extremities, where the high surface-area-to-volume ratio would cause heat to dissipate rapidly.

Countercurrent heat exchange is used by humans and many other animals to conserve heat. Major veins and arteries run parallel to one another deep in the muscles of the arms and legs, with blood flowing in opposite directions. When heat loss is not a concern, most blood returns from the skin through surface veins that are not parallel to the deep arteries. However, in cold conditions, blood returns through deep veins so that it may exchange heat with the arteries. Blood moving through the warm arteries cools as it travels toward the extremities. By the time it reaches the body's surface, it is much closer to the temperature of the outside air, minimizing the temperature gradient and helping to reduce heat loss. As the cold blood returns through the deep veins, it is reheated by the nearby warm arterial blood.

Source of peripheral resistance

Resistance to flow (peripheral resistance) is the impedance of blood flow in the arteries caused by blood entering the arteries faster than it can leave, resulting in the stretching of vessels from an increase in pressure. The elastic walls of the arteries contract during the diastole phase, but the heart contracts again before enough blood flows into the arterioles to completely relieve the pressure in the arteries. Peripheral resistance is a function of blood viscosity, blood vessel length, and blood vessel diameter. A higher concentration of blood cells and plasma proteins increases viscosity and creates a higher resistance to flow. Diseases that cause an increase in the number of blood cells are problematic due to the increase in resistance, which forces the heart to work much harder.

Total blood vessel length may impact flow resistance. An overweight individual with additional blood vessels to service fat cells has greater resistance, which is partly why they are prone to increased risk of heart problems. Blood vessel diameter is important as well. Vasoconstriction increases resistance, and vasodilation decreases resistance. Obstruction from plaque (atherosclerosis) inside blood vessels increases resistance since it reduces diameter.

Total peripheral resistance is the sum of resistance to flow offered by all systemic blood vessels, although the main determinant is resistance in the arterioles. Because TPR affects blood pressure, deviations from normal blood pressure elicit homeostatic reflexes that alter TPR to offset the difference. Hypertension causes blood vessels to dilate to relieve resistance, while hypotension causes vasoconstriction to increase resistance.

Oxygen and Carbon Dioxide Transport by Blood

Hemoglobin and hematocrit

Vertebrates utilize the red-colored pigment hemoglobin to increase the oxygen-carrying capacity of the blood. Hemoglobin includes a tetrameric *globin* protein that surrounds a *heme* group. Heme is an iron-containing molecule which loosely binds to a single O_2 molecule.

Hematocrit refers to the percentage of blood volume made of erythrocytes. It is a useful measure of a person's ability to transport oxygen. Low hematocrit indicates *anemia,* a reduced ability to carry oxygen in the blood. Anemia causes constant fatigue and weakness due to the decreased circulation of oxygen. It may be due to an abnormally low concentration of healthy, functional erythrocytes, insufficient hemoglobin or a combination of both. These issues can occur due to blood loss, iron or vitamin B_{12} deficiency, abnormal erythrocytes, insufficient erythrocyte production or excessive erythrocyte destruction.

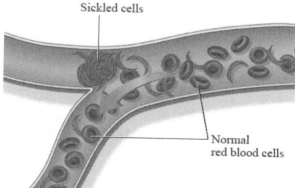

Sickled cells

Normal red blood cells

In sickle cell anemia, the red blood cells and are shaped like sickles (or crescent moon) and can become rigid and sticky. Due to their irregular shape, these cells can get stuck in small blood vessels (capillaries), slowing or blocking blood flow to parts of the body.

Oxygen content

Because hemoglobin is a tetramer, there are four heme groups per hemoglobin and therefore four O_2 binding sites. With millions of hemoglobin molecules per erythrocyte, the blood can carry 70 times more oxygen than if it was dissolved in plasma.

It is more difficult for animals to obtain O_2 from water than from the air. Water fully saturated with air contains a fraction of the O_2 as the same volume of air, and water is denser than air; therefore, aquatic animals must use more energy to breathe. Fish use up to 25% of their energy to breathe, while land mammals use 1–2%. The tradeoff is that air dries out wet respiratory surfaces, and as a result, humans lose 350 ml of water per day at 50% relative humidity.

Oxygen in the blood can be measured by its partial pressure (pO_2), its saturation, or its total content. *Partial pressure* describes oxygen's contribution to the total pressure of gases in the blood. *Saturation* refers to how many heme-binding sites are bound to oxygen in the blood. Content is calculated by taking into account the saturation, partial pressure and hemoglobin concentration to determine the total number of oxygen molecules in the blood.

Oxygen consumption increases in proportion to the magnitude of physical exercise up to the point of maximal oxygen consumption – *VO₂ max*. After VO₂ max is reached, any further increase in exertion can be sustained briefly by anaerobic metabolism. Normally, VO₂ max is determined by cardiac output. It may be limited by carbon monoxide content, the ability of the respiratory system to deliver oxygen to the blood and the ability of muscles to use oxygen.

Oxygen affinity

Oxygen that is inhaled diffuses into alveolar capillaries and binds to hemoglobin in erythrocytes. It is transported to tissues for cellular respiration. A small amount of blood oxygen is carried as dissolved O_2 in plasma, but the majority is reversibly combined with hemoglobin molecules in erythrocytes. Hemoglobin that has no bonded oxygen is *deoxyhemoglobin* (Hb), and that which is bonded to oxygen is *oxyhemoglobin* (HbO_2). The fraction of all Hb in the form of HbO_2 is the *percent of hemoglobin saturation*. This relationship is seen below:

$$\% \text{ hemoglobin saturation} = \frac{O_2 \text{ bound to Hb} \times 100}{\text{Maximal capacity of Hb to bind } O_2}$$

When pO_2 is high, hemoglobin binds to oxygen to form oxyhemoglobin. Blood is fully saturated with oxygen when all erythrocytes contain oxyhemoglobin. This occurs at the lungs, where oxygen content is highest. In body tissues deprived of oxygen, pO_2 is low, and erythrocytes release oxygen from oxyhemoglobin, diffusing into the plasma and tissue cells.

An *oxygen dissociation curve* shows the percent of oxyhemoglobin and non-bonded hemoglobin at various partial pressures of oxygen. Usually, a curve is labeled with the value at which the erythrocytes are fifty percent saturated with oxygen, the *p50 value*.

Hemoglobin has a sigmoidal oxygen dissociation curve because oxygen binding to one subunit somewhat relaxes the conformation of the other subunits, resulting in easier binding for additional oxygen. This is *cooperative binding*. When one O_2 molecule binds, the rest can bind with less difficulty. Likewise, when one O_2 is released, the remaining O_2 are released with ease. Thus, oxygen has the highest affinity to hemoglobin when three of its four polypeptide chains are already bound to oxygen.

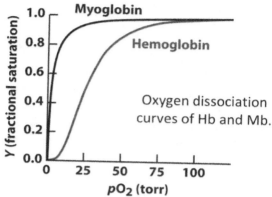

Myoglobin (hyperbolic) and hemoglobin (sigmoidal) dissociation curves

Myoglobin is a single chain protein subunit similar to hemoglobin, which is responsible for carrying oxygen in muscle tissue. It saturates quickly and is released in emergency situations of low oxygen when a burst of muscle movement is needed under reduced pO_2 conditions (e.g., swimming to the surface of a body of water after being submerged). Myoglobin binds to oxygen tighter than hemoglobin, as it has a greater affinity towards oxygen, but can bind one O_2 molecule. This does not allow for cooperative binding, giving myoglobin an oxygen dissociation curve that is hyperbolic. The myoglobin curve does not change over a wide range of pH. Myoglobin is in the bloodstream only after muscle injury and must be removed quickly since it can be toxic to the kidneys.

Biochemical characteristics of hemoglobin, modification of O_2 affinity

The oxygen dissociation curve is not static and can shift to the right or the left. A shift to the right means that for a given pO_2, less O_2 is bound to hemoglobin. Several conditions can produce a right shift, including an increase in temperature, an increase in pCO_2 or a decrease in pH. Increasing the temperature denatures the bond between oxygen and hemoglobin, which naturally decreases the concentration of oxyhemoglobin.

The Bohr Effect describes how the oxygen dissociation curve is altered by pH and CO_2. It was proposed in 1904 to explain how H^+ and CO_2 affect the affinity of O_2 for hemoglobin. The affinity of O_2 for hemoglobin is inversely related to both acidity ($\uparrow H^+$) and pCO_2 levels. Although CO_2 does not directly compete with O_2 for hemoglobin binding sites, it does bind to different areas of the hemoglobin molecule and encourages the release of O_2 molecules. Similarly, a decrease in pH is an increase in H^+ molecules, which bind to hemoglobin and promote the O_2 release. Both factors result in a Bohr shift, in which the O_2 dissociation curve shifts to the right.

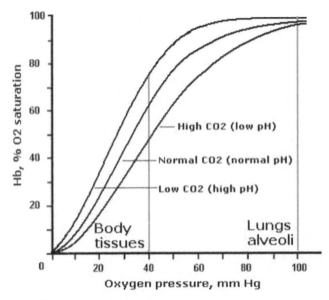

Bohr effect shifts hemoglobin dissociation curve to the right

High temperature, high CO_2 and low pH are all conditions that are observed during exercise, since temperature and CO_2 level rise due to increased metabolism, and pH is decreased due to lactic acid buildup.

2,3-Diphosphoglycerate (2,3-DPG), known as 2,3-Bisphosphoglycerate (2,3-BPG), is the primary organic phosphate in erythrocytes. Like CO_2 and H^+, 2,3-DPG is an allosteric effector that binds to hemoglobin and changes its conformation so that it decreases the affinity for oxygen. 2,3-DPG shifts the curve to the right (Bohr effect) to allow for the release of oxygen near the tissues that need it most.

Under normal physiology, the oxygen dissociation curve predictably shifts to the left in the lungs to maximize O_2 loading. As seen in the curve, as pO_2 increases, O_2 saturation of hemoglobin increases. The alveoli of the lungs are O_2 rich, causing hemoglobin of deoxygenated blood to be saturated with O_2. The diffusion gradient favoring oxygen movement from alveoli to blood is maintained because oxygen binds to Hb and keeps the plasma pO_2 low since only dissolved oxygen contributes to pO_2. In tissues, the procedure is reversed. The low pO_2 and high pCO_2 in the body tissues cause hemoglobin in the bloodstream to release O_2. There is a net diffusion of oxygen from blood into cells, and a net diffusion of carbon dioxide from cells into the blood.

Fetal hemoglobin binds to oxygen more tightly than adult hemoglobin to attract O_2 from maternal blood. Due to its higher binding affinity, the fetal hemoglobin curve shifts to the left of the adult curve.

Carbon monoxide (CO) is lethal since it can bind to hemoglobin around 200 times more tightly than oxygen. It interferes with the O_2 transport function of blood by combining with Hb to form *carboxyhemoglobin* (COHb). Due to the high binding affinity, small amounts of CO can occupy a large proportion of the Hb in the blood, thus making it unavailable for transporting O_2. If this happens, the Hb concentration and dissolved oxygen in the blood is normal, but total O_2 concentration is dangerously reduced. The presence of COHb shifts the O_2 dissociation curve to the left, thus interfering with the unloading of O_2 into oxygen-deprived tissues.

Carbon dioxide transport and level in the blood

Carbon dioxide is carried in blood in three different forms: dissolved in plasma as CO_2, dissolved in plasma as the bicarbonate ion (HCO_3^-) and combined with hemoglobin.

As carbon dioxide diffuses from the tissues into the blood, about 5% of it remains dissolved in plasma as CO_2. 10–20% combines with hemoglobin as *carbaminohemoglobin* ($HbCO_2$). As with oxygen, this complex can be unbound so that CO_2 is released back into the plasma. Unlike O_2, CO_2 binds to amino groups on hemoglobin rather than to the heme group of iron.

$$CO_2 + Hb \leftrightarrow HbCO_2$$

75 to 85% of CO_2 enters erythrocytes and combines with water to form carbonic acid (H_2CO_3), which dissociates into HCO_3^- and H^+. This reaction is catalyzed by the enzyme *carbonic anhydrase*. The ions are released back into the plasma to continue transport. The bicarbonate and hydrogen ions create a bicarbonate buffer system that helps maintain blood pH.

$$CO_2 + H_2O \leftrightarrow H_2CO_3 \leftrightarrow HCO_3^- + H^+$$

Some of the hydrogen atoms remain in erythrocytes and bind to oxyhemoglobin, releasing oxygen. This is an important process for oxygen transport and further prevents fluctuations in pH since excess H^+ can be removed from the blood.

Nervous and endocrine control

Both intrinsic and extrinsic stimulation controls the circulatory system. The intrinsic stimulation comes from pulses of the *sinoatrial* (SA) *node* as the "pacemaker" of the heart, in the upper dorsal wall of the right atrium; it initiates the heartbeat. The cardiac muscle cells in the SA node are particularly self-exciting and capable of initiating contraction in nearby cells. The capacity of the SA node for spontaneous, rhythmical self-excitation is a result of gradual depolarization, the *pacemaker potential*, of the cells.

When the node spontaneously generates an excitatory impulse, depolarization quickly spreads to the left atrium, and the two atria contract simultaneously. The action potential then spreads, after a small delay, through the *atrioventricular* (AV) *node* located at the base of the right atrium close to the septum. The delay in the action potential allows atrial contraction to be completed before the ventricles contract. *AV bundles* extend from the node toward the ventricles and carry the signal to the bottom of the heart, branching into *Purkinje fibers*. These fibers then cause rapid contraction of the ventricles.

The long refractory period of cardiac muscle cells limits the re-excitation of neurons and ensures that there is enough time for the chamber to be filled with blood before the next contraction occurs. This is why *tetanus*, when a muscle remains contracted and cannot relax, occurs only in skeletal muscle. However, the heart can experience many types of arrhythmias (abnormal heart rate patterns), which occur due to disruptions in the heart's electrical system. *Tachycardia* (high heart rate) and *bradycardia* (low heart rate) are two such arrhythmias. *Fibrillation* is a rapid, abnormal heartbeat that can affect both the atria and ventricles. Atrial fibrillation is concerning but typically not an emergency, while ventricular fibrillation is an emergency that usually accompanies a cardiac arrest. With the application of a strong electric current, the SA node may be able to reestablish a coordinated beat.

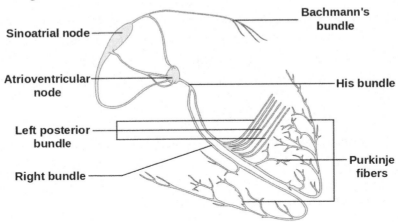

The innervation of the heart. Heartbeat originates in
SA node → AV node → bundle of His → Purkinje fibers for ventricular contractions

An *electrocardiogram* (ECG or EKG) is useful in diagnosing arrhythmias and other heart abnormalities. An EKG is a recording of the electrical changes that occur in the myocardium during a cardiac cycle. These contraction pulses generate currents in extracellular fluids that can be recorded at the skin's surface. An EKG typically consists of five major deflections: P wave, Q wave, R wave, S wave, and T wave.

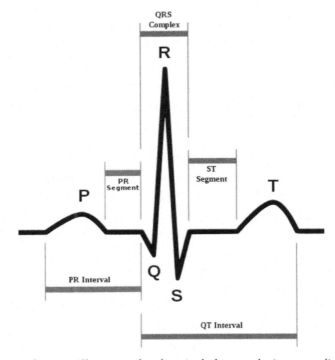

Electrocardiogram illustrates the electrical charges during a cardiac cycle

The cardiac cycle can be recorded by an electrocardiogram (EKG). The *P wave* represents depolarization of the atria; it occurs during atrial contraction. The *QRS complex* represents rapid depolarization of the ventricles, signaling that the ventricles are contracting. It has a much larger amplitude than the P wave. Atrial repolarization occurs at this time but is obscured by the activity of the QRS complex. Finally, the *T wave* represents repolarization of the ventricles as their muscle fibers recover.

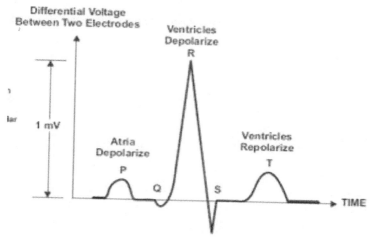

EKG with corresponding atria depolarization and ventricular depolarization and repolarization

The heart may spontaneously generate electrical impulses, but nervous and endocrine signals control the rate of these impulses. It is innervated by both the parasympathetic and sympathetic nervous systems. Stimulation of the parasympathetic fibers, such as the *vagus nerve,* decreases heart rate and force of contraction. This may occur in response to rest or hypertension. These neurons release the neurotransmitter *acetylcholine,* which increases the time required to reach the threshold voltage necessary for the initiation of an action potential. During times of rest, the parasympathetic nervous system dominates and diverts blood flow to the gastrointestinal tract to aid digestion.

Stimulation of the sympathetic fibers increases heart rate and force of contraction. These neurons release *norepinephrine,* which binds to receptors on the cardiac muscle cells of the SA node, increasing the uptake of calcium ions. Higher intracellular calcium levels allow the cells to reach the threshold voltage more rapidly, increasing heart rate and contractility. Sympathetic responses are activated during exercise; blood flow is redirected to muscles, cardiac output increases and constriction occurs in the veins to help return blood to the heart.

The adrenal medulla of the kidney releases *epinephrine* (i.e., adrenaline) in response to stress and other stimuli. This acts similarly to norepinephrine and causes a sympathetic heart rate increase. Circulatory responses take place not only in the heart but throughout the systemic and pulmonary circulation. Various receptors monitor conditions throughout the circulatory system. They may be *peripheral* (in blood vessels) or *central* (in the brain itself). Peripheral receptors often transmit the signal to the brain so that it may carry out responses. However, blood vessels can respond to stimuli with a reflex that occurs locally (*local control,* i.e., autoregulation). For example, the walls of arterioles automatically constrict in response to stretch.

Chemoreceptors monitor pH, oxygen, and carbon dioxide levels. Peripheral chemoreceptors are in the aorta and carotid arteries and transmit signals to the *cardiovascular control center* (CVCC) in the medulla oblongata of the brain. Central chemoreceptors are in the medulla itself, where they measure gas levels and pH in the brain. If any of these receptors detect that oxygen levels are low, carbon dioxide levels are high, or pH is low, the body attempts to increase gas exchange to compensate. This often occurs due to exercise. The sympathetic nervous system stimulates an increase in heart rate and stroke volume, resulting in higher cardiac output. This causes more rapid blood flow through the lungs, the more rapid release of excess carbon dioxide and uptake of oxygen. Selective vasoconstriction delivers oxygen to the tissues that need it most. In the long term, angiogenesis ensures these tissues receive greater oxygen.

Baroreceptors are in the carotid arteries, the aorta, and the medulla. They work with chemoreceptors to control blood pressure by continually adjusting vasoconstriction and vasodilation. A drop in arterial pressure causes the carotid and aortic baroreceptors to signal the medulla to immediately initiate a sympathetic response of vasoconstriction and higher cardiac output. In the long term, the kidneys retain more fluid to maintain blood volume. By contrast, a rise in arterial pressure stimulates the carotid baroreceptor reflex to initiate vasodilation. The medulla decreases sympathetic activity and increases parasympathetic activity, lowering cardiac output to reduce the pressure. Over time, the kidneys increase urination to lower blood volume.

Chapter 16

Lymphatic and Immune System

Lymphatic System

- **Major Functions**

- **Composition of Lymph**

- **Source of Lymph**

- **Lymph Nodes**

Immune System

- **Non-Specific and Specific Immunity**

- **Innate Immune System**

- **Adaptive Immune System**

- **Tissues: Bone Marrow, Spleen, and Thymus**

- **Concepts of Antigen and Antibody**

Lymphatic System: Major Functions

The lymphatic system is an open, unidirectional, secondary circulatory system. It contains a network of *lymph nodes, lymphatic vessels,* and *lymphatic capillaries.* The lymphatic system usually transports excess *interstitial fluids,* known as *lymph.* However, lymph can be moved by autonomic, smooth muscle contraction in larger lymph vessels. The lymph functions to return proteins to the bloodstream, redistribute body fluid, remove foreign bodies from the bloodstream and maintain the structural and functional integrity of tissues.

This system is associated with the cardiovascular system and has three main functions.

Equalization of fluid distribution

If the interstitial fluid pressure is greater than the lymphatic pressure, the lymph vessel open. When the lymph vessel opens, interstitial fluid enters the lymphatic capillaries. Lymphatic circulation eventually merges with the venous circulation, returning the lymph fluid to the blood. However, if the interstitial fluid pressure is less than the lymphatic pressure, the lymph vessel close, preventing lymph from leaking out.

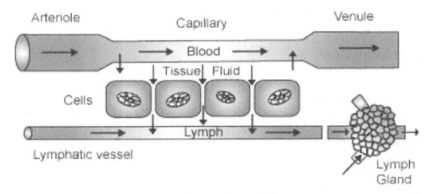

Relationship of fluid exchange between the circulatory and lymphatic systems

The movement of fluid is dependent upon skeletal muscle contraction. When muscles contract, fluid is squeezed past a valve that closes, preventing backflow.

Transport of proteins and large glycerides

The lymphatic system provides a pathway by which fat absorbed in the gastrointestinal tract can reach the blood. Fats are absorbed into the lacteals (lymphatic capillaries located in the small intestine) in the small intestine. Lacteals receive lipoproteins at the intestinal villi. The lymphatic vessels then transport these absorbed fats into the bloodstream. The lymph in the lacteals (chyle) has a milky appearance due to its high-fat content.

Proteins and large particles that cannot be taken up by capillaries are removed by lymph. Lymph monitors blood for bacterial or viral infection.

Return of materials to the blood

The lymphatic system takes up fluid that has diffused out of the blood capillaries and has not been reabsorbed. Lymphatic vessels carry interstitial fluid back to the cardiovascular system. This compensates for the net filtration out of the blood capillaries.

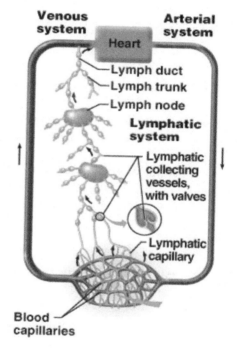

Interstitial fluid moves from the lymphatic to the cardiovascular system

The lymphatic capillaries collect cells and plasma proteins that leak out of the blood capillaries and returned to the venous system.

Composition of Lymph

Lymph is the tissue fluid that enters the lymphatic capillaries. Lymph, a fluid derived from interstitial fluid, runs through the lymphatic vessels, lymphatic nodes, and other lymphatic organs. Lymph tissue contains many *lymphocytes* (cells involved in immune response), which cleans and filters the fluid.

Peristalsis propels the lymph by the rhythmical contractions of smooth muscle lining the larger lymphatic vessels. The contractions are triggered by the stretching of the walls when lymph enters the system. Lymphatic vessels have valves that produce a unidirectional flow. The vessels are innervated by sympathetic neurons and are influenced by both the skeletal muscle and the respiratory pump.

The similarity to blood plasma

Lymph is a clear, colorless liquid that has a similar composition to blood plasma; however, lymph has a higher protein content (e.g., lymphocytes) compared to blood plasma.

The hydrostatic pressure of blood forces plasma fluid out of the capillary walls and into the surrounding tissues, which results in the formation of interstitial fluid. Some interstitial fluid then re-enters the capillaries, and some enter the lymphatic vessels to form lymph.

Substances transported

Lymph is mostly water, plasma proteins, chemicals, and white blood cells. The lymphatic system is responsible for the transportation of oxygen and carbon dioxide gas, along with chemical substances such as amino acids, glucose, and fats.

The lymphatic system transports other nutrients and cellular compounds. In particular, lymph contains various clotting factors including fibrinogen, globulin, an array of chemical elements (including calcium and iron) and some waste material (such as uric acid).

Source of Lymph

Diffusion from capillaries by differential pressure

Lymphatic capillaries have a single layer of endothelial cells resting on a basement membrane. Their water channels are permeable to interstitial fluid components, including protein molecules. Interstitial fluid enters these capillaries by bulk flow. The fluid then flows through the lymph nodes into two lymphatic ducts that drain into subclavian veins in the lower neck.

The basic flow of lymph is through the blood plasma from the capillaries into the interstitial fluid, where it becomes lymph, and is then returned to the blood.

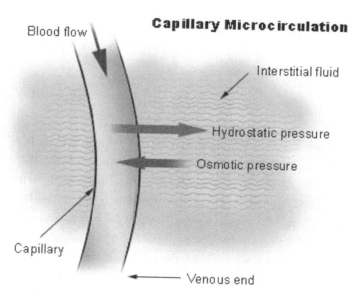

Blood flow becomes lymph due to a higher hydrostatic pressure that forces plasma into the interstitial space in forming lymph

Lymph Nodes

Lymph nodes are small (about 1–25 mm), ovoid or spherical masses of lymphoid tissue located along lymphatic vessels. They are responsible for filtering the lymph before it is returned to the blood. Lymph nodes are concentrated with phagocytic white blood cells. Lymph nodes are absent in the central nervous system.

A lymph node has two regions: the *outer cortex* and the *inner medulla*. A typical lymph node is surrounded by connective tissue compartments, known as *lymph nodules*. The cortex contains nodules where lymphocytes and macrophages congregate to fight pathogens. Macrophages are concentrated in the medulla and cleanse the lymph. Lymph sinuses separate the macrophages and lymphocytes.

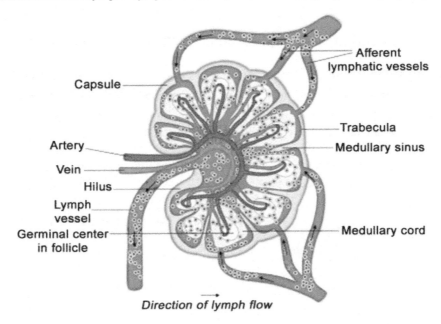

Capsule

Afferent lymphatic vessels

Artery

Vein

Hilus

Lymph vessel

Germinal center in follicle

Trabecula

Medullary sinus

Medullary cord

Direction of lymph flow

Afferent lymphatic vessels (only in the lymph nodes) enter all parts of the periphery of the lymph node. After branching and forming a dense plexus in the substance of the capsule, the afferent lymphatic vessels open into the lymph sinuses of the cortical part. Many afferent lymphatic vessels, which carry lymph to the nodes, enter via the convex side. The lymph travels through the lymph sinuses and eventually enters an efferent lymphatic vessel. The *efferent lymphatic vessels* (in the lymph nodes, the spleen, and tonsils) carry the lymph away from the node. In the lymph node, they begin at the lymph sinuses of the medullary part of the node and leave the nodes to either the veins or greater nodes.

Lymph nodes cluster in certain regions of the body, including the inguinal nodes in the groin, the axillary nodes in the armpits and the cervical nodes in the neck.

Lymphatic capillaries join as *lymphatic vessels* that merge before entering one of two ducts. The structure of the larger lymphatic vessels resembles veins, including the presence of valves.

The *thoracic duct* is larger than the right lymphatic duct. It serves the lower extremities, abdomen, the left arm, the left side of the head and neck and the left thoracic region. It delivers lymph to the left subclavian vein of the cardiovascular system.

The *right lymphatic duct* serves the right arm, the right side of the head and neck and the right thoracic region. It delivers lymph to the right subclavian vein of the cardiovascular system.

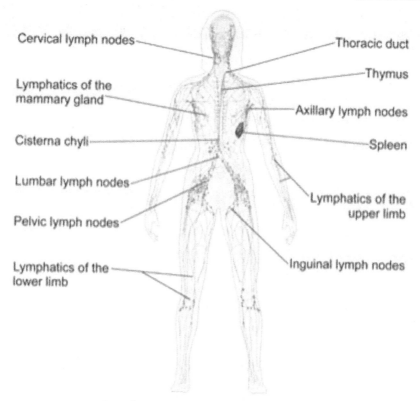

Lymphatic nodes and associated glands

Activation of lymphocytes

Lymphocytes produced in the red bone marrow by the differentiation of progenitor cells are the primary agents of an immune response. When pathogens or foreign antigens enter a lymph node, local lymphocytes are released into the bloodstream towards the site of the invasion. Once the lymphocytes are activated, they release chemicals that stimulate an immune response for proliferation, antibody production and the release of cytokines.

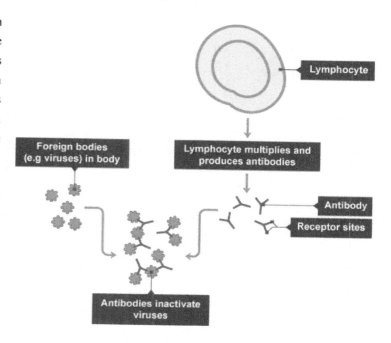

Immune System: Non-Specific and Specific Immunity

Immunity is the ability to defend against infectious agents, foreign cells and abnormal cancer cells. Immunity usually lasts for some time, and individuals do not ordinarily get the same illness twice. Immunity is acquired naturally through infection or artificially by medical intervention (e.g., vaccination). There are two types of *induced immunity*: active immunity and passive immunity. *Active immunity* is when an individual produces their antibodies. *Passive immunity* is when an individual receives prepared antibodies.

Active immunity can develop naturally after a person is infected. However, active immunity is often induced when a person is healthy to prevent future infection. An example of induced active immunity is a vaccine. Vaccinations are used to expose bodies to a particular antigen. These antigens are usually destroyed or severely weakened before they are administered to decrease their potency. After a vaccine, the immune response is measured by the antibody level present in the serum, the *antibody titer*. After the first exposure, a *primary response* occurs, going from no antibodies present to a slow rise in titer. After a brief plateau, a gradual decline follows as antibodies bind to an antigen or simply break down. After a second exposure to the same antigen, a *secondary response* occurs, and the antibody titer rises rapidly to a level much greater than before; this is a "booster." The high antibody titer is now expected to prevent any disease symptoms if the individual is infected. Active immunity depends on memory B and memory T cell responses (see adaptive immunity section). Active immunity is usually long-lasting, although a booster may be required every few years.

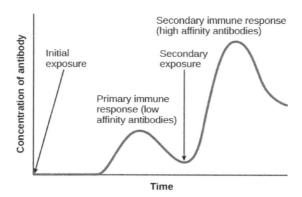

Antibody production profile comparing primary and secondary responses

Passive immunity defends against particular antigens by using antibodies from a foreign source. For example, an infant receives antibodies through the placenta or from the mother's breast milk, rendering the child immune to a specific pathogen. If in immediate danger of an infectious agent, a patient can receive antibodies from another individual through medical intervention. However, the effects of this type of passive immunity are short-lived, because antibodies are not made by an individual's B cells. For example, a person may be given a gamma globulin injection (a serum that contains antibodies against the agent) taken from an individual or animal who has previously been exposed to the antigen. If antibodies are derived from animals, some individuals may become ill with *serum sickness*, which is an inflammatory response that arises from the hypersensitivity of the individual to animal proteins, due to mistaking the animal's proteins for harmful antigens.

Immunity includes both *non-specific* and *specific* defenses (more detail below).

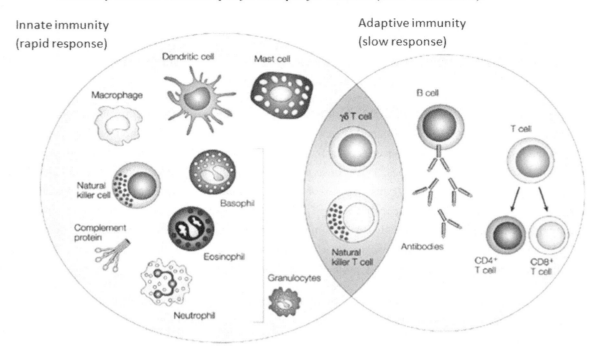

Innate immunity (rapid response) compared to adaptive immunity (slow response) on the right

Major Histocompatibility Complex

The *major histocompatibility complex* (MHC) is a group of genes that encode for specific proteins located on the surfaces of cells. These complexes, which are in all highly evolved vertebrates, aim to aid the immune system in recognizing foreign substances. In humans, these molecules are enclosed in many gene layers, all clustered on the same region of the chromosome. Human MHC proteins are called HLA (human leukocyte antigens).

There are two major classes of MHC molecules. Class one MHC molecules are along the membranes on all cell types. Class two MHC molecules are only in the immune system's macrophages and lymphocytes. No two individuals have the same MHC molecules, because genes have many different alleles (alternative forms of the gene). These molecules are collectively the tissue type. The MHC molecules contain a variety of genes that encode for other proteins, including complement proteins, chemical messengers of cytokines, and enzymes, sometimes called the third class of MHC molecules.

Initially, MHC molecules were recognized as antigens that stimulate an organism's immunological response to transplanted organs and tissues. After skin graft experiments on mice in the 1950s, it was found that graft rejection was an immune reaction mounted by the host organism against foreign tissue. The host was able to recognize the MHC molecules as foreign antigens and attacked them. Thus, the importance of the major histocompatibility complex (MHC) proteins was recognized to contribute to the difficulty of transplanting tissues from one organism (or person) to another.

Recognition of self vs. non-self and autoimmune disease

Each type of virus, bacteria or other foreign body has molecular markers that make it unique. Host lymphocytes (i.e., those in the body) can recognize and differentiate *self-molecules*, which belong to the body and are not foreign. *Non-self molecules* are lymphocytes. An example of a non-self molecule is an antigen, which triggers mitotic activity in B and T lymphocytes.

When a pathogen invades a body, MHC markers on the cells' plasma membranes distinguish between self and non-self cells. The pathogen displays a combination of self and non-self markers, and T cells interpret this as non-self. Cancer cells or tissue transplant cells are often recognized as non-self by T cells due to this combination. When T cells encounter non-self cells, they divide and produce four kinds of cells: cytotoxic T cells, helper T cells, suppressor T cells, and memory T cells. MHC molecules are vital parts of the immune system because they allow T lymphocytes to detect macrophages that have ingested a foreign microorganism. The partially digested microorganism displays a distinctive peptide, bound to an MHC complex, on the macrophage's surface. The T lymphocyte recognizes this foreign fragment attached to the MHC molecule and consequently initiates an immune response.

In an adaptive immune system, an antigen-specific response is initiated. This requires the identification and recognition of non-self antigens during antigen presentation.

Autoimmune diseases are when cytotoxic T cells or antibodies mistakenly attack the body's cells as a foreign antigen. Under these conditions, the differentiation between self and non-self molecules is lost. Autoimmune diseases may be genetic (i.e., heritable), or they may be caused by bacteria, viruses, drugs or various chemical agents.

Comparison of endogenous immune responses with autoimmune disease

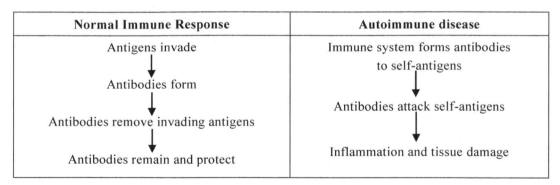

Normal Immune Response	Autoimmune disease
Antigens invade ↓ Antibodies form ↓ Antibodies remove invading antigens ↓ Antibodies remain and protect	Immune system forms antibodies to self-antigens ↓ Antibodies attack self-antigens ↓ Inflammation and tissue damage

A few examples of autoimmune disease are myasthenia gravis, multiple sclerosis, systemic lupus erythematosus, rheumatic fever, and type I diabetes. In myasthenia gravis, the neuromuscular junctions do not work properly and muscular weakness results. In multiple sclerosis (MS), the myelin sheath of nerve fibers is attacked. In systemic lupus erythematosus, many symptoms are elicited before dying from kidney damage. While there are no cures for autoimmune diseases, they can be controlled by drugs.

Innate Immune System

The first line of defense in the body is *surface coverage*. The skin and mucous membranes protect the body from pathogens. The skin is composed of a dry, acidic, dead cellular layer that allows for optimal, non-specific surface protection. The mucous membranes contain lysozymes which are enzymes that break down bacteria. Other cells in the mucous membranes contain cilia that filter the pathogens and particulates.

Nonspecific responses are generalized responses to pathogen infection and do not target a specific cell type. Individuals do not ordinarily become immune to their cells; the immune system can distinguish self from non-self cells. In this manner, the immune system aids, rather than counters, homeostasis. Lymphocytes recognize antigens because they have antigen receptors; the protein shape allows the receptors and antigens to combine like a lock and key. During maturation, enough differentiation occurs that there is a lymphocyte for various antigens. The non-specific response consists of white blood cells (WBCs) and plasma proteins. The four nonspecific defenses include a barrier to entry, inflammatory reaction, natural killer cells, and protective proteins.

The *barrier to entry* is the first non-specific response. Skin and the mucous membrane lining the respiratory, digestive and urinary tracts are mechanical barriers. Oil gland secretions inhibit the growth of bacteria on the skin. Ciliated cells are lining the upper respiratory tract sweep mucus and particles up into the throat to be swallowed. The stomach has a low pH (between 1.2 and 3) that inhibits the growth of bacteria. The harmless bacteria that reside in the intestines and vagina prevent pathogens from colonizing.

The *inflammatory reaction* is the second non-specific response, which is the series of events that occur if the tissue is damaged. The inflamed area often has four symptoms: redness, pain, swelling, and heat. Aspirin, ibuprofen, and cortisone are anti-inflammatory agents that counter the chemical mediators of inflammation.

Natural killer cells are the third non-specific response. Natural killer cells are a class of lymphocytes that recognize abnormal cells like cancerous cells. They attach to these abnormal cells and release chemicals that eventually destroy them.

Protective proteins are the fourth and final non-specific response. These proteins protect the cell nonspecifically. *Complement proteins* are plasma proteins, which have a role in both non-specific and specific defenses. The complement system contains some plasma proteins and is designated by the letter C and a subscript. One activated complement protein activates another protein in a set series of domino reactions. In this way, a limited amount of proteins can activate many other proteins, "complementing" certain immune responses. Additionally, the complement system amplifies an inflammatory reaction by attracting phagocytic cells to the site of infection. Some complement proteins bind to antibodies already on the surface of pathogens, thereby increasing the probability that a neutrophil or macrophage phagocytize pathogens. Complement proteins form a *membrane attack complex* that produces holes in bacterial cell walls and plasma membranes; from osmotic pressure, fluids and salts then enter the bacterium to the point where the cell bursts.

Lymphocytes

Lymphocytes are a type of leukocyte that produces antibodies. They are any of the three subtypes of white blood cells in a vertebrate's immune system. These types of white blood cells include the natural killer cells, T cells and B cells. Each lymphocyte makes one specific antibody. A large amount of different lymphocytes is needed so the body can produce different types of antibodies. The antibodies are on the surface of the plasma membrane of these lymphocytes, with the antigen-combining site projecting outwards. Pathogens have antigens on their surface which bind to the antigen-combining site of the antibodies of a specific lymphocyte. When this happens, the lymphocyte becomes active and starts to make clones of itself by dividing via mitosis. These clones then start to make more of this specific antibody needed to defend the body against the pathogen.

Macrophages

In general, *macrophages* are involved in engulfing foreign objects. A macrophage engulfs a pathogen and then acts as an antigen-presenting cell. When monocytes enter tissues, they differentiate into macrophages that ingest the pathogens. Connective and lymphoid tissues have resident macrophages that devour old blood cells and debris.

Macrophages trigger an explosive increase in leukocytes by releasing colony-stimulating factors; these chemicals then diffuse into the blood and are transported to the red bone marrow to stimulate the production of white blood cells (WBC). Inflammation promotes macrophage (phagocytic WBC) activity. Macrophages secrete *interleukins*, which are communication proteins among WBC. *Interleukin-1* increases body temperature, which causes a fever. The fever causes drowsiness and reduces the body's energy usage and stress, enhancing the WBC's ability to protect against infection.

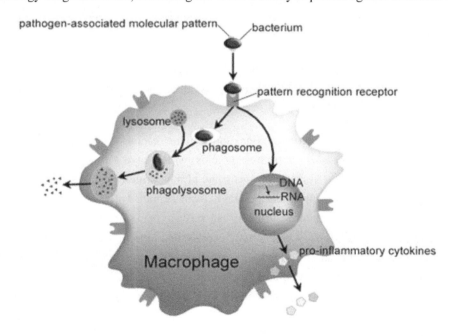

A *mast cell* is a specific example of a macrophage. Mast cells, which resemble basophils, reside in the connective tissue and mucous membranes. During an allergic response, they release histamine, bringing about inflammation. When tissue damage occurs, tissue cells and mast cells release chemical mediators, such as histamine and kinins. Histamine and kinins caused vasodilation and increased the permeability of capillaries to white blood cells. Enlarged capillaries produce redness and a local increase in temperature. The swollen area and the kinins stimulate free nerve endings, causing pain.

Phagocytes

Phagocytes are formed from stem cells (undifferentiated WBC) in bone marrow, and they consume foreign material to destroy it. They do this by recognizing pathogens and engulfing them by endocytosis. Enzymes (lysosomes) within the phagocytes then digest the pathogens. Phagocytes can ingest pathogens not only in the blood but within body tissue since they can pass through the pores of capillaries and into these tissues.

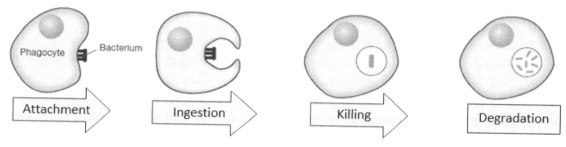

Stages of phagocytosis from attachment to degradation

Neutrophils phagocytize bacteria. *Eosinophils* are phagocytes that secrete enzymes to kill parasitic worms. *Basophils* secrete histamine to enhance inflammation.

Neutrophils are granulocytes due to the presence of granules in their cytoplasm. They are considered polymorpho-nuclear (PMN) leukocytes because of their unique, lobed nuclei. PMN leukocytes phagocytize pathogens and destroy them. Neutrophils and monocytes migrate by the amoeboid movement to the site of the injury and escape from the blood by squeezing through the capillary wall. They contain various toxic substances that kill or inhibit the growth of various bacteria and fungi. Just like macrophages, they activate a respiratory burst. The products of this respiratory burst are strong oxidizing agents like hydrogen peroxide, free oxygen radicals, and hypochlorite. Neutrophils are the most abundant type of phagocytes, and they are usually the first cells that go into action at the site of infection. Pus is the accumulation of dead neutrophils along with tissue, cells, bacteria and living WBC.

Dendritic cells are phagocytic cells present in tissues that are in contact with the external environment, mainly the skin, and the inner mucosal linings of the nose, small intestine, lungs, and stomach. They are important in the process of antigen presentation and serve as the connection between the innate and adaptive immune systems.

Plasma cells

Plasma cells (a type of B cell) secrete antibodies. They originate in bone marrow but leave the bone marrow to differentiate into plasma cells in the lymph nodes. They are transported by the blood plasma and the lymphatic system. When a B cell in a lymph node of the spleen encounters an appropriate antigen, it is activated to divide. B cells act as antigen-presenting cells and internalize various offending antigens, which are then taken by the B cell through receptor-mediated endocytosis and processed. The antigenic peptides (pieces of antigen) are then located on the surface of various MHC molecules and are presented to the T cells. The T cells bind to the MHC antigen molecule and activate the B cell, usually occurring in the spleen and lymph nodes. The B cell then starts differentiating into more specialized cells. For example, *germinal center B cells* differentiate into memory cells or plasma cells. This process of differentiation for B cells is affinity maturation. These B cells become plasmablasts (immature plasma). Eventually, they begin producing large volumes of antibodies in the lymph nodes and spleen. Once the threat of infection has passed, the development of new plasma cells ceases, while those that are present die.

Basic aspects of the inflammatory response

During the early stages of an infection, there is an *inflammatory response*, which is a nonspecific attack. Phagocytes become active and digest the pathogen. It causes localized redness, swelling, heat, and pain. White blood cells are more active at higher temperatures, and inflammation recruits white blood cells to the site of infection by sending out chemical signals. Changes in the capillary wall structure (higher permeability) allow more interstitial fluid and white blood cells to leak out into the tissues.

Neutrophils, lymphocytes, and monocytes are the cells involved in the inflammatory response.

Neutrophils are the first type of white blood cells that go to the injury site during acute inflammation. They are anti-bacterial cells that break down bacterial cells by releasing various lysosomal enzymes. The neutrophils recognize the bacterial cells as foreign agents by the antibody molecules that are attached to the surface of the bacteria. These antibody molecules are in the plasma of the blood and the interstitial fluid. Here, they bind to one specific antigen or foreign agent that the body has seen before.

Lymphocytes then start to accumulate during the inflammatory response process. If their presence is prolonged or is large in number, it may suggest that the antigen is still present and an actual infection has formed. The lymphocytes are responsible for producing large numbers of antibodies. These antibodies are unique to each type of lymphocyte. These specific antibodies provide a way to recognize foreign molecules and differentiate between self and non-self.

Monocytes are phagocytic cells that circulate in the blood along with macrophages. They are located in the connective tissue. The monocytes and macrophages engulf and digest foreign microorganisms, tissue debris, and various dead cells. They with lymphocytes to recognize and destroy foreign substances.

An *allergic reaction* is a hypersensitive immune response mechanism. It can cause itching, inflammation or even tissue injury. Allergies are hypersensitivities to foreign substances (e.g., pollen). A response to these antigens, *allergens*, usually involves tissue damage. Immediate and delayed allergic responses are two of the four possible responses. An *immediate allergic response* occurs within seconds of contact with an allergen. Cold-like symptoms are common. *Delayed allergic responses* are initiated by sensitized memory T cells at the site of the allergen in the body. The tuberculin skin test is an example: a positive test shows prior exposure to TB bacilli but requires some time to develop reddening of tissue.

Natural killer cells

Natural killer cells use cell-to-cell contact to destroy virus-infected cells and tumor cells; they lack any specificity or memory. They do not directly attack the invading microbes but instead destroy the host cells, such as the tumor cells. This concept is predominantly related to cells that have an abnormally low level of major histocompatibility complex markers, which can occur in viral infections of various host cells. The natural killer cells are activated when the MHC markers are altered, have the condition of "missing self" and are involved in the adaptive immune system. Many experiments show that natural killer cells can readily adjust to the immediate environment and then form an antigen-specific immunological memory, which is important to secondary infections with the same antigen. The natural killer cell's role in both innate and adaptive immunity is important in cancer therapy research.

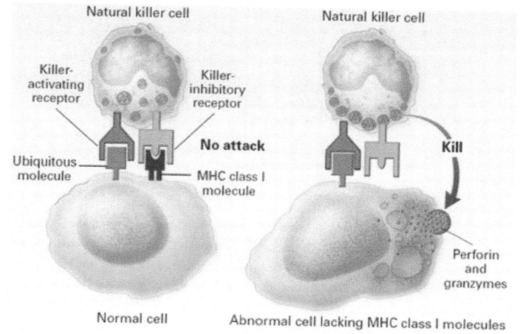

Natural killer cells mode of activation with action by killer-inhibitory receptor

Adaptive Immune System

Adaptive immunity is highly specified for a particular pathogen and antigen. Antigen-presenting cells have foreign antigens on their surface, and the antigens are recognized by T and B cells. Dendritic cells are an example of antigen presenting cells.

Specific defenses are physical, chemical, and cellular defenses against the invasion of viruses, bacteria and other agents of disease. Pathogens have antigens that can be components of foreign or cancer cells. The *specific response* is activated when nonspecific methods are insufficient, and infection becomes widespread. The *specific immune response* reacts to unique antigens and is primarily the result of the action of B lymphocytes and T lymphocytes. Some of these lymphocytes are capable of entrapping antigens on their surface. B lymphocytes (B cells) produce free-moving antibodies, while T lymphocytes (T cells) produce antibodies on their surface. When lymphocytes attach to these antigens, they can then begin to encode for unique antibodies, structures that are capable of catching these antigens.

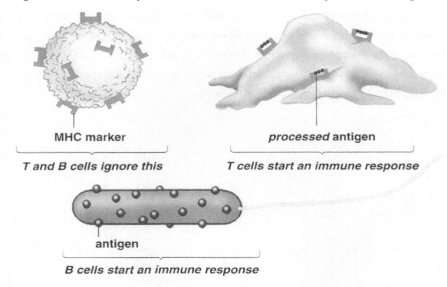

Molecular cues that stimulate lymphocytes to create an immune response

A *humoral response* (antibody-mediated response) responds to antigens that circulate in lymph or blood (e.g., bacteria, fungi, parasites, viruses, and blood toxins). The humoral response is a form of adaptive immunity. An example of a humoral response is a bacterial infection. Initially, inflammation persists, and macrophages and neutrophils engulf the bacteria. Then, interstitial fluid flushes into the lymphatic system where lymphocytes are in the lymph nodes. Macrophages process the foreign organism and present the appropriate bacterial antigen to the B lymphocytes. With assistance from helper T cells, the B cells differentiate into plasma and memory cells. Plasma cells produce antibodies that are released into the blood to attack the bacteria, while memory cells prepare for the same event if the bacteria ever attack again (secondary response).

Clonal Selection

Clonal selection is the response of lymphocytes to specific antigens. Clonal selection theory states that the antigen selects the B cell to produce a multitude of clones of the corresponding plasma cells. The selection portion is the B cell lymphocytes activating specific antigens. B cells are in the blood, and lymphoid tissues and are produced from lymphoid stem cells. The clonal portion of this concept deals with the multiplication of antibodies, which occurs in particular cells.

B cells present a specific antibody on the surface. The appropriate antigen forms a complex with the antibody on the cell, thereby activating the cell for further development. The ones that do not meet the specificity criteria are not activated. Antigen-antibody interaction occurs, and the complexes populate on the surface. These complexes are then internalized and begin to swell and divide rapidly. The B cells differentiate into plasma and memory cells. Plasma cells produce specific antibodies, while memory cells remain in circulation (but actively produce no antibodies).

T lymphocytes

T cells (T lymphocytes) arise from stem cells in the bone marrow. They travel to the thymus where they differentiate and mature. T cells have antigen receptors, but do not make antibodies; they check molecules displayed by non-self cells. In the thymus, if a T cell binds to a self-antigen, it is destroyed. If not, it is released for work in lymphoid tissue.

At maturity, T cells acquire receptors for self-markers, such as the major histocompatibility complex (MHC) molecules and antigen-specific receptors. They are then released into the blood as "virgin" T cells. T cells ignore other cells with MHC molecules, as well as free-floating antigens. The antigen must be presented to them by an *antigen-presenting cell* (APC). When an antigen-presenting cell presents a viral or cancer cell antigen, the antigen is first linked to an MHC protein; together they are presented to a T cell. Clinically, when a donor and recipient are histocompatible, it is likely that a transplant is successful. This binding promotes rapid cell division and differentiation into effector cells and memory cells (all with receptors for the antigen).

Cytotoxic T cells and helper T cells are responsible for cell-mediated immunity.

Effector cytotoxic T cells recognize infected cells with the MHC-antigen complex. They then destroy the cell with perforans (enzymes that perforate the cell membrane, allowing cytoplasm to leak out) and other toxins, which attack organelles and DNA. Once a cytotoxic T cell is activated, it undergoes clonal expansion and destroys cells that possess the antigen, if the cell bears the correct HLA antigen. As the infection disappears, the immune reaction wanes and few cytokines are produced. *Apoptosis* is the process of programmed cell death; it is critical to maintaining tissue homeostasis. Apoptosis occurs in the thymus if the T cell bears a receptor to recognize a self-antigen; if apoptosis does not occur, T cell cancers result (i.e., lymphomas and leukemias). The few T cells that do not undergo apoptosis survive as memory cells. The allergic response is regulated by the cytokines secreted by both T cells and macrophages.

Interferons and *interleukins* improve the ability of an individual's T cells to fight cancer. Interleukin antagonists may help to prevent skin or organ rejection, autoimmune diseases, and allergies when used as adjuncts for vaccines. Cancer cells with altered proteins on their cell surface should be attacked by cytotoxic T cells. Cytokines may awaken the immune system and lead to the destruction of cancer. Researchers withdraw T cells from a patient and culture them in the presence of interleukin. The T cells are then re-injected into the patient. The remainder of the interleukins is then used to maintain the killer activity of the T cells. They have storage vacuoles that contain *perforin molecules*, which perforate a plasma membrane and cause water and salts to enter, ultimately causing the cell to burst.

Effector helper T cells secrete interleukins that stimulate both T cells and B cells to divide and differentiate. Helper T cells stimulate the activation of B cells, cytotoxic T cells, and suppressor T cells. They regulate the immune system by improving the response of other immune cells. When exposed to an antigen, they enlarge and secrete cytokines. *Cytokines* stimulate the helper T cells to clone and stimulate other immune cells to perform their functions. Cytokines stimulate the macrophages to perform phagocytosis and stimulate B cells to become antibody-producing plasma cells. For example, HIV, which causes AIDS, infects helper T cells primarily and inactivates the immune response.

Memory T cells remain in the body after an encounter with an antigen and save time for an immune response to the same antigen. Like B cells, memory T cells have unique antigen receptors.

Suppressor T cells (*regulatory T cells*) use negative feedback in the immune system. They generally suppress the proliferation of effector T cells.

The cell-mediated response is effective against infected cells, using mostly T cells as a response to any non-self cell, including cells invaded by pathogens. A non-self cell binds to a T cell, which starts clonal selection. This initiates a series of events, including the production of cytotoxic T cells and helper T cells, the binding of helper T cells to macrophages that engulf pathogens, and the production of interleukins by the helper T cells, which stimulate the proliferation of T cells, B cells, and macrophages.

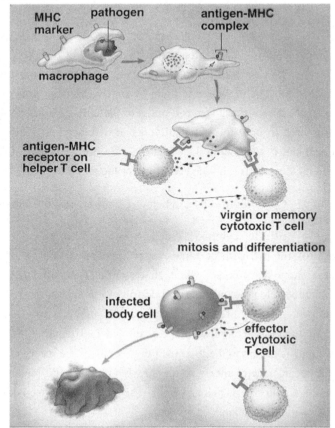

Cell-mediated immune response showing involvement by macrophages, antigen-MHC complex, helper T cells, and effector cytotoxic T cells

Tissue rejection occurs when cytotoxic T cells cause the disintegration of foreign tissue. This is a correct differentiation between self and non-self. The selection of compatible organs and the administration of immunosuppressive drugs prevents tissue rejection. Transplanted organs should have the same type of HLA antigens as the recipient. Cyclosporine and tacrolimus are immunosuppressant drugs that act by inhibiting the response of T cells to cytokines.

B lymphocytes

B lymphocytes (B cells) arise from stem cells in the bone marrow and mature in the spleen. They give rise to plasma cells that produce and secrete specific antibodies. Note that antibodies are released from plasma cells and are specific for an antigen. A single B lymphocyte produces one antibody type. Memory B cells are long-lived B cells that do not release antibodies in response to immediate antigen invasion; instead, they circulate in the body, proliferate and respond quickly (via antibody synthesis) to eliminate a subsequent invasion by the same antigen. These cells have a similar function to memory T cells.

Antibody-mediated immunity is the defense by B cells. It is *humoral immunity* because antibodies are present in the *humor*, which is any bodily fluid (i.e., blood and lymph). The plasma membrane of B cells contains *antigen receptor-antibodies* (immunoglobulins). *Immunoglobulins* are proteins that are specific to each antigen, and there are five classes (IgA, IgD, IgE, IgG, and IgM), based on variation in the Y-shaped protein-constant region and variable regions.

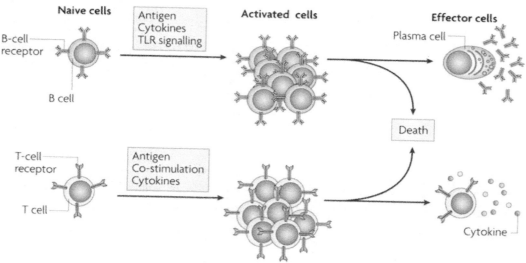

Roles of B cells and T cells in immunity

Each type of B cell carries its specific antibody as a membrane-bound receptor on its surface. A B cell does not clone until its antigen is present and it can recognize the antigen directly. However, B cells are stimulated by helper T cell secretions to clone. Some cloned B cells do not participate in antibody production but remain in the blood as memory B cells.

Antibodies are proteins that recognize antigens. The undifferentiated B cell produces antibodies that move to the cell surface and protrude. The B cell floats in the blood, and when it encounters the specific antigen, it becomes primed for replication. The B cell must receive an interleukin signal from a helper T cell, which has already become activated by a macrophage with an MHC-antigen complex. This promotes rapid cell division.

The B cell population then differentiates into effector and memory B cells. The effector B cells produce a staggering amount of free-floating antibodies. When these free-floating antibodies encounter an antigen, they tag it for destruction by phagocytes and complementary proteins. These types of responses only occur for extracellular toxins and pathogens; antibodies cannot detect pathogens or toxins located inside a cell. When mitosis occurs, the daughter populations become subdivided. Effector cells (or plasma cells), when fully differentiated, seek and destroy foreign cells.

Making memory cells is efficient because it does not necessitate the activation of T cells in proliferating antibodies if the same antigen is present. Memory cells allow the body to mount a greater and more sustained response against the same pathogen during secondary response. Thus, a secondary response usually takes less time. Immunological specificity and memory involve three events, including the recognition of a specific invader, repeated cell division that forms huge lymphocyte populations and the differentiation into subpopulations of effector cells and memory cells.

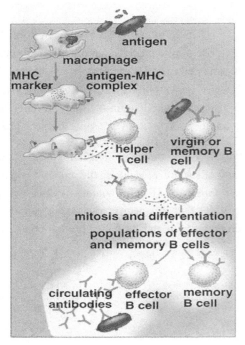

Antigen stimulates B cells to produce produce circulating antibodies

Interferons

Interferon is a protein that is produced by a virus-infected animal cell. Interferons bind to the receptors of non-infected cells, producing substances interfering with viral reproduction. Interferons are specific to a particular species (e.g., human interferons for humans).

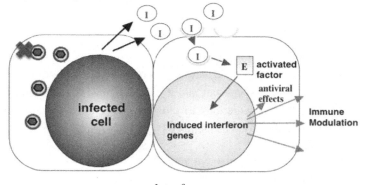

Interferons

Tissues

Lymphoid tissue is in the lymph nodes and thymus, as well as in various organs. It is where lymphocytes reside, proliferate and differentiate. Lymphoid tissues provide a site for white blood cells (WBC) to reside and proliferate.

Bone marrow

Human bone marrow consists of both red marrow and yellow marrow. Red (bone) marrow contains mostly hematopoietic tissue, whereas yellow marrow is predominantly fat cells. Both red and yellow marrows contain blood vessels and capillaries.

Red marrow is the origin of all blood cells, including all leukocytes, platelets and red blood cells. At birth, an individual's bone marrow is red. However, with age, a portion of the red marrow is converted to the yellow marrow, and by adulthood, about half of the bone marrow is red.

In adults, red marrow is located in the flat bones including the skull, sternum, ribs, clavicle, pelvic bones and vertebrae, and in the cancellous ("spongy") material at the epiphyseal ends of long bones (e.g., femur and humerus). Red marrow consists of reticular fibers that are produced by reticular cells packed around thin-walled sinuses. Differentiated blood cells enter the bloodstream at these bone sinuses.

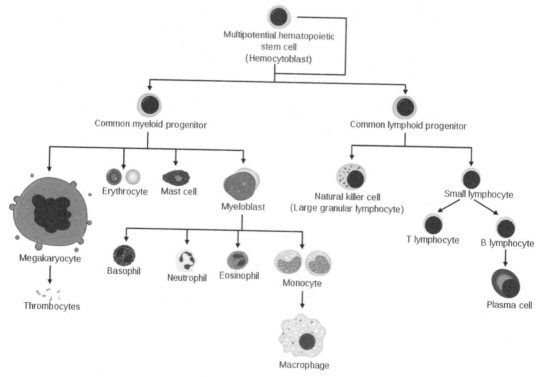

Blood stem cells give rise to cells of the immune system

Yellow bone marrow is in the medullary cavity (i.e., hollow interior in the center of long bones). When the body is exposed to trauma like extreme blood loss, the body can convert yellow bone marrow back into red bone marrow to increase blood cell production.

Spleen

The *spleen* is in the upper left abdominal cavity just below the diaphragm. The spleen is similar to a lymph node but is much larger (nearly the size of a fist). Whereas the lymph nodes cleanse the lymph, the spleen cleanses the blood. A capsule divides the spleen into nodules, which contain sinuses filled with blood. A *spleen nodule* has both red pulp and white pulp. The *red pulp* contains red blood cells, lymphocytes, and macrophages. It helps to purify the blood that passes through by removing microorganisms and worn-out or damaged red blood cells. *White pulp* mainly contains lymphocytes.

If the spleen ruptures due to injury, it can be removed, as other organs assume its functions. However, a person without a spleen is more susceptible to infections and may require antibiotic therapy.

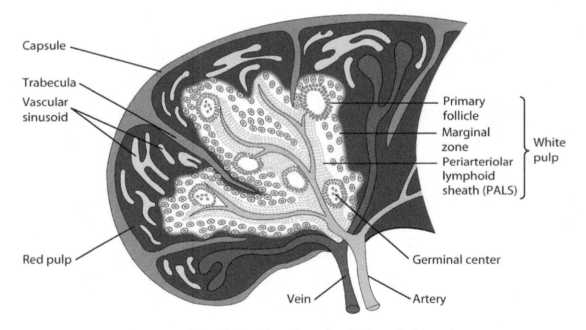

Anatomy of the spleen with regions of red pulp and white pulp

Thymus

The *thymus* is involved in the immune response. The *thymus gland* is located along the trachea behind the sternum in the upper thoracic cavity. It is larger in children than in adults and may disappear completely in old age. It is divided into lobules by connective tissue. The lobules are the site of T lymphocyte maturation. The interior or medulla of each lobule consists mostly of epithelial cells that produce thymic hormones (thymosin). Secreted *thymosins* stimulate lymphocytes to differentiate into T cells, which identify and destroy infected body cells.

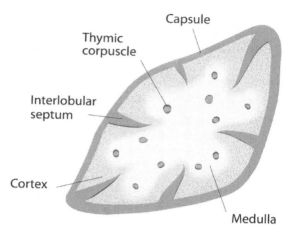

Structure of the thymus with cortex and medulla

Antigen and Antibody

Antigen-antibody recognition

An *antigen* is a non-self molecule that triggers an immune response. Antigens are foreign substances, usually proteins or polysaccharides, which stimulate the immune system to react, consequently stimulating the production of antibodies. *Antibodies* are molecules that bind to antigens so that lymphocytes can recognize the antigens. The antigen binds with a specific antibody at the *antigen-binding site*.

The *antigen-antibody complex,* or immune complex, marks the antigen for destruction via phagocytosis by neutrophils, macrophages, or complement activation. When antibodies bind to antigens, they bring about neutralization, where a pathogen cannot adhere to the host cell. This *opsonization* (i.e., when a pathogen is targeted for destruction by phagocytes) enhances phagocytosis-complement activation, where the antibody destroys the infected cell by creating holes in the cell membrane. Additionally, the antibodies mark for macrophage or natural killer cell phagocytosis by complement proteins, agglutination of antigenic substances or chemical inactivation (if a toxin).

Structure of the antibody molecule

Antibodies are large globular proteins that defend the body against pathogens by binding to antigens on the surface of these pathogens and destroying them. Antibodies are secreted into the blood, lymph, and other bodily fluids. Antibodies usually bind to one specific antigen.

Antibodies consist of two light chains and two heavy chains linked together by disulfide bonds. Antibodies are Y-shaped proteins, with each tip of the "Y" binding to an antigen. The tips of the "Y" are hypervariable regions because they are unique to each antigen-specific antibody.

Each arm of the antibody has a "heavy" polypeptide chain and a "light" polypeptide chain. *Heavy chains* have constant regions and variable regions, while *light chains* have constant and variable domains. The *constant regions* have amino acid sequences that do not change but are not identical among all antibodies.

The *variable regions* have portions of polypeptide chains with amino acid sequences that change, allowing for antigen specificity. The variable region forms the antigen binding sites of antibodies (i.e., tips of the "Y"), where their shape is specific to an antigen.

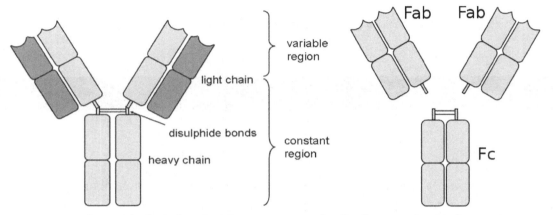

Structure of an antibody. Fab is the "fragment antigen binding" region that binds to antigens. Fc is the "fragment crystallizable" region that interacts with cell surface receptors, allowing antibodies to activate the immune system.

There are five classes of circulating antibodies, or immunoglobulins (Igs):

- *IgG antibodies* contain two Y-shaped structures and are the main type of antibody in the blood and extracellular fluid. They are not common in the lymph and tissue fluid. These antibodies protect the body from infection by binding to many types of pathogens and toxins.

- *IgM antibodies* are pentamers (contain five Y-shaped structures). They appear in blood soon after an infection begins and disappear before it is over. They are useful activators of the complement system.

- *IgA antibodies* contain two Y-shaped structures that attack pathogens before they reach the blood. They are the main antibody in bodily secretions.

- *IgD antibodies* contain two Y-shaped structures and serve as receptors for antigens on immature B cells.

- *IgE antibodies* contain two Y-shaped structures and are involved in immediate allergic reactions. IgE antibodies are attached to the plasma membrane of mast cells (in tissues) and basophils (in the blood). When an *allergen* attaches to IgE antibodies on these mast cells, they release large amounts of histamine and other substances, which cause cold symptoms or even anaphylactic shock. *Anaphylactic shock* is a severe systemic reaction that causes symptoms such as throat swelling, an itchy rash and a sudden drop in blood pressure. Allergy shots can sometimes prevent the onset of allergic symptoms. Injections of an allergen cause the body to build up high quantities of IgG antibodies because these antibodies do not cause allergy. The IgG antibodies combine with allergens received from the environment before reaching IgE antibodies located on the plasma membrane of mast cells and basophils.

Every plasma cell derived from the same B cell secretes antibodies against the same antigen; these are *monoclonal antibodies.* Monoclonal antibodies can be produced in vitro. B lymphocytes (usually harvested from mice) are exposed to a particular antigen. Activated B lymphocytes are then fused with myeloma cells (malignant plasma cells that divide indefinitely). *Hybridomas* are fused cells that result from two different cells—hybrid cells and cancerous cells (suffix *-oma*). Monoclonal antibodies are used for rapid, reliable diagnoses of various conditions (e.g., pregnancy). They identify infections, distinguish cancerous cells from normal cells and can be used to carry isotopes or toxic drugs to kill tumors.

Antigen presentation to stimulate antibodies

A pathogen enters an antigen-presenting cell (APC). Pieces of the pathogen are displayed at the surface of APCs. The T cell receptors recognize the presented antigen and activate various immune responses.

When a macrophage engulfs an extracellular pathogen, pieces of the pathogen become the antigen, and they are presented at the macrophage's cell surface. Helper T cells recognize the presented antigen and activate macrophages to destroy the pathogen. Helper T cells activate B cells to produce antibodies against the pathogen.

When an intracellular pathogen invades a host cell, pieces of the pathogen get presented on the host cell surface. Cytotoxic T cells recognize the presented antigen and signal the infected cell to self-destruct.

The Rh factor is an important antigen in human blood types. Rh positive (Rh^+) has the Rh factor on red blood cells (RBC); Rh negative (Rh^-) lacks the Rh antigen on the RBC. Rh negative individuals do not have antibodies to Rh factor but begin to make them if exposed to Rh positive blood. Rh positive is a genetically dominant trait.

Hemolytic disease of the newborn is possible if the mother is Rh negative and the father is Rh positive. The Rh factor is important during pregnancy; an Rh-negative mother and an Rh-positive father pose an Rh conflict. The child's Rh-positive RBC can leak across the placenta into the mother's circulatory system when the placenta breaks down. The presence of the "foreign" Rh positive antigens causes the mother to produce anti-Rh antibodies. Anti-Rh antibodies pass across the placenta and destroy the RBC of the Rh positive child. The Rh issue has been solved by giving Rh negative women an Rh immunoglobulin injection (Rho-Gam) either midway through the first pregnancy or no later than 72 hours after giving birth to an Rh positive child. The injection includes anti-Rh antibodies that attack a child's RBCs before they trigger the mother's immune system. The injection is not effective if the mother has already produced antibodies; hence, timing is important.

Blood types (blood groups) are classified by the presence (or absence) of inherited antigens on the surface of erythrocytes. These antigens include proteins, carbohydrates, glycolipids or glycoproteins. There are no O antigens. Type A$^-$ (A negative) blood has A antigens present on the erythrocyte, but no Rh factor (Rh$^-$) and no B antigens. Type A$^-$ blood produces anti-B antibodies that bind to B antigens and antibodies that bind to the Rh antigen.

RBC with a particular antigen agglutinate when exposed to corresponding antibodies. *Agglutination* is the clumping of red blood cells due to a reaction between antigens on the red blood cells. To receive blood, the recipient's plasma must not have an antibody that causes donor cells to agglutinate.

Patients with type AB blood can receive any type of blood; they are a universal recipient.

Patients with type O blood cannot receive A, B or AB, but they are a universal donor.

Patients with type A blood cannot receive B or AB blood.

Patients with type B blood cannot receive A or AB blood.

	Group A	Group B	Group AB	Group O
RCB Antigens	A antigen	B antigen	A and B antigens	None
Plasma Antibodies	Anti-B	Anti-A	None	Anti-A and Anti-B

Overview of blood types with associated antigens and corresponding antibodies

Please, leave your Customer Review on Amazon

Chapter 17

Digestive System

- **Introduction to the Digestive System**

- **Ingestion**

- **Stomach**

- **Liver**

- **Bile**

- **Pancreas**

- **Small Intestine: Duodenum, Jejunum, and Ileum**

- **Large Intestine**

- **Muscular Control**

- **Summary of GI Tract**

- **Endocrine Control**

- **Neural Control: Enteric Nervous System**

- **Nutrients and Human Nutrition**

Introduction to the Digestive System

Digestion provides the energy necessary for routine metabolic activities and to maintain homeostasis. The digestive tract ingests food, breaks it down into small molecules crosses plasma membranes, absorbs the nutrients, and eliminates indigestible remains.

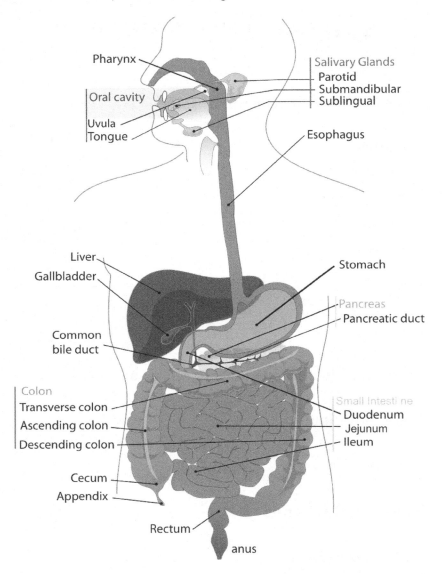

The human digestive system is a coiled, muscular tube (6-9 meters long when fully extended), beginning at the mouth and ending at the anus. Several specialized compartments occur along this length: mouth, pharynx, esophagus, stomach, small intestine, large intestine, and anus. Accessory digestive organs are connected to the main system by a series of ducts: salivary glands, parts of the pancreas, the liver and the gallbladder (*biliary system*).

The digestive tract performs six processes:

1. *Ingestion* – bringing food into the system

2. *Movement (peristalsis)* – moving food along the system

3. *Digestion* – breaking down food using mechanical and chemical processes

4. *Secretion* – release of enzymes and bile into the digestive tract

5. *Absorption* – moving food molecules from the digestive tract into the blood

6. *Defecation* – eliminating solid waste from the large intestine.

Four groups of macromolecules are digested and absorbed by the digestive tract. These include starches, proteins, triacylglycerols (fats) and nucleic acids.

Starches are broken down into glucose.

Proteins are broken down into amino acids.

Triacylglycerols (fats) are broken down into fatty acids and glycerol.

Nucleic acids are broken down into nucleotides.

Ingestion

Saliva as lubrication and source of enzymes

The gastrointestinal (GI) tract begins at the mouth, where a mechanical form of digestion begins (i.e., chewing or mastication). Saliva, a form of chemical digestion, contains mucus and the enzyme amylase. It is secreted from 3 pairs of salivary glands, located around the oral cavity. Mucus moistens the food, and amylase partially digests polysaccharides (starches). Swallowing moves food (bolus) from the mouth through the pharynx into the esophagus, past the esophageal sphincter and then into the stomach.

Human dentition (teeth structure) has many specializations because humans are omnivores (consume both plants and animals). Food is masticated (chewed) in the mouth and mixed with saliva. Food is manipulated in the mouth by a muscular tongue containing both touch and pressure receptors. *Taste buds,* receptors that are stimulated by the chemical composition of food, are located primarily on the tongue but on the surface of the mouth.

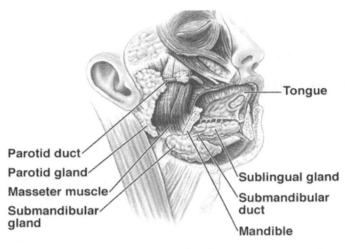

Salivary glands

Three pairs of *salivary glands* secrete saliva into the mouth by ducts. Saliva, a form of lubrication and a source of enzymes, dissolves food and contains mucin. *Mucin* is a protein that lubricates the bolus. Saliva contains amylase, an enzyme that breaks down polysaccharides (starch and glycogen), as well as antibodies and lysozymes that kill pathogens. *Salivary amylase* is the enzyme that begins digesting starch (α-linked glucose polysaccharide). It breaks down the complex carbohydrates into the simple sugar maltose (a disaccharide of two glucose monomers). Mucus (composed of mucin) moistens food and lubricates the esophagus. Additionally, bicarbonate ions in saliva neutralize the acids in foods.

$$\text{Starch } + H_2O \xrightarrow{\textit{Salivary Amylase Mechanism}} \text{maltose}$$

Pharynx functions in swallowing

The pharynx (throat) is between the mouth and the esophagus and is where the food and air passage cross. It is a muscular tube that squeezes and routes food to the *esophagus* when swallowing. It closes off pathways to the nasal cavity and airway to prevent choking. In the pharynx, the bolus triggers an involuntary swallowing reflex that prevents food from entering the trachea and directs the bolus into the esophagus. The digestive and respiratory passages come together in the pharynx and then separate. Air moves into the rigid trachea while food descends the flexible esophagus into the stomach.

Epiglottal action

During swallowing, the bolus is moved to the back of the mouth by the tongue. If food were to enter the trachea, the pathway of air to the lungs could be blocked. The epiglottis is a flap of cartilage that closes off the airway during swallowing. It covers the opening into the trachea as muscles move bolus through the pharynx into the esophagus.

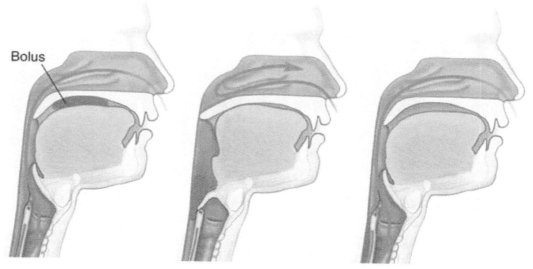

Swallowing where the epiglottis blocks food from entering the trachea

Esophagus functions in transport

After the food and saliva are swallowed, they are pushed into the esophagus, a tube that uses muscular contractions to propel food toward the stomach. *Peristalsis* is the involuntary muscle contractions that move food via the esophagus past the esophageal sphincter and into the stomach.

Stomach

The stomach, an important component of the digestive system, is an elastic, muscular sac that can stretch to store food. The bolus passes through the *gastroesophageal sphincter* (lower esophageal or cardiac sphincter), a ring of smooth muscle fibers that connect the esophagus and stomach. The stomach secretes hydrochloric acid (HCl), which destroys bacteria and other harmful organisms, ultimately preventing food poisoning. When gastric (stomach) juices leak through the cardiac sphincter, irritation of the esophagus causes acid reflux or heartburn. The stomach's acidic environment (pH 1.5–3.0) provides the optimum conditions for the enzyme pepsin to function. The stomach secretes pepsin, which begins digesting proteins into smaller polypeptides and amino acids. These smaller amino acid molecules can then be absorbed by the villi (small finger-like projections) that protrude from the epithelial layer of the small intestine.

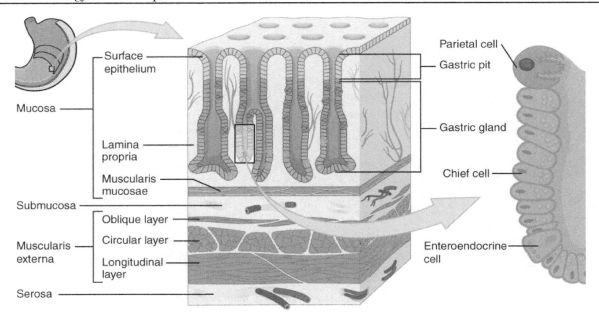

Storage and churning of food

The stomach stores partially digested food, freeing humans from the burden of constantly eating for energy. During a meal, the stomach gradually fills from an empty capacity of 50-100 milliliters to a capacity of 1,000 milliliters (1 liter) with discomfort, the stomach can distend to hold 2,000 milliliters (2 liters) or more.

The muscles in the stomach are responsible for the mechanical breakdown of food. This process is churning, where food is mechanically digested and mixed. Walls of the stomach contract vigorously, mixing food with juices secreted when the food enters. The mixing of food with water and gastric juice generates a creamy medium, called *chyme.* Food is mixed in the lower part of the stomach by peristaltic waves that propel the acid-chyme mixture against the pyloric sphincter. Over a 1 to 2-hour period, increased contractions of the stomach push the food through the pyloric sphincter and into the duodenum of the small intestine as the stomach empties. High-fat diets generate satiation by significantly increasing the period for food to remain in the stomach.

Low pH, gastric juice, protection by mucus against self-destruction

Gastric glands are *exocrine glands* (secreted by duct) within gastric pits. *Gastric pits* are indentations in the stomach that denote entrances to the gastric glands, which contain secreting epithelial cells (chief cells, parietal cells, and mucous cells). These epithelial cells line the inner surface of the stomach, secreting about 2 liters of gastric juice per day. Gastric juice contains several components for digestion (e.g., HCl, pepsinogen, and mucus).

Chief cells secrete *pepsinogen,* a zymogen (i.e., precursor) to pepsin. Pepsinogen is activated to form *pepsin* by low pH in the stomach. Once active, pepsin begins protein digestion. *Parietal cells* secrete HCl and intrinsic factor, which is important in vitamin B-12 absorption. *G cells* secrete *gastrin*, a large peptide hormone that is absorbed into the blood, stimulating parietal cells to secrete HCl. Acetylcholine increases secretion of all cell types, while gastrin and histamine increase HCl secretion.

Mucous cells (Goblet cells) secrete mucus that lubricates and protects the stomach's epithelial lining from the acidic environment. The mucus forms a protective barrier between the cells and the stomach acids. Additionally, pepsin is inactivated (i.e., pepsinogen) when it contacts the mucus. Bicarbonate ions reduce acidity near the cells lining the stomach by increasing the pH. For protection, tight junctions link the epithelial, stomach-lining cells together, preventing stomach acids from affecting other structures.

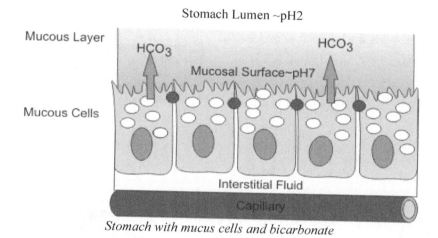

Stomach with mucus cells and bicarbonate

Peptic ulcers result from the failure of the mucosal lining to protect the stomach. Ulcers are eroded areas of gastric surface or breaks in the mucosal barrier. Ulcers expose the underlying gastric muscle tissue to the corrosive action of acid and pepsin. Bleeding ulcers result when tissue damage is so severe that blood enters the stomach. Perforated ulcers are life-threatening emergencies that occur when a hole forms in the stomach wall. About 90% of all peptic ulcers are caused by *Helicobacter pylori,* a species of bacteria that burrows into the mucous lining of the stomach, exposing the underlying epithelial cells to the acidic environment of the stomach. Other factors that contribute to ulcers are stress, excess HCl and aspirin.

Production of digestive enzymes, site of digestion

Carbohydrate digestion begins with salivary amylase in the mouth and continues as the bolus passes to the stomach. The bolus is further broken down into acidic chyme in the lower third of the stomach. Hydrochloric acid does not directly function in digestion, but it lowers the pH of the gastric stomach contents to about 2. The low pH stops the activity of salivary amylase and promotes pepsin activity, allowing protein digestion to begin. Pepsin is an enzyme that controls the hydrolysis of proteins into peptides. Chyme leaves the stomach and enters the small intestine.

$$\text{protein} + H_2O \xrightarrow{\text{pepsin}} \text{peptides (small chains of amino acids)}$$

Structure (gross)

The stomach has an inner membrane made of dense folds as rugae, allowing it to accommodate stretching. The stomach is sealed off at the top by the cardiac (gastroesophageal) sphincter. On the bottom, it is sealed by the pyloric sphincter, a circular muscle that controls the release of chyme into the small intestine.

When the pyloric sphincter relaxes, a portion of chyme exits the stomach and enters the duodenum (first part of the small intestine). A neural reflex causes the sphincter to contract, closing off the opening. The slow, rhythmic pace with which chyme exits the stomach allows for thorough digestion.

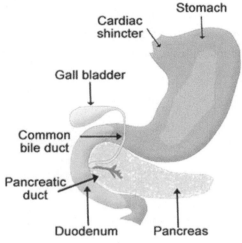

Stomach and duodenum are shown with accessory organs for digestion

Liver

Structural relationship of the liver within the gastrointestinal system

The liver is a large glandular organ that occupies the top of the abdominal cavity, just below the diaphragm. Being the largest gland in the body, it spans both sides of the abdomen (right side occupies

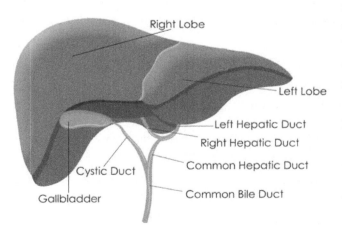

much more space) and has ducts that drain into the duodenum and gallbladder. The liver is responsible for the detoxification of blood (by removing and metabolizing poisonous substances), the synthesis of blood proteins (making plasma proteins, including albumin and fibrinogen), the production of bile, the destruction of old erythrocytes and the storage and regulation of glucose in the blood, as well as the production of urea from amino acids and ammonia.

The liver stores glucose as glycogen and breaks down glycogen to maintain a constant blood glucose concentration. Blood vessels from the large and small intestines lead to the liver as the *hepatic portal vein*. The liver maintains the blood glucose level at 0.1% by removing glucose from the hepatic portal vein to store as glycogen. When needed, glycogen is broken down, and glucose reenters the hepatic vein.

Amino acids can be converted to glucose by the liver, but deamination (removal of amino groups) must occur beforehand. The liver produces urea from these amino groups and ammonia. Using complex metabolic pathways, the liver converts amino groups to urea. Urea is the most common human nitrogenous waste; the blood transports it to the kidneys. (See "Excretory System")

Roles in nutrient metabolism and vitamin storage

The liver provides storage for many important nutrients, vitamins (A, D, E, K and B_{12}) and minerals (e.g., iron, copper), releasing these essential substances to the tissues as needed. These compounds are absorbed from the blood as they go through the hepatic portal system.

Glucose is stored in the liver in its polysaccharide form, glycogen. Under the influence of insulin, glucose is transported to the liver cells (hepatocytes). *Hepatocytes* absorb and store fatty acids in the form of digested triglycerides. The storage of these nutrients allows the liver to maintain homeostasis of blood glucose levels.

Roles in blood glucose regulation and detoxification

The liver acts as a storehouse for glycogen, the storage form of glucose. Through gluconeogenesis, glycogenolysis, and glycogenesis, the liver regulates the level of glucose in the blood. When blood sugar is too low, the liver conducts gluconeogenesis (synthesizing glucose and increasing blood glucose count), and glycogen lysis (breaking down stored glycogen and increasing blood glucose count). When blood glucose is too high, the liver engages in glycogenesis, converting the extra glucose in the blood into glycogen as a form of storage.

Insulin, an important hormone in glucose regulation, is released in response to an increase in glucose levels. Therefore, insulin promotes the conversion of glucose into glycogen, whereby the excess glucose can be stored in the liver for later use.

Glucagon, a hormone antagonistic to insulin, is released in response to decreased glucose levels, and therefore promotes the conversion of glycogen into glucose. The lack of glucose can be compensated for by the new supply of glucose brought about by glycogen.

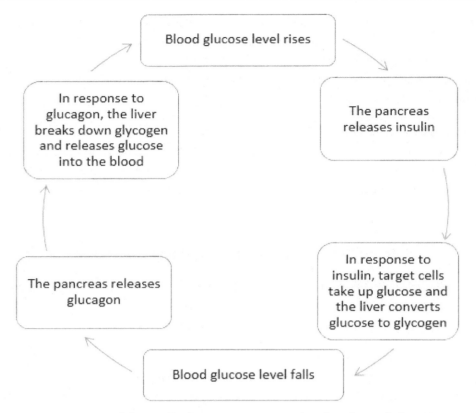

Homeostasis of glucose by the antagonistic actions of insulin and glucagon

The liver is responsible for the detoxification of the body. It can metabolize alcohol (alcohol dehydrogenase), remove blood ammonia and inactivate other drugs or toxins. The liver excretes detoxified chemicals as part of bile (or polarized to be excreted by the kidneys). *Kupffer cells* are macrophages in the liver that phagocytize the bacteria from intestines. Although most red blood cells are destroyed in the spleen, Kupffer cells assist in the destruction of irregular erythrocytes.

Bile

Liver produces bile

The liver makes bile from cholesterol and stores it in the gallbladder. To produce bile, hepatocytes in the liver destroy old red blood cells. Bile is a green byproduct resulting from the breakdown of hemoglobin, converted into the two key components of bile, bilirubin, and biliverdin. Bile subsequently enters the duodenum from the gallbladder to emulsify fats.

Gallbladder stores bile

Bile reaches the gallbladder through hepatic ducts and is stored there for later use. When needed, the bile flows through the *cystic duct*, which merges with the *pancreatic duct* to form the *common bile duct*.

The sphincter of Oddi is a muscular valve that controls the entry of digestive juices (bile and pancreatic juice) from the liver and pancreas into the duodenum of the small intestine. When the sphincter is closed, secreted bile is shunted into the gallbladder. The presence of fat in the intestine releases CCK, which relaxes the sphincter to discharge bile salts into the duodenum.

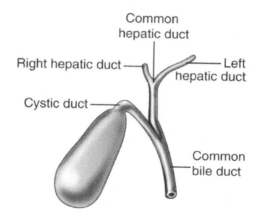

Gallbladder and associated ducts

Excess, unused bile is concentrated and stored in the gallbladder. The excess bile is secreted when needed. During a meal, for example, bile is secreted from the gland by smooth muscle contraction. It reaches the duodenum as the first portion of the small intestine by the common bile duct.

Function

Bile salts solubilize fats, while bicarbonate ions neutralize stomach acids. *Emulsification* breaks fat globules into microscopic droplets. Bile salts, secreted by hepatocytes (liver cells), enter the GI tract, are reabsorbed by transporters in the intestine, and are returned to the liver via the *portal vein*. This recycling pathway is the *entero-hepatic circulation*.

$$\text{fat} \xrightarrow{\text{bile salts}} \text{fat droplets (by emulsification)}$$

This process increases fat digestion by increasing the surface area of fat globules exposed to lipase enzymes. Bile emulsifies fats, facilitating their breakdown into progressively smaller fat globules until they can be acted upon by lipases. Bile contains cholesterol, phospholipids, bilirubin and a mix of salts. Fats are completely digested in the small intestine, unlike carbohydrates (i.e., salivary amylase in the mouth) and proteins (predigested in the stomach).

Due to their hydrophobic nature, digested fats are not soluble. To compensate for this, bile salts surround fats, forming micelles that pass into the epithelial cells. Afterward, bile salts are recycled back into the lumen to repeat the process. Fat digestion is usually completed by the time the food reaches the ileum (lower third) of the small intestine. Bile salts are absorbed in the ileum and are recycled by the liver and gallbladder. Fats pass from the epithelial cells to the small lymph vessels that run through the villi.

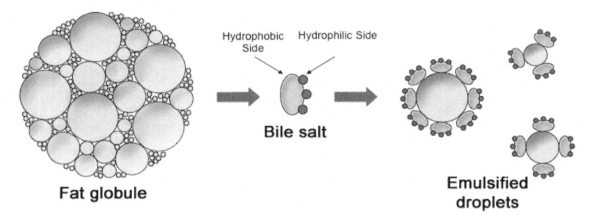

Bile salts emulsify fat into smaller goblets for absorption in the small intestine

Excessive secretion of water-insoluble cholesterol in bile results in crystals, as *gallstones*. Gallstones can obstruct the opening of the gallbladder or the bile duct. If a gallstone prevents bile from entering the intestine, the rate of fat digestion and absorption decreases. If a gallstone blocks the entry of the pancreatic duct, pancreatic enzymes cannot enter the intestine, thus preventing the digestion of other nutrients. Blocked bile secretion results in accumulation of bilirubin in tissues. This produces a yellowish coloration called jaundice. *Jaundice* is common in newborns and patients with liver disease.

Pancreas

The *pancreas*, an elongated, tadpole-shaped organ, lies deep within the abdominal cavity. It is located just below the stomach along the posterior abdominal wall and leads to the duodenum. The endocrine functions of the pancreas include secreting the glucose regulatory hormones (glucagon and insulin) into the bloodstream using two different types of cells. Depending on the blood glucose concentration, one of these two hormones targets the liver. From its exocrine function, the pancreas secretes digestive enzymes and a fluid rich in HCO_3^- ions to neutralize the acid from the stomach.

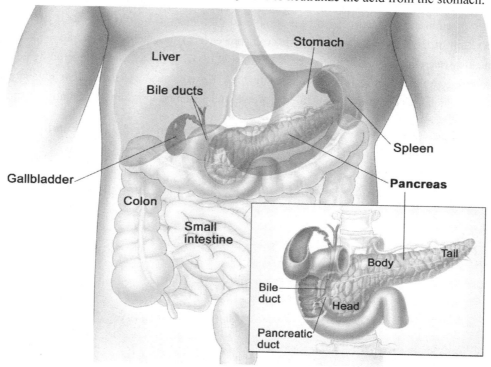

Pancreas as an endocrine and exocrine organ aiding in digestion

The pancreas produces enzymes and bicarbonate

The pancreas secretes pancreatic juice, a mixture of enzymes and ions. The pancreas releases its major enzymes from pancreatic cells, *acinar cells*, into the duodenum via the pancreatic duct. These major enzymes include *trypsin* and *chymotrypsin* (digest proteins), lipase (digests fats), pancreatic amylase (digests starch), and deoxyribonucleases and ribonucleases (digest nucleic acids). These enzymes exist first as zymogens / proenzymes (inactive forms). Once trypsin becomes activated by the intestinal enzyme *enterokinase*, it activates the other digestive enzymes released earlier by the pancreas. However, certain enzymes (e.g., pancreatic amylase and lipase) are secreted in active forms.

- Pancreatic amylase digests starch to maltose:

  starch + H_2O + *pancreatic amylase* → maltose

- Trypsin (and other enzymes) digest protein to peptides:

  protein + H_2O + *trypsin* → peptides

- Lipase digests fat droplets into glycerol and fatty acids:

  fat droplets + H_2O + *lipase* → glycerol + fatty acids

Secretion of pancreatic enzymes is stimulated by *cholecystokinin* (CCK), a GI peptide hormone. The secretion of CCK is triggered by the detected presence of fatty acids and amino acids in the small intestine. The secretion of bicarbonate ions is stimulated by *secretin*, which is triggered by acidity in the small intestine. Acinar cells secrete pancreatic enzymes. The enzymes organize into *acini* (singular acinus structures), which are composed of several acinar cells and a duct (empties into the small intestine).

The pancreas secretes bicarbonate ions (HCO_3^-) to neutralize the HCl from the stomach. The enzymes of the pancreas function best in alkaline (pH > 7) solutions.

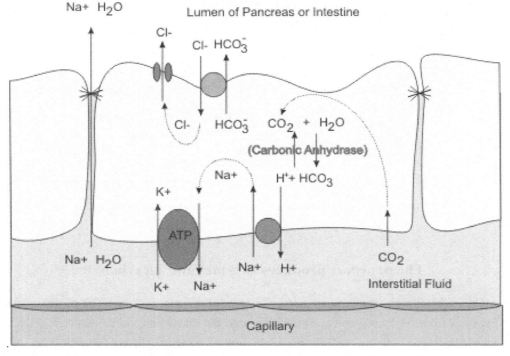

Bicarbonate secretion in pancreas or intestine to aid in digesting lipids

Transport of enzymes to the small intestine

The enzymes produced by the pancreas travel to the small intestine, where digestion is continued. These enzymes enter the small intestine through the *pancreatic duct* (i.e., the *duct of Wirsung*). This duct joins the pancreas to the common bile duct so that the pancreatic juices can be used in the small intestine.

The hormone *cholecystokinin* (CCK), produced in the small intestine, is secreted in response to the presence of nutrients, and this hormone regulates the transport of enzymes into the duodenum.

Secretin is another hormone produced in the small intestine, and it controls the transport of bicarbonate into the small intestine to regulate pH level.

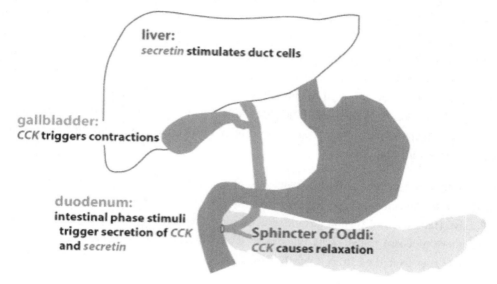

Organs and enzymes associated with digestion

Small Intestine

Most digestion and absorption of nutrients occurs in the small intestine. The intestinal wall secretes enzymes and receives enzymes from the pancreas. The main function of the small intestine is to complete digestion and absorb the monomers of food particles after chemical digestion. The small intestine contains many villi, which are structures that increase the surface area for absorption.

Absorption of food molecules and water

Most animals need to digest food into small molecules to cross the plasma membranes. There are two reasons why the complete digestion of large food molecules is essential for proper absorption. First, consumed food is made of many compounds synthesized by other organisms. Not all of the ingested compounds are suitable to be used by human tissues. Therefore, these compounds had to be broken down and reassembled so that the body can use them appropriately. Second, the food molecules have to be small enough to be absorbed by the villi in the intestine by diffusion, facilitated diffusion or active transport. Therefore, large food molecules need to be broken down for absorption. Intestinal folds, villi, and microvilli increase the surface area for absorption of digested compounds into circulation (fats into lacteals, all others into capillaries).

Active transport occurs to absorb against the concentration gradient. Since the intestinal lumen has less glucose than the enterocyte, secondary active transport is by the Na$^+$ / K$^+$ pump and Na$^+$ / glucose symport. Passive (facilitated) diffusion absorbs glucose down its concentration gradient. The enterocyte (now having more glucose) transfers its glucose to the extracellular fluid using facilitated diffusion. Then, the glucose goes from the extracellular fluid to blood.

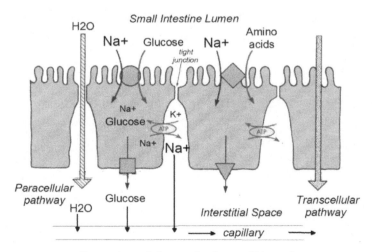

Absorption of glucose in the small intestine

Function and structure of villi

Most absorption occurs in the duodenum and jejunum (second third) of the small intestine. The inner surface of the intestine has circular folds that more than triple the surface area for absorption. Finger-like projections as villi cover the mucous membrane in these ridges and furrows. Villi increase the surface area for greater digestion and absorption. They are covered with epithelial cells that increase the surface area by another factor of 10. The epithelial cells are lined with microvilli that further increase the surface area; a 6-meter-long tube has a surface area of 300 square meters. The small intestine is specialized for absorption by the huge number of villi that line the intestinal wall. If the small intestine were merely a smooth tube, it would have to be 500–600 meters long to have a comparable surface area.

The structure of the villus is specific. Firstly, there are a great number of them, so this increases the surface area for absorption in the small intestine. Also, the villi have their projections (microvilli). Microvilli are minute projections (collectively as the brush border) on the surface of cells of the intestinal villi. These microvilli have protein channels and pump in their membranes to allow the rapid absorption of food by facilitated diffusion and active transport. The villi contain an epithelial layer (one cell layer thick) so that food can pass through easily and be absorbed quickly. The blood capillaries in the villus are associated with the epithelium so that the distance for the diffusion of the food molecules is small. This thin layer of cells contains mitochondria to provide the ATP needed for the active transport of certain food molecules. Finally, there is a lacteal branch at the center of the villus that carries away fats after absorption.

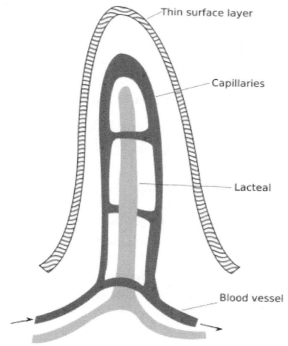

Intestinal villus with lacteals for fat absorption and capillaries for other nutrients

Sugars and amino acids enter villi cells and are absorbed into the bloodstream. Glycerol and fatty acids enter villi cells, reassemble into fat molecules, and move into lacteals. Each villus has a *lacteal*, which is a lymphatic vessel surrounded by a capillary network in an intestinal villus that aids in the absorption of fats. Absorption involves diffusion and active transport requiring the expenditure of cellular energy. Villi have cells that produce intestinal enzymes which complete the digestion of peptides and sugars. The villi contain large numbers of capillaries that take the amino acids and glucose produced by digestion to the hepatic portal vein and the liver.

Absorption occurs in the small intestine's specialized villi when food molecules pass through a layer of cells and into the body's tissues. Absorption is followed by assimilation, which occurs when the food molecules become part of the body's tissue.

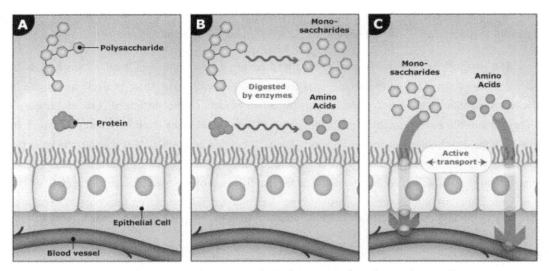

Digestion and absorption of amino acids and monosaccharides in the small intestine

Production of enzymes and site of digestion

As chyme enters the duodenum, proteins and carbohydrates are partially digested. At this stage, no fat has been digested yet, because the secretions from the liver and the pancreas secreted into the duodenum aid fat digestion.

Digestion of carbohydrates, proteins, and fats continues in the small intestine. Starch and glycogen are broken down into maltose. Proteases (enzymes secreted from the pancreas) continue the breakdown of proteins into small peptide fragments and some amino acids.

The upper part of the small intestine, the duodenum, is the most active in digestion. It uses secretions from the liver and the pancreas to break down ingested compounds. Epithelial cells of the duodenum form Bruner's glands by secreting watery mucus. The mucus aids in protecting the duodenal epithelium from the acidic chyme (from the stomach), lubricating the intestinal walls and providing an alkaline environment for pancreatic enzymes to operate.

Epithelial cells of villi produce the intestinal enzymes attached to the plasma membrane of microvilli. Intestinal secretions complete the digestion of peptides and sugars; peptidases digest peptides into amino acids:

$$\text{peptides} + H_2O + \textit{peptidases} \rightarrow \text{amino acids}$$

Maltose (from the first step in starch digestion) is converted by maltase to glucose:

$$\text{maltose} + H_2O + \textit{maltase} \rightarrow \text{glucose} + \text{glucose}$$

Maltose, sucrose, and lactose are the main carbohydrates in the small intestine; the microvilli absorb them. Starch is broken down into two-glucose units (maltose). Enzymes in the epithelial cells convert these disaccharides into monosaccharides that then leave the cell and enter the capillary. Lactose intolerance results from the genetic lack of the enzyme lactase (lactose is milk sugar) produced by the intestinal cells. The lack of lactase results in the incomplete digestion of lactose to glucose and galactose.

While the pancreas is the major source for the enzymes used, the small intestine does make some of its enzymes, including *protease, amylase, lipase,* and *nuclease.*

Cellulose, a polysaccharide similar in structure to starch, cannot be digested by humans. Humans lack cellulase needed to hydrolyze β glycosidic bonds of cellulose.

Pancreatic juices neutralize stomach acid

Pancreatic juice is secreted by the pancreas and contains sodium bicarbonate [$NaCO_3$], which neutralizes the acidity of chyme (due to the HCl from the stomach). The pH of the small intestine is therefore slightly basic. This neutralization process facilitates enzymes in the small intestine, which would normally be denatured by stomach pH.

Anatomic subdivisions of the small intestine

The small intestine is divided into three segments: the duodenum, the jejunum, and the ileum. The small intestine is a coiled tube up to 6 meters long and 2 to 3 cm wide. Coils and folding, along with villi, give this 6−meter tube the surface area of a tube 500-600 meters long. Food moves from the stomach to the small intestine through the *pyloric sphincter* (into the first 25 cm of the small intestine, the duodenum). The duodenum has a pH of 6, mainly due to bicarbonate ions secreted by the pancreas.

In the duodenum, the breakdown of starches, proteins and remaining food types (fats and nucleotides) continue. The ileocecal valve separates the small intestine from the large intestine. The duodenum is mostly responsible for digestion, with the jejunum and ilium mostly responsible for absorption. 90% of digestion and absorption occurs in the small intestine. *Goblet cells* secrete mucus to lubricate and protect from mechanical and/or chemical damage.

While the length of the entire intestinal tract contains lymphoid tissue, only the ileum has abundant *Peyer's patches* (i.e., unencapsulated lymphoid nodules) that contain large numbers of lymphocytes and immune cells.

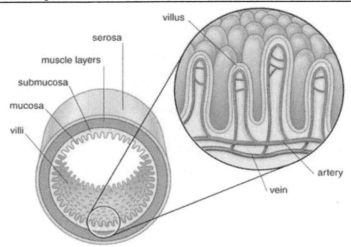

The luminal surface of the small intestine is covered by a single layer of epithelium containing exocrine and endocrine cells. The epithelia, with an underlying layer of *lamina propria* (connective tissue), and *muscularis mucosa* (muscle) are *mucosa*. Below the mucosa is a layer of inner circular, and an outer longitudinal smooth muscle is the *muscularis externa*, which provides the force for moving and mixing the GI contents. *Serosa* is the outermost layer of the tube made of connective tissue.

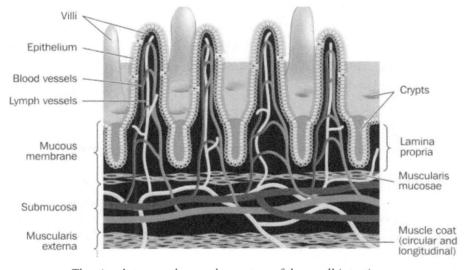

The circulatory and muscular system of the small intestine

Large Intestine

Structure of large intestine

The large intestine is the region of the gastrointestinal tract following the small intestine. It has four parts: cecum (blind pocket containing the appendix), colon, rectum, and anal canal. Chyme enters the cecum through the ileocecal sphincter, which relaxes and opens by the gastroileal reflex. The large intestine terminates at the anus, an external opening. The large intestine has lobes (pockets) along its length due to muscle tone. Unlike the small intestine, the large intestine has no folds or villi because its primary function is to store and concentrate fecal material for elimination.

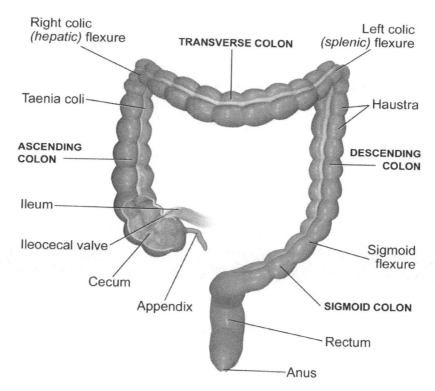

Large intestine with colon, rectum and anal canal

The *appendix* is a fingerlike projection extending from the cecum, a blind sac at the junction of the small and large intestines. It may play a role in fighting infections. If an infected appendix bursts, the result is a general abdominal infection as peritonitis.

The large intestine secretes an alkaline mucus (HCO_3^- ions) into the lumen, which protects epithelial tissues and neutralizes acids produced by bacterial metabolism.

Absorption of water

Material entering the large intestine is mostly indigestible residue and liquid. Undigested chyme is passed to the large intestine, where it is temporarily stored and concentrated by reabsorption of salts and water.

About 1.5 liters of water enter the digestive tract daily from drinking, and another 8.5 liters enter from various secretions. The small intestine reabsorbs about 80% of this total liquid. The large intestine absorbs remaining water not absorbed by the small intestine. If the water is not reabsorbed, it causes diarrhea, which can cause serious dehydration and ion loss. However, if too much water is reabsorbed, the result can be constipation. Sodium ions (Na^+) are absorbed whenever water is reabsorbed.

Rectum for storage and elimination of waste

The large intestine moves material that has not been digested from the small intestine. Water, salts, and vitamins are absorbed in the large intestine. However, the remaining contents in the lumen form feces, which are stored in the rectum and are then released through the anus. Feces consist of about 75% water and 25% solid matter. One-third of the solid matter is intestinal bacteria. The remainder is undigested wastes, fats, organic material, mucus and dead cells from the intestinal lining.

Following a meal, there is a wave of intense contraction, called mass movement. The mass movement of fecal material into the anus initiates the defecation reflex. Contractions of the rectum expel the feces through the anus. During defecation, the anal sphincter opens, and feces are released through the anus.

Muscular Control

Sphincter muscle

The *cardiac sphincter* (gastroesophageal sphincter) is between the esophagus and the stomach. It prevents backflow of food.

The *pyloric sphincter* is between the stomach and small intestine. It releases food into the small intestine, a small amount at a time.

The *anal sphincter* is at the end of the rectum and ties the end of the rectum. The *internal anal sphincter* is made of smooth muscle and closes the anus, while the *external anal sphincter* is made of skeletal muscle and is under voluntary control. Both sphincters regulate the anal opening and closing.

The defecation reflex, which is mediated by mechanoreceptors, causes the two anal sphincters to open to expel the feces. If defecation is delayed, rectal contents are driven back into the colon by reverse peristalsis until the next mass movement.

Peristalsis

Peristalsis is regular contractions of the circular smooth muscle that produces a slow, rhythmic, bidirectional segmentation movement similar to the unidirectional movement of material. The undigested material moves back and forth slowly to provide resident bacteria time to grow and multiply. The bacteria of the large intestine (e.g., *E. coli*) are symbiotic organisms that ferment undigested nutrients, making gas as a byproduct. These bacteria produce vitamins (vitamin K), as an important component in the formation of blood clots. Undigested polysaccharides (fiber) are metabolized to short-chain fatty acids by the residing bacteria and are absorbed by diffusion. Bacterial metabolism produces *flatus*, a mixture of gases.

Peristalsis is the involuntary movement of smooth muscles that squeeze food along the digestive tract. Chyme moves through intestines via peristalsis. Layers of circular and longitudinal smooth muscle enable the chyme (partly digested food and water) to be pushed along the ileum by waves of muscle contractions is peristalsis. The remaining chyme is passed to the colon.

The stomach produces peristaltic waves in response to the arrival of food. The pyloric sphincter between the stomach and duodenum opens to release small amounts of chyme into the duodenum with each wave. These waves are generated by pacemaker cells in the longitudinal smooth muscle layer and are spread by gap junctions. Gastrin secretion distension of the stomach and other factors stimulate gastric motility, while distension of the duodenum inhibits it.

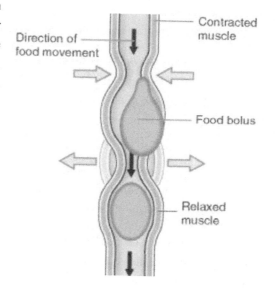

The most common motion of the small intestine is stationary contraction and relaxation, called segmentation. Segmentation mixes chyme with digestive juices but results in little net movement. The chyme is mixed and brought into contact with the intestine wall, and is then moved slowly toward the large intestine. The movements are initiated by pacemaker cells in the smooth muscle layer.

After most of the materials are absorbed, segmentation is replaced by a peristaltic activity called migrating motility complex, which moves any undigested material to the large intestine. The intestinal hormone, *motilin*, initiates migrating motility. Movements in the large intestine (due to involuntary contractions) shuffle contents back and forth, and propulsive contractions move material through the large intestine.

Summary of GI Tract

The human digestive tract is a complete tube-within-a-tube system. Each part of the digestive system has a specific function. In humans, the digestion of food is an extracellular process. Enzymes are secreted in the digestive tract by nearby glands. Food is never within the accessory glands, only within the tract itself. Digestion requires a cooperative effort by the production of hormones and the actions of the nervous system.

- *Mouth*: grinds and moistens food by mastication (chewing for mechanical digestion); begins digestion of starch by amylase and lipase contained in saliva (chemical digestion by enzymes).

- *Esophagus*: moves food to the stomach.

- *Stomach*: is responsible for churning (mechanical digestion), acid digestion by HCl, protein digestion by pepsin (chemical digestion) and food storage.

- *Small intestine*: the longest and most extensively folded part of the GI tract is the site of almost all of the digestion and absorption of nutrients and water. Chyme is subjected to bile from the liver (to emulsify fats) and digestive enzymes (amylase, protease, lipase, and nuclease) for chemical digestion (enzymes are predominantly from the pancreas).

- *Large intestine*: does not assist in digestion; water is reabsorbed from the chyme, leaving behind the solid indigestible waste product (feces); bacteria in the large intestine produce vitamin K.

Enzymes are needed in the process of digestion. They are the biological catalysts that break down the large food molecules into smaller ones for absorption. Digestion can occur naturally at body temperature; however, this process takes a long time because it happens at such a slow rate. Enzymes are vital because they speed up the digestive process by lowering the activation energy required for the reaction to occur at body temperature.

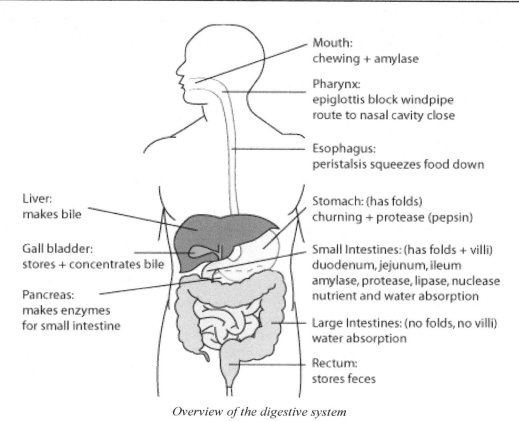

Mouth:
chewing + amylase

Pharynx:
epiglottis block windpipe
route to nasal cavity close

Esophagus:
peristalsis squeezes food down

Liver:
makes bile

Stomach: (has folds)
churning + protease (pepsin)

Gall bladder:
stores + concentrates bile

Small Intestines: (has folds + villi)
duodenum, jejunum, ileum
amylase, protease, lipase, nuclease
nutrient and water absorption

Pancreas:
makes enzymes
for small intestine

Large Intestines: (no folds, no villi)
water absorption

Rectum:
stores feces

Overview of the digestive system

The source, substrate, products and optimum pH conditions for amylase, protease and lipase enzymes.

	Amylase	Protease	Lipase
Enzyme	Salivary Amylase	Pepsin	Pancreatic Lipase
Source	Salivary Glands	Chief cells in the stomach lining	Pancreas
Substrate	Starch	Proteins	Triglycerides such as fats and oils
Products	Maltose	Small polypeptides	Fatty Acids and Glycerol
Optimum pH	pH 7	pH 1.5 – 2	pH 7

Endocrine Control

Hormones

- *Gastrin* is produced by gastric glands in the stomach lining when food reaches the stomach or upon sensing food. Gastrin stimulates increases gastric motility. Its secretion is stimulated by meals rich in protein.

- *Secretin* is produced by cells lining the duodenum when food enters the duodenum. Secretin stimulates the pancreas to secrete fluids rich in $NaCO_3$ into the duodenum. This secretion is stimulated by acidic chyme.

- *Cholecystokinin* (CCK) is produced in the duodenal wall of the small intestine in response to fats. CCK stimulates the pancreas to increase pancreatic juice. It induces the liver to increase the output of bile and causes the gallbladder to release bile.

- *Gastric Inhibitory Peptide* is produced by the duodenal wall in response to fat and protein in chyme as it enters the duodenum. It causes a mild decrease in the rate of digestion in the stomach by inhibiting both gastric gland secretion and stomach motility.

Endocrine cells are scattered throughout the GI epithelia, and the surface of these cells is exposed to the lumen. Chemical substances in the chyme stimulate endocrine cells to release hormones into the blood.

Also scattered throughout the lining of the stomach are *enterochromaffin-like* (ECL) cells and other cells that secrete *somatostatin*, an inhibitory protein that helps the body release glucagon and insulin. *ECL cells* are neuroendocrine cells in the digestive tract; gastrin stimulates them to release histamine, which stimulates parietal cells to produce gastric acid. The *pyloric antrum*, a lower portion of the stomach, secretes gastrin.

Increased protein content in a meal stimulates the release of gastrin and histamine, which stimulates HCl secretion. Somatostatin inhibits acid secretion by inhibiting the release of gastrin and histamine. *Enterogastrone* in the duodenum inhibits gastric acid secretion.

The precursor pepsinogen, which is produced by chief cells, is converted to pepsin by excess acid in the stomach.

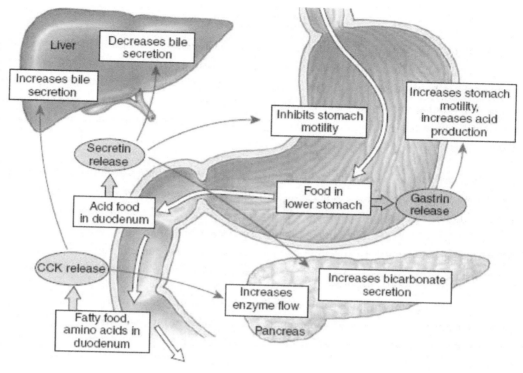

Digestive tract with glands and associated hormones

The *Islets of Langerhans* of the pancreas are irregularly shaped patches of endocrine tissue. The islets contain specific cell types including *alpha cells*, *beta cells*, *delta cells*, *F-cells*, and *C-cells*.

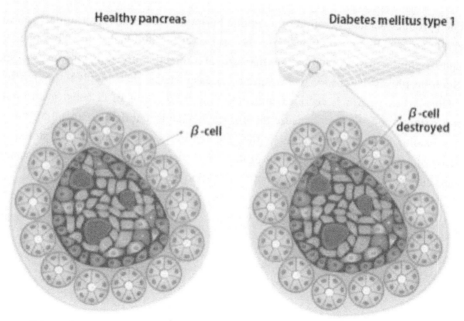

Islets of Langerhans with β-cells producing insulin in nondiabetic patients

Target Tissues

Vitamin D is a fat-soluble nutrient that promotes calcium absorption in the digestive system. Vitamin D is involved in modulating cell growth, supporting the immune system, and reducing inflammation. Cells in the digestive system have vitamin D receptors that are necessary for protein synthesis. Vitamin D provides nerve cells with the calcium they need to send adequate signals, which is important because nerves contained in the digestive tract must communicate with one another to control the digestive process.

Eighteen different endocrine cells can be identified within the gastrointestinal tract. Peptides that bind with target cell receptors and stimulate the cell to react are *agonists*. Others that fit the receptor but do not cause a particular reaction are *antagonists*. The ability of an antagonist to occupy a receptor and prevent access to an agonist is the principle behind treating a peptic ulcer disease with histamine receptor blockers. By occupying the receptors on the parietal cells, antagonists do not allow the histamine the opportunity to produce hydrochloric acid, the primary cause of peptic ulcers.

The discharge of granules of gastrin from the G cells occurs when a meal is consumed. While the concentration of hydrogen ions remains low because of the buffering effect of the food, the release of gastrin continues. As digestion takes its due course and the stomach starts to empty, the acidity increases because of the diminishing neutralizing effect of the food. When the contents of the stomach, in contact with the mucosa of the antrum, reach a certain level of acidity, the release of gastrin stops. Failure of this process causes inappropriate secretion of acid when the stomach is empty and may cause peptic ulcers in the duodenum. Some endocrine cells contain microvilli on their surface that project into the lumen of the gland or the main channel of the stomach or intestine. These cells can sample the luminal contents in their vicinity continuously.

When production and secretion of a peptide hormone are excessive, it initiates an increase in the number of target cells and may increase the size of the individual cells. This phenomenon is trophism and is similar to the increase in the size of skeletal muscle in response to appropriate exercise. Such trophism is observed in certain disease states that involve the gastrointestinal hormones. Thus, when gastrin is secreted into the blood by a tumor of G cells (gastrinoma) of the pancreas, it is a continuous process because there is no mechanism at that site to inhibit the secretion. This process results in a great increase in the number of parietal cells in the stomach and the overproduction of acid. Because the defenses of the upper gastrointestinal tract mucosa cannot handle autodigestion, intractable and complicated peptic ulceration occurs as a result.

Neural Control: Enteric Nervous System

Neural Regulation of the GI tract

Impulses to the GI muscles and exocrine glands are supplied by the enteric nervous system, the local nervous system of the GI tract. The neural regulation allows local, short reflexes are independent of the CNS. Long reflexes through the CNS are possible via sympathetic and parasympathetic nerves, which innervate the GI tract.

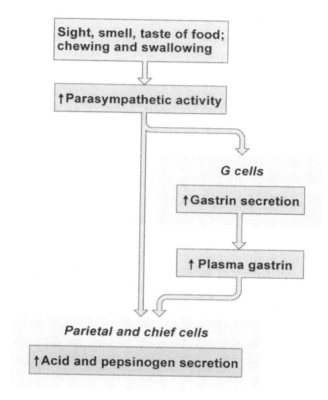

Neural control of the GI tract and hormone release

Somatic nerves control chewing to the skeletal muscles and the reflex activation of mechanoreceptors on the palate, gums, and tongue.

Autonomic nerves stimulate saliva secretion in response to chemoreceptors and pressure receptors in the mouth.

Swallowing is mediated by pressure receptors on the walls of the pharynx, which send impulses to the swallowing center in the *medulla oblongata*. The swallowing center activates muscles in the pharynx and esophagus. Multiple responses occur in a temporal sequence. The palate is elevated to prevent food from entering the nasal cavity, respiration is inhibited, and the epiglottis covers the glottis to prevent food from entering the trachea (windpipe).

The upper esophageal sphincter opens, and food enters the esophagus and moves toward the stomach by muscle contractions are peristaltic waves. Food then moves to the stomach when the lower esophageal sphincter opens. A less efficient (or faulty) lower esophageal sphincter results in the reflux of gastric contents into the esophagus (gastroesophageal reflux); this reversal results in heartburn and over time contribute to ulceration of the esophagus.

The *vomit reflex* results in the forceful expulsion of toxic gastric contents and is coordinated by the vomiting center in the medulla oblongata. Various mechano- and chemoreceptors in the stomach and elsewhere can trigger this reflex. Increased salivation, sweating, heart rate, pallor, etc. accompany the reflex. Abdominal muscles contract to raise abdominal pressure, while the lower esophageal sphincter opens and gastric contents are forced into the esophagus (retching). If the upper esophageal sphincter opens, contents are then expelled from the mouth (vomiting). Excessive vomiting can lead to loss of water and salts, which ultimately results in dehydration.

Phases of GI control

Each phase is named according to the location of the receptor for a reflex. These phases do not occur in a temporal sequence.

1. The *cephalic phase* is initiated when sight, smell, taste, chewing, and emotional states stimulate receptors in the head. Reflexes mediated by sympathetic and parasympathetic fibers activate secretory and contractile activity.

2. The *gastric phase* is initiated by distension, acidity and the presence of amino acids and peptides in the stomach. This phase is mediated by short and long reflexes and activates the secretion of gastrin.

3. The *intestinal phase* is initiated by distension, acidity, and osmolarity of digestive products in the intestine. The phase is mediated by GI hormones and short and long neural reflexes.

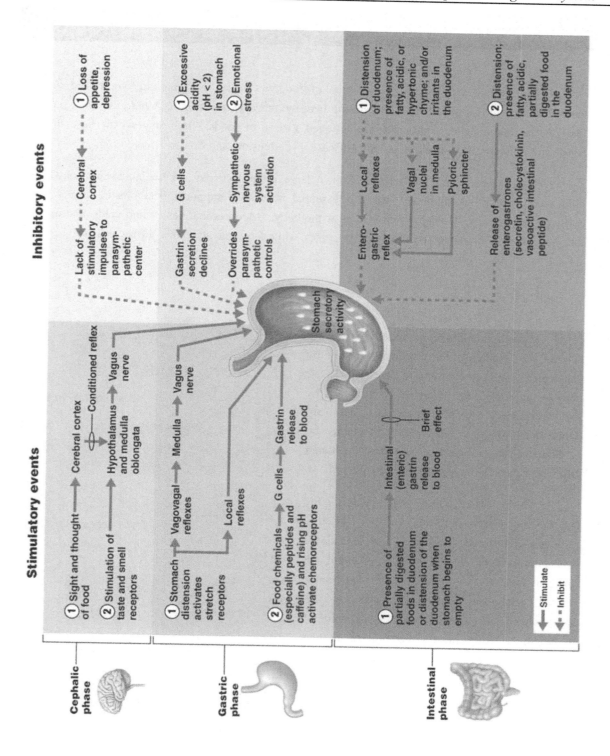

Stimulatory events

Inhibitory events

Cephalic phase

Gastric phase

Intestinal phase

① Sight and thought of food → Cerebral cortex → Conditioned reflex

② Stimulation of taste and smell receptors → Hypothalamus and medulla oblongata → Vagus nerve

① Stomach distension activates stretch receptors → Vagovagal reflexes → Medulla → Vagus nerve

Local reflexes

② Food chemicals (especially peptides and caffeine) and rising pH activate chemoreceptors → G cells → Gastrin release to blood

① Presence of partially digested foods in duodenum or distension of the duodenum when stomach begins to empty → Intestinal (enteric) gastrin release to blood → Brief effect

Stomach secretory activity

① Lack of stimulatory impulses to parasympathetic center ⟵ Cerebral cortex ⟵ ① Loss of appetite, depression

① Gastrin secretion declines ⟵ G cells ⟵ ① Excessive acidity (pH < 2) in stomach

Overrides parasympathetic controls ⟵ Sympathetic nervous system activation ⟵ ② Emotional stress

① Distension of duodenum; presence of fatty, acidic, or hypertonic chyme; and/or irritants in the duodenum → Local reflexes → Enterogastric reflex

Vagal nuclei in medulla → Pyloric sphincter

② Distension; presence of fatty, acidic, partially digested food in the duodenum → Release of enterogastrones (secretin, cholecystokinin, vasoactive intestinal peptide)

→ = Stimulate
--→ = Inhibit

Nutrients and Human Nutrition

Nutrition deals with the composition of food, its energy content and synthesized organic molecules. There is a quantitative relationship between nutrients and health. A *balanced diet*, required for good health, includes a properly proportioned variety of foods. Imbalances can cause disease. Nutrition is a major factor in cardiovascular disease, hypertension, and cancer.

Macronutrients are foods required on a large scale each day. These include carbohydrates, lipids, and proteins. Additionally, water is essential because it supports metabolism. Correct water balance is a must for the proper functioning of the body. Monosaccharides, amino acids, and mineral salts are absorbed by transporter-mediated processes, while fatty acid and water diffuse passively.

Carbohydrates

Digestion begins in the mouth by salivary amylase and is completed in the small intestine by pancreatic amylase. Monosaccharides (e.g., glucose, galactose and fructose) are produced by the breakdown of polysaccharides and are transported to the intestinal epithelium by facilitated diffusion or active transport. Facilitated diffusion moves the sugars into the bloodstream.

During the absorptive state, glucose is the major energy source, and some of it is converted to glycogen and stored in skeletal muscle and the liver. In adipose tissue, glucose is transformed and stored as fat. *Glucose sparing* is the reduction of glucose catabolism and the increase in fat utilization by most tissues. Glucose is spared for the brain, resulting in minimization of protein breakdown.

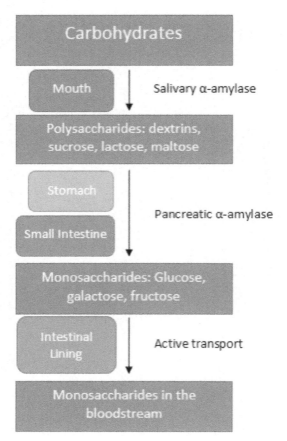

Carbohydrate digestion and absorption

Proteins

Proteins are polymers composed of amino acids. Proteins are in meat, milk, poultry, fish, cereal grains and beans. They are needed for cellular growth and repair. Twenty common amino acids are in proteins, of which humans can naturally synthesize eleven. The remaining nine are the essential (must be in the diet) amino acids.

Proteins are not normally used for energy; however, muscle proteins are broken down for energy during starvation. Excess protein can be used for energy or converted to fats. Proteins are broken down to peptide fragments by pepsin in the stomach, and by pancreatic trypsin and chymotrypsin in the small intestine. The fragments are then digested to free amino acids by *carboxypeptidase* from the pancreas and *aminopeptidase* from the intestinal epithelium.

Free amino acids enter the epithelium by secondary active transport and leave it by facilitated diffusion. Small amino acid chains can enter interstitial fluid by endo- and exocytosis. Most amino acids enter cells and are used to synthesize proteins. Any excess amino acids are converted to carbohydrates or fat.

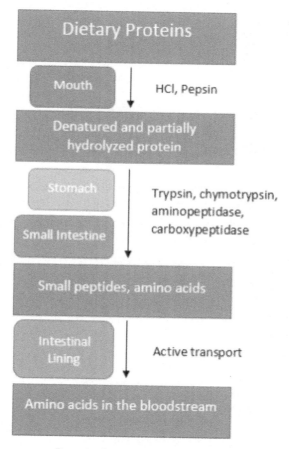

Protein digestion and absorption

Lipids

Lipids generate the greatest energy yield. Therefore, many plants and animals store energy as fats (a type of lipid). Lipids are present in oils, meats, butter and plants (e.g., avocado and peanuts). Some fatty acids, such as linoleic acid, are essential and must be included in the diet. When present in the intestine, lipids promote the uptake of fat-soluble vitamins A, D, E and K.

Fatty acids of plasma chylomicrons are released within the adipose tissue capillaries and form triacylglycerols.

Fat digestion occurs by pancreatic lipase in the small intestine, producing a monoglyceride and two fatty acids from this process. Large lipid droplets are first broken down into smaller droplets by the process of emulsification. Emulsification is driven by mechanical disruption (by contractile activity of the GI tract) and by emulsifying agents (amphipathic bile salts). Pancreatic colipase binds the water-soluble lipase to the lipid substrate.

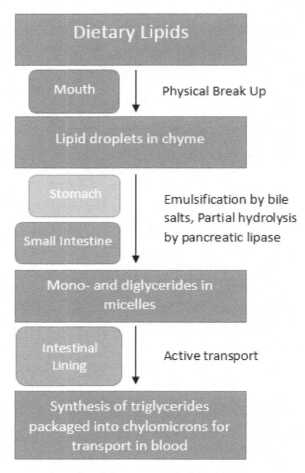

Lipid digestion and absorption

Digested products and bile salts form amphipathic micelles. These micelles keep the insoluble products insoluble aggregates, from which small amounts are released and absorbed by epithelial cells via diffusion. Free fatty acids and monoglycerides then recombine into triacylglycerols at the smooth ER. They are processed in the Golgi and enter the interstitial fluid as chylomicron droplets, which are then taken up by the lacteals in the intestine.

Absorptive and post-absorptive states

The *absorptive state* is when ingested nutrients enter the blood. Some of these nutrients supply the energy need of the body, while the remainder is stored. Carbohydrates and proteins are absorbed into the blood primarily as monosaccharides and amino acids, respectively, while fat is absorbed into the lymph as triacylglycerols.

The *post-absorptive state* is when the GI tract is depleted of nutrients, and the body must supply the required energy. In this state, the net synthesis of glycogen, fat, and protein ceases, and net catabolism of these substances begins. Plasma glucose level is maintained by:

- *Glycogenolysis*, the hydrolysis of glycogen in the liver and skeletal muscles

- *Lipolysis*, catabolism of triacylglycerols into glycerol and fatty acids in adipose tissues. Glycerol reaching the liver is converted to glucose

- *Gluconeogenesis*, catabolism of non-carbohydrate substances to glucose.

Vitamins

Vitamins are necessary organic compounds required for metabolic reactions. They usually cannot be made by the body and are needed in trace amounts. Vitamins may act as enzyme cofactors, known as *coenzymes*. Some vitamins are soluble in fats, while others are soluble in water.

Fat-soluble vitamins are absorbed and stored along with fats. Most water-soluble vitamins are absorbed by diffusion or mediated transport. Fat-soluble vitamins necessitate more complex mechanisms to cross the phospholipid bilayer. For example, due to its large size and charged nature, Vitamin B-12 must first bind to the intrinsic factor protein to be absorbed via endocytosis.

The body is unable to make some vitamins required for metabolic activities. Lack of vitamins results in vitamin deficiencies. The 13 vitamins are divided into fat-soluble (A, D, E, K) and water-soluble vitamins (8 B vitamins and vitamin C). Many vitamins are portions of coenzymes; for example, niacin is part of NAD^+ and riboflavin is a part of a FAD. Coenzymes are needed in small amounts because they are used over and over again. Vitamin A is not a coenzyme but a precursor for the visual pigment that prevents night blindness.

Vitamin A is part of a group of hydrophobic unsaturated nutritional organic compounds that include retinol, retinal, retinoic acid and several carotenoids (e.g., beta-carotene). Vitamin A has multiple functions. It is important for growth and development, for the maintenance of the immune system, and for good vision. Vitamin A is needed by the retina of the eye in the form of retinal, which combines with protein opsin to form rhodopsin (i.e., light-absorbing molecule). Rhodopsin is necessary for low-light (night vision) and color vision.

Skin cells contain a precursor cholesterol molecule converted to vitamin D by UV light exposure. Only a small amount of UV is needed to cause this change. Vitamin D leaves the skin and is modified in the kidneys and then in the liver to become calcitriol. *Calcitriol* circulates throughout the body, regulating calcium uptake and metabolism. It promotes absorption of calcium by the intestines. The lack of vitamin D leads to rickets in children, a condition in which poor mineralization of the skeleton causes bowing of the legs. Most milk is fortified (added during processing) with vitamin D to prevent rickets.

Cell metabolism generates *free radicals*, unstable molecules with an extra electron. O_3^- is a common free radical. Free radicals stabilize by donating electrons to another molecule; however, doing so damages cellular molecules, such as DNA and proteins. This damage may cause plaque in arteries or cancer. Vitamins C, E, and A—abundant in fruits and vegetables—are antioxidants that defend against free radicals. While vitamin supplements can assist in fighting vitamin deficiencies, they do not replace fruits and vegetables, which contain many phytochemicals and other beneficial compounds.

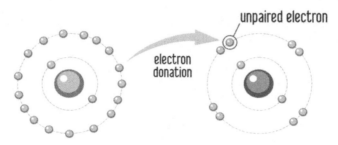

Antioxidants reduce free radicals

Minerals

Minerals are trace elements required for normal metabolism, cell and tissue structure, nerve conduction and muscle contraction. They are essential and can only be obtained from the diet. Iron, iodine, calcium, and sodium are examples of minerals.

Humans require *macrominerals* (e.g., calcium, phosphorus) in amounts of over 100 mg per day. They are constituents of cells and body fluids and are structural components of tissues. Calcium is needed to build bones and tooth enamel, as well as to generate nerve conduction and muscle contraction. Sodium is a key macromineral used to control blood pressure and blood volume and to keep nerves and muscles functioning correctly.

Microminerals are elements (e.g., zinc, iron) recommended in amounts less than 20 mg per day. Microminerals are likely to have specific functions. For example, iron is needed to produce hemoglobin, of which adult females need more due to menstrual blood loss. Iodine is used to produce thyroxin, a hormone of the thyroid glands. Minute amounts of molybdenum, selenium, chromium, nickel, vanadium, silicon and arsenic (among others) are essential to keep the body functioning properly.

Chapter 18

Excretory System

- **Introduction to Excretory System**

- **Kidney Structure**

- **Nephron Structure**

- **Roles in Homeostasis**

- **Formation of Urine**

- **Storage and Elimination of Urine: Ureter, Bladder, and Urethra**

Introduction to Excretory System

The *human excretory system* is an organ system that removes excess fluids and materials to prevent damage to the body. It consists of several parts. The human *kidneys* are two bean-shaped, reddish-brown organs, each about the size of a fist. They are located on each side of the vertebral column (below the diaphragm) and are partially protected by the lower rib cage. The kidneys are the sites of urine formation, and each is connected to a *ureter*, which moves urine from a kidney to the *urinary bladder*.

The urinary bladder stores urine from the kidneys until it is voided from the body through the *urethra*. The male urethra runs through the penis and conducts semen. In females, the urethra opens ventral to the vaginal opening.

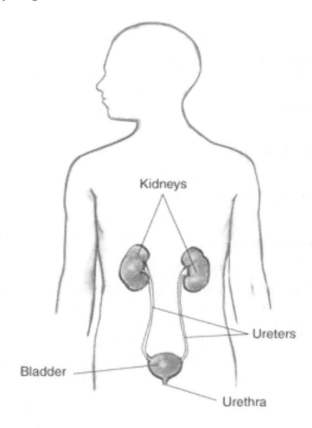

Two kidneys connect by ureters to the urinary bladder.
Urine passes through the urethra as it leaves the body.

Kidney Structure

The kidney has three regions: the renal cortex (outer part), medulla (inner part) and renal pelvis (innermost, the hollow structure of the kidney).

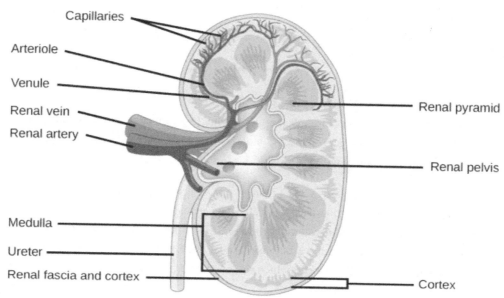

Anatomy of the kidney with renal cortex, medulla and renal pelvis

Cortex

The *renal cortex* is the thin, outer region (or shell) of the kidney and it is composed of many convoluted tubules that give it a granular appearance.

Medulla

The *renal medulla* is the inner part of the kidney consisting of the striped, pyramid regions that lie on the inner side of the cortex and the loop of Henle.

Renal pelvis

The *renal pelvis* is a hollow, funnel-shaped structure in the innermost part of the kidney that collects urine. It receives the urine from collecting ducts and papillary ducts, and it releases urine into the ureters, which lead to the urinary bladder.

Nephron Structure

Each kidney has approximately one million subunits, called *nephrons*. Nephrons are considered the functional units of the kidney because they reabsorb nutrients, salts, and water. Nephrons are composed of the *renal corpuscle* and the *renal tubule*.

Glomerulus

The *glomerulus* is a ball of fenestrated capillaries that acts like a sieve (or colander). Small molecules dissolved in the fluid pass through the glomerulus (e.g., glucose, which is later reabsorbed), while large molecules such as plasma proteins and blood cells do not. If blood cells or proteins are in the urine, this likely indicates a problem with the glomerulus.

The *juxtaglomerular apparatus* (JGA), named for its proximity to the glomerulus, consists of the macula densa cells and juxtaglomerular (JG) cells. The *macula densa* cells are important in blood flow regulation and can affect the glomerulus filtration rate. They are located in the part of the ascending limb passing between the afferent and the efferent arterioles. The wall of the afferent arteriole near the Bowman's capsule has *juxtaglomerular* cells, which secrete the hormone renin.

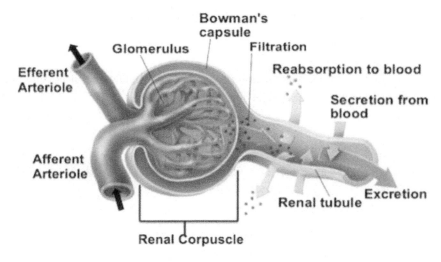

The nephron as the functional unit of the kidney.
Blood enters via the afferent arteriole and exits via the efferent arteriole

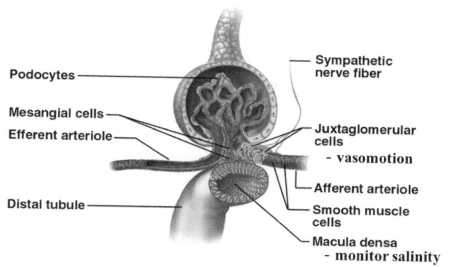

Podocytes

Mesangial cells

Efferent arteriole

Distal tubule

Sympathetic
nerve fiber

Juxtaglomerular
cells
- vasomotion

Afferent arteriole

Smooth muscle
cells

Macula densa
- monitor salinity

Juxtaglomerular apparatus of the kidney

Bowman's capsule

Bowman's capsule is a capsule conformation that surrounds the glomerulus. It is responsible for performing the first step in the filtration of blood for the formation of urine. Bowman's capsule encloses the glomerulus involved in filtering liquids and microscope components from the blood.

Proximal tubule

The *proximal tubule* is a convoluted tubule on the side of the Bowman's capsule and is the major site of active reabsorption of glucose, ions and amino acids. Additionally, it is responsible for the secretion of ions, except potassium. The proximal tube is divided into an initial convoluted portion and a straight (descending) portion.

Fluid in the filtrate entering the proximal convoluted tubule is reabsorbed into the peritubular capillaries. This reabsorption includes approximately two-thirds of the filtered salt as well as water filtered out by the glomerulus. The proximal convoluted tubule secretes chemicals, such as ammonium, which is formed by the deamination processes that convert glutamine to alpha-ketoglutarate. Finally, the proximal convoluted tubule drains into the loop of Henle.

Loop of Henle

The majority of each nephron is the *loop of Henle*, a U-shaped tube that extends from the proximal tubule and consists of a *descending limb* and an *ascending limb*. The loop begins in the cortex, receiving filtrate from the proximal convoluted tubule. It then extends into the medulla as the descending limb and returns to the cortex as the ascending limb to continue into the distal convoluted tubule.

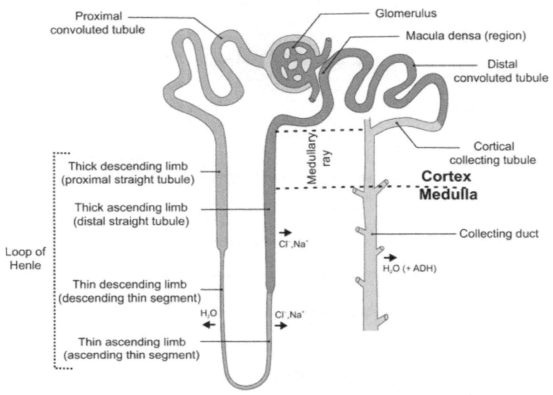

Cross section of a nephron including glomerulus and loop of Henle

The primary role of the loop of Henle is to concentrate salt in the surrounding tissues (interstitium). This is accomplished using a countercurrent multiplier mechanism.

The descending limb of the loop of Henle is the site of water reabsorption. The ascending limb of Henle is the site of salt absorption. Their considerable differences can easily distinguish the ascending and descending limbs of the loop of Henle. The thin descending limb is permeable to H_2O and noticeably less permeable to salt. This part of the loop of Henle indirectly contributes to the concentration of the interstitium. Excess water is picked up by the *vasa recta* and returned to the blood. As the filtrate descends deeper into the hypertonic interstitium of the renal medulla, H_2O flows freely out of the descending limb by osmosis until the tonicity of the filtrate and interstitium equilibrate. Longer descending limbs allow more time for H_2O to flow out of the filtrate; thus, longer limbs make the filtrate more hypertonic than shorter limbs.

The thick ascending limb of the loop of Henle is impermeable to H_2O, a critical feature of the countercurrent exchange mechanism. The ascending limb actively pumps Na^+ out of the filtrate, generating the hypertonic interstitium that drives countercurrent exchange. This is done through secondary active transport via Na-K-Cl cotransporters. In passing through the ascending limb, the filtrate becomes hypotonic, since it has lost much of its sodium content. This hypotonic filtrate is passed to the distal convoluted tubule in the renal cortex.

Distal Tubule

The loop of Henle leads to the *distal convoluted tubule*, located on the side of the collecting ducts. The distal convoluted tubule is hormone-controlled and fine-tunes the work done by the proximal tubule by continuing the reabsorption of salts and water. It is partially responsible for the reabsorption of glucose, ions, and water. The cells of the distal convoluted tubule contain numerous mitochondria, which produce enough energy (ATP) for active transport. The endocrine system regulates much of the ion transport in the distal convoluted tubule. Additionally, the distal convoluted tubule regulates the pH by secreting protons, absorbing protons and secreting bicarbonate ions.

Calcium reabsorption occurs in the distal convoluted tubule in response to low blood calcium levels. When blood calcium levels are low, parathyroid hormone is released. In the presence of parathyroid hormone, the distal convoluted tubule reabsorbs more Ca^{2+}, osteoclast activity is stimulated (Ca^{2+} released from bone), and additional phosphate is secreted.

The hormone aldosterone controls sodium levels (through absorption) and potassium levels (through secretion). When aldosterone is present, more Na^+ is reabsorbed (along with H_2O), and more K^+ is secreted. The atrial natriuretic peptide causes the distal convoluted tubule to secrete more Na^+.

Collecting duct

Many of the distal convoluted tubules drain into *collecting ducts*, which are comprised of the connecting tubules, the cortical collecting ducts, and the medullary collecting ducts. *Medullary collecting ducts* from numerous nephrons merge and drain into the renal pelvis, which is connected to the ureter. The tubules are connected to another set of blood vessels, the *peritubular capillaries*.

Each collecting duct is shared by many nephrons and is the place where the *antidiuretic hormone* (ADH)-controlled reabsorption of water and the hormone-controlled reabsorption / secretion of sodium occurs. Collecting ducts are normally impermeable to water. If the hormone ADH is present, however, collecting ducts become permeable to water and water reabsorption occurs. ADH causes blood pressure to rise and plays an important role in homeostasis by regulating glucose, sodium, and water levels in the body.

Roles in Homeostasis

The kidneys function to excrete waste, control plasma pH and maintain homeostasis of body fluid volume and solute composition.

Blood pressure

Since sodium is the major extracellular solute, changes in total body sodium result in changes in the volume of extracellular fluid. These changes then lead to changes in plasma volume and blood pressure, which is detected by baroreceptors. Usually, more than 99% of the sodium filtered out at the glomerulus is returned to the blood. Most are reabsorbed at the proximal tubule, 25% is extruded by the ascending limb of the loop of the nephron, and the remaining sodium is reabsorbed from the distal convoluted tubule and collecting duct.

Blood pressure is constantly monitored within the *juxtaglomerular apparatus*. If the blood pressure is insufficient to promote glomerular filtration, the afferent arteriole cells secrete *renin*. Renin catalyzes the conversion of *angiotensinogen* (a protein produced by the liver) into *angiotensin I*. Angiotensin I is then converted to *angiotensin II* by *angiotensin-converting enzyme* (ACE); the rate-limiting step of this reaction is controlled by renin from JG cells. JG cells act as internal baroreceptors and receive sympathetic inputs from external baroreceptors.

Angiotensin II increases the blood pressure as a vasoconstrictor by promoting the retention of sodium ions. This hormone stimulates cells in the adrenal cortex to produce aldosterone, which stimulates sodium reabsorption by *cortical collecting ducts* (and large intestine, sweat, and salivary glands). Aldosterone acts on the distal convoluted tubules to increase the reabsorption of Na^+ and the excretion of K^+. Increased Na^+ in the blood causes water to be reabsorbed, increasing blood volume and pressure. Aldosterone production is triggered by the *renin-angiotensin-aldosterone system*.

Thirst is stimulated by lower extracellular volume, higher plasma osmolarity or angiotensin II. The brain centers for thirst are located in the hypothalamus.

Atrial natriuretic hormone (ANH) is produced by the atria of the heart when cardiac cells stretch. When blood pressure rises, the heart produces ANH to inhibit the secretion of renin and the release of ADH, which decreases blood volume and pressure. Additionally, the heart can release atrial natriuretic peptide (ANP), which is the antagonist for aldosterone and causes the kidney to excrete more sodium ions and water, causing vasodilation.

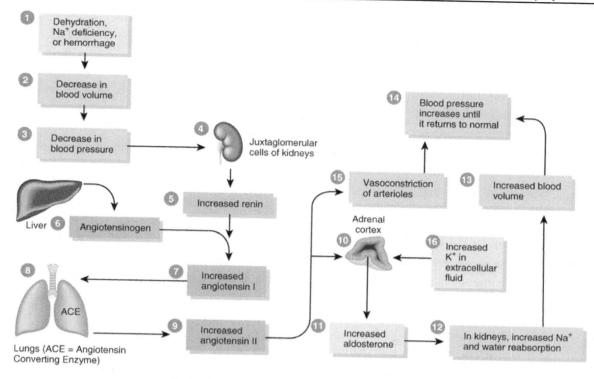

Renin-Angiotensin-Aldosterone (RAA) pathway

Osmoregulation

If osmolarity differs between two regions, water moves into the region with the higher solute concentration. Marine environments are high in salt and promote the loss of water and the gain of ions when drinking saltwater. Freshwater promotes a gain of water by osmosis and a loss of ions as this excess water is excreted. For example, the body of a marine fish living in saltwater is hypotonic relative to its environment, resulting in water being constantly lost by osmosis. To account for this, marine fish constantly consume water, rarely urinate and secrete accumulated salts through their gills.

Conversely, freshwater fish live in an environment where their bodies are hypertonic relative to the environment. Water, therefore, is being constantly gained. This characteristic leads freshwater fish to rarely consume water, constantly urinate and readily absorb salts through gills.

The excretory system regulates ions and water in body fluids. Regulation depends on the concentration of mineral ions (i.e., Na^+, Cl^-, K^+, and HCO_3^-). Body fluids gain mineral ions when animals eat food and drink fluids, and lose ions through excretion. Water enters organisms when they eat food or drink water, and by metabolism where cellular respiration produces water. Water is lost by evaporation from the skin and lungs, and by excretion (i.e., in urine or feces). For balance, the volume of water entering the body must equal the volume of water lost.

The kidneys function to eliminate wastes (urea, H^+) generated by the metabolic activity while reabsorbing important substances (glucose, amino acids, sodium) for reuse by the body. The generation of a solute concentration gradient from the kidney's cortex to its medulla allows a considerable amount of water to be reabsorbed. Excretion of concentrated urine serves to limit water loss from the body and helps to preserve blood volume.

The long loop of a nephron is comprised of a descending limb (cortex → medulla) and an ascending limb (medulla → cortex). Salt (NaCl) passively diffuses out of the lower portion of the ascending limb, but the upper, thick portion of the limb actively transports salt out into the tissue of the renal medulla. Less salt is available for transport from the tubule as fluid moves up the thick portion of the ascending limb. Urea leaks from the lower portion of the collecting ducts, causing the concentrations in the lower medulla to be highest.

Because of the solute concentration gradient within the renal medulla, water leaves the descending limb of the loop of Henle along its length. The decreasing water concentration in the descending limb encounters an increasing solute concentration; this is a countercurrent mechanism. Fluid received by a collecting duct from the distal convoluted tubule is isotonic to cells of the cortex. As this fluid passes through the renal medulla, water diffuses out of the collecting duct into the renal medulla. The urine finally delivered to the renal pelvis is usually hypertonic to the blood plasma.

Some terrestrial animals near oceans drink seawater despite its high osmolarity. Such birds and reptiles have a nasal salt gland that excretes concentrated salt solution. Terrestrial animals, in general, tend to lose both water and ions to the environment.

In humans, water and sodium are gained from ingestion and oxidation of organic nutrients. Water is then lost via sweat glands, respiratory passageways, and the gastrointestinal and urinary tracts.

Blood plasma mainly contains sodium ions and chloride ions, whereas cells themselves mainly contain potassium and hydrogen ions. Blood osmolarity is determined predominantly by the concentrations of sodium and potassium ions. If the osmolarity is too low, aldosterone is released so that reabsorption can take place. Aldosterone controls potassium concentrations by allowing the renal system to reabsorb sodium ions and excrete potassium ions in urine. Osmolarity is regulated by the secretions and reabsorptions of the kidney tubules.

Antidiuretic hormone (ADH) is released from the posterior lobe of the pituitary gland. ADH acts on the collecting ducts by increasing its permeability to H_2O, thereby increasing H_2O retention. When ADH is released, more water is reabsorbed, and there is less urine. When ADH is not released, more water is excreted and more urine forms. Thus, if an individual does not drink, the pituitary releases ADH; if hydrated, ADH is not released. *Diuresis* is increased urine production, while *antidiuresis* is a decreased amount of urine.

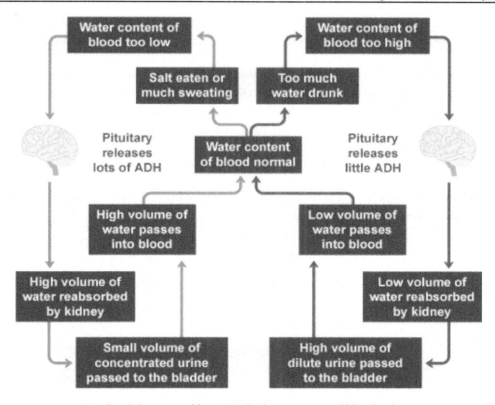

Feedback loops used by ADH for homeostasis of blood volume

While aldosterone and ADH ultimately do the same thing (i.e., increase water reabsorption in the kidneys), they have different mechanisms of action: ADH directly increases water reabsorption from the nephron's collecting duct, while aldosterone indirectly increases water reabsorption by increasing sodium reabsorption from the collecting duct.

Acid-base balance

The balance between acids and bases within the blood keep the blood pH steady. One of the primary ways this balance is regulated is through the bicarbonate HCO_3^- buffer system in the blood and extracellular fluid, and the phosphate buffer system inside the cells.

During bicarbonate regulation, HCO_3^- is filtered at the renal corpuscle. It then undergoes reabsorption in the proximal tubule and can be secreted in the collecting ducts.

$$HCO_3^- \text{ excreted} = HCO_3^- \text{ filtered} + HCO_3^- \text{ secreted} - HCO_3^- \text{ reabsorbed}$$

Inside red blood cells, CO_2 and H_2O combine to form H_2CO_3, which dissociates to yield H^+ and HCO_3^- ions. HCO_3^- moves to the interstitial fluid by diffusion, while H^+ ion is secreted into the lumen by an active process involving H-ATPase pumps. The secreted H^+ ion combines with filtered HCO_3^- in the lumen and generates CO_2 and H_2O. These two compounds diffuse into the cell, and the process is repeated. If an excess of H^+ ions is secreted, it combines with a nonbicarbonate buffer, usually HPO_4^{2-}, in the lumen, and is excreted. Then, the HCO_3^- generated within the cell and entering the plasma produces a net gain of HCO_3^-.

The bicarbonate buffer system and the respiratory system work together to maintain blood pH. Breathing out CO_2 decreases the acidity in the blood. Bicarbonate ions lower the pH of the blood. Then, the excretion of H^+ ions and NH_3 and the reabsorption of bicarbonate ions (HCO_3^-) is adjusted. If the blood is basic (pH is high), fewer H^+ ions are excreted, and fewer (HCO_3^-) ions are reabsorbed to lower the pH.

$$H^+ + HCO_3^- \rightleftarrows H_2CO_3 \rightleftarrows H_2O + CO_2$$

Ammonia is produced by the deamination of amino acids in the tubule cells. Reabsorption or excretion of ions is homeostasis by the kidneys to regulate blood pH and osmolarity.

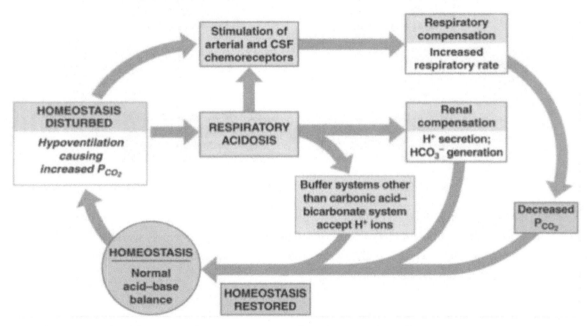

Respiratory and excretory systems function to maintain homeostasis

Removal of soluble nitrogenous waste

The breakdown of nucleic acids and amino acids produces nitrogenous wastes. Amino acids derived from protein are needed for the synthesis of body proteins or nitrogen-containing molecules. Unused amino acids are oxidized to generate energy or are stored as fats or carbohydrates. The resulting *amino groups* ($-NH_2$) must be removed.

Depending on the organism, nitrogenous wastes are excreted as ammonia, urea or uric acid. Amino groups removed from amino acids form *ammonia* (NH_3) by adding a third hydrogen ion. This requires little or no energy. Ammonia, while quite toxic, is water soluble and requires a significant amount of water to wash away from the body (e.g., aquatic animals).

Urea ($CO(NH_2)_2$) is a harmless form of ammonia. Amino acids are converted into ammonia, which is then converted to urea. Urea is excreted by urination in the form of urine, which is simply concentrated urea in ionized water (e.g., humans).

Mammals and terrestrial amphibians usually excrete urea as their main nitrogenous waste, because urea is much less toxic than ammonia. Urea is excreted in a moderately concentrated solution, which conserves body water. Urea is produced in the liver as a product of the energy-requiring *urea cycle*. In the urea cycle, carrier molecules take up carbon dioxide and two molecules of ammonia, resulting in the release of urea.

Some animals have adapted to survival in dry arid environments, and can completely avoid drinking water. They form concentrated urine and fecal matter that is almost completely dry and are able to meet their water requirements with the metabolic water generated through aerobic respiration. Insects, reptiles, and birds excrete *uric acid* as their main nitrogenous waste. Uric acid is not very toxic and is poorly soluble in water, but is concentrated for water conservation. Uric acid is synthesized through enzymatic reactions using even more ATP than urea synthesis, demonstrating the tradeoff between water conservation and energy expenditure.

Formation of Urine

Glomerular filtration

Urine formation follows filtration, secretion, and reabsorption.

Filtration of plasma from glomerular capillaries into the Bowman's capsule is *glomerular filtration*. The glomerular filtrate passes the glomerulus (afferent arteriole → glomerulus → efferent arterioles) to the rest of the nephron. It is referred to as a bulk flow process because water and solutes are moved together due to a pressure gradient. The glomerular filtrate contains all plasma substances in the same concentrations as plasma, except plasma proteins and the molecules bound to these proteins.

The *afferent arterioles* carry blood into the glomerulus, while the efferent arterioles carry blood away from the glomerulus. Efferent arterioles exit the glomerulus and web around the nephron as *peritubular capillaries*. These peritubular capillaries act to surround the proximal convoluted tubule (PCT) and distal convoluted tubule (DCT). They reabsorb the nutrients and ions filtered out by the glomerulus. The efferent arterioles continue to form the *vasa recta*, surrounding the loop of Henle (maintaining the concentration gradient) before merging with the renal branch of the renal vein. The peritubular capillaries drain into a venule. Then the venules from many nephrons drain into a small vein. Many small veins then join to form the renal vein, a vessel that enters the inferior vena cava.

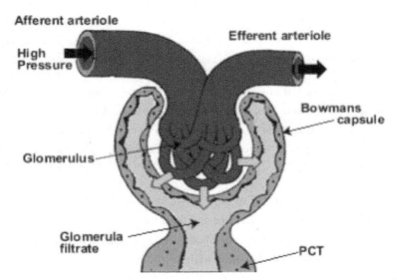

Glomerulus and Bowman's capsule with afferent arteriole and efferent arteriole

During passage through tubules, substances move from tubules to peritubular capillaries as *tubular reabsorption.* In *tubular secretion*, substances move from peritubular capillaries to tubules. Both waste (e.g., urea, creatine and uric acid) along small useable ions and nutrients are filtered out. The nutrients and usable ions are reabsorbed, while the waste is excreted. Particles that are too large to filter through (e.g., blood, albumin) remain in the circulatory system. This is a passive process that is driven by the hydrostatic pressure of blood. In the filtration process, the filtrate is pushed (hydrostatic pressure) from the glomerulus into Bowman's capsule.

The *juxtaglomerular apparatus* monitors filtrate pressure in the distal tubule via granular cells. It is responsible for the secretion of renin, which starts a signal transduction cascade involving angiotensin. The result is that the adrenal cortex is stimulated to create more aldosterone, which stimulates the retention of sodium in the body.

When blood enters the glomerulus (cluster of capillaries), hydrostatic pressure (blood pressure) forces small molecules from the glomerulus across the inner membrane of the glomerular capsule into the lumen of the glomerular capsule, a process of pressure filtration. The glomerular walls are a hundred times more permeable than the walls of most capillaries. *Glomerular filtrate* is the molecules that leave the blood and enter the glomerular capsules. Plasma proteins and blood cells are too large to be part of the glomerular filtrate. Failure to restore fluids would soon cause death due to loss of water and nutrients and lowering of blood pressure.

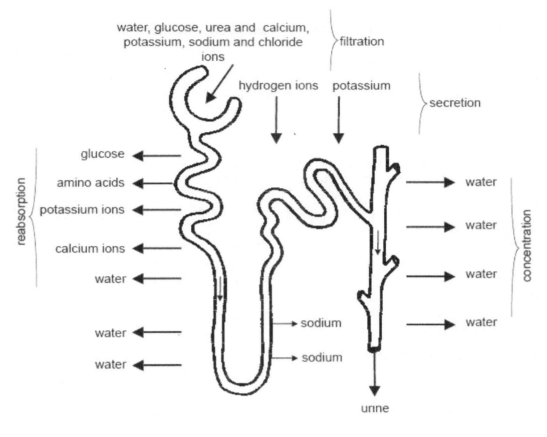

$$A_E \ = \ A_F + A_S - A_{RA}$$

where A is amount: A_E = excreted, A_F = filtered, A_S = secreted, A_{RA} = reabsorbed.

For a given substance:

Net glomerular filtration pressure = $P_{GC} - P_{BS} - \pi_{GC}$

where P_{GC} = glomerular capillary hydrostatic pressure is favoring filtration from the capillaries to Bowman's capsule, P_{BS} = the Bowman's capsule hydrostatic pressure favoring movement from Bowman's capsule to the capillaries and π_{GC} = osmotic pressure resulting from the presence of protein in glomerular capillary plasma and the absence of protein in Bowman's capsule.

Normally, net glomerular filtration pressure is positive (fluid moves into Bowman's capsule).

Glomerular filtration rate (GFR) is the volume of liquid filtered from glomeruli into the Bowman's capsule per unit time. GFR is determined by net filtration pressure, the permeability of the corpuscular membranes and the surface area available for filtration. GFR is subject to physiological regulation by neural and hormonal inputs to afferent and efferent arterioles. Constriction of afferent arterioles decreases P_{GC}, while constriction of efferent arterioles increases it. *Mesangial cells*, modified smooth muscle cells, are involved in this constriction process.

The *filtered load* is the total amount of non-protein substance filtered into the Bowman's space. It is given by multiplying GFR with the plasma concentration of the substance.

Filtered load = GFR × plasma concentration.

If the quantity of a substance excreted in urine is less than the filtered load, tubular reabsorption has occurred; if it is more, tubular secretion has occurred.

Secretion and reabsorption of solutes

During *secretion,* substances such as acids, bases, and ions (e.g., K^+) may be secreted by both passive and active transport in *peritubular capillaries,* which are formed from efferent arterioles from the glomerulus.

The *proximal convoluted* tubules reabsorb all the nutrients and most of the ions.

Soluble waste products are left in the filtrate (urea) to be excreted. Also, NH_4^+, creatinine and organic acids are excreted.

The *loop of Henle* reabsorbs water and salt using the countercurrent mechanism.

The *distal convoluted* tubules selectively reabsorb or secrete ions and compounds based on hormonal control. The collecting duct reabsorbs water to concentrate urine if ADH is present.

The *collecting duct* can secrete and reabsorb substances based on hormonal control. The regulation of blood occurs in the following manner: H^+ is secreted out of the blood when blood pH is too acidic, whereas HCO_3^- is secreted out of the blood when blood pH is too basic.

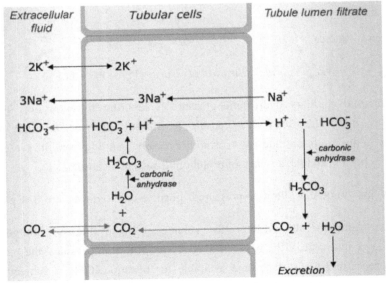

Ion exchange in the tubular cells of the kidney

Tubular secretion moves substances (e.g., H^+ and K^+ ions) from peritubular capillaries into the tubular lumen by diffusion or transcellular mediated transport. Some of the movements are coupled with reabsorption of Na^+ ions. Secretion back into the filtrate is primarily associated with the distal convoluted tubule. This helps rid the body of potentially harmful compounds that were not filtered into the glomerular capsule (e.g., uric acid, hydrogen ions, ammonia, and penicillin).

During *reabsorption*, glucose, salts, amino acids and water are reabsorbed from the filtrate and returned to the blood. *Tubular reabsorption* of fluids from the nephron into the blood occurs through the walls of the proximal convoluted tubule. Reabsorption recovers much of the glomerular filtrate. The osmolarity of the blood and filtrate are equal, so osmosis of water does not occur. Sodium ions are actively reabsorbed, pulling along chlorine, which changes the osmolarity of the blood, and water now moves passively from the tubule back to the blood.

About 60–70% of salt and water is reabsorbed at the proximal convoluted tubule. Cells of the proximal convoluted tubule have numerous microvilli, which increases the surface area available for absorption. They have numerous mitochondria, which supply the energy needed for active transport. Only molecules with carrier proteins are reabsorbed. For example, if there is more glucose than carriers, excess glucose appears in the urine. In diabetes mellitus, there is high plasma glucose due to a lack of insulin or insulin-resistant receptor cells. When diabetic, the liver does not properly convert glucose to glycogen.

Waste products are mostly excreted in the urine. Useful products are reabsorbed completely and are mostly not excreted. Mediated transport is the reabsorption mechanism of many substances (e.g., glucose molecules are coupled to the reabsorption of sodium).

The limit to which these mediated transport systems can move materials per unit time is the *transport maximum* (T_m). This limit is a result of the saturation of binding sites on membrane transport proteins.

Sodium moves out of the lumen into the epithelium by diffusion (via ion channels) by cotransport with glucose (which is being reabsorbed), or by countertransport with H^+ ions (which are being secreted). The sodium-potassium pump (Na^+ / K^+–ATPase) transports sodium out of the epithelium into the interstitial fluid.

The removal of sodium lowers the osmolarity of the lumen and raises that of the interstitial fluid. This causes a net diffusion of water from the lumen into the interstitial fluid through the epithelium. Water permeability of the proximal tubule is high, but only that of collecting ducts is under the control of vasopressin (ADH). ADH stimulates insertion of aquaporin channels, increasing water permeability. Low ADH leads to water diuresis or *diabetes insipidus*, a condition caused by the inability to produce ADH and promote the retention of water. As a result, the person excretes more urine than necessary. Increased urine flow due to increased solute excretion is *osmotic diuresis*.

Concentration of urine

The *concentration of urine* occurs when there is a low volume of fluid in the bloodstream. To counteract this, the body produces small amounts of concentrated urine. Concentrated urine production is achieved through the hormone ADH, which reduces water loss by making the distal convoluted tubule (DCT) and the collecting ducts permeable to water. When blood pressure is low, aldosterone increases reabsorption of Na^+ by the DCT, which increases water retention (serum $[Na^+]$ increases blood pressure).

The distal convoluted tubule contains a dilute solution of urea. The collecting duct concentrates this dilute solution by water reabsorption (facilitated diffusion) using ADH-controlled action. Water reabsorption in the collecting duct is possible because the loop of Henle has high osmolarity (concentrated) at the bottom.

The area where the distal convoluted tubules join the collecting duct becomes more concentrated as the filtrate descends the collecting duct (because the surrounding medulla is salty and water leaves).

The collecting duct leads to the renal calyx, which empties into the renal pelvis and drains into the ureter.

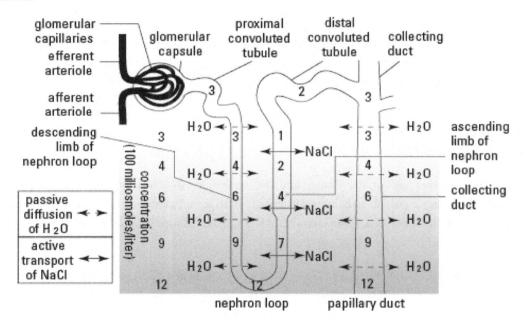

Regulation of urine concentration.
Passive and active transport of substances for sections of the nephron.

Countercurrent multiplier mechanism

The countercurrent multiplier creates an osmotic gradient down the loop of Henle, which is used by the collecting duct to concentrate urine. The sodium-potassium pump on the ascending limb drives this gradient. Countercurrent is the water flow out of the filtrate on the descending limb and the salt flow out of the filtrate on the ascending limb. The descending limb is impermeable to salt, and the ascending limb is impermeable to water.

The multiplier is the product of the gradient-producing power of each sodium-potassium pump. This gradient-producing power multiplies down the length of the loop of Henle. The longer the loop of Henle, the greater the osmotic gradient. The greater the osmotic gradient, the more concentrated the urine.

Urea recycling contributes to the high osmolarity at the bottom of the loop of Henle. Urea recycling occurs when the urea at the bottom of collecting ducts leaks into the interstitial fluid and returns to the filtrate.

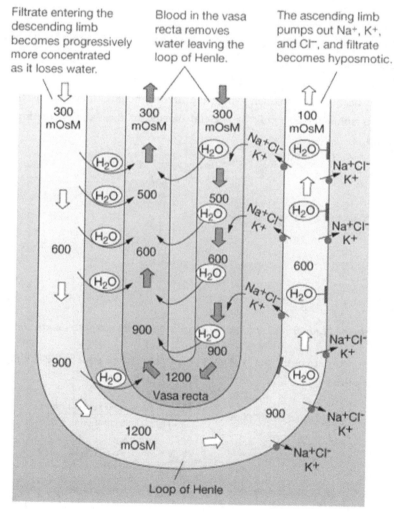

Counter-current multiplier mechanism in the loop of Henle that creates high interstitial osmolarity. The thin descending loop is on the left, and the thick ascending loop is on the right. Note the molecules and ions which pass each section.

Fluid from the proximal tubule has the same osmolarity as plasma since it absorbs sodium and water equally. In the ascending limb, sodium, but not water, is actively reabsorbed from the lumen, making the interstitial fluid of the medulla hyperosmotic. Due to hyperosmolarity, there is passive diffusion of water from the lumen into the interstitial fluid in the descending limb. Fluid in the distal tubule becomes progressively dilute as sodium is transported out. Then, in the cortical and medullary ducts, water diffuses out of the tubule into the hyperosmotic interstitial fluid, and urine is concentrated.

Storage and Elimination of Urine: Ureter, Bladder, and Urethra

The collecting ducts empty into the ureter, which then drains into the bladder. The bladder stores the urine. Its special epithelium (transitional epithelium) distends to accommodate the storage of large amounts of urine. The urine is excreted from the bladder through the urethra.

The smooth-muscle contractions of the ureter wall allow for urine to flow through. Urine is stored in the bladder and ejected during urination (micturition). The bladder is a chamber with walls made of smooth muscle (detrusor muscle). The contraction of detrusor muscles produces urination. Part of the muscle at the base of the bladder, where the urethra begins, functions as the internal urethral sphincter. Below this sphincter is a ring of skeletal muscle of the external urethral sphincter, which surrounds the urethra.

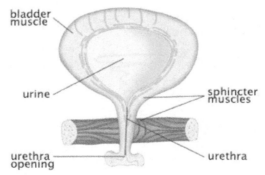

Urinary bladder and sphincter muscles as urine pass into the urethra

Renal clearance measures the volume of plasma from which the kidneys per unit time completely remove a substance.

$$\text{Clearance of substance S} = \frac{\text{Mass of S excreted per unit time}}{\text{Plasma concentration of S}}$$

$$C_S = \frac{U_S V}{P_S}$$

where C_S = clearance of S, U_S = urine concentration of S, V = urine volume per unit time, P_S = plasma concentration of S. C_S of a substance equals glomerular filtration rate (GFR) if the substance is filtered but not reabsorbed, secreted or metabolized.

Chapter 19

Muscle System

- **Muscle Cell**

- **Important Functions**

- **Structural Characteristics**

- **Mechanics of Muscle Contraction**

- **Regulation of Cardiac Muscle Contraction**

- **Oxygen Debt: Fatigue**

- **Neural Control**

Muscle Cell

Muscular Cell Structure

A muscle consists of muscle cells bound together by connective tissue. A single muscle fiber is a multinucleated cell formed from myoblasts during development. Muscle cells contain several parallel *myofibrils*, each of which is composed of *myofilaments* (primarily actin and myosin). These myofibrils contain the *sarcomeres*, which are the basic units of the contraction in the skeletal muscle. Myofibrils are packed together within the multinucleated skeletal muscle cells. The *sarcolemma* (plasma membrane) of the muscle cell contains the myofibrils and keeps them packed together. The nuclei of the muscle fibers are located at the edges of the diameter of the fiber, adjacent to the sarcolemma.

The *sarcoplasm* is the cytoplasm of muscle fibers and contains numerous mitochondria that produce ATP for muscle contraction. The *sarcoplasmic reticulum* is similar to the smooth endoplasmic reticulum; it extends throughout the sarcoplasm of the muscle cell and is involved in storing calcium ions used in muscle contraction.

Muscle is attached to a bone by collagen bundles as tendons. After infancy, new fibers are formed from undifferentiated satellite cells and generally do not undergo mitosis to create new muscle cells after development, *hyperplasia*. However, muscle cells increase in size and thus increase the overall volume of the muscle, *hypertrophy*. In adulthood, any compensation for lost muscles occurs mostly through an increase in the size of fibers.

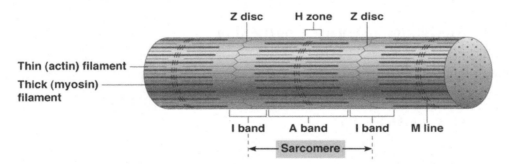

The function unit of a muscle cell is the sarcomere

Transverse tubules (*T-tubules*) are tunnel-like extensions of the sarcolemma that pass through muscle cells from one side of the cell to another, forming a network around myofibrils. They are referred to as transverse because of the way they are oriented. The transverse tubules play a vital role in muscle contraction. A muscle action potential, which is the movement of electrical charge, travels along the transverse tubules and stimulates the release of calcium ions from the sarcoplasmic reticulum. This allows the calcium ions to flood back into the sarcoplasm and binds to troponin.

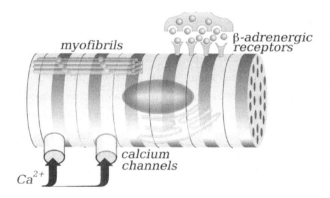

Calcium triggers the movement of various protein filaments (including *actin, myosin,* and *tropomyosin*) within the myofibrils, which results in muscle contraction. The general function of the T-tubules is to conduct impulses from the surface of the cell to the sarcoplasmic reticulum where Ca is released.

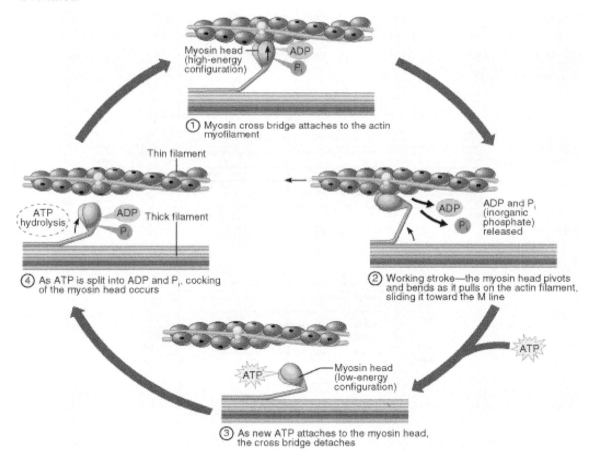

Calcium binds to troponin to cause a conformational change in tropomyosin and expose the myosin binding sites on the actin myofilament (step 1)

The sarcoplasm in muscle cells is equivalent to the cytoplasm of other cells. The sarcoplasmic reticulum in muscle cells is homologous to the endoplasmic reticulum. Unlike the endoplasmic reticulum, the sarcoplasmic reticulum stores and secretes Ca^+, the ion essential in muscle contraction.

The sarcoplasmic reticulum forms a sleeve around each myofibril, with enlarged *lateral sacs* that store Ca^{2+}. It is abundant in skeletal muscle cells and related to myofibrils. The membrane of the sarcoplasmic reticulum contains active pumps involved in moving calcium into the sarcoplasmic reticulum from the sarcoplasm. The sarcoplasmic reticulum contains specialized gates for calcium. Action potentials lead to depolarization of the sarcoplasmic reticulum membrane, leading to depolarization of the T-tubules. This opens the Ca^{2+} channels of the lateral sacs, causing the contraction to begin. To end contraction, Ca^{2+} is pumped back into the lateral sacs by active transport proteins, called plasma membrane Ca^{2+} ATPase (PMCA).

Sarcomeres with I and A bands, M and Z lines, and the H zone

Skeletal muscle cells have longitudinal bundles of myofibrils. Each myofibril consists of thin (actin) and thick (myosin) filaments, which repeat along the myofibril in units as sarcomeres. Organelles within the sarcomeres resemble the form and function of other types of eukaryotic cells. In the *H zone*, the central region of the sarcomere, there is no overlap between thin and thick filaments. The *M line*, in center of the H zone, links the center regions of thick filaments and divides the sarcomere vertically.

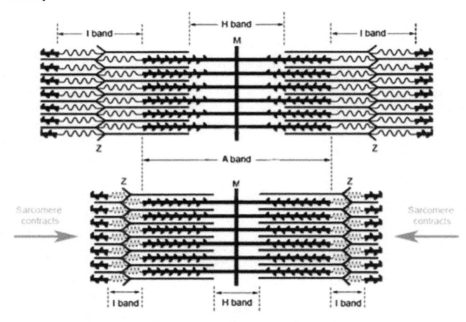

Sarcomere with associated regions

Each sarcomere has a band of thick filaments in the middle as the *A band*. Sarcomeres are flanked on both sides by thin filaments. Vertical borders between sarcomeres are *Z lines*, which anchor thin filaments.

I bands represent thin actin filaments, and *H bands* represent thick myosin filaments. Titin protein fibers from the Z line are linked to the M line and the thick filaments. Due to the banded pattern provided by thin and thick filaments, skeletal muscle is striated muscle.

Fiber type

Fibers within muscle tissue can be organized into fast and slow fibers. *Fast fibers* contain myosin with high ATPase activity and have high shortening velocity. *Slow fibers* contain myosin with low ATPase activity and have low shortening velocity. Fast fibers fatigue rapidly, while slow fibers fatigue gradually.

Oxidative fibers have numerous mitochondria and therefore a high capacity for oxidative phosphorylation. ATP production is dependent on oxygen. Oxidative fibers contain myoglobin, an oxygen-binding protein, which increases the rate of oxygen diffusion into the fiber. Myoglobin gives oxidative fibers a red color, as red muscle fibers.

Glycolytic fibers have few mitochondria but a high concentration of glycolytic enzymes and glycogen. Therefore, it is glycolysis, rather than oxidative phosphorylation, which fuels the contractions. These fibers are white muscle fibers due to their pale color. Glycolytic fibers can develop more tension than oxidative fibers because they are larger and contain more filaments, both thick and thin. However, they fatigue rapidly.

The three major types of skeletal muscle fibers, determined by their myosin ATPase activity and energy source, are 1) slow oxidative, 2) fast oxidative / glycolytic, and 3) fast glycolytic. Most muscles contain all three fiber types.

Slow oxidative fibers are one of the two main skeletal muscle fibers with abundant mitochondria and myoglobin. They generate energy predominantly through the aerobic conditions. Slow oxidative fibers twitch at a slow rate and are resistant to fatigue. The peak force exerted by these muscles is low. Slow muscle fibers have a lot of oxidative enzymes but are low in ATP activity.

Fast oxidative / glycolytic fibers can contract at a fast rate and produce a large peak force while being resistant to tiring even after many cycles. These fibers have a large amount of ATP activity and are high in oxidative and glycolytic enzymes. They are used for anaerobic activities that need to be sustained for a prolonged time.

Fast glycolytic fibers can exert a large force and contract at a fast rate. However, this comes at the expense of the fibers tiring easily. After a small amount of exertion, the muscle requires rest to recover. These fibers have low oxidative capacity while ATP and glycolytic activity is high. These fibers are used during anaerobic activity for short durations of time.

Abundant mitochondria in red muscle cells as ATP source

Mitochondria are abundant in muscle cells because of the need for a quick source of energy at any time. However, certain types of muscles have more mitochondria than others. *Type I muscle* (slow twitch or red muscle) contains more mitochondria than other types of muscle. Type I muscles are desirable for long-distance running. This type of muscle uses mitochondria to produce ATP from oxygen aerobically.

Type IIB muscles are common in weightlifters. These types of muscles require short bursts of energy and therefore are not as dependent on mitochondria. Type IIB muscles are mitochondria-poor and appear white. These types of muscles are more reliant on short bursts of glycolysis and therefore have greater stores of glycogen. While they can generate greater force than type I muscles, type IIB muscles experience muscle fatigue at a much faster rate.

Type IIA muscles are those intermediate between types I and IIB. They have fewer mitochondria than type I, but more than type IIB. Type IIA muscles fatigue less easily than type IIB but cannot replenish ATP as efficiently as type I. Type IIA muscle is often pink.

Red muscle (type I)	White muscle (type IIB)
High endurance, but slow	*Fast, but fatigue easily*
• Predominantly aerobic respiration	• Anaerobic respiration (glycolysis)
• Many mitochondria because red muscles undergo aerobic respiration	• Few mitochondria because white muscles mainly undergo glycolysis
• Equipped to receive abundant oxygen supply: many capillaries and much myoglobin	• Equipped for short bursts of glycolysis: store high amounts of glycogen

Long-distance runners typically have a greater percentage of red fibers than white fibers. Short-distance runners typically have a greater percentage of white than red fibers.

Many muscle cells rely on ATP to perform their function. Muscle cells contain myoglobin, which stores oxygen. Cellular respiration does not immediately supply the ATP needed. Since ATP availability is essential, the body has adapted additional mechanisms than only glycolysis to generate it. For example, when in need of ATP, muscle fibers rely on *creatine phosphate* (phosphocreatine), a stored form of high energy phosphate. Creatine phosphate does not directly participate in muscle contraction; however, it contributes by regenerating ATP rapidly: Creatine − P + ADP → ATP + Creatine.

When all creatine phosphate is depleted, and O_2 is limited, fermentation produces a small amount of ATP to compensate. Over time, however, this results in a buildup of lactic acid that leads to muscle fatigue due to oxygen debt. Lactic acid is transported to the liver, where 20% is broken down to CO_2 and H_2O via aerobic respiration. The ATP gained from this respiration is then used to reconvert 80% of the lactate to glucose. For example, those that train for marathons increase the number of mitochondria, allowing aerobic respiration for longer periods of time. Rigor mortis, the stiffness seen in a recently deceased corpse, occurs because nonliving organisms do not produce ATP. Therefore, the mechanisms behind muscle contraction are unable to allow the muscles to relax; muscles remain contracted until the enzymatic breakdown of cross-linking of actin and myosin filaments occurs.

Increased amounts of contractile activity (exercise) increases the size (hypertrophy) of muscle fibers and capacity for ATP production. Low-intensity exercise affects oxidative fibers, increasing the number of mitochondria and capillaries. High-intensity exercise affects glycolytic fibers, increasing their diameter by an increased synthesis of actin, myosin filaments, and glycolytic enzymes.

Muscle glycogen is the major fuel in the initial stages of exercise. After the initial stages, blood glucose and fatty acids are used. Muscle fiber generates ATP by one phosphorylation of ADP with the use of creatine phosphate, which is the source of ATP during the initial phase of contraction. This is followed by the slower pathways of second oxidative phosphorylation in mitochondria, or by glycolysis in the cytosol. At the end of muscle activity, creatine phosphate and glycogen levels are restored by energy-dependent processes, leading to a continued elevated level of oxygen consumption, *oxygen debt*, even after exercise finishes.

Important Functions

Support and mobility

The muscular system consists of contractile fibers held together by connective tissue. Muscle contraction can result in movement, stabilization of position, movement of substances throughout the body and generation of body heat. Muscles provide *support* for stabilizing joints, maintaining posture while sitting or standing. Muscles provide *mobility*. For example, skeletal muscles facilitate body movement, and smooth muscles move substances through the gut.

Peripheral circulatory assistance

The heart is a muscle that pumps blood. The pumping action of cardiac muscle in the heart causes blood to flow through blood vessels. Other muscles in the body provide peripheral assistance outside of the heart to help keep the circulatory system flowing. For example, contraction of the skeletal muscles of the diaphragm not only draws air into the lungs but squeezes on the abdominal veins to draw blood back to the heart.

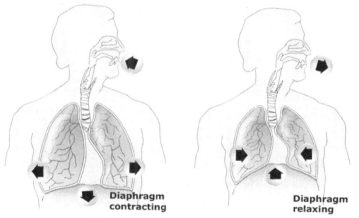

Contraction of the diaphragm: inhaling (left) and exhaling (right). During inhalation, the diaphragm contracts as the lungs expand. During expiration, the diaphragm relaxes as air is expelled from the lungs.

Thermoregulation by shivering reflex

Thermoregulation allows an organism to keep its body temperature within a certain range. One method of thermoregulation is the "*shivering reflex*," the generation of heat due to rapid skeletal muscle contractions (hydrolysis of ATP generates heat). These contractions are important for homeostasis to maintain body temperature in cooler environments.

Structural Characteristics

All muscles are contractile fibers. There are three major types, each with distinctive properties: skeletal muscle, cardiac muscle, and smooth muscle.

Comparison of muscle types:

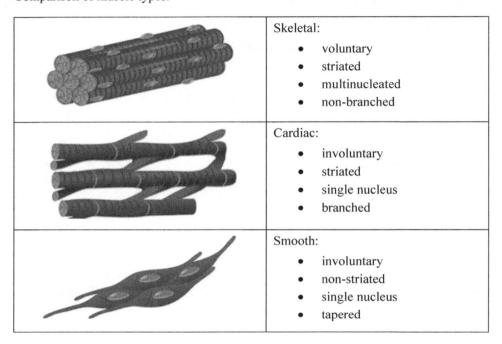

	Skeletal: • voluntary • striated • multinucleated • non-branched
	Cardiac: • involuntary • striated • single nucleus • branched
	Smooth: • involuntary • non-striated • single nucleus • tapered

Skeletal Muscle

Skeletal muscle cells are multinucleated. When observed under a microscope, skeletal muscle cells appear as long, non-branched fibers with *striations* or vertical stripes. The striations occur due to the presence of sarcomeres.

The *sarcomere*, an individualized, organized structure that allows for muscle contraction, is the basic functional unit of a myofibril. The myofibrils are contractile portions of fibers that lie parallel and run the length of the fiber. Sarcomeres communicate using *transverse tubules* (T-tubules). T-tubules penetrate the cell and make contact with, but do not fuse to, the *sarcoplasmic reticulum*. Ions are exchanged through the myofibril using this transverse system.

The ends of skeletal muscles attach to bones via tendons, allowing skeletal muscle contractions to move the skeleton. Skeletal muscle, under voluntary control, is responsible for the everyday movements of the skeleton.

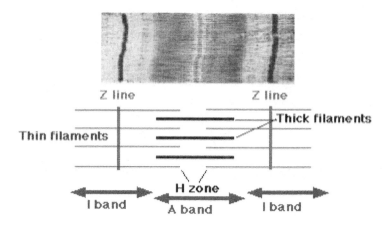

Sarcomere with electron micrograph showing striated pattern and illustration of thin (actin) and thick (myosin) filaments with associated A and I band and H zone

In a healthy state, skeletal muscle never fully relaxes to its full length. There is always *muscle tone* (some degree of tension) in the muscle, which provides important protection to the fibers. The muscle tone gives the muscle fibers a passive resistance to stretching. This is driven by the natural viscoelastic properties of the muscle fibers and a degree of alpha motor neuron activation (lower motor neurons of the brainstem and spinal cord). *Hypertonia* is abnormally high muscle tone that can result in brief spasms, prolonged cramps or constant rigidity.

A skeletal muscle with multinucleated myofibrils

Cardiac Muscle

Cardiac muscle has a striated appearance due to sarcomeres, and cells in cardiac muscle contain either one or two central nuclei. Unlike skeletal muscle, cardiac muscle is branched. Cardiac muscle is under involuntary control and is exclusively in the heart.

The *myocardial* (cardiac muscle) cells are separated by intercalated discs that have gap junctions to allow action potentials to flow through electrical synapses. Cardiac muscle has a high mitochondria concentration and is stimulated by autonomic innervation. Both cardiac muscle and smooth muscle are myogenic and are capable of contracting without stimuli from nerve cells. Some myocardial cells do not function in contraction; instead, they form the conducting system, which initiates the heartbeat and spreads it throughout the heart. Blood is supplied to cardiac muscle cells by coronary arteries and drained by coronary veins. Blood pumped through the chambers does not exchange substances with the cells of the heart muscle.

The vital cardiac muscle cells are innervated with a rich supply of sympathetic fibers, which release norepinephrine, and parasympathetic fibers, which release acetylcholine.

A cardiac muscle with branching between myocardial cells

Smooth Muscle

Smooth muscle fibers are composed of spindle-shaped cells with a single nucleus. Smooth muscle fibers are *nonstriated* because they do not have the highly-organized sarcomeres that form the basic units of skeletal and cardiac muscle. However, they have intermediate filaments involved in the sliding filament mechanism for contraction using myosin and actin. The thick (myosin) and thin (actin) filaments run diagonal to the long axis of the cell, and they are attached either to the plasma membrane or *dense bodies* as cytoplasmic structures. Like skeletal muscle, smooth muscle requires Ca^{2+} ions for contraction, which (like skeletal muscle) is released from the sarcoplasmic reticulum from inside the cell.

While cardiac and muscle cells do not generally divide in adult humans, smooth muscle cells maintain the ability to divide throughout life. They are controlled by the autonomic nervous system under involuntary control. For example, smooth muscle is in the lining of the bladder, the uterus, the digestive tract, and blood vessel walls.

A smooth muscle with spindle-shaped cells and a single nucleus

The plasma membrane of smooth muscles receives both excitatory and inhibitory inputs, and the contractile state of the muscle depends on the relative intensity of both. However, some smooth muscle fibers generate action potentials spontaneously. The potential change occurring during such spontaneous depolarizations is the *pacemaker potential*.

Unlike skeletal muscles, smooth muscles do not have motor end plates. The postganglionic autonomic neuron divides into branches in the smooth muscle fibers, with each branch containing a series of *varicosities* (swollen regions). The varicosities contain vesicles filled with a neurotransmitter, which is released from an action potential. The same neurotransmitter can produce excitation in one fiber and inhibition in another. Varicosities from a single axon may innervate several fibers, and a single fiber may receive signals from varicosities of both sympathetic and parasympathetic neurons.

Smooth muscle plasma membranes bind and respond to hormones. Paracrine agents, acidity, oxygen concentration, osmolarity, and ion composition can influence smooth muscle tension, providing a response mechanism to local factors.

Single unit smooth muscle (or visceral muscle) cells are connected by gap junctions, which are intracellular channels that allow the passage of molecules from cell to cell. Therefore, an action potential can influence many single-unit smooth muscle cells, causing a unified contraction. In single-unit smooth muscle, all the fibers undergo synchronous activity due to adjacent fibers being linked by gap junctions; an action potential occurring on any of the fibers propagates to the other cells. This allows the whole muscle to respond to stimulation as a single unit. Some of the fibers may consist of pacemaker cells, which can control the contraction of the entire muscle. However, a majority of the smooth muscle fibers consist of non-pacemaker cells.

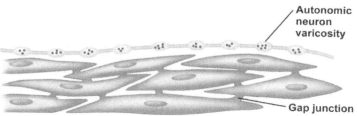

A single unit smooth muscle with autonomic neural innervation and gap junction

Multiunit smooth muscle is made of cells that contract independently from one another. This is because each multiunit smooth muscle cell is directly attached to a neuron. In addition to the neuronal response, smooth muscle cells can respond to hormones, changes in pH, O_2 and CO_2 concentration, temperature and ion concentration.

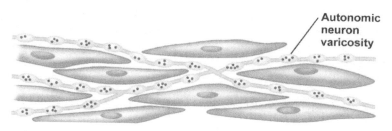

Multiunit smooth muscle which contracts independently

Mechanics of Muscle Contraction

Actin and Myosin Filaments

Under a microscope, a skeletal and cardiac muscle cell shows dark and light bands. An electron microscope shows that placement of protein filaments within sarcomeres form these striations of myofibrils. The light and dark bands correspond to two different types of proteins, actin, and myosin.

Actin filaments (thin filaments) are long actin protein chains in a spherical conformation that include the proteins troponin and tropomyosin, which wrap around the actin protein. Each thin myofilament is comprised of two chains that coil around each other. Actin molecules have a unique combination of strength and sensitivity. These molecules are constantly destroyed and renewed as needed so that they can contribute to the correct function of the muscle tissue. This is controlled by the ATP which is attached to each actin monomer. The state of the ATP determines the stability of each actin molecule.

Myosin filaments (thick filaments) consist of myosin protein molecules arranged in a bipolar structure. They are made of protruding club-like heads that lie towards the ends of the thick filaments, while their shafts lie towards the middle.

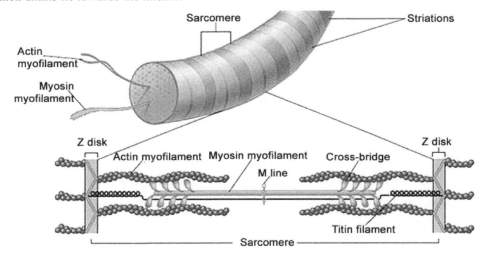

Skeletal and cardiac muscle have striated appearance from sarcomeres as the functional unit

Each myosin protein thick filament has a central bare zone and an array of protruding heads of opposite polarity at each end. Each myosin molecule possesses a tail which forms the core of the thick myofilament, while the head projects from the core filament. These myosin heads are cross bridges. Each myosin head has ATP binding sites and actin-binding sites. The myosin head contains a hinge at the point where it leaves the core of the thick myofilament to allow the head to swivel, and this swiveling action is the cause of any type of muscle contraction. The swiveling action occurs when actin combines with the myosin head, leading to the ATP associated with the head hydrolyzing ATP → ADP + energy. Each actin molecule contains a binding site for myosin; each myosin molecule contains a binding site for both actin and ATP.

Troponin and tropomyosin in muscle contraction

In a resting muscle fiber, tropomyosin, a long protein that spirals along actin, blocks the myosin head or the binding sites on actin, preventing the formation of cross-bridges with actin. An action potential releases Ca^{2+} from the sarcoplasmic reticulum of the muscle cell and increases the cytosolic concentration of Ca^{2+}. When Ca^{2+} binds to troponin, it shifts tropomyosin from the myosin-binding site and makes it possible for myosin cross bridges to attach with actin. Tropomyosin can be thought of as protecting actin's binding sites from the advances of the myosin head. In the presence of Ca^{2+}, troponin changes its conformation and moves tropomyosin away from its guard position, thereby allowing myosin heads to bind to actin and muscle contraction to ensue.

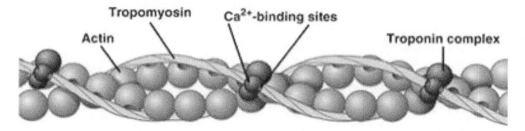

Myosin-binding sites blocked by tropomyosin on actin filament

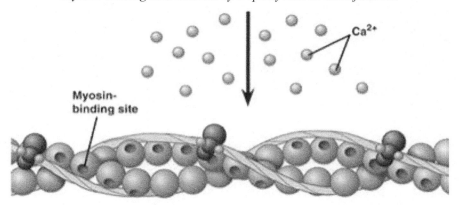

Myosin-binding sites exposed when Ca^{2+} binds to troponin, a conformation change in tropomyosin exposes the myosin binding sites on actin.

Cross Bridges

The overlapping thick and thin filaments form *cross bridges* from the thick myosin filaments. In a cross bridge, each thick filament is surrounded by a hexagonal array of six thin filaments, and each thin filament is surrounded by a triangular array of three thick filaments. Thick filaments contain the contractile protein myosin. Thin filaments contain the contractile protein actin, and troponin and tropomyosin proteins.

Power stroke of muscle contraction

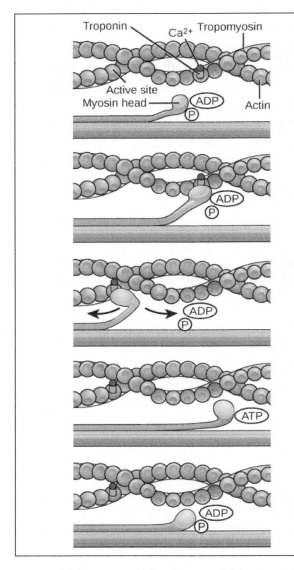

1 - The active site on actin is exposed as Ca^{2+} binds troponin and causes a conformational change in tropomyosin on actin.

2 – Movement of tropomyosin allows the myosin head forms a cross-bridge with actin.

3 – During the power stroke, the myosin head bends, the muscle shortens, and ADP and phosphate are released.

4 – A new molecule of ATP attaches to the myosin head, causing the cross-bridge from the myosin head to actin to detach.

5 – ATP hydrolyzes to ADP and phosphate, which returns the myosin head to the "cocked" position in preparation for the next contraction.

Without new ATP, the cross bridges remain attached between the myosin head and the actin filament. This explains why corpses are stiff; non-living bodies do not produce ATP. The strength of contraction of a single muscle fiber cannot be increased, but the strength of the overall contraction can be increased by recruiting more muscle fibers. The cross bridges break due to enzymatic activity in a deceased individual, which gives rise to cadaver movement.

Sliding Filament Theory

The sliding filament model proposes that a muscle fiber contracts, causing the sarcomeres within the myofibrils to shorten. As a sarcomere shortens, actin filaments slide past the myosin; the I band shortens, and the H zone disappears.

The swivel movement of the cross-bridges makes the overlapping thick and thin filaments slide past each other. Each actin and myosin fiber does not change in length during a contraction.

The *Z line* is the boundary of a single sarcomere and is involved in anchoring thin actin filaments. Visually under an electron microscope, it is a zigzag line on the sides of the sarcomere that connects the filaments of adjacent sarcomeres. The *M line* is a line of myosin in the middle of the sarcomere that is linked by accessory proteins. The *I band* is the region containing thin filaments (actin) only. The *H zone* is the region containing thick filaments (myosin) only. The *A band* is made of an actin end overlapping with a myosin end. The H zone and I band reduce during contraction, while the A band remains constant.

In the sliding filament model, cross-bridge forms and the myosin head bends, the power stroke. The power stroke causes actin to slide toward the M line, which allows the muscle fiber to contract. Next, ATP binds to the myosin head and is converted to ADP + Pi, which remains attached to the head. The release of myosin Ca^{2+} binds to troponin and results in a conformational change in tropomyosin.

Now, tropomyosin exposes myosin attachment sites, and cross bridges are formed between myosin heads and actin filaments. Subsequently, ADP + Pi are released, causing a sliding motion of actin to bring the sarcomere's Z lines together (contraction, power stroke). A new ATP molecule attaches to the myosin head, and the cross bridges unbind, resulting in sarcomere relaxation, as phosphorylation breaks the cross bridge. Though counterintuitive, ATP is not directly needed for the power stroke. ATP binding is needed for the detachment of the myosin head to actin and is needed to remove the power stroke by cocking the myosin head (in preparation of the next power stroke).

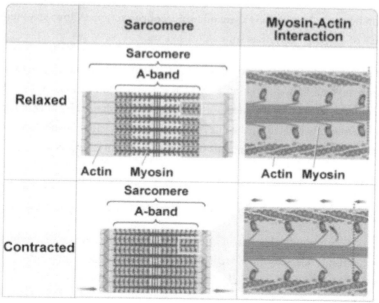

Sliding filament model: thick (myosin) and thin (actin) filaments

slide past each other during muscle contraction, shortening of the sarcomere

Calcium in sarcoplasmic reticulum for muscle contraction

Nerves generally stimulate muscle cells during muscle contraction. An action potential runs along the muscle cell membrane and deep into the muscle cell via T-tubules. This stimulates the *sarcoplasmic reticulum* (terminal cisternae) to release calcium ions. Calcium causes muscles to contract via the sliding filament mechanism.

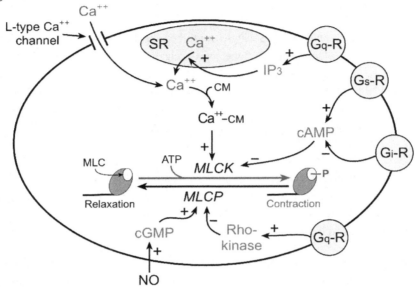

Smooth muscle stores Ca^{2+} in the sarcoplasmic reticulum and receive extracellular Ca^{2+} for muscle contractions. Overview of enzymatic cascade for Ca^{2+} and nitrous oxide.

Extracellular calcium enters the cell via the opening of Ca^{2+} channels in the plasma membrane. The rate of Ca^{2+} removal is slower, resulting in a muscle twitch. The concentration of Ca^{2+} in response to a single action potential is sufficient to activate a proportion of the cross bridges. Therefore, tension in smooth muscle can be graded by varying cytosolic Ca^{2+} concentration. A low-level tension is maintained, even in the absence of external stimuli as smooth muscle tone.

Cytosolic Ca^{2+} binds to the calmodulin protein. This newly formed calcium-calmodulin complex binds to the enzyme, myosin light-chain kinase (MLCK). The activated kinase uses ATP to phosphorylate the myosin cross bridges. The phosphorylated cross bridges bind to actin, resulting in muscle contraction. In response, myosin is dephosphorylated by the enzyme myosin light chain phosphatase (MLCP), antagonistic to MLCK. This enzyme inhibits contraction, thus causing smooth muscle relaxation.

Muscular Response

Muscle spindles monitor changes in muscle length and the rates by stretch receptors, which are present in modified muscle fibers as *intrafusal fibers*. The other fibers responsible for skeletal movement are *extrafusal fibers*. Stretching the muscle fires these muscle spindles, while the contraction of the muscle slows the firing.

Passive tension in a relaxed fiber increases with increased stretch due to the elongation of *titin filaments*. However, the maximum tension during contraction depends primarily on the levels of overlap between the thick and thin filaments, which depends on the *resting length* of muscles.

Two main factors allow for the highest maximum tension during contraction. *Increased resting length* increases the maximum tension, because it allows the thick and thin filaments more room to slide, allowing more cross-bridge cycling to occur. Additionally, the *decreased resting length* can increase the maximum tension. Tension can arise due to increased overlap between thick and thin filaments, allowing for greater interaction during contraction.

These two factors that allow for maximum tension in muscle are opposed: 1) increasing the resting length too much does not permit enough interaction between filaments, and 2) decreasing the resting length too much causes the thick filaments to hit the Z lines and stop contraction. Hence, there is an optimal resting length between the two extremes that result in the greatest possible maximum tension during contraction. Most fibers are near this length, l_0, and are relaxed.

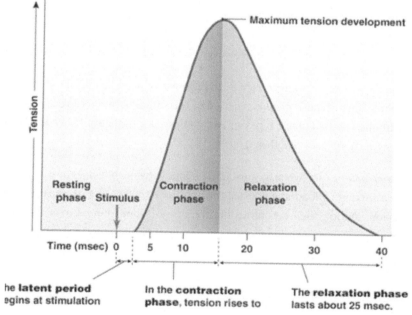

Muscle contraction with tension vs. time. The contraction phase produces a maximum force of tension before the relaxation phase returns muscle to the resting phase for a 40 msec twitch

Skeletal muscles are attached to the skeleton by tendons made of fibrous connective tissue. When muscles contract or shorten, they cannot inherently lengthen on their own. Instead, skeletal muscles work in antagonistic pairs for skeletal muscles to return to their original length. One muscle of an antagonistic pair bends the joint and brings a limb toward the body. The other straightens the joint and extends the limb. In this mechanism, the shortening of one muscle leads to the lengthening of another.

Force exerted on an object by a contracting muscle is *muscle tension*, while the force exerted on the muscle by the object (weight) is a *load*. These are opposing forces, and whether the exertion of force leads to a change in fiber length depends on the relative magnitudes of the tension and the load.

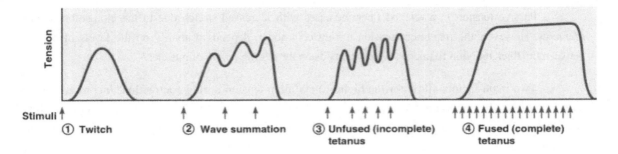

Summation occurs when contractions combine and become stronger and more prolonged (i.e., increase in muscle tension from successive action potentials). The summation can be temporal (successive in time) or spatial (separate in location).

Tetanus occurs when continuous contractions are sustained, and the muscle cannot relax. In tetanus, the rate of muscle stimulation is so fast that twitches blur into one smooth constant motion. If tetanus oscillates, it is unfused or incomplete tetanus, while tetanus without oscillations is fused or complete tetanus.

Tonus is a state of partial contraction. However, the muscle never completely relaxes.

A *twitch contraction* refers to the mechanical response of a single muscle fiber to a single action potential. They are not tetanic because there is a period of relaxation between separate stimuli, preventing the summation of stimuli. The two major types of contractions are isometric (fiber length is unchanged) and isotonic (fiber length shortens or lengthens).

Isometric contractions are when a muscle develops tension but does not change in length. It occurs when a muscle supports a load in a constant position or attempts to move a supported load that is greater than the tension. During such a contraction, the cross bridges bound to actin do not move. In isometric contractions, the *latent period* is the time between when the muscle receives the stimulus and when tension starts developing in the fiber. The following phase is the *contraction time*; the time required for the fiber to reach maximum tension. Lastly, the *relaxation period* releases the fiber's tension. The fiber rests until a new stimulus is received.

Isotonic contractions occur when the load remains constant while the fiber length changes. If the fibers shorten, the contraction is *concentric*. In such a contraction, the cross bridges bound to actin move, shortening the fibers. Before an isotonic shortening, there is a period of *isometric* contraction, when the tension of the fiber increases to meet the load and move it. In isotonic contractions, the latent period is the time between when the stimulus is received and when shortening takes place in the fiber. The time from the beginning of shortening to the maximum shortening is the contraction time. During the relaxation period, the fiber relaxes to a greater length. The fiber rests until a new stimulus is received. If the fibers lengthen, the isotonic contraction is an *eccentric contraction*. This occurs when the load on a muscle is greater than the tension, forcing the muscle to lengthen.

Isometric and isotonic twitches are measured in slightly different ways. Since isometric contractions do not change the muscle length, plots of this process are shown as tension versus time. Isotonic contractions involve a change in muscle length, so plots of isotonic contractions are shown as distance shortened versus time.

Isometric contractions *do not* move a load; they develop tension in a muscle fiber without any change in fiber length. Therefore, the isometric contractions do not perform physical work, and it is appropriate that they be compared to isotonic twitches. Isometric twitches have a *short latent period* and a *short contraction time*.

The components of isotonic twitches can be compared at loads of different forces. At a heavier load, a longer time is required in the isometric component of the twitch to build tension to meet the load. Thus, increasing the load leads to a longer latent period. Heavier loads decrease the shortening distance that a single stimulus can provide. Although the contraction time is slightly shorter with a heavier load, the decreased shortening distance leads to a slower velocity of contraction (or distance shortened) per unit time. The relaxation phase depends on the load helping the fiber return to original length; increased loads lead to a shorter relaxation phase.

Isotonic twitches *do* move a load and *do* involve a change in fiber length. Compared to isometric twitches, they have a long latent period and long contraction time. Even though eccentric isotonic contractions involve muscle lengthening over time, they involve temporary fiber shortening to give a controlled movement of the load. For example, slowly lowering one's body from a pull-up requires eccentric muscle effort from the latissimus dorsi, whereas quickly dropping from pull-up to a dead-hang position while maintaining a grip on the bar would not involve isotonic contraction from the lats (and would be painful).

Isometric twitches have a shorter contraction time but have a longer relaxation period, a *longer overall duration* compared to isotonic twitches. Isotonic twitches have a shorter relaxation period due to the presence of a load, which brings the muscle to its original length.

The contractile velocity of different muscle types

Contractile velocity is the force that a muscle creates and depends on the length, volume, fiber type and shortening velocity of the muscle.

Muscle cross-sectional area (CSA) divides the volume by the length.

$$CSA = V / l$$

Physiological CSA $= m \cdot \cos \theta / l \cdot \rho;$ where m is muscle mass, θ is the fiber angle, l is the length of the fiber, and ρ is muscle density.

Short fibers have a higher physiological muscle volume per unit of muscle mass. There is greater force produced, and it shortens at slower speeds.

Long fibers have a lower physiological muscle volume per unit of muscle mass, meaning there is a lower force production, and it shortens at faster speeds.

Type 1 fibers have a slow increase in force and a low force production. They have a smaller diameter, which leads to slower contraction.

Type IIA fibers have a fast contraction and a faster force production.

Type IIB fibers have a fast contraction, faster increase in force, and large force production. The type IIB fibers are fatigued quickly, and maintaining that force is difficult.

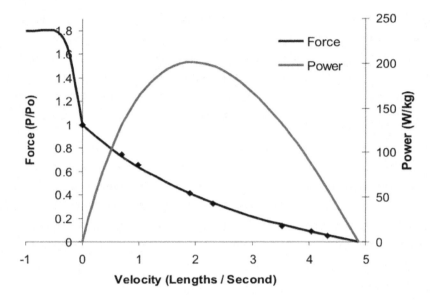

In the force-velocity relationship, changes in muscle lengths affect force generation. Force decreases with increasing contractile velocity.

Regulation of Cardiac Muscle Contraction

Cardiomyocytes (cardiac muscle cells) coordinate contractions, regulated through intercalated discs, which spread action potentials to support the synchronized contraction of the myocardium. The contraction in a cardiac muscle, as well as in smooth and skeletal muscles, occurs via *excitation-contraction coupling* (ECC). ECC is where an electrical stimulus originating in a neuron is converted into a mechanical response. In muscle cells, the electrical stimulus is the action potential, and the desired mechanical response is a contraction.

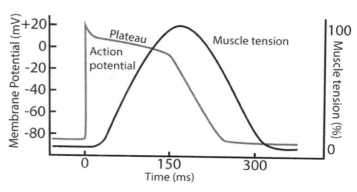

In particular cardiac muscle, ECC is dependent on calcium-induced calcium release. This involves the conduction of calcium ions into the cell, triggering further release into the cytoplasm. Like skeletal muscle, the initiation and upshoot of the action potential in ventricular cardiac muscle cells are derived from the entry of sodium ions across the sarcolemma in a regenerative process. However, in cardiac muscle, an inward flux of extracellular calcium ions through the calcium channels on the T-tubules sustains the depolarization of cardiac muscle cells for a longer duration. Contraction in cardiac (like skeletal) muscle occurs via the sliding filament model. The contraction pathway for cardiac muscle is summarized in five steps.

1. An action potential, induced by pacemaker cells, is conducted to contractile cardiomyocytes through intercalated discs (gap junctions).

2. As the action potential travels between sarcomeres, it activates the calcium channels in the T-tubules, resulting in an influx of calcium ions into the cell along with the release of Ca^{2+} from the sarcoplasmic reticulum.

3. Calcium in the cytoplasm then binds to cardiac troponin-C, which moves the troponin complex from the actin binding site. This movement of the troponin complex frees the myosin to bind actin and activates contraction.

4. The myosin head pulls the actin filament towards the center of the sarcomere, contracting the muscle.

5. Intracellular calcium is then removed by the sarcoplasmic reticulum, dropping the intracellular calcium concentration, returning the troponin complex to its inhibiting position on the actin and effectively ending contraction.

Neural Control

Motor neurons

Motor neurons are *efferent neurons* that send signals to muscles and organs. Motor neurons have cell bodies in the CNS and contain large, myelinated axons that can propagate action potentials at high velocities. *Autonomic motor neurons* control the sympathetic and parasympathetic branches of the nervous system, while *somatic motor neurons* control skeletal muscles. Muscle fibers of single muscle do not all contract at once. A single motor neuron innervates multiple muscle fibers, collectively a *motor unit*. Motor units have varying amounts of muscle fibers. Usually, smaller motor units are activated first, then larger ones are activated as needed. This leads to smooth increases in force. The fine movement uses smaller motor units.

The total tension a muscle develops depends on the tension in each fiber and the number of fibers contracting at a particular time. The number of fibers contracting depends on the recruitment (or activation) of these muscle fibers.

Sensory neurons, or *afferent neurons*, are the opposite of motor neurons. Afferent fibers from receptors can take one of four pathways. Some fibers go back directly to motor neurons of the same muscle without the interposition of any interneurons, as *monosynaptic stretch reflex arcs*. Some fibers end on interneurons that inhibit the antagonistic muscles, as *reciprocal innervation*. Some fibers activate motor neurons of synergistic muscles. Lastly, some fibers continue to the brainstem.

A *motor program* is the pattern of neural activities required to perform a movement. It is created and transmitted via neurons that are organized hierarchically and is continuously updated. A skill can be learned if the program is repeated frequently enough. Local control of motor neurons is important in keeping the motor program updated by gathering information from local levels through afferent nerve fibers.

Alpha motor neurons are the larger motor neurons that control the extrafusal fibers, which are responsible for skeletal movement. *Gamma motor neurons* are the smaller motor neurons that control the intrafusal fibers. The neurons in these intrafusal fibers are excited or co-activated with the neurons in the extrafusal fibers to get continuous information about muscle length.

Withdrawal reflex is a stimulus that activates flexor motor neurons and inhibits extensor motor neurons to move the body away from an external stimulus. The effect is produced on the same (ipsilateral) side of the body where the stimulus arose. An opposite effect (*crossed extensor reflex*) may be produced on the other side, the contralateral side, to compensate for any lost support due to the withdrawal.

Interneurons are synapses that integrate inputs from both higher centers and peripheral receptors. Afferent inputs to local interneurons bring information such as the tension of muscles or movement of joints, which influence movements. *Golgi tendon organs* are receptors located in tendons to monitor the tension of a muscle.

Complex muscular activities (e.g., maintenance of posture and balance) require a carefully coordinated effort from several muscles. The afferent vestibular apparatus pathways of the eyes and the somatic receptors must first relay sensory information to the brain centers. The information is then compared with an internal representation of the body's geometry, and corrections to skeletal muscles are made through alpha motor neurons in the efferent pathways.

Walking is a coordinated effort of multiple muscles. On one leg, extensor muscles are activated to support the body's weight. At the same time, contralateral extensors are inhibited through reciprocal inhibition, allowing flexors to swing the non-supporting leg forward.

Neuromuscular junctions and motor end plates

Muscles contract when stimulated by motor nerve fibers. Before a signal crosses the neuromuscular junction, an action potential must reach the axon terminal. The axon terminals of a motor neuron contain *acetylcholine* (ACh) vesicles. Neurotransmitters are released into the synapse, where they travel across the gap and reach receptors on the motor end plate, which is a part of the sarcolemma (sheath around fibers of skeletal muscle) on the muscle cell. A graded potential is created across this junction, and if the potential reaches a threshold, an action potential is created.

The action potential then travels down the sarcolemma and causes the muscle to contract. An action potential opens Ca^{2+} channels in the nerve plasma membrane, allowing Ca^{2+} to diffuse in and enable neurotransmitter vesicles to fuse with the plasma membrane. After the vesicles fuse, they release ACh into the extracellular space. ACh opens ion channels in the motor end plate and produces a depolarization as *endplate potential* (EPP). The enzyme acetylcholinesterase on the motor end plate degrades ACh to close the ion channels and return the plate to resting potential.

A *neuromuscular junction* is the meeting of a nerve (motor end axon terminal) with a muscle at the motor end plate. The *motor end plate* is the sarcolemma that synapses with the motor neuron and has receptors for the neurotransmitters. When nerve impulses travel down a motor neuron to the axon bulb, ACh-containing vesicles merge with the presynaptic membrane and ACh is released into the synaptic cleft. ACh rapidly diffuses to and binds with receptors on the sarcolemma. The sarcolemma generates impulses spreading down the T-tubule system into the sarcoplasmic reticulum, where it triggers the release of Ca^{2+} ions from the sarcoplasmic reticulum. The Ca^{2+} ions bind to troponin to initiate muscle contraction.

Sympathetic and parasympathetic innervation

Both the sympathetic and parasympathetic systems contain motor neurons that innervate involuntary muscles.

The *sympathetic nervous system* controls the "fight or flight" responses, when heart rate and blood pressure increase, the pupils dilate, less blood is directed to the digestive system, and more blood is directed to the muscles. The *parasympathetic nervous system* controls the "rest and digest" responses. Contrary to the sympathetic responses, heart rate, and blood pressure decrease, the pupils constrict, and more blood is directed to the digestive system rather than the muscles.

Voluntary and involuntary muscles

Voluntary muscles are consciously controlled muscles with cylindrical fibers. These muscles are generally attached to bones (i.e., skeletal muscles), and the brain is involved in the movement of these muscles. An example of voluntary muscles is the biceps in the upper arm.

Involuntary muscles cannot be consciously controlled. They are spindle-shaped fibers associated with the autonomic nervous system. Smooth muscles of the gut and cardiac muscles of the heart are involuntary muscles.

Contractions of voluntary muscles are usually rapid and forceful, while contractions of involuntary muscles are usually slow and rhythmic. Motor behaviors are a continuum of these two types of contractions, having components of both voluntary and involuntary muscles to differing degrees.

Oxygen Debt: Fatigue

Muscle fatigue occurs when there is a decline in muscle tension from the previous contractile activity. A fatigued muscle has decreased shortening velocity and a slower rate of relaxation. Onset and rate of fatigue depend on the type of skeletal muscle and the duration of contractile activity. If a fatigued muscle is allowed to rest, it recovers. The rate of recovery depends on the duration and intensity of the previous exercise. Fatigue is not due to low ATP; a fatigued muscle still has a high concentration of ATP but is an adaptation to prevent the rigor that results from a low ATP level.

High-frequency fatigue accompanies high-intensity, short-duration exercise is due to failure in the conduction of action potential in the T-tubule. Recovery from such fatigue is rapid. Low-frequency fatigue (low-intensity, long-duration exercise) is due to the buildup of lactic acid, which changes the conformation of muscle proteins. Recovery from such fatigue is slow. The basic molecular mechanism for muscle fatigue is: continuous synaptic activity causes depletion of the required neurotransmitter, leading to fatigue.

Please, leave your Customer Review on Amazon

Chapter 20

Skeletal System

- **Skeletal System Functions**

- **Skeletal Structure**

- **Bone Structure**

- **Endocrine Control**

Skeletal System Functions

Structural rigidity and support

The bones form the rigid framework of the body. The strength of the bones allows for movement of the muscles that are anchored to the bone. For example, the large leg bones support the body against the pull of gravity, the leg and arm bones permit flexible body movement, the pelvis bones support the trunk and the atlas (the 1st vertebra) supports the skull.

Calcium storage

Bones store many essential minerals needed to sustain life (e.g., calcium, phosphorous and others). Bones are critical in maintaining calcium homeostasis. When blood calcium is low, parathyroid hormones (PTH) signal osteoclasts of bone to break down and release calcium. Most of the Ca^{2+} in the body is stored in the bone matrix as *hydroxyapatite*.

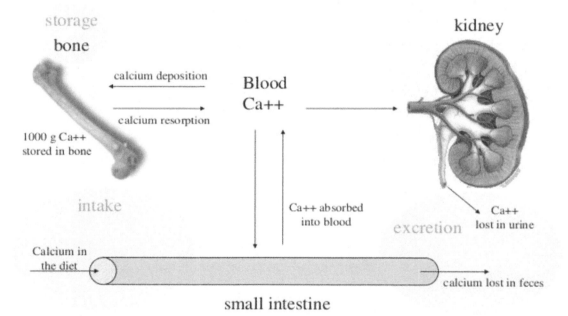

Calcium homeostasis with calcium stored in bone

Physical protection

One of the functions of the skeletal system is to protect the internal organs. The rib cage, in particular, protects delicate internal organs, including the heart and the lungs. The skull protects the brain, and the spine protects the spinal cord. Additionally, many large bones shelter bone marrow, which contains stem cells needed to produce blood.

Skeletal Structure

Basic anatomy of the skeletal system

The skeletal system is made of three main components: bone, cartilage, and joints. *Bone* is a tough, rigid form of connective tissue that gives the human skeleton its strength. *Cartilage* is a form of connective tissue but is not rigid and tough in comparison to the bone. *Joints* are where bones connect, and they allow for the skeletal system to be mobile.

The human vertebrate skeletal system is divided into the axial (midline) skeleton and appendicular skeleton.

The *axial skeleton* lies at the midline of the body and consists of the skull, vertebral column, sternum, and rib cage.

The cranium and the facial bones form the skull. Additionally, newborns have membranous junctions as *fontanels* that usually close by the age of two. The bones of the cranium contain *sinuses*, air spaces lined with mucous membranes that reduce the weight of the skull. Sinuses give a resonant sound to the voice. Two mastoid sinuses drain into the middle ear; *mastoiditis* is an inflammation of these sinuses that can lead to deafness.

The cranium is composed of eight bones: one frontal, two parietal, one occipital, two temporal, one sphenoid and one ethmoid bone.

The spinal cord passes through the *foramen magnum*, an opening at the base of the skull in the *occipital bone*. Each *temporal bone* has an opening that leads to the middle ear. The *sphenoid bone* completes the sides of the skull and forms the floors and walls of the eye sockets. The *ethmoid bone*, in front of the sphenoid, is part of the orbital wall and is a component of the nasal septum.

There are fourteen facial bones, including one mandible, two maxillae, two palatine, two zygomatic, two lacrimal, two nasal and one vomer. The *mandible* bone (lower jaw) is the movable portion of the skull; it contains tooth sockets. The *maxilla* bone forms the upper jaw and the anterior of the hard palate; it contains tooth sockets. The *palatine* bones make up the posterior portion of the hard palate and the floor of the nasal cavity. The *zygomatic* bone gives the cheekbones their prominence. *Nasal* bones form the bridge of the nose. Other bones make up the nasal septum, which divides the nasal cavity into two regions. The ears are elastic cartilage and lack bone, whereas the nose is a mixture of bone, cartilage, and fibrous connective tissue.

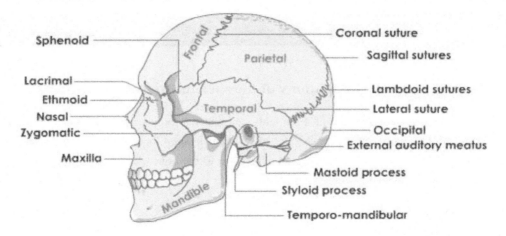

The *vertebral column* supports the head and trunk and protects the spinal cord and roots of the spinal nerves. The vertebral column serves as an anchor for the other bones of the skeleton. There are seven *cervical vertebrae* located in the neck. The twelve *thoracic vertebrae* are in the thorax (or chest). The *lumbar vertebrae* are in the small of the back. One *sacrum* is formed from five fused *sacral vertebrae*. One *coccyx* is formed from four fused *coccygeal vertebrae*.

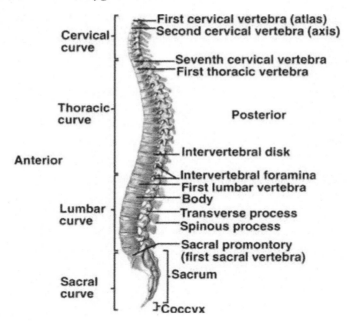

Normally, the spinal column has four normal curvatures that provide strength and resiliency in posture. *Scoliosis* is an abnormal sideways curvature; *hunchback* and *swayback* are abnormal conditions of the spinal column. *Intervertebral discs* between the vertebrae act as padding to prevent the vertebrae from grinding against each other and to absorb shock during physical activities. Intervertebral discs weaken with age. In contrast, *vertebral discs* allow for motion between vertebrae for movement such as bending forward.

The *rib cage* protects the heart and lungs, yet is flexible enough to allow breathing. All twelve pairs of ribs connect directly to the *thoracic vertebrae* in the back; seven pairs of ribs attach directly to the sternum, and three pairs connect to the sternum indirectly via cartilage at the front of the sternum.

There are two pairs of ribs (floating ribs) that are completely unattached to the sternum and only attach to the vertebrae.

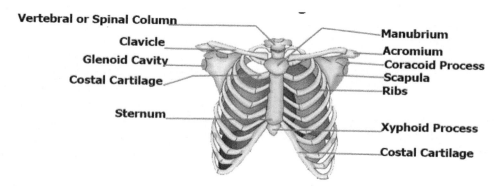

Vertebral or Spinal Column

Clavicle

Glenoid Cavity

Costal Cartilage

Sternum

Manubrium

Acromium

Coracoid Process

Scapula

Ribs

Xyphoid Process

Costal Cartilage

Rib cage with twelve pairs of ribs

The *appendicular skeleton* consists of the bones within the pectoral girdle, the pelvic girdle, and the upper and lower limbs.

The *pectoral girdle* is specialized for flexibility and built for strength. Ligaments loosely link the components of the pectoral girdle. The *clavicle* (collarbone) connects with the *sternum* in the front and the *scapula* in the back. The scapula is held in place by muscles and can move freely. The *humerus* is the long bone of the upper arm; its rounded head fits into a socket of the scapula. The *radius* is the more lateral of bones in the lower arm (i.e., the forearm); it articulates with the humerus at the elbow joint (a hinge joint), and the radius crosses in front of the ulna for easy twisting. The *ulna* (elbow bone) is the more medial of the two bones in the lower arm. The larger end of the ulna joins with the humerus to make the elbow joint.

The flexibility of the hand is attributable to the presence of many bones. The wrist has eight *carpal bones* which look like small pebbles. Five *metacarpal bones* fan out to form the framework of the palm. The *phalanges* are the bones of the fingers and thumb.

The *pelvic girdle* consists of two heavy, large coxal (hip) bones. The *coxal bones* are anchored to the *sacrum*; together with the sacrum, they form a hollow cavity that is wider in females than in males. It transmits weight from the vertebral column via the sacrum to the legs.

The *femur* is the largest, longest and strongest bone of the body; however, it is limited in the amount of weight that it can support. The *patella* is the kneecap and is a thick, roughly triangular bone that allows for knee extension. The *tibia* has a ridge called the "shin"; its end forms the inside of the ankle. The shin bone is strong. The *fibula* is the smaller of the two bones; its end forms the outside of the ankle.

There are seven *tarsal bones* in each ankle. In standing or walking each tarsal bone receives the weight and passes it to the heel and ball of the foot. The *metatarsal bones* form the arch of the foot and provide a springy base. The *phalanges* are the bones of the toes, which are stouter than those of the fingers.

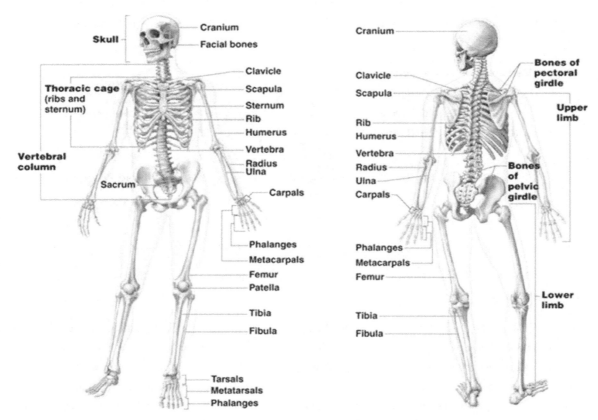

Human adult skeletal system: anterior view (left) and posterior view (right)

Specialization of bone types and structures

Bones are classified by their shape.

Long bones consist of a *diaphysis* (shaft) with two ends that are shaped like rods. Examples include the thigh (femur), upper arm (humerus), and finger bones (phalanges).

The *diaphysis* of the bone consists of a central medullary cavity that is filled with *fatty tissue* (yellow bone marrow), which is surrounded by a thick collar of compact bone. Yellow marrow in the shaft of long bones has fat molecules and serves as an important energy reservoir.

The *epiphysis* (expanded end of the bone) consists mainly of spongy bone surrounded by a thin layer of compact bone. *Spongy bone* has numerous plates separated by irregular spaces. Spongy bone is lighter but designed for strength. The solid portions of the bone follow the lines of stress. Bone spaces are often filled with *hematopoietic tissue* (red bone marrow), which is specialized tissue that produces blood cells. The epiphysis contains *articular cartilage*, which is a pad of hyaline cartilage. This is where long bones articulate, or join, and acts as a "shock absorber."

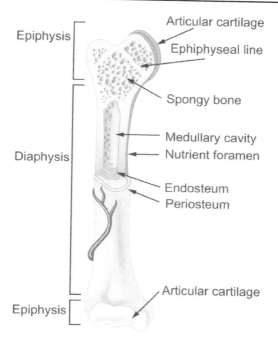

Epiphysis

Articular cartilage

Ephiphyseal line

Spongy bone

Medullary cavity

Nutrient foramen

Endosteum

Periosteum

Diaphysis

Articular cartilage

Epiphysis

Long bone with central diaphysis and terminal epiphysis

The *epiphyseal line*, or the remnant *epiphyseal disc* or *plate*, is cartilage at the junction of the diaphysis and the epiphysis is the growth plate.

The *periosteum* (outer fibrous) protective covering of the diaphysis is richly supplied with blood vessels, lymph vessels, and nerves. It allows for the insertion of tendons and ligaments to the bone. The *nutrient foramen* is a perforating canal that allows blood vessels to travel in and out of the bone.

Osteons are the functional unit of *compact bone*. The bone cells are located in small chambers as lacunae, which are arranged in concentric circles around central canals. *Lacunae* are separated by a matrix that contains protein fibers of collagen and mineral deposits. The *osteogenic layer* contains the *osteoblasts,* the bone-forming cells, and the *osteoclasts,* the bone-destroying cells. The *endosteum layer* is the inner lining of the medullary cavity and contains a layer of both osteoblasts and osteoclasts.

Short bones are cube-like. The outer surface is a thin layer of compact bone and internally contains spongy bone. The short bones are located in the hands and feet. They are predominantly made of spongy bone. They contain a tubular shaft and articular surfaces at each end but are much smaller than a long bone. The articulations that they are joined by allows for increased flexibility and decreased mass. These bones provide stability and strength. Examples include wrist (carpals) and bones (tarsals) bones.

Flat bones are thin and usually curved with an outer layer of periosteum-covered compact bone that surrounds an inner core of endosteum-covered spongy bone. Essentially, the spongy bone is sandwiched between two layers of thin, compact bone. The spongy bone within the epiphyses of flat bone and long bone contains hematopoietic tissue (red marrow). Their structure provides a flat and broad surface area for tendon attachment. The strong structure of these bones offers protection to many internal organs (e.g., most skull bones, breastbone (sternum), shoulder blades (scapulae), and the ribs).

Irregular bones are not long, short or flat. Their complicated shape is due to their specialized function of providing mechanical support for the body and protecting the spinal cord. Their structure consists of a thin layer of compact bone with an internal component of spongy bone (e.g., the vertebrae, hips and auditory ossicles).

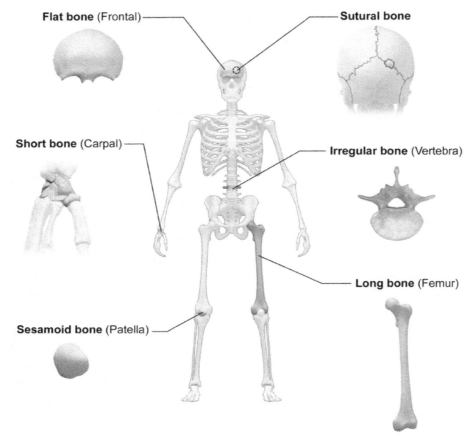

Bone classification by shape

Sesamoid bones develop within a tendon. Since they act to hold the tendon further away from the joint, the angle of the tendon is increased, and thus the force of the muscle is increased (e.g., patella and pisiform).

Wormian bones (sutural bones) are small bones that lie within the major skull bones. These are irregular isolated bones that appear in addition to the usual centers of ossification of the cranium. They are predominantly in the lambdoid suture (posterior aspect of the skull), which is more tortuous than other sutures.

Cartilage: structure and function

Cartilage is an avascular connective tissue that has a dense matrix of collagen and elastic fibers embedded in a rubbery ground substance. The matrix is produced by *chondroblasts* cells that become embedded in the matrix as *chondrocytes* (mature cartilage cells). They occur, either singly or in groups, within spaces of *lacunae* (sing. *lacuna*) in the matrix. There are three types of cartilage: hyaline cartilage, fibrocartilage, and elastic cartilage.

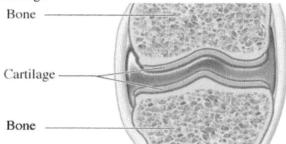

Bone

Cartilage

Bone

Cross section of a joint with cartilage cushioning the apposition of two bones

The cartilage cells secrete into the extracellular matrix, which contains fiber meshwork that gives the cartilage its characteristic properties of flexibility and resilience. The surface of most cartilage in the body is surrounded by a membrane of dense, irregular connective tissue as *perichondrium*. Unlike other connective tissues, cartilage contains no blood vessels or nerves.

Cartilage is softer and more flexible than bone. For example, the ear, nose, larynx, trachea, and joints are made of cartilage. It possesses the properties of compressibility and resilience (ability to resume its original shape after deformation), which is important in locations such as the ends of bones in joints, knees and between vertebrae.

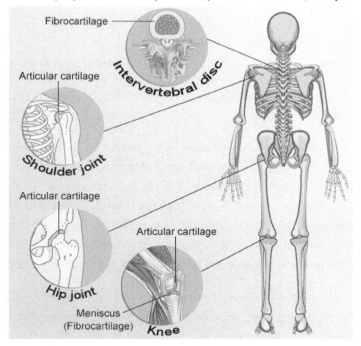

Hyaline cartilage is the most abundant of the three types. Hyaline cartilage consists of a bluish-white, shiny, ground elastic material with a matrix of chondroitin sulfate into which many fine collagen fibrils are embedded. Mesenchyme tissue initiates the formation of chondrocytes, which then produce collagen. Collagen is present in tissue as a triple helix with hydroxyproline and hydroxylysine, ground substance and elastin fibers.

Hyaline cartilage covers the surface of bones at joints (especially in osteoarthritis-prone areas that are vulnerable to damage due to wear), including the ends of long bones and the anterior ends of the ribs. It facilitates smooth movements at joints, provides flexibility and support, reduces friction and absorbs shock in joints. Hyaline cartilage in the embryonic skeleton provides smooth surfaces, enabling tissues to move and slide easily over each other. For example, hyaline cartilage is in bronchi, bronchial tubes, costal cartilages, the larynx (voice box), the nose and the trachea.

Fibrocartilage (meniscus) is a tough form of cartilage that consists of chondrocytes scattered among clearly visible dense bundles of collagen fibers within the matrix. Fibrocartilage lacks a perichondrium. Fibrocartilage tissue provides support and rigidity to attached and surrounding structures and is the strongest of the three types of cartilage. An example of fibrocartilage is *calli*, which is the tissue formed between the ends of the bone at the site of a healing fracture. When there is a blood clot, granulation tissue forms into cartilage, and then eventually into full-fledged bone. Other examples include intervertebral discs (between the vertebrae of the spine), the menisci (cartilage pads of the knee joint), the pubic symphysis (hip bones join at the front of the body), and in the portions of the tendons that insert into cartilage tissue, especially at the joints.

Knee bones are a capped crescent-shaped piece of meniscus cartilage. The knee joint contains 13 fluid-filled sacs as *bursae*, which ease the friction between the tendons, ligaments, and bones. Inflammation of the bursae (*bursitis*) is the cause of "tennis elbow."

Elastic cartilage is yellowish with cartilage cells (chondrocytes) located in a threadlike network of elastic fibers within the matrix of the cartilage. A perichondrium is present. Elastic cartilage provides support to surrounding structures and helps define and maintain the shape of the area in which it is present (e.g., external ear). Other examples include the auditory (Eustachian) tubes, external ear (the auricle) and the epiglottis (flap on the larynx).

Joint structures

A *joint* is where bones meet each other. Joints connected to bones are classified as synovial, fibrous or cartilaginous.

Synovial joints (mobile joints) have a fluid-containing cavity that lubricates bone movement. They are usually involved with bones that move relative to each other. Most joints are synovial joints, with the two bones separated by a cavity (e.g., the carpals, wrist, elbow, humerus and ulna, shoulder, hip joints, and knee joints). Synovial joints are subject to arthritis. In osteoarthritis, the cartilage at the ends of bones disintegrate, and the bones then become rough and irregular from mechanical "wear and tear." In rheumatoid arthritis, the synovial membrane becomes inflamed and thickens. The joint degenerates and becomes immovable and painful. This is likely caused by an autoimmune reaction.

Types of synovial joints include ball and socket (e.g., shoulder and hip), hinge (e.g., fingers, elbow), gliding (e.g., scaphoid and lunate bones of the wrist) and immobile (e.g., plates of the skull and rib-to-sternum connection). The ball and socket joint allows the most freedom of motion.

Fibrous joints (immovable) connect bone to bone with cartilage (or fiber). An example of an immovable joint is a suture, which is usually holding together the bones of a skull. Examples include the skull, pelvis, spinous process and the vertebrae.

Cartilaginous joints (slightly moveable) include joints between the vertebrae, spine, and ribs. For example, the two hipbones are slightly movable, because they are ventrally joined by cartilage and respond to pregnancy hormones.

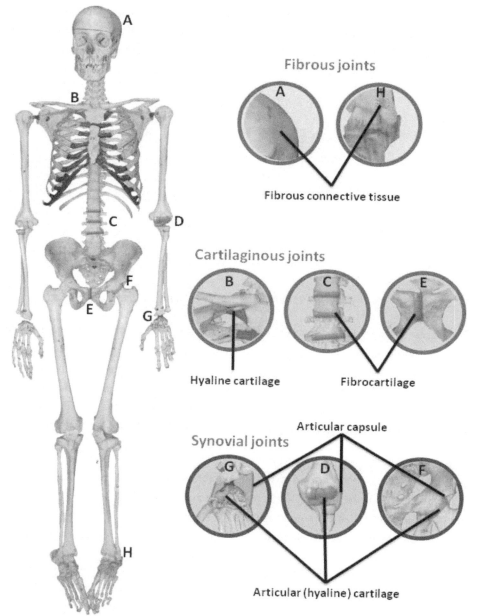

The three types of joints are synovial, fibrous and cartilaginous, with examples of cartilage and connective tissue. Note the reference to examples and locations on the skeletal structure.

Ligaments and tendons

Ligaments and tendons are both soft collagenous tissues. *Ligaments* connect bone to bone, while *tendons* connect muscle to bone. Ligaments and tendons play a significant role in musculoskeletal biomechanics, and they represent an important area of orthopedic treatment in which many medical challenges remain. A challenge is restoring the normal mechanical function of these tissues. Ligaments and tendons have a hierarchical structure that affects their mechanical behavior. Also, ligaments and tendons can adapt to changes in their mechanical environment due to injury, disease or exercise.

Unlike bone, there are not as many quantitative structure-function relationships for ligaments and tendons. For one, the hierarchical structure of ligaments and tendons is much more difficult to quantify than bone. Secondly, the ligaments and tendons exhibit both nonlinear and viscoelastic behavior even under physiologic loading, which is more difficult to analyze than the linear behavior of bone.

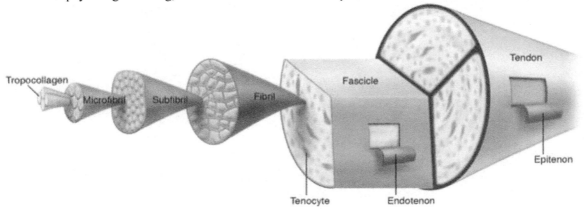

Ligaments and tendons have a hierarchical structure

The largest structure for these soft collagenous tissues is the ligament (or tendon). The ligament (or tendon) then is split into smaller entities as *fascicles*. The fascicle contains the basic *fibril* of the ligament (or tendon), and the *fibroblasts*, which are the biological cells of the ligament (or tendon). There is a structural characteristic at this level that plays a significant role in the mechanics of ligaments (and tendons): the crimp of the fibril. The *crimp* is the waviness of the fibril; this contributes significantly to the nonlinear stress-strain relationship for ligaments, tendons and essentially all soft collagenous tissues.

Ligaments connect bone to bone, forming a *joint capsule*, which stabilizes and strengthens the joints. The joint capsule is lined with a synovial membrane that produces lubricating synovial fluid. Ligaments are made of dense bundles of connective tissue made of collagenous fibers, which are surrounded and protected by dense irregular connective sheaths. Blood is supplied to ligaments through microvascularity from insertion sites to supply the nutrition needed for growth, matrix synthesis, and repair.

Capsular ligaments are a part of the articular capsule that surrounds synovial joints. They act as mechanical reinforcements. *Extra-capsular ligaments* join with other ligaments and provide joint stability. *Intra-capsular ligaments* are less common and promote stability but allow a larger range of motion. *Cruciate ligaments* occur in pairs of three.

The collagen fibrils in ligaments have slightly less volume and organization compared to the fibrils in tendons. Ligaments have a higher percentage of proteoglycan matrix than tendons. Fibroblasts are present in ligaments. In "double-jointed" individuals, the ligaments are unusually loose, allowing them to stretch their ligaments more than average.

A *tendon* is a tough band of fibrous connective tissue that can withstand tension. Tendons connect muscle to bone at moveable joints and anchor the muscle. Tendons contain collagen fibrils (Type I), a proteoglycan matrix and fibroblasts arranged in parallel rows. Type I collagen constitutes about 86% of dry tendon weight; glycine (~33%), proline (~15%) and hydroxyproline (~15%, to identify collagen because it is almost unique to it). Tendons carry tensile forces from muscle to bone, and they carry compressive forces when wrapped around bone like a pulley. The tendons procure blood through the vessels in the *perimysium*, a sheath of connective tissue which covers the tendon, through the *periosteum*, the membrane covering the outer surface of the bone, and from the surrounding tissues.

Origin is the point of attachment of the muscle to stationary bone, and *insertion* is a point of attachment of the muscle to a bone that moves.

Viscoelasticity is another important aspect of ligament and tendon behavior, which indicates time-dependent mechanical behavior. Thus, the relationship between stress and strain is not constant but depends on the time of displacement (or load). There are two major types of behavior characteristic of viscoelasticity. *Creep* is increasing deformation under constant load. This contrasts with an elastic material which does not exhibit increased deformation no matter how long the load is applied. *Relaxation stress* is the second significant behavior, as the stress is reduced or relaxes under a constant deformation.

Ligaments are viscoelastic as they gradually strain under tension, and then return to their original shape when the tension is released. However, a joint cannot retain its original shape when extended past a certain point or extended for a prolonged period. A joint becomes *dislocated* when this occurs, often due to trauma. Once a joint has become dislocated, it must be manually moved back to its original position as soon as possible. If the ligaments are lengthened for a prolonged period, the joint is weakened, making it more susceptible to future dislocations.

Hysteresis (energy dissipation) is a characteristic of the viscoelastic material. If a viscoelastic material is loaded and unloaded, the unloading curve does not follow the loading curve. The difference between the two curves represents the amount of energy that is dissipated (or lost) during loading.

There are three major regions of the stress-strain curve: 1) toe region, 2) linear region and 3) yield and failure region. In the physiologic activity, most ligaments and tendons exist in the toe and somewhat in the linear region. These constitute a nonlinear stress-strain curve since the slope of the toe region is different from that of the linear region.

Regarding structure-function relationships, the toe region represents "un-crimping" in the collagen fibrils.

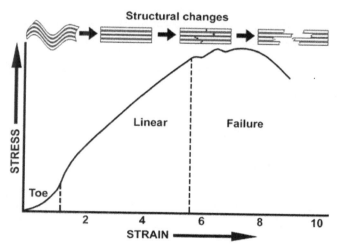

Since it is easier to stretch out the crimp of the collagen fibrils, this part of the stress-strain curve shows a low stiffness. As the collagen fibrils become uncrimped, the collagen fibril backbone itself becomes stretched, which gives rise to a stiffer material. As individual fibrils within the ligament or tendon begin to fail, damage accumulates, stiffness is reduced, and the ligaments and tendons begin to fail. Thus, the overall behavior of ligaments and tendons depends on the individual crimp structure and the failure of the collagen fibrils.

Endoskeleton versus exoskeleton

Humans possess an *endoskeleton,* which is located internally and contains both bone and cartilage. The endoskeleton does not limit the space for internal organs, and it supports greater weight. Soft tissues surround the endoskeleton to protect it because injuries to soft tissue are easier to repair. Usually, an endoskeleton has elements that protect vital internal organs.

Insects, crustaceans and some other animals have an external *exoskeleton.* Mollusks have exoskeletons that are predominantly calcium carbonate ($CaCO_3$). *Chitin*, present in insects and crustaceans, have jointed exoskeletons and a strong, flexible, nitrogenous polysaccharide. The exoskeleton protects against damage and keeps tissues from drying out. Although stiffness provides support for muscles, the exoskeleton is not as strong as an endoskeleton. The exoskeletons of clams and snails grow with the animals; their thick non-mobile $CaCO_3$ shell is used for protection. The chitinous exoskeleton of arthropods is jointed and moveable. Arthropods must molt when their exoskeleton becomes too small; a molting animal is vulnerable to predators. The jointed endoskeleton of vertebrates and exoskeleton of arthropods allows for flexibility. The exoskeleton is what made arthropods well-adapted for colonizing land because it protected against desiccation and allowed for locomotion without water.

Bone Structure

The bone's chemical composition is organic components (~25% by weight), inorganic components (~70% by weight) and water (~5% by weight). Bone is considered a connective tissue and is characteristically hard, strong, elastic and lightweight. Bones can be made from a combination of compact and spongy bone.

Macroscopically, bone is a solid structure with internal canals where blood vessels run and holes where cells can reside. The structure as a whole is surrounded by membranes that contain stem cells, including osteoblast (bone building) and osteoclast (bone degrading) cells.

Microscopically, bone is composed of cells, with the extracellular matrix arranged in cylinders as *osteons*, which contain blood vessels and nerves running through the middle.

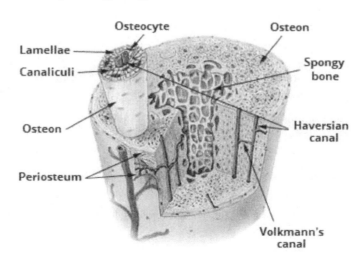

Four types of cells make up the bone matrix.

Osteoprogenitor cells are derived from mesenchyme and may undergo mitosis and differentiate into osteoblasts.

Osteoblasts are stem cells that form the bone matrix by secreting collagen and organic compounds from which bone is formed. They cannot undergo mitosis. As the collagenous matrix is released around them, they are enveloped by the matrix and differentiate into osteocytes. Osteoblasts secrete a matrix material of osteoid.

Osteoids are primarily made of collagen, which gives bone its high tensile strength. They contain glycolipids and glycoproteins.

Osteocytes are mature bone cells derived from osteoblasts. They are principal bone cells that cannot undergo mitosis. They maintain daily cellular activities such as the exchange of nutrients and waste with blood.

Osteoclasts are multinucleated cells that function in bone resorption, including the destruction of the bone matrix and are important in the development, growth, maintenance, and repair of bone. They develop from monocytes.

Compact bone is a highly organized, solid, smooth, dense type of bone that does not appear to have cavities from the outside. In compact bone, osteoclasts burrow tunnels, as *Haversian canals*.

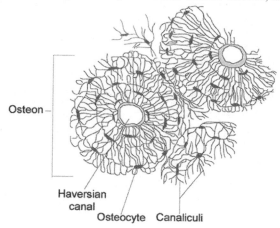

Compact bone from a transverse section of a long bone's cortex

The Haversian system is comprised of osteocytes, the star-shaped bone cells that lie in the *lacunae*. The osteoblast secrets the matrix of collagen and calcium salts in concentric *lamellae* (layers) around the central Haversian canal, which contains blood vessels and nerves. The elongated cylinders are bonded together to form the long axis of a bone. The *canaliculi* is a communication canal within the bone that connects the lacunae of the osteons. *Volkmann's canals*, small channels that run perpendicular to the surface of the bone, connect the blood and nerve supplies of adjacent Haversian canals together.

Spongy bone (cancellous) is less dense and consists of poorly-organized *trabeculae*, which are small, needle-like pieces of bone with much of open space between them. Spongy bone is nourished by diffusion from nearby Haversian canals. Spongy bone provides support of soft tissue, protection of internal organs, assistance in body movement, mineral storage, blood cell production and energy storage in the form of adipose (fat) cells in the marrow.

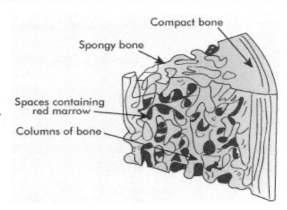

Bone with spongy bone in the medulla and compact bone along the cortex

Calcium–protein matrix

Calcium is the most abundant mineral in bone and the most abundant mineral in the human body. Calcium is an element that cannot be produced by any biological processes and must enter the body through diet. Bones act as a storage site for calcium. The extracellular matrix of bone consists of calcium salts, primarily calcium phosphate [$Ca_3(PO_4)_2(OH)_2$], which gives bone its hardness (or rigidity), collagen fibers and ground substance (glue). Calcium ions are vital for physiology and are needed for bone mineralization, tooth health, regulation of the heart rate, blood coagulation, contractions of smooth and skeletal muscle cells and regulation of nerve impulse conduction. The normal level of calcium in the blood is highly regulated at about 9-10 mg/dL. When the body cannot maintain this level, a person experiences hypocalcemia (or hypercalcemia), as described below.

Bone has a mineral content of roughly 50% by volume. The mineral content of the matrix consists mostly of calcium phosphate, hydroxyapatite, with minimal amounts of magnesium, carbonate and acid phosphate. Hydroxyapatite bone crystals are small and are soluble, and vital in mineral metabolism.

The organic matrix of bone is composed of about 90% collagen protein. Collagen is formed through chains that resemble short threads that twist into triple helices (resembling strings). They then line up and bind together, forming fibrils. The fibrils are arranged in layers, and the mineral crystals then deposit between the layers.

Matrix maturation expresses alkaline phosphatase and many noncollagenous proteins, (i.e., osteocalcin, osteopontin, and sialoprotein). Calcium and phosphate-binding proteins aid in the ordered deposition of minerals by the regulation of the amount and size of the hydroxyapatite crystals formed. The mineral portion of the matrix provides for the mechanical rigidity and strength of the bone, whereas the organic portion of the matrix contributes to the flexibility and elasticity of the bone.

Bone growth: osteoblasts and osteoclasts

The human fetal skeleton is cartilaginous that serve as scaffolds for bone construction. The fetal skeleton is formed from mesenchyme and hyaline cartilage that is loosely shaped like bones. This "skeleton" provides supporting structures for ossification (hardening into bone). At about 6-7 weeks gestation, ossification begins and continues throughout adulthood. The cartilaginous models are converted to bones when calcium salts are deposited in the matrix, first by cartilaginous cells and later by bone-forming osteoblast cells.

The main components of the ossification are: cartilage cells (chondrocytes), precursor cells (osteoprogenitor cells), bone deposition cells (osteoblasts), bone resorption cells (osteoclasts) and mature bone cells (osteocytes).

During the process of ossification, blood vessels invade the cartilage and transport osteoprogenitor cells to the center of ossification. At the center of ossification, the cartilage cells die, which form small cavities. Osteoblast cells form the progenitor cells, which then begin depositing bone tissue outwards from the center. From this process, both spongy textured calcaneus bone and smooth outer compact bone form.

Endochondral ossification is the conversion of cartilaginous scaffolds to bones. However, some bones, such as facial bones, are formed without a cartilaginous scaffold, and this is *intramembranous ossification*. The replacement of preexisting connective tissue with bone occurs through both intramembranous ossification and endochondral ossification.

Endochondral ossification occurs when a bone is formed from hyaline cartilage. Most bones in the skeleton are formed in this manner. During endochondral ossification, there is a primary ossification center at the middle of a long bone; secondary centers later form at the ends. *Primary ossification centers* harden in a fetus and during infancy. *Secondary ossification centers* develop in a child and harden during adolescence and early adulthood. A cartilaginous growth plate forms between the primary and secondary ossification centers. As the growth plate remains between the two centers, bone growth continues.

The perichondrium becomes the periosteum, which contains layers of undifferentiated cells, including osteoprogenitor cells that later develop into osteoblasts. *Appositional growth* is when osteoblasts secrete osteoid against the shaft of the cartilage scaffold to provide support for the new bone.

Hypertrophy occurs when chondrocytes in the primary ossification center begin to grow. They stop secreting collagen and other proteoglycans and begin secreting alkaline phosphatase and other enzymes essential for mineral deposition.

After the calcification of the matrix, hypertrophic chondrocytes start to die off to form cavities within the bone. The hypertrophic chondrocytes start to secrete vascular endothelial cell growth factors that induce the sprouting of blood vessels from the perichondrium. Blood vessels forming in the periosteal bud invade the cavity left by the chondrocytes and branch in opposite directions along the length of the shaft. These blood vessels carry hematopoietic cells, osteoprogenitor cells, and various other cells within the cavity. The hematopoietic cells eventually form the bone marrow.

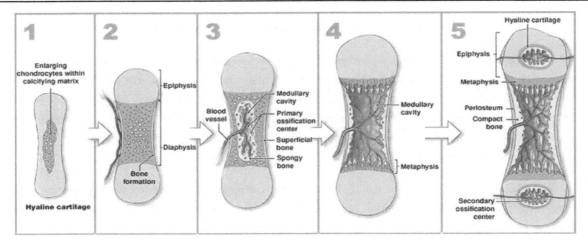

1) Enlargement of chondrocytes, 2) formation of a superficial layer of bone, 3) production of spongy bone at primary ossification center, 4) growth in length and diameter, 5) formation of secondary ossification centers

Osteoblasts, differentiated from the osteoprogenitor cells that enter the cavity via the periosteal bud, use the calcified matrix as a scaffold and begin to secrete osteoid, which forms the bone *trabecula*. The osteoclasts formed from macrophages break down spongy bone to form the medullary (or bone marrow) cavity. *Hematopoiesis* is blood cell formation. All blood cells are formed in the red marrow of certain bones. The flat bones of the skull, ribs, and breastbone (i.e., sternum) contain red bone marrow that manufactures blood cells.

During intramembranous ossification, a bone forms on (or within) a fibrous connective tissue membrane. The connective tissue membrane eventually forms the periosteum, which is made of fibers and granular cells in a matrix. The peripheral portion is predominantly fibrous, whereas the internal environment is predominantly osteoblasts. The tissue as a whole is heavily supplied with blood vessels.

As the ontogenetic fibers move out of the periphery, they continue to calcify and give rise to fresh bone spicules. A network of bone is then formed from meshes that contain blood vessels, and the delicate connective tissue is populated with osteoblasts. The bony trabeculae are thickened by the additional fresh layers of bone formed by the osteoblasts on its surface, and the meshes are simultaneously encroached upon. Many layers of bony tissue are continuously added under the periosteum and around the large vascular channels that eventually become the Haversian canals, which thickens the bone.

During infancy and childhood, *longitudinal growth* occurs, where long bones lengthen entirely by growth at the epiphyseal plates. Also, bones grow in thickness by appositional growth. The epiphyseal plates are replaced by bone in adulthood.

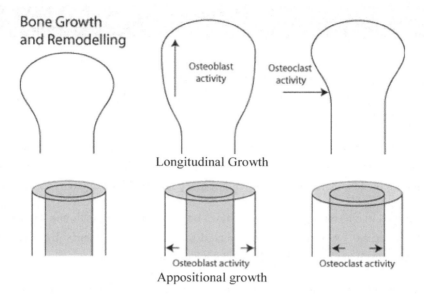

The structure of the epiphyseal plate contains four zones: the *zone of resting cartilage*, which anchors the plate to epiphysis, the *zone of proliferating cartilage* (where the chondrocytes divide to replace those that die at the diaphyseal surface of the epiphysis), the *zone of hypertrophic cartilage* (site of maturing cells) and the *zone of calcified cartilage* (which consists of dead cells, because the matrix around them is calcified).

Ossification of most bones is completed by age 25. As a child grows, cartilage cells are produced by mitosis on the epiphyseal side of the plate. The cartilage of the epiphyseal plate is replaced and destroyed by bone-forming on the diaphyseal side of the plate. For this reason, the thickness of the plate remains almost constant, while the bone on the diaphyseal side increases in length. Lengthwise bone growth occurs at the ends of long bones. The role of osteoblasts in lengthwise bone growth is to add bone tissue at the bone ends. Osteoblasts lengthen the knobs at the ends of the bone, while osteoclasts remodel bone tissue by degrading (chipping away) the bone ends (i.e., the knobs) until they are the right size and shape.

Along with the increase in length, bones increase in thickness (or diameter). Appositional growth occurs in an osteogenic layer of periosteum; osteoblasts lay down matrix (compact bone) on the outer surface. This is accompanied by osteoclasts destroying the bone matrix at the endosteal surface. Osteoblasts' role in diameter growth of bones is to add bone tissue to the outside of the bone. Osteoclasts' role in diameter growth of bones is to remove some bone tissue from the inside of the bone (bones are hollow). Without osteoclasts, diameter growth results in bones that are too thick and too heavy. Even with osteoclasts, bones become thicker with age.

Age-related skeletal changes are apparent at the cellular and whole body level. Height begins to decrease incrementally at around age 33. Bone loss gradually exceeds bone replacement. After menopause, females lose bone more rapidly than males. By age 70, bone loss is similar in both sexes. The likelihood of fractures increases as bones age. Adults need more calcium in the diet than do children to promote the work of osteoblasts. The rate of remodeling varies. For example, the distal femur is replaced every four months, while the shift of the femur is not replaced during one's lifetime.

Endocrine Control

The rate of bone growth is controlled by hormones, including growth hormones and sex hormones. Eventually, the epiphysis plates become ossified, the bone stops growing, and a person reaches final adult height.

In adults, bone is continually being remodeled (broken down and built up again). This involves the action of osteoblasts, osteoclasts and the hormones calcitonin and parathyroid hormone (PTH) through the negative feedback mechanism, which ultimately affects blood calcium homeostasis.

Osteoclasts (bone-absorbing cells) break down bone, remove worn cells and deposit calcium in the blood. Osteoclasts secrete lysosomal enzymes that digest the organic matrix by secreting acids that decompose calcium salts into Ca^{2+} and PO_4^- ions, which then enter the blood.

Osteoblasts (blood-forming cells) form new bone, taking calcium from the blood. Osteoblasts become entrapped in the bone matrix and become osteocytes in the lacunae of osteons. This continual remodeling allows the bone to change in thickness gradually. Osteoclasts determine calcium level in blood, which is important for muscle contraction and nerve conduction.

Minerals needed for bone growth and remodeling include calcium, phosphorus (a component of the hydroxyapatite matrix), magnesium (normal osteoblast activity), boron (inhibiting calcium loss) and manganese (formation of a new matrix).

Several vitamins are needed for bone growth, remodeling, and repair. Vitamin D greatly increases intestinal absorption of dietary calcium and slows its urine loss. Vitamin D deficiency causes rickets in children and osteomalacia in adults. Vitamin C helps maintain the bone matrix and collagen synthesis; deficiency of vitamin causes scurvy. Vitamin A is required for bone resorption and controls the activity, distribution, and coordination of osteoblasts and osteoclasts during development. Vitamin B_{12} plays a role in osteoblast activity.

Hormones needed for bone growth and remodeling include the human growth hormone (HGH), sex hormones, thyroid hormones, parathyroid hormones, and calcitonin. The pituitary gland secretes HGH. It is responsible for the general growth of all tissues. It stimulates reproduction of cartilage cells at the epiphyseal plate.

Sex hormones include estrogen and androgens (e.g., testosterone), which aid in osteoblast activity by promoting new bone growth. They degenerate cartilage cells in the epiphyseal plate by closing the plate. Estrogen's effect is greater than androgen's effect.

Thyroid hormones include triiodothyronine (T_3) and thyroxine (T_4). These hormones stimulate the replacement of cartilage by bone in the epiphyseal plate. Calcium homeostasis (blood calcium level of about 10 mg/dL) is critical for normal bodily functions. Calcium homeostasis is controlled by PTH, vitamin D, calcitonin and the interactions of the skeletal, endocrine, digestive and urinary systems. Together, these body systems work to maintain calcium level in the blood.

PTH is secreted by the parathyroid glands when blood calcium levels are low. It stimulates osteoclast proliferation, and the resorption of bone occurs. The demineralization releases calcium into the blood. PTH causes kidney tubules to reabsorb Ca^{2+} into the blood. It causes intestinal mucosa to increase dietary absorption of Ca^{2+} and causes an increase in blood calcium levels (homeostasis). PTH stimulates the synthesis of vitamin D, which stimulates calcium absorption from any digested food in the small intestine. The body deposits calcium in the bones when blood levels get too high, and it releases calcium when blood levels drop too low. This is regulated by PTH, vitamin D and calcitonin. When these processes return blood calcium levels to normal, there is enough calcium to bind the receptors on the surface of the cells of the parathyroid glands, and this cycle of events is turned off.

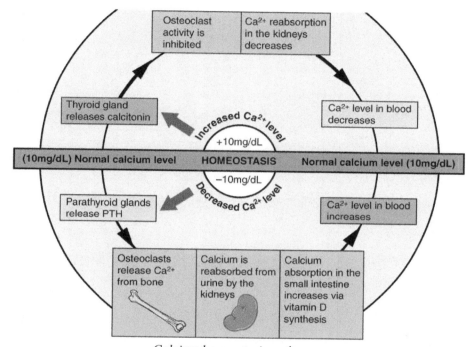

Calcium homeostasis pathways

Calcitonin is secreted by the thyroid gland when blood calcium levels are too high (hypercalcemia). It inhibits bone resorption and increases osteoblast activity (i.e., deposition of bone matrix). Calcitonin causes the kidney tubules to secrete excess Ca^{2+} into the urine and therefore results in a decrease in blood calcium levels (back to normal). Parafollicular cells of the thyroid gland secrete calcitonin, leading to a decreased plasma calcium concentration by reducing bone resorption. These actions lower the blood levels of calcium. When blood calcium levels return to normal, the thyroid gland stops secreting calcitonin. *Hypocalcemia* is where there is an abnormally low level of blood calcium.

Notes

Please, leave your Customer Review on Amazon

Chapter 21

Skin System

- Introduction

- Structure

- Functions in Homeostasis and Osmoregulation

- Functions in Thermoregulation

- Physical Protection

- Hormonal Control: Sweating, Vasodilation, and Vasoconstriction

Introduction

The *integument* is an organ system comprised of the skin, glands and accessory structures (e.g., hair, nails, blood vessels and nerves). All four of the basic tissue types are well-represented in this organ system: 1) epithelium in the hair, skin, nails and blood vessels, 2) muscle tissue surrounding blood vessels and attached to hair follicles, 3) nervous tissue in the lower layers of the skin and 4) connective tissue (including blood) throughout the entire organ system.

Dermatology is the study of the integumentary system. The integumentary system has several functions, including regulation of body temperature, acting as a physical barrier and protection against injury, dehydration, chemicals, UV radiation, and infectious agents.

The integument is a sensory organ that gathers information about temperature, pressure, and touch from the outside environment and transmits it to the central nervous system. The integument has a storage function, keeping glucose and fat as energy stores and holding water and blood. Vessels in the dermis hold up to 10% of the blood in the resting adult.

The blood content of the skin and the process of sweating are important for thermoregulation. Sweating regulates the body's water content and excretes waste. Finally, the skin helps synthesize and store vitamin D, essential in the regulation of blood calcium levels, as well as other vital processes.

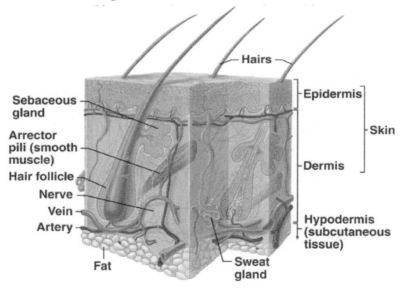

Skin with epidermis, dermis, fat cells and accessory structures

Structure

The integument (skin) is a *cutaneous membrane* that covers the outer surface of the body. It is the largest organ by surface area and weight. The average adult has a skin surface area of about 2 square meters and a weight of 5 kg, 16% of the entire body weight. Thickness varies from 0.5 millimeters on the eyelids to 6 millimeters on the heels of the feet. Humans lose almost a kilogram of skin epithelium per year that becomes a major part of household "dust." Cutaneous glands are any of the glands in the skin: merocrine sweat glands, apocrine sweat glands, sebaceous glands, ceruminous glands, and mammary glands. Other structures such as hair, nails, blood vessels, and nerves are in the skin.

Layer differentiation, cell types, and tissue

The integument is stratified into three major layers. The *epidermis* is the outer, thinner layer and consists of epithelial tissue. The *dermis* is the middle, thicker layer and consists of epithelial tissue, muscle tissue, nervous tissue, and connective tissue. Below the dermis is the *subcutaneous* or *subQ layer*, the *hypodermis*. The hypodermis is not technically part of the skin but is a supporting layer of loose connective tissue and fat that supplies blood to the dermis and attaches the skin to the underlying tissues and organs.

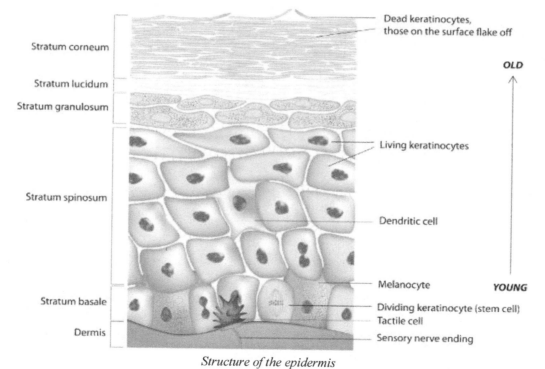

Structure of the epidermis

Most of the body is covered with thin skin, which has hair and glands. The *volar surfaces*, which are on the palm and sole, are covered with thick skin. Thick skin has an additional level of stratification in the epidermis and lacks hair and sebaceous glands.

The epidermis is a superficial, stratified, avascular epithelial tissue, and depends on the dermis for nutrients. Most oxygen enters the epidermis not from the dermal blood supply but the atmosphere. The epidermis is composed of *keratinized stratified squamous epithelium*, an epithelial tissue that contains four major types of cells: keratinocytes, melanocytes, Langerhans (dendritic) cells, and Merkel cells. *Keratinocytes* are the majority of the epidermal cells (nearly 90%) and produce *keratin*, a tough fibrous protein that provides protection. Keratinocytes flatten as they die and migrate upward to the surface of the skin, forming a tight layer of dead keratinocytes that give skin its waterproof quality. The keratinocytes are held together by *desmosomes* (adhesion cell junctions) that resist stretching and tearing of the skin but still allow some passage of materials between cells.

The layers of the epidermis include (from the superficial to the deepest): stratum corneum, stratum lucidum, stratum granulosum, stratum spinosum, and stratum basale.

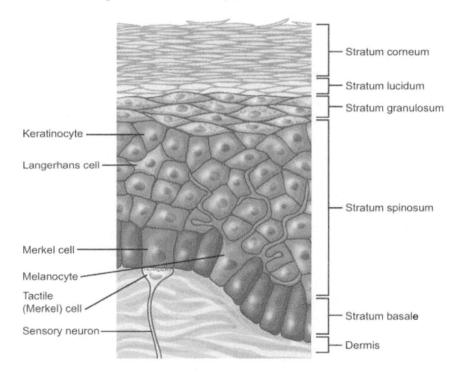

Skin layers

The *stratum corneum* is the outermost layer of the epidermis and consists of 25 to 30 layers of dead keratinocytes, the *corneocytes*. These keratin plates are sealed with lipids secreted by *lamellar granules*, organelles in living keratinocytes in the deeper layers of the epidermis. The keratin packed in the corneocytes and the lipids sealing the gaps between cells make skin impenetrable to many environmental agents and prevent water from entering or leaving. The average keratinocyte flakes off the epidermal surface about one month after it is created, a process of exfoliation.

Below the corneum is the *stratum lucidum,* only in the thick skin at the fingertips, palms of the hands and soles of the feet. The stratum lucidum has 3 to 5 layers of dead keratinocytes that add an extra layer of protection to the thick skin.

The *stratum granulosum* (granular layer) is an intermediate layer containing 3 to 5 layers of keratinocytes in the process of dying and migrating upward. Keratinocytes in the stratum granulosum contain *lamellar bodies*, known as *membrane-coating vesicles,* which secrete lipids that seal the epidermis. They are in the process of converting their cytoskeletal filaments into keratin, so that they may become keratin-packed cells in the stratum corneum and/or stratum lucidum.

The *stratum spinosum* (spinous layer) lies under the stratum granulosum and provides strength and flexibility to the epidermis. It is made of primarily of 8 to 10 layers of keratinocytes. *Langerhans cells* are all layers of the epidermis, but most predominantly in the stratum spinosum. These cells are macrophages that originate in red bone marrow and are involved with immune responses. They interact with the helper T-cells of the immune system.

The *stratum basale* (basal layer) lies below the stratum spinosum and is the *stratum germinativum* or *germinal layer*. The stratum basale is the deepest layer of the epidermis. Continuous division of stem cells occurs here and is the source of all the upper layers. Keratinocytes produced in the stratum basale are pushed up to the top layer, losing their cytoplasm, nucleus and other organelles as they reach the stratum corneum, before sloughing off the body. The stratum basale contains *Merkel cells*, which associate with sensory neurons and function in the sensation of touch. *Melanocytes* in the stratum basale produce *melanin* pigment, which protects against damage by UV radiation.

Below the epidermis is the *dermis*, which is about 0.2 to 4 millimeters thick and made mostly of connective tissue. The connective tissue is mainly collagen but includes elastic fibers, reticular fibers, fibroblasts, and other cell types. *Fibroblasts* are cells that produce fibrous proteins for the extracellular matrix of the connective tissue. The dermis contains blood vessels, sweat glands, sebaceous glands, nerves, hair follicles, nail roots, and smooth muscle.

In most places, *dermal papillae* (upward projections of the dermis) interlock with downward *epidermal ridges* to form a wavy boundary. Within the dermal papillae are the tiny *capillary loops* that supply them with blood. The papillae form the *friction ridges* of the fingertips and toes, which function in increasing firmness of grip by increasing friction.

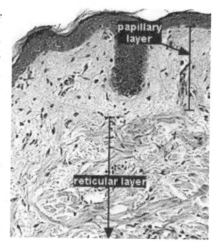

Papillary layer is the top 20% of the dermis. The papillary region lies just below the epidermis and consists of areolar connective tissue with thin collagen and elastic fibers, dermal papillae, and nerve endings. The papillary layer forms an arena for the mobilization of macrophages to defend against pathogens that breach the epidermis.

The deeper *reticular layer* of the dermis is a dense connective tissue (collagen and elastic fibers) that is packed with oil glands, nerves, sweat gland ducts, fat and hair follicles. It provides strength and elasticity to the skin. Dermal tears cause stretch marks. *Cleavage lines* (*Langer lines*) are "tension lines" on the skin that indicate the predominant direction of collagen fibers in the dermis.

The dermis contains *thermoreceptors*, which detect temperature, especially in the skin of the ears and face. *Mechanoreceptors*, which recognize sensations such as texture, pressure, and vibration, are in all layers of the integumentary system. Merkel cells and Meissner corpuscles are the most sensitive and are on the top layer of the dermis and epidermis, usually on non-hairy parts of the body (e.g., lips, palms, soles of the feet and tongue). The other mechanoreceptors are Ruffini corpuscles and Pacinian corpuscles, both in the hypodermis. *Ruffini corpuscles* are mechanoreceptors that have a thermoreceptor-like function.

Nociceptors are located in the dermis, as receptors detect any stimulus that may cause damage to the skin or other parts of the body. They signal an individual experiencing pain to avoid the particular stimulus that is causing the pain. Nociceptors can cause dull pain in an area where injury has taken place to signal the individual not to use that particular limb.

The subcutaneous layer (hypodermis) is not part of the skin. It is composed of areolar and adipose tissue, in larger quantity compared to the reticular layer of the dermis. This layer attaches the dermis and epidermis securely to the deeper tissues of the body. Additionally, blood vessels and nerves in the hypodermis extend to the dermis. One important function of the hypodermis is to store fat. The adipose tissue absorbs shock and provides insulation to protect the inner organs. Areas of the hypodermis composed mainly of adipocytes are *subcutaneous fat*.

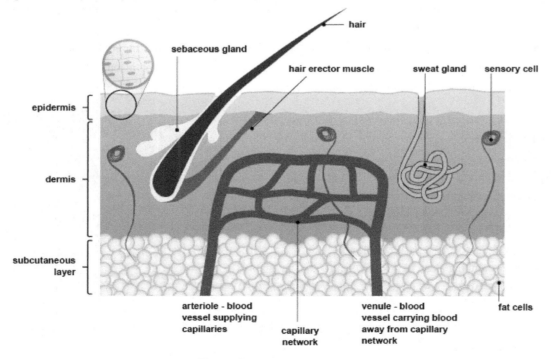

Skin and accessory structures

Skin color is from various proportions of different skin pigments, including eumelanin, pheomelanin, hemoglobin of the blood, the white collagen of the dermis and dietary carotene. Pathological conditions resulting in abnormal skin coloration include cyanosis, erythema, pallor, albinism, jaundice, and hematomas. Variations in hair color are attributable to the relative amounts of eumelanin and pheomelanin.

Melanocytes are specialized skin cells that produce the protective skin-darkening pigment of *melanin*. Melanin is located in the stratum basale of the epidermis and the matrix of the hair. Melanocytes are *dendritic* (branched), and their dendrites are used to transfer pigment granules to adjacent cells.

Skin markings include the friction ridges of the fingertips, flexion lines of the palms, wrists, and other places, and freckles, moles, and *hemangiomas* (bright red birthmarks). *Freckles* are clusters of concentrated melanin triggered by exposure to sunlight, and their amount is determined by genetics. *Nevi* are chronic lesions of the skin and are, by definition, benign. A melanocytic nevus is a mole or birthmark. Some nevi can become cancerous; *malignant melanoma* is a cancer of melanocytes.

Melanin normally decreases as an individual ages and loses melanocytes, resulting in gray hair and atypical skin pigmentation. However, certain disorders cause an abnormal lack of melanin. *Vitiligo* is a chronic disorder that causes depigmentation patches in the skin. The precise pathogenesis (cause) is not known but is most likely a combination of genetic factors coupled with a disorder of the immune system. *Albinism* is a congenital condition characterized by the complete (or partial) absence of pigment in the skin, hair, and eyes due to a defect of the enzyme involved in melanin production.

Relative impermeability to water

The epidermal water barrier is created by keratinocytes and the desmosomes which link them, a lipid coating that fills the gaps between keratinocytes, and a thick protein layer on the inner surface of the keratinocyte plasma membranes. The water barrier greatly reduces water loss from the body. Keratin is water-insoluble, and hydrophobic glycolipids seal the space between the dead keratin-packed cells. Sebum (oil) from the sebaceous glands contributes partly to the impermeability to water. These glands are everywhere except the palms and the soles of the feet.

Functions in Homeostasis and Osmoregulation

Homeostasis is a state in which the internal environment remains relatively constant despite changes in the external environment. Organisms usually strive for homeostasis, and organ systems of the human body are dedicated to maintaining homeostasis, including the integumentary system. The integumentary system has a particularly important role in water homeostasis via *osmoregulation*. Osmoregulation is the process by which the osmotic pressure of an organism's fluid is actively regulated to maintain the homeostasis of the body water content.

Water homeostasis involves insulating the body against losing water to the outside (or taking it in). The intercellular spaces of the epidermis are sealed with an array of lipids and amino acids that counteract the loss of water and salts from the skin and the penetration of water-soluble materials into the skin.

The loss of water from the stratum corneum into the environment is *transepidermal water loss*, a normal process. The integumentary system ensures that excessive transepidermal water loss does not occur. However, it may occasionally promote water loss through the sweat glands to regulate body water content and temperature. Perspiration excretes salts and nitrogenous wastes (e.g., urea, uric acid and ammonia). Increased sodium retention in the blood must offset sodium loss through perspiration. *Aldosterone* is a steroid hormone which promotes sodium reabsorption in the kidneys with associated water retention and therefore is an important regulator for osmoregulation.

Functions in Thermoregulation

Homeothermy is the ability to maintain body temperatures within a normal range. The total heat content of the body is the net difference between heat production and heat loss. In steady state, heat production equals heat loss. Water loss through the skin, sweat and respiratory tract can contribute to heat loss. The body surface can lose heat by radiation, conduction, convection and by the evaporation of water.

Sensory information from *thermoreceptors* in the skin travels to the hypothalamus, which then integrates and responds to this information to maintain homeothermy. The hypothalamus signals various parts of the body (e.g., sweat glands and the skin arterioles) to initiate thermoregulation responses.

Changes in body temperature are detected by two types of thermoreceptors: *peripheral thermoreceptors* (in the skin) and *central thermoreceptors* (in the brain and spinal cord). Central thermoreceptors provide the essential negative feedback for maintenance of core body temperature, while peripheral thermoreceptors provide feedforward information.

Thermoreceptors can be classified as cold receptors and hot receptors. *Cold receptors* start to perceive a sensation of cold when the temperature of the surface of the skin drops below 95 °F, and are most stimulated when the surface of the skin is at 77 °F, but are no longer stimulated when the temperature drops below 41 °F. This is why feet and hands go numb when they are submerged in ice-cold water for an extended period. *Hot receptors* start to perceive hot sensations when the temperature of the surface of the skin rises above 86 °F; they are most stimulated at 113 °F. Pain receptors take over at above 113 °F to avoid damage to the skin and underlying tissues.

Hair and erectile musculature

In animals, hairs help insulate the body by trapping air. Normally, hair lies at an angle to the skin, with a bundle of the smooth muscle of the *arrector pili* or *piloerector* attached to the follicle. When body temperature drops, the piloerector contracts and the hair stands up. In most mammals, this erect position fluffs up the fur and traps a layer of air between the fur and skin. This provides better insulation and thus increases body temperature. However, because humans are mostly hairless, this response is the *goosebumps* on the skin.

Fat layer for insulation

The adipose tissue in the hypodermis, along with its functions of providing protective padding and serving as energy storage, can act as an insulator. Fat is an ideal insulating tissue because heat travels poorly through this medium. There are two types of mammalian fat: brown fat and white fat. *Brown fat* generates heat due to its mitochondria and is abundant in infants as well as in the neck and around the large blood vessels of adults. It can be in hibernating animals.

White fat is the primary insulator and energy storage in adults. The thickness of the fat layer varies widely depending on the location of the body. For example, the fat layer on the eyelids is just a fraction of an inch, while the fat layer on the buttocks can be several inches.

Sweat glands and location in the dermis

There are two types of skin (sweat glands). Both are simple, coiled tubular glands. Sudoriferous glands produce sweat, which cools the body through evaporative cooling. Increased sweating is a corrective response to reduce the temperature of the organism.

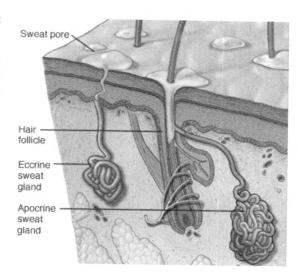

Merocrine (or *eccrine*) sweat glands are the major sweat glands of the body; they are most numerous on the palms and soles of feet. These glands secrete a watery solution (600 ml per day) that cool the body and eliminates small amounts of wastes like urea. Eccrine glands can release sweat in response to emotional stress (e.g., fear or embarrassment); this type of sweating is emotional sweating or a "cold sweat."

Apocrine sweat glands are larger than eccrine sweat glands and are associated with hair follicles mainly on the skin of the axillae (armpit), the groin, the areolae and bearded facial regions of adult males. They become active during puberty along with the appearance of hair in these regions and secrete pheromones (or pheromone-like) substances. The sweat of apocrine glands is slightly viscous and usually odorous (unlike eccrine sweat). Much of body odor is due to apocrine sweat. The secretory portion of the apocrine sweat glands is located mostly in the hypodermis, with the excretory duct directing into hair follicles. *Myoepithelial cell* contractions around the base of a follicle squeeze the secretion up the duct to the skin surface.

Vasoconstriction and vasodilation in surface capillaries

The skin's effectiveness as an insulator is subject to physiological control by a change in the blood flow to the skin. Blood vessels reduce the insulating capacity of the skin by carrying heat to the surface to be lost to the external environment. *Vasodilation* is the distention of blood vessels. Vasodilation relaxes the smooth muscle around a blood vessel, increasing its diameter and promoting blood flow. When the temperature is high, the hypothalamus initiates vasodilation of arterioles near the skin. This increases blood supply to the capillaries in the skin, which leads to more heat loss at the skin surface.

At low temperatures, the hypothalamus initiates the opposite response. *Vasoconstriction* of arterioles causes blood to bypass the skin capillaries and redirects blood to the veins, reducing blood supply to the skin capillaries and thus reducing heat loss at the skin surface.

The *thermoneutral zone* is the range of environmental temperatures over which the basal metabolic rate can maintain body temperature and adjusted by vasoconstriction or vasodilation alone. When the temperature is below (or above) this zone, the body must use an amount of energy which exceeds the basal metabolic rate to alter heat production (and heat loss). In temperatures below the thermoneutral zone, an increase in heat production may be accomplished by behaviors such as shivering and exercise. Other behavioral mechanisms, such as curling up into a ball when cold, reduce the surface area exposed to the environment and decrease heat loss by radiation and conduction. In temperatures above the thermoneutral zone, a decrease in heat production can be accomplished by lowering the metabolic rate. An increase in heat loss can be promoted by evaporative cooling via sweating.

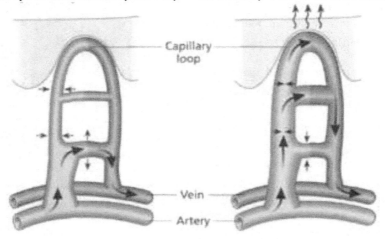

The heat kept in and retained by the body   Heat loss by convection and radiation

Cold environment   Warm environment

Vasoconstriction in cold environments and vasodilation in warm environments are physiological responses to regulate body temperature.

Physical Protection

The skin is an excellent line of defense because it provides an almost impenetrable physical barrier protecting the internal environment. Multiple layers in the skin allow for specialization. The epidermis is adapted to fast cell turnover, heals quickly and resists damage. It offers waterproofing and protection from pathogens to underlying tissues. The dermis provides temperature stability and prevents dehydration, but unlike the epidermis, it is capable of limited healing. Nails and hair originate in the dermis and provide additional protection.

Nails, calluses, and hair

Keratinization replaces viable cells in the stratum basale with more waxy, water-resistant keratin. Constant friction can stimulate this process and produce a *callus*, which is a thick buildup of keratinocytes in the stratum corneum. Calluses are a response to repeated physical stress that function to protect the dermis from this stress. Excessive friction or pressure in too short a time usually results in blisters.

Nails and hair are formed of hard keratin and are tougher than the soft keratin in the skin. Hard keratin is more compact and extensively cross-linked than soft epidermal keratin.

A *pilus* (hair) is a slender filament of the dead, keratinized epidermal cells growing from an oblique tube, the *hair follicle,* which is composed of both epidermal and dermal tissue. Hair is on all surfaces of the skin except the palms, the anterior surfaces of the fingers and the soles of the feet. Genetics determine hair color, thickness, and distribution. The functions of hair vary with type and location and include thermal insulation, protection from the sun and foreign objects, sensation, facial expression, signaling of sexual maturity and regulation of pheromone dispersal.

There are three types of hair: lanugo hairs, vellus hairs, and terminal hairs. *Lanugo hairs* are fine, unpigmented, downy hairs that cover the body of a fetus. These hairs are shed before or shortly after birth. *Vellus hairs* are short, fine, pale hairs that are barely visible to the naked eye. They develop in childhood and are on all humans. *Terminal hairs* are long, coarse, heavily pigmented hairs.

Deep in the follicle, a hair begins with a dilated *bulb,* continues as a narrower *root* below the skin surface, and extends above the skin as the *shaft.* The bulb contains a *dermal papilla* of vascularized connective tissue, from which the hair receives its only nourishment. The *hair matrix* just above the papilla is the site of hair growth by the division of the matrix cells. In cross-section, a hair exhibits a thin outer cuticle, a thicker layer of keratinized cells (forming the hair cortex), and a *medulla* core. A hair follicle has an inner *epithelial root sheath,* (an extension of the epidermis) and an outer *connective tissue root sheath* (condensed dermal tissue). It exhibits a *bulge,* which is a site of stem cells for follicle growth. A hair follicle is associated with *hair receptors* (nerve endings) to detect hair movements.

Differences in hair texture are attributable to differences in cross-sectional shape: straight hair is round, wavy hair is oval and curly hair is relatively flat. A hair has a life cycle consisting of a growing *anagen* stage of 6 to 8 years, a shrinking *catagen* stage of 2 to 3 weeks, and a resting *telogen* stage of 1 to 3 months. Hairs usually fall out during the catagen or telogen stage. A scalp hair typically lives 6 to 8 years and grows at a rate of 10 to 18 centimeters per year.

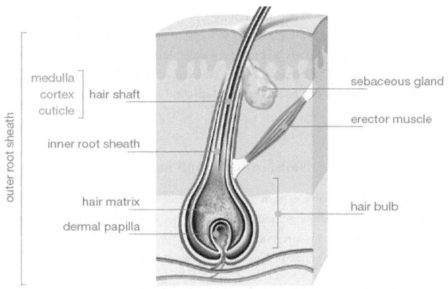

Anatomy of hair with bulb, root, and shaft

Nails are composed of hard, densely keratinized epidermal cells located over the dorsal surfaces of the ends of the fingers and toes.

The nail structure has a free edge (a transparent nail body) with a whitish *lunula* at its base, and a nail root embedded in a fold of the skin. The skin underlying the nail is the nail bed; its epidermis is the *hyponychium*. The nail matrix is a growth zone composed of epidermal stratum basale at the proximal end of the nail bed.

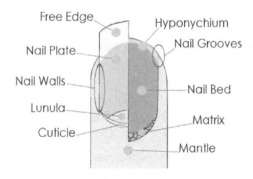

Anatomy of a nail

Protection against abrasion and disease organisms

Generally, the skin forms a physical barrier that prevents pathogens from entering the body. Besides its physical toughness, the skin has chemical defenses. *Sebaceous glands,* associated with hair follicles, produce *sebum* (an oily secretion), which does two important things: prevents the dehydration of hair and skin and inhibits the growth of certain bacteria. Lactic acid and fatty acids in sebum create an acidic environment on the surface of the skin that discourages the growth of pathogens. The sebum contains antibodies that fight infection. Sebaceous glands are *holocrine glands*, which break down to form the secretion.

Ceruminous glands are modified sweat glands located in the auditory canal. Along with nearby sebaceous glands, they are involved in producing *cerumen* (a waxy secretion). Cerumen, or earwax, is a mixture of ceruminous gland secretion, sebum, and dead epidermal cells. It keeps the eardrum pliable, waterproofs the auditory canal and kills bacteria.

Lymphocytes (immune cells) are abundant throughout the dermis, and all have various functions in the suppression of infection. The epidermis contains lymphocytes and other cells with immune function (e.g., Langerhans cells).

Mucous membranes form another type of barrier against pathogens. These are soft and moist areas in internal organs exposed to external environments (the trachea) and on certain areas of the skin (e.g., nose, ears and genitals). Mucous membranes are not strong enough to create a physical barrier, but they do produce mucus, which contains antibodies, lysosomal enzymes and other agents of the immune system. Mucus in the respiratory tract traps pathogens and irritants which are then expelled out of the nose or the trachea.

Senescence (cell aging) of the integumentary system is marked by thinning and graying of the hair, dryness of the skin and hair due to atrophy of the sebaceous glands, and thinning and loss of elasticity in the skin. The loss of subcutaneous fat reduces the capability for thermoregulation. UV radiation accelerates the aging of the skin, promoting wrinkling, age spots, and skin cancer. Aged skin is more vulnerable than younger skin to trauma and infection and heals more slowly.

Depending on the depth of the injury, there are two kinds of wound-healing processes. *Epidermal wound healing* follows superficial wounds that affect only the epidermis. This healing serves to restore normal function. *Deep wound healing* occurs when an injury extends to the dermis and subcutaneous layer. This process is more serious and requires the formation of scar tissue (fibrosis), which reduces some function of the skin of the repaired wound.

A burn is tissue damage caused by excessive heat, electricity, radioactivity or corrosive chemicals that denature the proteins in skin cells. Burns destroy some of the skin's important contributions to homeostasis like resistance to microbial invasion, protection against desiccation and thermoregulation. They are graded according to their severity.

A *first-degree burn* involves only the epidermis and is characterized by mild pain and *erythema* (redness) but no blisters; the skin quickly regains function. A *second-degree burn* destroys the epidermis and part of the dermis; some skin functions are lost. Indications of a second-degree burn include erythema, blister formation, edema, and pain. A *third-degree burn* is a full-thickness burn, meaning it destroys the epidermis, dermis, and hypodermis. After a third-degree burn, most (if not all) skin functions are lost, and the region becomes numb because sensory nerve endings have been destroyed.

According to the American Burn Association's classification of burn injury, a major burn includes third-degree burns over 10% of body surface area; second-degree burns over 25% of body surface area; or any third-degree burns on the face, hands, feet or perineum (which includes the anal and urogenital regions). A burn exceeding 70% of body area results in death for more than half of victims.

Burn Type	Layers affected	Sensation	Healing time	Prognosis
Superficial (1°)	Epidermis	Painful	5–10 days	Heals well; repeated sunburns increase the risk of skin cancer
Superficial partial thickness (2°)	Extends into superficial (papillary) dermis	Very painful	Less than 2–3 weeks	Local infection/cellulitis but no scarring
Deep partial thickness (2°)	Extends into deep (reticular) dermis	Pressure and discomfort	3–8 weeks	Scarring, contractures (may require excision and skin grafting)
Full thickness (3°)	Extends through entire dermis	Painless	Prolonged (months) and incomplete	Scarring, contractures, amputation (early excision recommended)
Fourth degree (4°)	Extends through the entire skin and into underlying fat, muscle, and bone	Painless	Requires excision	Amputation, significant functional impairment and, in some cases, death

Skin cancer is classified as one of 3 types and is distinguished by the cells of origin and the appearance of the lesions: *basal cell carcinoma, squamous cell carcinoma,* and *malignant melanoma.* Malignant melanoma is the least common form but is the most dangerous because of its tendency to quickly metastasize (move to other regions in the body by circulatory or lymphatic systems).

Hormonal Control: Sweating, Vasodilation, and Vasoconstriction

Hormones are chemical messengers that relay information throughout the body, usually to maintain homeostasis. One hormone is estrogen, which is the primary sex hormone in females but is present in males. Estrogen affects skin thickness, elasticity, and fluid balance. Estrogen can increase the production of glycosaminoglycans (GAGs) such as hyaluronic acid, which maintain the skin's structure and fluid balance. Additionally, estrogen can increase collagen production to maintain epidermal thickness and smooth the appearance of wrinkles. Estrogen insufficiency, which occurs in menopausal women, can cause excessive perspiration.

Thyroid hormones affect body temperature and skin dryness. Excess thyroid hormone causes the skin to become warm, sweaty and flushed. Conversely, a deficit of thyroid hormone causes the skin to become dry, coarse and thick. An imbalance of serotonin, a hormone involved in mood, sleep, digestion, and memory, can lead to excessive perspiration.

Testosterone is in both sexes but is the primary male sex hormone. Coarser hair, oily skin, and general skin aging are due to the activity of testosterone. Females can experience increased oiliness and acne when their hormones are not in balance.

Vasodilation and vasoconstriction control the distribution of blood flow in the body, which controls heat loss in the body. They are influenced by chemical factors (CO_2, H^+, and K^+), sympathetic nerves and autonomic nerves. Hormones play an important role in regulating this system. Vasodilation involves the relaxation of smooth muscle around the blood vessels. Muscle relaxation is dependent on the intracellular concentration of calcium ions and is related to the phosphorylation of the light chain of the *myosin* contractile protein. Vasodilation occurs when there is a decrease in intracellular calcium levels or when myosin is dephosphorylated. Various hormones promote vasodilation through specific pathways by decreasing calcium content within the cells. These endogenous vasodilators include epinephrine, histamine, prostacyclin, and prostaglandins. Vasoconstriction involves the constriction of blood vessels. This process works to increase intracellular calcium levels. Some of the hormones involved in vasodilation include norepinephrine, dopamine, thromboxane, and vasopressin (ADH).

Although epinephrine and norepinephrine have the same effect in the heart, these two hormones have vastly different effects on the blood vessels. Epinephrine causes vasodilation, while norepinephrine causes vasoconstriction. This is because the heart contains only beta-2 receptors, while blood vessels contain both alpha receptors and beta-2 receptors. Epinephrine preferentially activates beta-2 receptors, so if there are sufficient beta-2 receptors, vasodilation occurs.

Notes

Chapter 22

Reproductive System

- **Female and Male Reproductive Structures and their Functions**

- **Gametogenesis by Meiosis**

- **Sperm and Ovum**

- **Hormonal Control of Reproduction**

Female and Male Reproductive Structures and Their Functions

The *female reproductive system* includes the ovary, oviduct, uterus, and vagina.

The *male reproductive system* consists of the testes, epididymis, vas deferens, prostate gland, bulbourethral glands, and the penis.

Genitalia

Genitalia is the external genital organs of the reproductive system.

The major difference between the male and female reproductive structures is that male structures are mostly external, for the delivery of sperm, whereas female structures are mostly internal, to nurture a growing fetus. In males, there is one opening for both urine and sperm, while females have separate openings for urine and menstruation and sexual intercourse.

Female Genitalia

The external genitalia of females is collectively the *vulva*. The urethra opens into the vulva. The *vulva* (pudendum or external genitalia) includes the *clitoris, mons pubis, labia majora,* and *labia minora*. The labia are located on each side of the vaginal and urethral openings. The *clitoris*, a short shaft of erectile tissue capped by a pea-shaped gland, is located at the front juncture of the labia minora. This structure is homologous to the penis in males. Additionally, the vulva contains the vaginal orifice, *greater* and *lesser vestibular glands, paraurethral glands* and *vestibular bulbs* (erectile tissues).

The *vagina* is a tubular organ at a 45° angle with the small of the back. The vagina is connected to the uterus by the cervix, which is a cylinder-shaped neck of tissue about 1 inch across. The *cervix* is cartilage covered by smooth, moist tissue. The vagina tilts posteriorly between the urethra and rectum. It has no glands but is moistened by transudation of serous fluid through the vaginal wall and by mucus from glands in the cervical canal. The adult vagina is lined with a stratified squamous epithelium with antigen-presenting *dendritic cells*.

The vagina's mucosal lining lies in folds, and it can extend as necessary in childbirth. The vagina receives the penis during copulation. *Copulation* is a sexual union that facilitates the reception of sperm by a female. During birthing, the vagina is where the fetus passes out of the body (i.e., the birth canal).

The *uterus* has an upper *fundus,* middle *corpus* (body), and a lower *cervix* (neck), where it meets the vagina. A narrow *cervical canal* connects the uterine lumen with the vaginal lumen. *Cervical glands* in the canal secrete mucus, which prevents vaginal microbes from spreading into the uterus.

The *uterus* is a hollow, thick-walled muscular organ located superior to the urinary bladder. Its size and shape are comparable to that of an inverted pear; it is where the fertilized ovum develops until birth. The uterine wall has three layers: an outer serosa *perimetrium,* thick muscular *myometrium,* and an inner mucosa *endometrium*. The endometrium contains numerous tubular glands and is divided into two layers – a thick superficial *stratum functionalis* (shed during each menstrual period) and a thinner basal *stratum basalis* (retained from cycle to cycle).

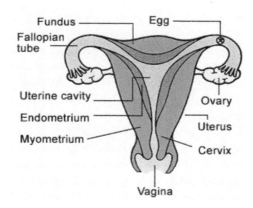

A female reproductive system with associated structures

The uterus is supported by a pair of lateral wing-like *broad ligaments* and cordlike *cardinal, uterosacral,* and *round ligaments.* It receives blood from a pair of *uterine arteries.*

The *Fallopian tubes* (oviducts) are two tubes that branch from the uterus and provide a passage to the uterus for the ovum released by the ovary. Fallopian tubes, uterine tubes and salpinges (singular salpinx), are the normal site of fertilization. They are lined with ciliated epithelia; after fertilization, the embryo is slowly moved by ciliary movement toward the uterus. The flared distal end of the Fallopian tube, near the ovary, is the *infundibulum* and has feathery projections of *fimbriae* to receive the ovulated egg. Its long midportion is the *ampulla,* and the short constricted zone near the uterus is the *isthmus.* A ligamentous sheet of mesosalpinx supports the tube.

Male Genitalia

The external genitalia of men consists of the penis, male urethra and the scrotum.

The *penis* is a cylindrical copulatory organ that introduces semen (a fluid containing spermatozoa) and secretions into the female vagina. The penis is divided into an internal *root* and an external *shaft* and *glans.* It is covered with loose skin that extends over the glans as the *prepuce* (foreskin). Internally, the penile shaft consists mainly of three spongy long *erectile tissues*—a pair of dorsal *corpora cavernosa* (engorge with blood and produce most of the effect of erection), and a single ventral *corpus spongiosum* (contains the urethra). All three tissues have *lacunae* (blood sinuses) separated by *trabeculae* composed of connective tissue and *trabecular muscle* (smooth muscle).

At the proximal end of the penis, the corpus spongiosum dilates into a *bulb* that receives the urethra and ducts of the bulbourethral glands. The corpora cavernosa diverge into a pair of *crura* that anchor the penis to the pubic arch and perineal membrane.

A pair of internal *pudendal arteries* supplies the penis. Each branch into a *dorsal artery*, which travels dorsally under the skin of the penis. The *deep artery* travels through the corpus cavernosum and supplies blood to the lacunae. The dorsal arteries supply most of the blood when the penis is flaccid, and the deep arteries supply blood during an erection.

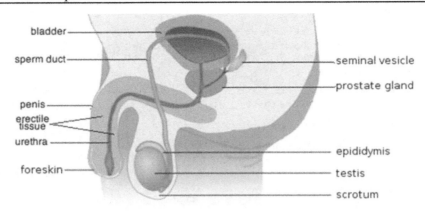

A male reproductive system with anatomy and glands

The *scrotum* is part of the external male genitalia located behind and underneath the penis. It is a pouch of skin that contains and protects the testicles.

The scrotum contains the testes and the *spermatic cord*, which is a bundle of connective tissue, testicular blood vessels and *ductus deferens* (sperm duct). The spermatic cord passes up the back of the scrotum and through the external inguinal ring into the inguinal canal.

Spermatic ducts carry sperm from the testes to the urethra. They include *efferent ductules* (leaving the testes); *duct of the epididymis* (highly coiled structure adhering to the posterior side of the testes); muscular *ductus deferens* (travels through the spermatic cord and inguinal canal into the pelvic cavity); and a short *ejaculatory duct* (carries sperm and seminal vesicle secretions towards the last 2 cm to the urethra).

The *urethra* completes the path of the sperm to the outside of the body. The urethra is the duct by which urine is conveyed out of the body from the bladder, and by which male vertebrates convey semen.

Gonads

Gonads are primary sex organs that are specialized to produce gametes (haploid cells of egg and sperm). There are two types of gonads: testes, which produce spermatozoa (sperm) and ovaries, which produce ova (singular ovum, or egg).

Female Gonads

The female gonads are the *ovaries*, which house immature eggs that mature one (or more) at a time (on a monthly basis). The *ovary* is where the ova (or eggs) are produced (ovaries produce a secondary oocyte each month) along with the female sex hormones, estrogen, and progesterone, during the ovarian cycle. The ovaries are located in the abdominal cavity. Each female has two ovaries.

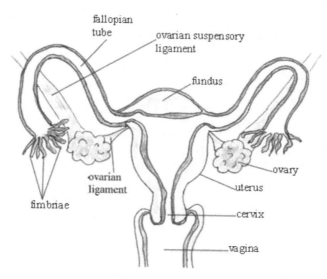

Female ovary with fallopian tube and ovarian ligament

The *oviduct* (Fallopian tube or uterine tube) is a tube through which eggs move from the ovary to the uterus. Each ovary is associated with an oviduct. The ovary has a central *medulla*, a surface *cortex* and an outer fibrous capsule of the *tunica albuginea*. Fimbriae are fingerlike projections that sweep over the ovaries and waft the egg into the Fallopian tubes when it is released from the ovary. Generally, the ovaries alternate in producing one oocyte each month.

The ovary is supported by a medial *ovarian ligament*, a lateral *suspensory ligament,* and an anterior *mesovarium*. The ovary receives blood from a branch of the *uterine artery* medially and the *ovarian artery* laterally.

Male Gonads

The male gonads are the *testes*, which produce sperm and testosterone.

Paired testes are suspended in the scrotal sacs of the scrotum. Shortly before birth, the fetal testes descend through the inguinal canal into the scrotum. The low temperature in the scrotum is vital to normal sperm production. If the testes do not descend, surgery or hormonal therapy is required; otherwise, sterility results.

The testie each consist of *seminiferous tubules* for the production of sperm and *interstitial cells* (Leydig cells) for the production of testosterone. The epithelium of a seminiferous tubule consists of *germ cells* and *sustentacular cells*. The germ cells develop into sperm. The sustentacular cells support and nourish the germ cells by forming a *blood-testis barrier* between them and the nearest blood supply.

Like the ovary, the testis has a fibrous capsule of the *tunica albuginea*. Fibrous septa extend from the tunica and divide the interior of the testis into 250−300 compartments as *lobules*. Each lobule contains one to three sperm-producing seminiferous tubules. Testosterone-secreting interstitial cells lie in clusters in between the tubules.

A long, slender *testicular artery* supplies each testis and drained by veins of the *pampiniform plexus*, which converge to form the *testicular vein*. The testis is supplied with *testicular nerves* and lymphatic vessels.

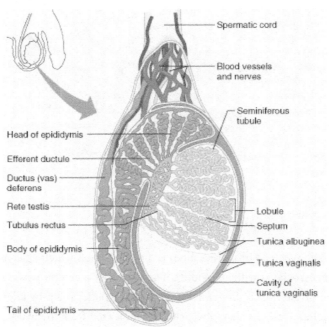

Male reproductive system

Semen (seminal fluid) is a thick, whitish fluid that contains about 10% sperm, 30% glandular secretions from the prostate vesicles and 60% secretions from the seminal vesicles and bulbourethral glands. It contains about 50–120 million sperm/mL, as well as seminogelin (seminal vesicle protein), a serine protease (prostate-specific antigen), fructose, prostaglandins and other substances.

The male has three sets of accessory glands: a pair of *seminal vesicles* posterior to the urinary bladder, a single *prostate gland* inferior to the bladder (enclosing the prostatic urethra) and a pair of small *bulbourethral glands* that secrete into the proximal end of the penile urethra. The seminal vesicles and prostate secrete most of the semen. The bulbourethral glands produce a small amount of clear, slippery fluid that lubricates the urethra and neutralizes its pH.

The seminal vesicles lie at the base of the urinary bladder, which contains two glands that join the *vas deferens* (pl. *vasa deferentia*) to form an ejaculatory duct that enters the urethra. A thick fluid containing nutrients, including mucous (liquid for the sperm), fructose (for ATP) and prostaglandins, is secreted into the ejaculatory duct.

The prostate gland is located below the urinary bladder and surrounds the upper portion of the urethra. It secretes a milky, slightly alkaline solution that promotes sperm motility and viability. This fluid neutralizes the acidity of urine that may still be in the urethra and neutralizes vaginal acidity. The prostaglandin secretions neutralize seminal fluid, which is too acidic from the metabolic waste of sperm. The bulbourethral glands are located below the prostate gland and on either side of the urethra; they release mucous secretions that provide lubrication.

The *epididymis* (pl. *epididymides*) is a coiled tube that attaches to each testicle for the site of final maturation and storage of sperm. The vas deferens and epididymis store the sperm until ejaculation. Sperm is non-motile at this time. During the passage through the epididymis, they are concentrated by fluid absorption. When a male is sexually aroused, the sperm enters the urethra, part of which extends through the penis. Sperm travels through the vas deferens into the ejaculatory duct, which leads to the urethra and penis. The urethra transports urine from the bladder during urination. The mnemonic for the path of sperm is "Seven Up" (Seminiferous tubules, Epididymis, Vas deferens, Ejaculatory duct, nothing, Urethra, Penis).

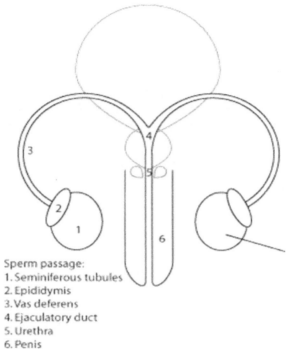

Sperm passage:
1. Seminiferous tubules
2. Epididymis
3. Vas deferens
4. Ejaculatory duct
5. Urethra
6. Penis

Path of sperm from the seminiferous tubules and through the penis

Sperm cannot develop at the core body temperature of 37 °C. The testes are kept about 2 °C cooler than normal body temperature by three structures in the scrotum: *cremaster muscle* of the spermatic cord, *dartos muscle* in the scrotal wall and the *pampiniform plexus* of veins in the spermatic cord. The cremaster muscle relaxes when it is warm and contracts when it is cool, lowering or raising the scrotum and testes. The dartos muscle contracts and tautens the scrotum when it is cool. The pampiniform plexus acts as a countercurrent heat exchanger to cool the blood on its way to the testes.

SAT Biology E/M: Complete Content Review

Gametogenesis by Meiosis

Gametogenesis is the meiotic cell division that produces eggs (oogenesis) and sperm (spermatogenesis). *Meiosis* is a nuclear division that is broken into two broad stages (meiosis I and meiosis II), which reduces the chromosome number from diploid (2n) to haploid (n). The haploid number is half of the diploid number of chromosomes. Meiosis of one diploid cell results in the production of four haploid cells. These daughter cells, containing half the genetic information, are *gametes.*

Before fertilization, the building blocks to create that cell must be available, so that the initial fertilized cell (zygote) can develop into an embryo, fetus and a mature adult. Developing gametes (egg and sperm) are germ cells, and the first stage in their development is the proliferation of primordial germ cells by mitosis (replication via cell division), in which each daughter cell receives a full set of chromosomes identical to those of the original cell. In females, mitosis activity of germ cells occurs only during embryonic development, while in males it begins at puberty and continues throughout life. For many years, it was believed that at birth, females contain all the egg cells that they will ever have. However, new findings are coming to light with regards to mitotic activity in female gonads, and the idea that females do not produce new eggs during their lifetime is beginning to change with new research.

The next stage of development is meiosis, in which each daughter cell receives half of the chromosomes of the original cell. During meiosis, chromosomes stay in their homologous pairs. In humans, for example, instead of 46 individual chromosomes lining up, there are 23 pairs of chromosomes. Gamete production via meiosis occurs in all sexually reproducing eukaryotes, including animals, plants and fungi.

Sexual reproduction results in the formation of gametes. Human males produce small, motile sperm in the testes, and human females produce a large, immobile, nutrient-laden egg or ovum in the ovarian follicles. These gametes fuse to form a *zygote.* A zygote has the full or diploid (2n) number of chromosomes. If gametes contained the same number of chromosomes as somatic (body) cells, the zygote would have twice the correct number of chromosomes. When the gametes fuse, the chromosomes from both parents naturally combine. Each fertilized egg (zygote) cell has a pair of homologous chromosomes, one 1N homolog from the mother (egg) and one 1N homolog from the father (sperm).

During meiosis, the chromosomes align within the daughter cells in many possible combinations. $(2^{23})^2$ or about 70 trillion combinations are possible without crossing over. This allows for genetic variability in sexually reproducing organisms. Crossing over is the process during meiosis I, where homologous chromosomes are paired with another (synapsis) and can exchange their genetic material to form genetically unique (recombinant) chromosomes. If crossing over occurs once, then $(4^{23})^2$ or 70 trillion squared genetically different zygotes are possible for one couple. Crossing over is unique to prophase I of meiosis.

The Phases of Meiosis

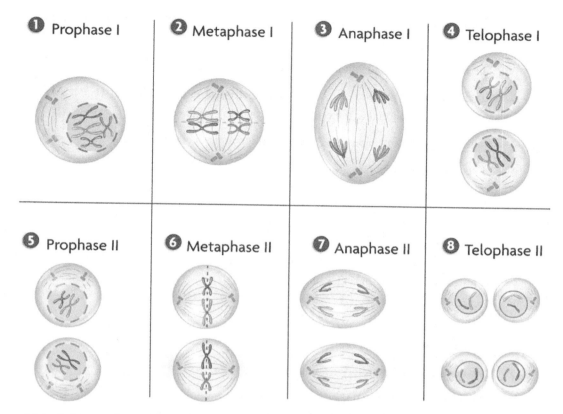

Meiosis I is a reduction stage (2n → 1n), meiosis II separates sister chromatids at the centromere

Both meiosis I and meiosis II have four phases: prophase, metaphase, anaphase, and telophase. Before meiosis I, DNA replication occurs in S phase of interphase, and each chromosome has a pair of sister chromatids; attached at the centromere (similar to mitosis for somatic cells).

Meiosis I: reduction phase of diploid to haploid (2n → 1n)

In general, during meiosis I, homologous chromosomes come together and line up at the synapsis. During synapsis, the two sets of paired chromosomes lay alongside each other as bivalents (tetrads). Bivalents are a pair of related homologous (one from the mother and one from the father) chromosomes that are held together by a chiasma complex.

Prophase I

1. Nuclear division occurs: nucleolus disappears, nuclear envelope fragments, centrosomes migrate away from each other and spindle fibers assemble.

2. Homologous chromosomes undergo synapsis, forming bivalents; crossing over may occur, in which case sister chromatids exchange genetic material by recombination.

3. Chromatin condenses, and chromosomes become microscopically visible.

Metaphase I

1. During prometaphase I, bivalents held together by chiasmata move toward the metaphase plate at the equator of the cell.

2. The fully formed spindle aligns the bivalents at the metaphase plate.

3. Kinetochores are proteins associated with centromeres; they attach to *kinetochore spindle fibers* anchored to the centrioles at each pole of the cell.

4. Homologous chromosomes independently align at the metaphase plate.

5. Maternal and paternal homologs may be oriented toward either pole.

Anaphase I

1. The homologs separate and move toward opposite poles.

2. Each chromosome is attached with a centromere and two sister chromatids (replicated previously during S phase).

Telophase I

1. In animals, this occurs at the end of meiosis I.

2. The nuclear envelope reforms and nucleoli reappear.

3. This phase may or may not be accompanied by cytokinesis for portioning the two nuclei with separate plasma membranes (i.e., two daughter cells).

Interkinesis

1. This period between meiosis I and meiosis II is similar to the interphase between mitotic divisions.

2. However, no DNA replication (as in S phase of mitosis) occurs; the chromosomes are 1n (each chromosome has a sister chromatid).

Meiosis II: 1n → 1n

Before meiosis II begins, the DNA does not replicate, and the centromere still attaches the sister chromatids. During meiosis II, the centromeres split and the sister chromatids separate. Chromosomes in the four daughter cells contain one chromatid. Counting the number of centromeres verifies the number of chromosomes. Fertilization then restores the diploid number (2n) in the zygote and subsequent somatic cells originating from this zygote.

1. During metaphase II, the haploid chromosomes (with sister chromatids) aligns at the metaphase plate.

2. During anaphase II, sister chromatids separate at the centromeres and the two daughter chromosomes move toward the poles.

3. Due to crossing over in prophase I, each gamete can contain chromosomes with different gene combinations than either parent.

4. At the end of telophase II and cytokinesis, there are four haploid cells (1 sperm for males and 1 egg and 3 polar bodies for females).

5. In animals, the haploid cells mature and develop into gametes, which may eventually fuse into a zygote (2n) from a 1n sperm and 1n egg.

6. In plants, the daughter cells become spores and divide to produce a haploid adult generation.

7. In some fungi and algae, a zygote results from gamete fusion and immediately undergoes meiosis; therefore, the adult is always haploid.

DNA is replicated once before both mitosis and meiosis I and II; in mitosis, there is one nuclear division, while in meiosis there are two nuclei divisions between syntheses of DNA.

In humans, meiosis occurs in reproductive organs to produce gametes, while mitosis occurs in somatic cells (not germline cells) for growth and repair.

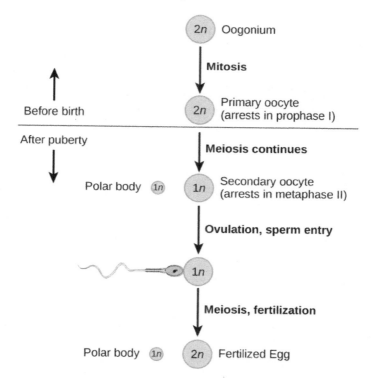

Meiosis generates 4 unique haploid cells while mitosis generates two identical diploid cells.

•   During prophase I of meiosis, homologous chromosomes pair for crossing over, which increases genetic variation. Crossing over does not occur during mitosis.

•   During metaphase I of meiosis, homologous chromosomes align at the metaphase plate; in mitosis, individual chromosomes align.

- During anaphase I in meiosis, homologous chromosomes, with centromeres intact, separate and move to opposite poles; in mitosis at this stage, sister chromatids separate and move to opposite poles.

- Events of meiosis II are now the same stages as in mitosis; however, the nuclei contain the haploid number of chromosomes in meiosis.

- Mitosis produces two genetically identical diploid daughter cells; meiosis produces four genetically different haploid daughter cells (4 sperm or 1 egg and 3 polar bodies).

Sperm and Ovum

In human males, meiosis is part of *spermatogenesis*, the production of sperm, and occurs in the testes. In human females, meiosis is part of *oogenesis*, the production of egg cells, and occurs in the ovaries.

Differences in the formation of egg and sperm

Spermatogenesis occurs in the seminiferous tubules in the testes and produces sperm from primary spermatocytes. *Spermatogonia* are undifferentiated germ cells that divide by mitosis and differentiate into *primary spermatocytes*. Primary spermatocytes grow and undergo the first meiotic division to form *secondary spermatocytes*. In the secondary spermatocyte stage, cells undergo a second meiotic division to form *spermatids*.

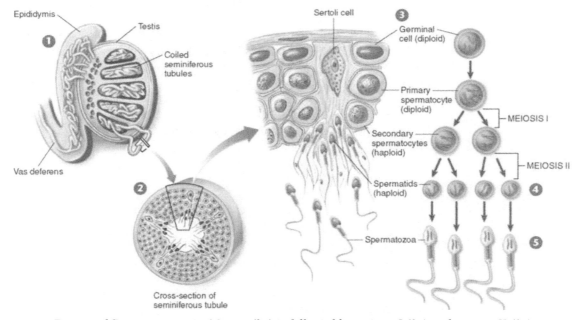

Testis and Spermatogenesis. Mitosis (2n) is followed by meiosis I (1n) and meiosis II (1n)

The spermatogonium replicates all of its chromosomes during interphase. It has 46 primary spermatocyte chromosomes, and each chromosome is made of two sister chromatids joined by a centromere. The cell undergoes prophase I, where the homologous chromosomes line up on spindles at the equator. The cell undergoes anaphase I (reduction phase of meiosis), but the centromeres do not divide. Instead, the homologous chromosome pairs separate. During telophase I, two cells are formed. Each cell is a secondary spermatocyte that now contains 23 chromosomes, and each chromosome has two chromatids joined by a centromere. These cells are haploid. Each of these cells then goes through meiosis II: prophase II, metaphase II, anaphase II and telophase II. This second set of divisions DOES resemble mitosis. Chromosomes condense (but do not pair up) during prophase II. The chromosomes align by spindle fibers during metaphase II, and the centromeres divide during anaphase II. The cells finish dividing during telophase II. At the end of meiosis II, there are four haploid cells with 23 chromosomes.

Spermatids, formed from the division of the secondary spermatocytes, develop into mature spermatozoa (sperm). *Sertoli cells* are stimulated by follicle-stimulating hormone (FSH) in the seminiferous tubules and surround and provide nourishment to spermatids during differentiation. They complete maturation (e.g., gain of motility) in the epididymis. *Sertoli cells* secrete the peptide hormone *inhibin* (acts on the pituitary gland to inhibit FSH release) and *androgen-binding protein* (binds testosterone and acts as an intermediary between germ cells and hormones). The Sertoli cells divide the tubules into compartments with separate environments where different stages of spermatogenesis continue. *Leydig cells* between the tubules produce testosterone in the presence of luteinizing hormone (LH).

Sperm produced in the testes mature within the *epididymides*. These are tightly-coiled tubules outside of the testes. The maturation time in the epididymis is required for sperm to develop the ability to swim. Once sperm has matured, they are propelled into the *vasa deferentia* by muscular contractions. Sperm is stored in the epididymides and the vasa deferentia.

Oogenesis results in the production of a single ovum from a single primary oocyte. Unlike spermatogenesis, the *ovarian cycle* occurs in a monthly rhythm and usually produces one gamete (ovum as an unfertilized egg) per month. Each egg develops in its bubble-like *follicle*, located primarily in the cortex. Each month, about 20 to 25 *primordial follicles* resume their development. The single layer of squamous follicular cells around the oocyte thicken into cuboidal cells. The follicle is then a *primary follicle*. As the egg enlarges, the follicular cells multiply and pile up into multiple strata; the follicle is then a *secondary follicle*, and the follicular cells are *granulosa cells*.

Oogonia are primitive germ cells that undergo mitosis and develop into *primary oocytes* (with 46 chromosomes), which are the initial cells in oogenesis. Primary oocytes remain in meiotic arrest (i.e., begin first meiotic development but do not complete it). All of the germ cells in a female are believed to be at this developmental stage at birth. However, this is an area currently under investigation by researchers.

Primary oocytes are in the ovaries of the female reproductive system. Oogenesis occurs on a monthly basis, beginning at puberty and ending at menopause. At puberty, primary oocytes destined for ovulation complete meiosis, and each daughter cell receives 23 chromosomes. Although some primary oocytes undergo *atresia* (immature and degraded) during childhood, there are about 300,000 to 400,000 oocytes at puberty. When the primary oocyte divides, one of the two daughter cells, the *secondary oocyte*, retains most of the cytoplasm. The *first polar body* is the other daughter cell, is nonfunctional and (often) does not proceed to meiosis II. The secondary oocyte proceeds to the metaphase II of meiosis and then suspends. Meiosis II resumes if fertilization occurs. The completion of meiosis II allows the secondary oocyte to become a fertilized egg (2n zygote when fused with the sperm).

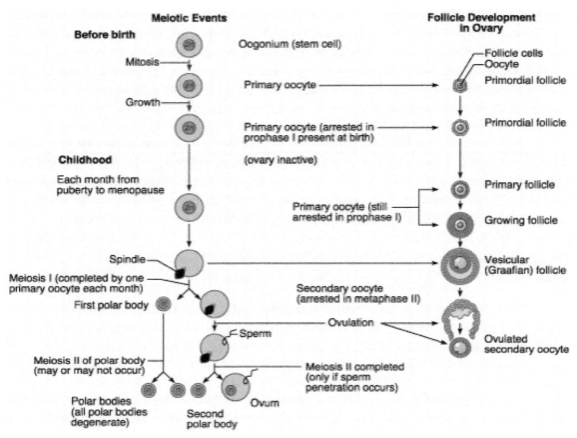

Oogenesis and the ovarian cycle

The meiotic division produces two *second polar bodies* that disintegrate because they receive insufficient cytoplasm. The body ultimately absorbs second polar bodies, and serve to retain most of the cytoplasm in the egg. The cytoplasm serves as a source of nutrients for the developing embryo. This is different from spermatogenesis, in which four functional, mature sperm are produced from a single spermatogonium.

A single ovum is produced per month. Normally, one of the tertiary follicles becomes a fully mature *vesicular* (Graafian) *follicle* or the *tertiary follicle* destined to ovulate. Ovulation occurs around day 14 of a typical cycle. The follicle swells and bursts, releasing the egg and *cumulus oophorus* (a mass of follicular cells surrounding the ovum in the vesicular ovarian follicle) into the mouth of the uterine tube.

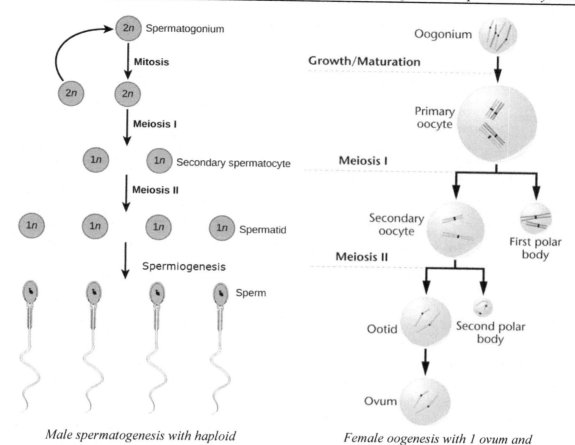

Male spermatogenesis with haploid secondary spermatocytes (1n) and spermatids (1n)

Female oogenesis with 1 ovum and (up to) 3 polar bodies

Male and female gametogenesis		
Male	**Female**	**Difference**
Spermatogonium (2n)	Oogonium (2n)	Spermatogonium renews its population by mitosis throughout life. Oogonium stops renewing its population before birth
Primary spermatocyte (1n)	Primary oocyte (1n)	Primary oocyte arrests at prophase I
Secondary spermatocyte (1n)	Secondary oocyte (1n)	Secondary oocyte arrests at metaphase II
Sperm (1n)	Ovum (1n)	Between the secondary spermatocyte and sperm is the spermatid

Differences in morphology

Sperm is compact cells of DNA with flagella that provide motility.

Eggs are non-motile and filled with cytoplasm.

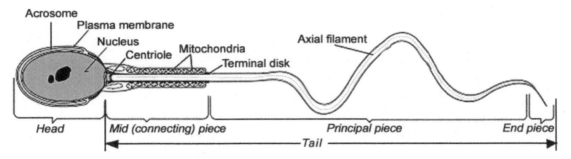

Human spermatozoon with head, midpiece, and tail

Spermatozoa (mature sperm) have three parts. The sperm *head* (haploid with 23 chromosomes), contains a nucleus and DNA covered by an acrosome. The *acrosome* is a cap-like covering over the anterior end of the nucleus that stores enzymes to help the sperm penetrate the several layers of cells and thick membrane enclosing the egg. The *middle piece* contains multiple mitochondria for energy wrapped around microtubules of the flagellum in a 9 + 2 microtubule array. The *tail* contains microtubules as components of a flagellum for its whip-like movement propelling the sperm. The ejaculate of a human male contains several hundred million sperms. Fewer than 100 ever reach the vicinity of an egg, and usually, one sperm ultimately enters an egg.

The ovum (unfertilized egg cell) is the female gamete. Unlike sperm, the egg is not capable of active movement. The egg is much larger than the sperm (it is visible to the naked eye). Human sperm is about 55 micrometers (μm) in length (head is 5 μm, and flagellum is 50 μm). A mature ovum is between 120-150 μm in diameter. Therefore, the ratio of the length of the sperm to the diameter of the egg is about 1:3. Comparing the width of a sperm cell (~3 μm) to an egg's diameter, the ratio is about 1:50.

The *granulosa cells* produce a glycoprotein gel layer as the *zona pellucida* around the egg; the connective tissue around the granulosa cells condenses into a tough fibrous *theca folliculi*. The granulosa cells secrete *follicular fluid*, which forms small pools. The follicle is then a *tertiary follicle*. The fluid pools eventually coalesce to form a single cavity, the *antrum*. The egg is now held against one side of the antrum by a collection of cells of the *cumulus oophorus*; the innermost layer of these cells is the *corona radiata*.

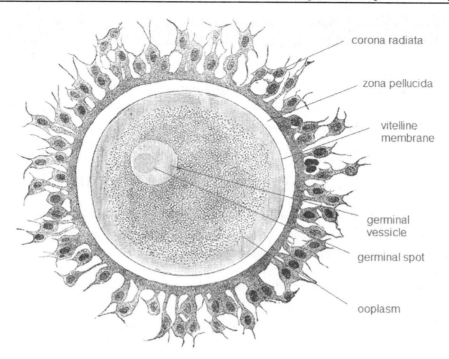

Human ovum with the protective glycoprotein layer of zona pellucida
and inner vitelline membrane enclosing the ooplasm.

The relative contribution to the next generation

Sexual reproduction produces offspring that combine genes from each parent. It entails the union of two gametes (ovum and sperm) to form a zygote (fertilized egg). The gametes, each containing half the genetic information needed to produce the 2n progeny.

Sperm contributes to the chromosomal DNA only, as the egg actively destroys any mitochondria in the sperm. The egg contributes chromosomal DNA and everything else (mitochondria, organelles, and biomolecules within the ooplasm) within the large volume of cytoplasm (i.e., 1 large egg and 3 small, non-functional polar bodies). The egg contains the cytoplasm nutrients and biomolecules needed by a developing embryo.

A diploid zygote (haploid sperm and haploid egg) represents the first stage in the development of a genetically unique organism. The zygote contains the essential components needed for development as the genes of chromosomes. The genes of the zygote are not activated to produce proteins until after several cleavage divisions. During cleavage, the large zygote subdivides into cells of a more conventional size via mitosis. After initial cleavage, blastomeres are the initial cells of an organism's development.

Hormonal Control of Reproduction

Ovaries produce the hormones *estradiol* and *progesterone*. Estradiol is an estrogen. Testes produce *testosterone*, which is the main sex hormone in males. Testosterone is an androgen, which has masculinizing effects. Androgens are not unique to males and estrogens are not unique to females.

Male and female sexual development

Many organs of the male and female reproductive systems develop from the same embryonic organs. Organs with the same embryonic precursor are *homologous*. For example, the scrotum and labia majora are homologous, because both develop from the labioscrotal folds.

Secondary sex organs are other anatomical structures needed to produce offspring, such as the male glands, ducts and penis, and the female uterine tubes, uterus, and vagina. *Secondary sex characteristics* are features not essential to reproduction but which help to attract mates by indicating sexual maturity (e.g., breasts). Secondary sex characteristics comprise the external differences between males and females.

Puberty is when reproductive organs mature, and reproduction becomes possible. Puberty is initiated by the secretion of *gonadotropin-releasing hormone* (GnRH) by the hypothalamus. GnRH stimulates the release of *follicle-stimulating hormone* (FSH) and *luteinizing hormone* (LH) by the anterior pituitary. FSH and LH act upon the gonads to stimulate the development of sperm and ova and the secretion of sex hormones. The sex hormones exert negative feedback on the secretion of GnRH, FSH, and LH.

The visible phenotypic and physiological changes at puberty result from many hormones (e.g., testosterone, estrogens and growth hormone). At the conclusion of puberty, the individual attains fertility. Adolescence continues until the full adult height is attained.

The earliest visible sign of male puberty is enlargement of the testes and scrotum; the ejaculation of motile sperm marks the completion of puberty.

Testosterone has three main roles: 1) stimulating the development of prenatal genitalia, 2) stimulating the development of the male secondary sexual characteristics, such as growth of the skeletal muscle (longer legs and broader shoulders), development of pubic hair, facial hair and chest hair, and 3) the maintenance of sex drive during adulthood. Testosterone stimulates the secretion of oil and sweat glands (i.e., attributed to body odor and acne). Additionally, testosterone prompts the larynx and vocal cords to enlarge, resulting in a deeper voice. It is involved in triggering baldness if "baldness genes" are present and regulates testosterone synthesis by acting on Leydig cells to stimulate testosterone secretion.

Female puberty is marked by *thelarche* (onset of breast development), *pubarche* (appearance of pubic and axillary hair; also in males), and *menarche* (onset of menstruation). Regular ovulation and fertility are attained about a year after menarche.

Estrogens maintain the normal development of the related organs and for the secondary sex characteristics of females. Compared to males, females have less body and facial hair and more fat beneath the skin (i.e., a more rounded appearance). In females, the pelvic girdle enlarges, and the pelvic cavity is larger for wider hips. Estrogen and progesterone are required for breast development.

The female breast has a conical or *pendulous body* and a narrower *axillary tail* extending toward the armpit. The breast contains 15–24 lobules, each with a mammary duct. The *nipple* is at the apex and is surrounded by a zone of darker skin as the *areola*. The areola has small *areolar glands* that may appear as bumps around the nipple. They produce a secretion that prevents chafing and cracking of the skin in a nursing mother.

The mammary duct begins at the nipple and divides into numerous ducts that end in alveoli (blind sacs). Prolactin hormone is for *lactation* (milk production) to begin. The feedback inhibition suppresses production of prolactin that estrogens and progesterone have on the anterior pituitary during pregnancy. Therefore, it takes a couple of days after delivery of a baby for milk production to begin. Before this, the breasts produce a watery, yellowish-white fluid (*colostrum*) similar to milk but containing more protein and less fat, which is rich in IgA antibodies that provide immunity to a newborn. Breast cancer is a common form of cancer in females; women should have regular breast exams and mammograms as recommended.

At midlife, both sexes go through a period of hormonal and physical change of *climacteric*. This is marked by a decline in testosterone or estrogen secretion and a rise in the secretion of FSH and LH. In females, climacteric is accompanied by *menopause*, the cessation of ovarian function and fertility.

Menopause is the cessation of menstrual periods with age from a decrease in the number of ovarian follicles and their hypo-responsiveness to gonadotropins. Plasma estrogen levels decrease, resulting in high gonadotropin secretion. A decrease in a bone mass called osteoporosis occurs, as well as hot flashes or the sudden dilation of arterioles, which increases body temperature and sweating.

Male reproductive cycle

The hypothalamus has ultimate control of the testes' sexual function through the secretion of GnRH. This hormone stimulates the pituitary to produce the gonadotropic hormones FSH and LH in the anterior pituitary gland. FSH promotes spermatogenesis in males by stimulating primary spermatocytes to undergo meiosis I (forming secondary spermatocytes). FSH enhances Sertoli cells (nurse cell that helps develop sperm), by causing them to bind to androgens more effectively. In males, LH is *interstitial cell-stimulating hormone* (ICSH).

LH acts on the cells of Leydig and stimulates the production and secretion of testosterone. The *sustentacular cells* of the seminiferous tubules release the hormone *inhibin*, which regulates the rate of sperm production and produces androgen-binding protein, making the testes responsive to testosterone. The hypothalamus-pituitary-testis system uses a negative feedback relationship that maintains a fairly constant production of sperm and testosterone. Although hormone and gamete production is constant in males, this is not true for females.

Female reproductive cycle

The Ovarian Cycle

In a longitudinal cross-section, an ovary shows many cellular follicles, each containing an oocyte (egg). A female is born with as many as two million follicles. The number is reduced to 300,000–400,000 by puberty, and a small number of follicles (about 400) ever fully mature. As a follicle matures, it develops from a *primary follicle* to a *secondary follicle* to a vesicular follicle (Graafian). As oogenesis occurs, a secondary follicle contains a secondary oocyte that is pushed to one side of the fluid-filled cavity. The vesicular follicle fills with fluid until the follicle wall balloons out on the surface and bursts, releasing a secondary oocyte surrounded by a *zona pellucida* and *follicular cells.*

During the *follicular phase*, FSH and LH stimulate primary follicles (containing primary oocytes) to grow and stimulate *theca cells*, express receptor for LH to produce androstenedione (androgen). Androgens, as a response to LH, are converted into estrogen by follicle stimulating (FSH-induced) hormone by granulosa cells. Estrogen leads to the thickening of the endometrium (uterine epithelium). As the estrogen levels rise, it exerts feedback control over the anterior pituitary secretion of FSH, causing the follicular phase to end.

As FSH decreases, the follicles cannot be maintained, and all but one follicle degenerates. The one dominant follicle (Graafian follicle) survives because 1) it is hyperresponsive to FSH and can maintain itself even under low FSH, and 2) it becomes sensitive to LH. Estrogen levels in the blood rise, causing the hypothalamus to secrete more GnRH, which causes a surge in LH secretion. LH does not drop but shoots up (LH surge), because increased estrogen exerts positive feedback on the LH-releasing mechanism of the pituitary. The LH spike triggers ovulation.

Ovulation is the rupture of the vesicular follicle with the discharge of the secondary oocyte into the pelvic cavity. The *secondary oocyte* completes a second meiotic cell division when fertilization occurs. Meanwhile, the follicle develops into the *corpus luteum* (promoted by LH), which secretes progesterone. Progesterone is the hormone responsible for maintaining the endometrium. If pregnancy does not occur, the corpus luteum degenerates in about 14 days, and estrogen and progesterone levels recede. The lack of estrogen and progesterone leads to the collapse of the vascular endometrium, which leads to menstruation. If pregnancy does occur, the corpus luteum persists for about 3 months. The ovarian cycle is under the control of gonadotropic hormones FSH and LH, just as in males. The gonadotropic hormones are not present constantly but are secreted at different rates during the cycle. The *luteal phase* in the ovary (corresponding to the secretory phase of the uterus) is the second half of the ovarian cycle following ovulation.

Progesterone and estrogen from the corpus luteum inhibit the normal functioning of GnRH, which slows the production of FSH and LH. Progesterone converts the endometrium into a secretory tissue full of glycogen and blood vessels, ready to receive a fertilized egg. As the level of progesterone in the blood rises, negative feedback decreases the anterior pituitary's secretion of LH, and the corpus luteum degenerates.

Emergency postcoital contraceptives (e.g., Plan B) include progesterone antagonists that prevent progesterone from binding to their receptors, leading to erosion of the endometrium. When oral contraceptives (i.e., birth control pills) are used, a combination of synthetic progesterone and estrogen inhibit pituitary gonadotropin release, thereby preventing ovulation.

The Menstrual Cycle

The endometrium undergoes cyclic histological changes are the *menstrual cycle*, governed by the shifting hormonal secretions of the ovaries. FSH, LH, estrogen, and progesterone are in a complex interaction to regulate menstruation. An average 28-day uterine cycle is divided into four phases. The *proliferative phase* is the mitotic rebuilding of tissue lost in the previous menstrual period and is primarily regulated by estrogens. The *secretory phase*, regulated primarily by progesterone, consists of a further thickening of the endometrium by secretions (not by mitosis). The *premenstrual phase* is ischemia and necrosis of the endometrium. The *menstrual phase* begins when endometrial tissue and blood is first discharged from the vagina and marks day 1 of a new cycle. It is triggered by the decline in ovarian secretions of progesterone and estrogens.

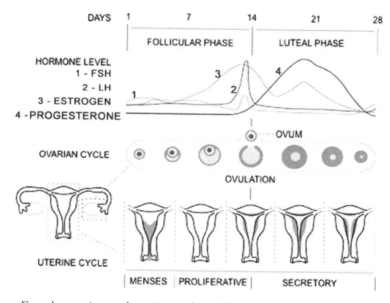

Female ovarian and uterine cycles with associated hormone levels

The menstrual cycle begins on day 1, with the onset of *menstruation* (sloughing of the endometrium resulting in bleeding). During days 1 to 5, low levels of estrogen and progesterone cause menstruation. *Menstruation* is the periodic shedding of tissue and blood from the endometrium; this lining disintegrates, and the blood vessels rupture. *Menses* is the flow of blood and tissues discharged from the vagina. FSH from the anterior pituitary stimulates the growth of a follicle in the ovary (the follicular phase), which secretes estrogen as it grows.

After day 5, the rising estrogen levels stimulate the uterus to grow a new inner lining. Between days 6 and 13, increased production of estrogens by an ovarian follicle causes the endometrium to thicken and become vascular and glandular (proliferative phase).

By day 14, the endometrial lining is thick. A surge in LH from the anterior pituitary gland releases the oocyte and some of the follicular cells from the ovary (ovulation). Ovulation usually occurs on day 14 of the 28-day cycle. LH causes the remaining follicular cells to become the corpus luteum. Days 15 through 28 see increased production of progesterone by the corpus luteum, which causes the endometrium to double in thickness, and uterine glands mature, producing a thick mucoid secretion (secretory phase). The endometrium is now prepared to receive an embryo. If no pregnancy occurs, the progesterone and estrogen levels decline, and the corpus luteum degenerates. With low levels of progesterone, the uterine lining begins to degenerate. During menstruation, the anterior pituitary increases FSH production, and a new follicle begins to mature. The ovarian cycle controls the uterine cycle.

If fertilization and implantation do not occur, the corpus luteum degenerates after about 14 days, and the drop in progesterone and estrogen causes the lining to degrade and shed, initiating the next cycle.

Pregnancy, parturition, and lactation

Events Following Fertilization

The cells of the placenta produce human chorionic gonadotropin (HCG), which maintains the corpus luteum. This hormone can be detected about 11 days after conception. Home pregnancy tests rely on the presence of HCG to confirm pregnancy. In general, the level of HCG doubles every 48 hours during the first four weeks of pregnancy, every 72 hours by 7-8 weeks, reaches a peak at 11-12 weeks of pregnancy. The corpus luteum produces progesterone and estrogen to maintain the uterus during the first trimester of pregnancy, and HCG maintains the corpus luteum itself until the placenta produces its progesterone and estrogen, at which time the corpus luteum regresses.

The progesterone and estrogen have two effects at this stage. They inhibit the anterior pituitary, so no new follicles mature. They maintain the lining of the uterus, so the corpus luteum is not needed, thus eliminating menstrual cycles during pregnancy.

Major causes of female infertility are 1) blocked oviducts, 2) failure to ovulate due to low body weight (<10-15%) and 3) *endometriosis*, the spread of uterine tissue beyond the uterus.

The common causes of male sterility and infertility are low sperm count and abnormal sperm from disease, radiation, chemical mutagens and excessive heat near the testes.

Pregnancy

Mammals are *viviparous*, as the embryo remains in the female's body during development. There are many forms of viviparity, but the most developed form is *placental viviparity*, where the mother constantly supplies the nutrients needed for development.

Parturition (birthing)

Normal human pregnancy lasts for about 40 weeks from the first day of the woman's last menstrual period. When the fetal brain matures, the hypothalamus causes the pituitary to stimulate the adrenal cortex so that androgens are released. The placenta uses androgens as precursors for estrogens that stimulate the production of prostaglandin and oxytocin. The hormones estrogen, prostaglandin and oxytocin cause the uterus to contract rhythmically and expel the fetus. *Labor* is a series of strong uterine contractions and has three stages:

1) cervix thins and dilates; amniotic sac ruptures and releases fluids;

2) rapid uterine contractions, followed by the birth of a newborn;

3) uterus contracts and expels umbilical cord and placenta.

The cervix dilates, and the infant moves through the vagina. When the infant is born, the following changes occur:

- oxygen from mother's blood to breathing with functional lungs

- nutrients from mother's blood to suckling for lactation

- fetal circulation (which bypasses the lungs and liver) to normal circulation (closing ducts and openings)

Lactation

Lactation is the secretion of milk by the mammary glands. Estrogen and progesterone secretion at the onset of puberty leads to breast enlargement due to the development of the duct system. During pregnancy, estrogen stimulates the secretion of prolactin and placental lactogen, which stimulate the development of the glands (the alveoli) in breasts.

During pregnancy, milk secretion is inhibited by estrogen and progesterone. The inhibitory effect stops after childbirth once the placenta is removed. Milk secretion is maintained by the release of prolactin from afferent input from nipple receptors to the hypothalamus during suckling. The secretion of dopamine (inhibiting prolactin) regulates milk secretion.

Milk ejection from the alveoli to the ducts is stimulated by oxytocin, released by the suckling reflex. Suckling inhibits the hypothalamic-pituitary-ovarian hormone chain, which, in effect, blocks ovulation. The breast is internally divided into lobes, each with a *lactiferous duct* that conveys milk to the nipple. Other than during pregnancy or lactation, the breast is mostly adipose and fibrous tissue with small traces of mammary glands. During pregnancy, the ducts grow, and branch and secretory *acini* (cell clusters) develop at the ends of the smallest branches. Each duct expands into a *lactiferous sinus* just beneath the skin of the nipple. The acini have contractile *myoepithelial cells* that respond to oxytocin and cause milk to flow down the ducts.

Integration with neural control

Erection

The penis contains vascular compartments and arteries. Normally, these vessels are constricted so that there is little blood in them, causing the penis to remain flaccid. During sexual arousal, nervous reflexes cause an increase in the arterial blood flow to the penis. Nerves of the penis converge on a pair of *dorsal nerves*, which lead via the *internal pudendal nerves* to the sacral plexus and then the spinal cord. The penis receives sympathetic, parasympathetic and somatic motor nerve fibers.

Sexual excitation in higher brain centers, stimulation of mechanoreceptors in the penis, inhibition of sympathetic fibers and release of nitric oxide all contribute to the dilation of these arteries. Dilation then causes these compartments to engorge with blood at high pressure. The increased blood flow fills and distends the erectile tissue, making the penis elongated and rigid, like an *erection*. Erectile dysfunction is the inability to achieve an erection due to various physiological or psychological causes. Viagra and related products release nitrous oxide (NO) and block the breakdown of cGMP, a messenger involved in the relaxation of the arterial smooth muscle, which promotes erection.

Ejaculation

The stimulation of sympathetic nerves contracts the smooth muscles lining the ducts and discharges semen through the urethra. The sphincter at the base of the urinary bladder is closed so that sperm cannot enter the bladder and urine is not expelled.

Ejaculation is the expulsion of semen and is achieved at the peak of sexual arousal. *Emission* is the first phase of ejaculation. Nerve impulses from the spine trigger the epididymides and vasa deferentia to contract. Subsequent motility causes the sperm to enter the ejaculatory duct. Secretions are released from the seminal vesicles, the prostate gland, and the bulbourethral glands. A small amount of secretion from the bulbourethral glands may leak from the end of the penis to clean the urethra of acid but may contain sperm.

Expulsion is the second phase of ejaculation. Rhythmical contractions at the base of the penis and within the urethral wall expel the semen in spurts. Rhythmical contractions are a release from *myotonia* (muscle tenseness), an important sexual response. Ejaculation lasts for a limited time, and the penis returns to a flaccid state following ejaculation. A *refractory period* follows when stimulation does not result in an erection. An *orgasm* is the physiological and psychological sensations that occur at the climax of sexual stimulation. During an orgasm, heart rate and blood pressure increase and skeletal muscles contract throughout.

The *clitoris* in females contains many sensory receptors as a sexually sensitive organ. An orgasm involves the release of neuromuscular tension in the muscles of the genital area, vagina, and uterus.

Please, leave your Customer Review on Amazon

Chapter 23

Development

- **Embryogenesis**

- **Mechanisms of Development**

Embryogenesis

Development describes the changes in the life cycle of an organism. *Embryogenesis* begins the process of growth and development of an embryo.

Fertilization

In sexually reproducing organisms, *fertilization* is the first step of embryogenesis. It marks the combination of sperm and egg to form a zygote. Reproduction may be sexual or asexual. *Asexual reproduction* does not involve fertilization, as only one parent is required. Asexual reproduction is the primary form of reproduction for single-celled organisms (e.g., bacteria, many protists, and fungi). Rarely are animals exclusively asexual, but most animals that can reproduce asexually use a combination of sexual and asexual reproduction. In *sexual reproduction, ovum* (egg) of one parent is fertilized by the sperm of the other. The ovum and sperm are *gametes* and are produced by sexually mature organisms in *gametogenesis*. Sexually-reproducing animals use various strategies to ensure that their gametes find each other. Fertilization of the ovum by the sperm may occur externally (e.g., in water) or internally (i.e., within the organism).

Animals that use external fertilization (e.g., fish and amphibians) must release thousands (or even millions) of gametes (egg or sperm) since the chance of fertilization is low. However, the advantages of external fertilization are a greater degree of genetic variation and a low chance of disease transmission. Most externally fertilizing animals synchronize the release of their gametes using environmental signals. Tides, which fluctuate with the lunar cycle, are one such trigger.

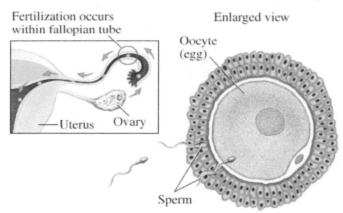

Animals that use internal fertilization, including all terrestrial vertebrates, require one (or a few) ova and have the advantage of being more selective about reproductive partners. The disadvantage of this reproductive strategy is that it reduces genetic variation and requires direct contact between male and female, which promotes the transmission of disease. Most mammals, including humans, use internal fertilization, and their young develop within the mother's uterus until birth. Reptiles, birds and a class of mammals called monotremes fertilize internally, but subsequently, lay eggs in which their young develop.

To preserve the existence of the species and pass on their genetic material, it is essential that mature adults are capable of producing fertile offspring. Many animals are only fertile during certain periods. In human females, the ovum is available for fertilization by sperm once per cycle, during *ovulation*. During ovulation, an ovary releases an ovum, which is then swept by ciliary cells into the *uterine tube* (the oviduct or Fallopian tube). During intercourse, millions of sperm are ejaculated and then move through the vagina, uterus and uterine tube towards the ovum using their *flagella*, whip-like tails. Temperature and chemical signals direct the sperm towards the ovum. They are aided in their journey by rhythmic propulsions of the vagina and uterus. However, most sperms die before reaching the ovum due to their depleted energy supply (ATP) and the acidic vaginal environment.

A sperm secretes proteins that bind to receptors on the glycoprotein layer surrounding the plasma membrane of a same-species ovum, preventing cross-species fertilization. This glycoprotein layer is the *vitelline layer* or *zona pellucida* in mammals. Upon binding, the *acrosome* in the sperm head digests a path through the zona pellucida with the release of hydrolytic enzymes. The sperm body follows into the ovum.

Most of the sperm cells that reach the ovum are viable and have the potential to fertilize it. Many sperm cells attach to the ovum, but only one should succeed in penetrating the ovum. *Polyspermy* occurs when more than one sperm penetrates the ovum. Usually, a fertilized ovum has 2 copies of each chromosome, but an ovum affected with polyspermy has 3 or more copies, which usually leads to the formation of a non-viable zygote. Down's Syndrome is trisomy 21.

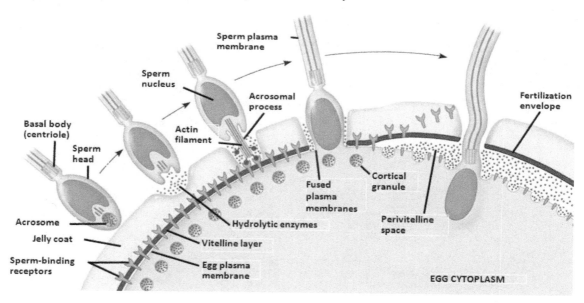

Fertilization of the egg by sperm

Prevention of polyspermy depends on certain changes that occur when a sperm cell first binds and fuses to an ovum. During this time, the ovum is vulnerable to other sperm attempting to fertilize it, so immediate action must be taken to avoid this scenario by two mechanisms: fast block and slow block to polyspermy.

In marine invertebrates (e.g., sea urchins), fertilization triggers an influx of sodium ions into the ovum, which causes its membrane potential to depolarize rapidly. Sperm cells are immediately unable to bind to the positively charged ovum, so this is the *fast block* because it quickly prevents polyspermy.

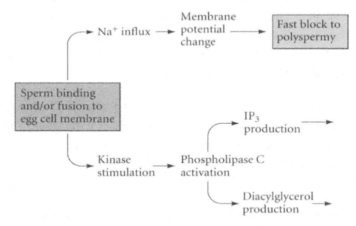

Fast block to polyspermy involves a change to the membrane potential
that inhibits additional sperm from penetrating the fertilized egg

The *slow block* is a slower and more long-term process. Many species other than mammals utilize *both* a fast block and slow block because although the fast block is effective, it is transient and another countermeasure to polyspermy is often needed. The slow block is initiated upon fertilization, as the ovum secretes various hormones to prevent it from being overwhelmed by the hundreds of sperm attempting fertilization. In most species that use this technique, the slow block is the *cortical reaction.*

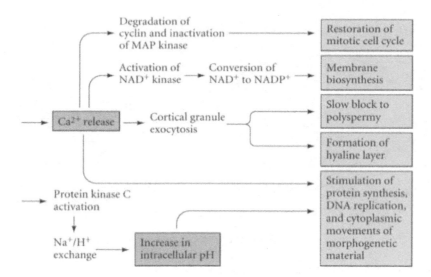

Slow block to polyspermy uses cortical granule to create a physical barrier around the zygote

The cortical reaction involves calcium release from the ovum, which causes *cortical granules* in the ovum to secrete calcium, which modifies the zone pellucida. Proteases lift the vitelline layer as hyaluronic acid pushes the vitelline layer away from the surface of the ovum, forming a barrier that prevents entrance of other sperm. Hyalin and peroxidases harden the fertilization envelope and inactivate the sperm-binding sites to prevent additional sperm cells from fusing. The vitelline layer is the *fertilization envelope.*

In mammals, the vitelline layer is the zona pellucida, so the cortical reaction is the *zona reaction.* The zona pellucida hardens, and its binding sites become inactivated. The zona reaction of mammals is less effective than the cortical reaction in other species. Research shows that in the zona reaction, contents of the cortical granules inactivate ZP2 and ZP3 sperm receptors on the zona pellucida. Note that mammals use only a slow block to polyspermy, and not a combination fast block and slow block as in other species (e.g., sea urchins).

Even after the sperm binds to and fuses with the ovum, fertilization is not complete until the gametes fuse their two haploid nuclei, as *pronuclei.* The pronuclei form a single diploid nucleus, which contains the unique genome of the zygote.

In most animals, including humans, the sperm contributes more than just its pronucleus to the zygote. The sperm centriole enters the ovum along with the pronucleus and flagellum. Recent evidence suggests that sperm mitochondria also enter, but are destroyed by the ovum. The centriole replicates and is assembled into a centrosome, which then replicates to organize the assembly of the first mitotic spindle in the zygote. This explains why extra mitotic spindles form in cases of polyspermy since several sperm cells contribute their centrioles to the ovum.

Cleavage

As the zygote is being swept by ciliary movement through the uterine tube into the uterus, it begins a series of cell divisions as *cleavage.* Once the zygote begins mitotic division, it is now an *embryo.* Cleavage involves rapid cell divisions without cell growth; therefore, with each division, *blastomeres* (cells) become smaller.

Cleavage may be either *indeterminate* or *determinate.* When cleavage is indeterminate, blastomeres can individually complete normal development if separated. Blastomeres formed by determinate cleavage do not develop if separated; each is a necessary part of the embryo. This is typical of *protostomes,* a superphylum of animals (e.g., arthropods, nematodes, platyhelminths, rotifers, mollusks and annelids). They are the counterpart to the *deuterostomes,* the most well-known of which are the chordates and the echinoderms.

Deuterostomes typically display indeterminate cleavage, which gives rise to identical twins. Identical twins are *monozygotic* twins because they originate from a single zygote that divided into two completely separate organisms. Fraternal twins result from two separately fertilized ova which both implant and develop in the uterus independently.

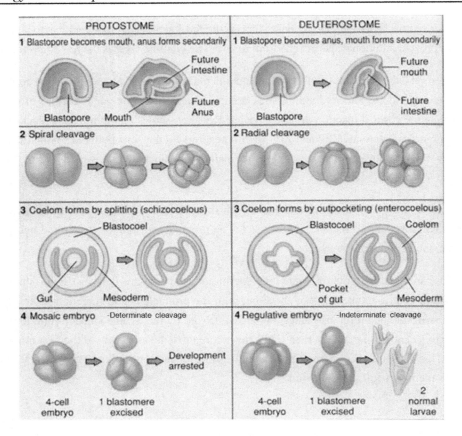

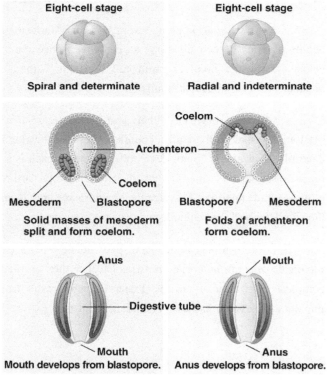

In many species, the ovum has an animal pole (upper) and a vegetal pole (lower hemisphere), known as embryo polarity. The *vegetal pole* contains larger cells filled with *yolk*, a nutritious substance that feeds the embryo. The yolk is denser than the cytoplasm, which causes it to settle at the bottom. In mammals, it later differentiates into *extraembryonic membranes* that protect and nourish the embryo. The *animal pole* is made of smaller, more rapidly dividing cells, and forms the embryo itself.

Determinative vs. intermediate cleavage

The *archenteron* is the center cavity formed by gastrulation, and its opening is the *blastopore*. The fate of the blastopore defines deuterostomes and protostomes. The term protostomes ("first mouth") has the blastopore first form the mouth. The term deuterostome ("second mouth") has the blastopore first form the anus, and the mouth develops second.

For the zygote, the first cleavage is *polar* and divides the ovum into two segments along the vegetal-animal axis. It is followed by cleavages perpendicular to the vegetal-animal axis, which is *equatorial*. *Radial cleavage* is the alternation of polar and equatorial cleavage and is a hallmark of deuterostome development. In contrast, protostomes typically display *spiral cleavage*.

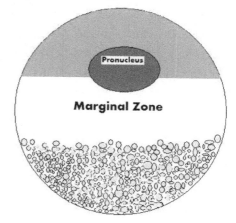

The amount of yolk in the zygote influences the manner of cleavage. Species that lay external eggs (e.g., birds, some fish, reptiles and most insects) have a large amount of yolk. This leads to *meroblastic* (incomplete cleavage). Mammals, worms, insects, and some other fish have eggs with much less yolk. This leads to *holoblastic* (complete cleavage).

Animal pole is on the upper portion while vegetal pole is the lower portion of the egg; pronucleus is at the boundary above marginal zone

Blastula formation

In humans, a few days after fertilization, cleavage creates a *morula*, a solid ball of cells. By the fifth day, the morula is transformed into a hollow *blastula* (*blastocyst* in humans). The blastocyst is formed as blastomeres migrate to the outside of the morula, leaving behind the *blastocoel*, a fluid-filled cavity. A blastocyst consists of two parts: an outer ring of cells as a *trophoblast* and an inner cell mass, the *embryoblast*.

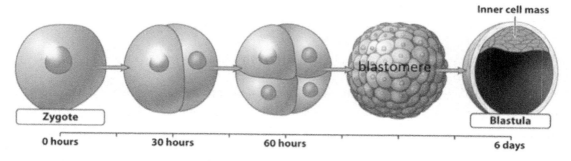

The trophoblast accomplishes implantation of the blastocyst by embedding into the nourishing epithelium of the uterus, the *endometrium*. Upon implantation, the trophoblast releases *human chorionic gonadotropin* (HCG) to maintain estrogen and progesterone production from the *corpus luteum* of the ovary, which maintains the endometrium.

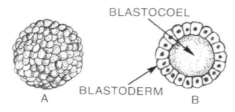

Blastula with the surface (A) and a cross-sectional view (B)

The amount of yolk affects the first three stages of development (cleavage, blastulation, and gastrulation). The lancelet and frog develop quickly in water and travel into the larvae to feed themselves. The chick provides a significant amount of yolk inside a hard shell, and its development continues until the chick can exist on its own. The early stages of human development resemble the chick due to our common evolutionary history. Frog embryo cells at the animal pole have little yolk; cells at the vegetal pole contain more yolk.

The presence of yolk causes cells to cleave more slowly, so cells at the animal pole are smaller. Chick cell cleavage is incomplete; only those cells lying on top of the yolk cleave and spread out over the yolk surface, in contrast to the ball-like morula of the lancelet. The frog blastocoel is formed at the animal pole.

The cells containing yolk do not participate in gastrulation and do not invaginate; instead, when the animal pole cells begin to evaginate from above, a slit-like *blastopore* is formed. Other pole cells move down over the yolk; the blastopore then becomes rounded. These yolk cells temporarily left in the region form a *yolk plug*. Cells from the *dorsal lip* of the blastopore migrate between the ectoderm and endoderm, forming the *mesoderm*. Later, the *coelom* is created by splitting off from the mesoderm.

For example, in a chick, a blastocoel is created when cells lift from the yolk and leave a space between the cells and the yolk. There is so much yolk that endoderm formation does not occur by invagination. Instead, the upper layer of the cells differentiates into the ectoderm, and the lower layer differentiates into the endoderm. The mesoderm arises by an invagination of cells along the edges of the longitudinal furrow in the embryo midline, named the primitive streak. Later, the newly formed mesoderm splits to form the coelomic cavity.

Gastrulation

After fertilization (3 weeks in humans), *gastrulation* occurs with an invagination of some cells into the blastocoel to form the *primary germ layers*. These are the initial cell layers from which all the body tissues develop.

Vertebrates and other higher animals are *triploblastic* with three primary germ layers. The most primitive animals (cnidarians and sponges) are *diploblastic*, with two germ layers.

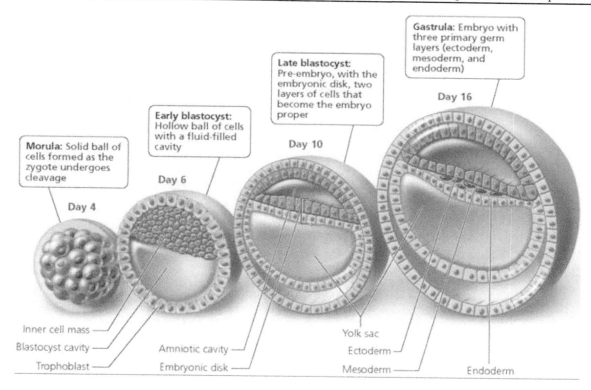

Morula: Solid ball of cells formed as the zygote undergoes cleavage

Day 4

Early blastocyst: Hollow ball of cells with a fluid-filled cavity

Day 6

Late blastocyst: Pre-embryo, with the embryonic disk, two layers of cells that become the embryo proper

Day 10

Gastrula: Embryo with three primary germ layers (ectoderm, mesoderm, and endoderm)

Day 16

Inner cell mass

Blastocyst cavity

Trophoblast

Amniotic cavity

Embryonic disk

Yolk sac

Ectoderm

Mesoderm

Endoderm

First cell movements

Gastrulation involves a series of orchestrated cell movements, modulated by complex cell-signaling pathways.

The exact manner of gastrulation varies by species. In humans (and other mammals), before gastrulation, the inner cell mass flattens into the *embryonic disc*, which is divided into two layers: the *epiblast* and the *hypoblast*. The beginning of gastrulation is marked by the appearance of the *primitive streak*, a line of cells along the midline of the embryo. The epiblast then ingresses along the primitive streak, pushing the hypoblast along with it. This first ingression is the blastopore. As invagination continues, the blastopore deepens to form the *archenteron*, the primitive gut. The *hypoblast* becomes the amnion, one of the extraembryonic membranes.

Formation of primary germ layers: endoderm, mesoderm, and ectoderm

The first epiblast cells that migrate inwards differentiate into the *mesoderm* (middle germ layer). Diploblastic animals like sponges and cnidarian develop *mesoglea*, a non-cellular layer, instead of a mesoderm. Epiblastic cells that continue to invaginate the blastocyst become the *endoderm* (inner germ layer), while those that remain outside are the *ectoderm* (outer germ layer). After gastrulation, the germ layers begin to develop into the entire array of structures and organs in the body.

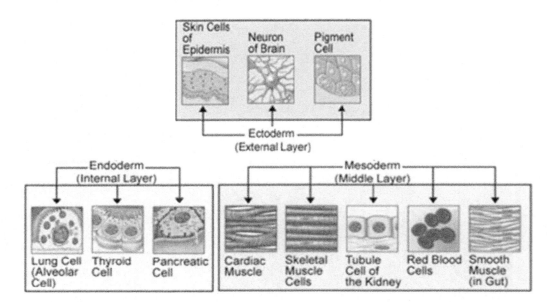

Primary germ layers and some representative tissue types formed from each layer

Neurulation

The newly formed mesoderm cells of the gastrula coalesce along the main axis to form a dorsal *notochord* as a stiff rod that provides support in lower chordates and is replaced in amphibians, birds, and mammals by the vertebral column. The portion of the ectoderm located just above the notochord develops into the nervous system in *neurulation.*

At first, the ectoderm cells on the dorsal surface of the embryo thicken, forming the *neural plate.* A *neural groove* develops down the midline of the neural plate; on either side of the neural groove, *neural folds* begin to fold upward and fuse. This forms the *neural tube,* which eventually becomes the central nervous system. At this point, the embryo is a *neurula.*

Neurulation with the formation of the neural tube

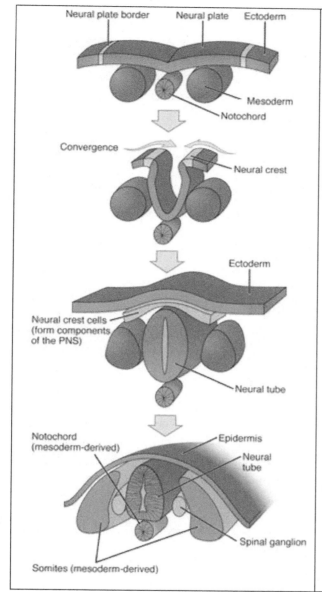

1 – Neuroectodermal tissues differentiate from the ectoderm and thicken into their neural plate. The neural plate border separates the ectoderm from the neural plate.

2 - The neural plate bends dorsally, with the two ends eventually joining at the neural plate borders, which are now referred to as the neural crest.

3 – The closure of the neural tube disconnects the neural crest from the epidermis. Neural crest cells differentiate to form most of the peripheral nervous system.

4 – The notochord degenerates and only persists as the nucleus pulposus of the intervertebral discs. Other mesoderm cells differentiate into the somites, the precursors of the axial skeleton and skeletal muscle.

Neural crest

The *neural crest* is the ectoderm at which the two neural folds close together to form the neural tube. Neural crest cells migrate and differentiate into a variety of cell types (e.g., neurons and glial cells of the peripheral nervous system, melanocytes of the epidermis, epinephrine-producing cells of the adrenal gland, portions of the heart and the connective tissue of the head).

Although derived from ectoderm, the neural crest is referred to as the fourth germ layer because of its importance.

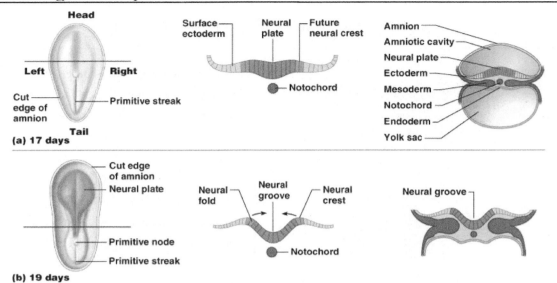

The neural crest is ectoderm that forms during neurulation

Major structures arising from primary germ layers

Aside from the nervous system, the ectoderm gives rise to the epidermis of the skin, the epithelial lining of the mouth and rectum, and several glands. The ectoderm is the origin of the nails, hair and skin glands, as these are all epidermal derivatives.

During neurulation, midline mesoderm cells that did not contribute to the formation of the notochord now become two longitudinal masses of tissue as *somites*, a process of *somitogenesis*. In vertebrates, the somites give rise to the dermis, skeletal muscle, tendons, cartilage and bone of the axial skeleton. The other portions of the mesoderm develop into the other connective, muscular and skeletal portions of the body. Mesoderm forms the excretory and reproductive system, along with the heart, blood vessels and blood cells, and the linings of all body cavities (coeloms).

The endoderm gives rise to the epithelial linings of the digestive tract, respiratory tract and excretory tract, along with associated organs and glands of the digestive and respiratory systems.

Structures forming from primary germ layers		
Ectoderm	Mesoderm	Endoderm
Epidermis Nervous system Epithelial linings: mouth and rectum Glands: adrenal medulla, pineal, pituitary	Dermis Skeletal system Muscular system Excretory system Reproductive system Circulatory system Lymphatic system Epithelial linings: coelom Glands: adrenal cortex	Epithelial linings: respiratory, digestive and excretory system Glands: pancreas, thymus, thyroid, parathyroid Liver

Mechanisms of Development

Development is the changes in the life cycle of an organism due to growth, differentiation, and morphogenesis. Most animals go through the same embryonic stages: zygote, morula, blastula, and gastrula. Gastrulation is succeeded by neurulation, which is the first event of *organogenesis* when body organs begin to develop. Neurulation marks the point at which the embryo leaves the embryonic stage and becomes a *fetus*. Organogenesis is related to *morphogenesis,* the development of an organism's body plan and shape.

Cell specialization

Cells are specialized when they have a specific function. For example, epidermal cells produce keratin to protect the skin against abrasion, myocytes produce actin and myosin to make muscles contract, and neurons produce neurotransmitters that transmit electrochemical impulses. *Stem cells* are undifferentiated cells which have yet to specialize. As an embryo develops, its stem cells differentiate to give rise to the different tissues of the body.

Cell specialization is achieved in two stages: *determination* and *differentiation.* Determination of a cell is the period when the cell begins to commit to a certain cell type, and it can be divided into two states: a cell is *specified,* and then it is *determined.* If a specified cell is placed into different body tissue, the cell alters its differentiation pathway to match that tissue. In this regard, the specification is reversible. However, once the cell is determined, the cell is irreversibly committed to its differentiation pathway and does not adapt to a different environment by altering its differentiation.

Determination

Determination begins with the specification (reversible) and ends with an irreversible commitment to a certain cell type. A cell is determined if influences from the cytoplasm cannot change its final form. These cytoplasmic influences are narrowed with successive cell divisions; complete determination usually occurs late in the cell specialization process.

Differentiation

Cellular differentiation is when cells become specialized in their structure and function. Although the two terms are sometimes used interchangeably, differentiation is the process by which cells commit to a certain cell type, while specialization refers to what that particular cell type is. Differentiation is an action, while specialization is a state of being.

Differentiation is an important component of *morphogenesis* when the overall form of the organism and its body parts are shaped; this includes both early cell movement and later *pattern formation,* the process by which cells assume different functions in different parts of the body. As cells differentiate, their functions change and they interact with one another, eventually organizing into body tissues. Cells can be traced during differentiation to build a lineage map of their development by *fate mapping.*

Each body cell contains a full set of chromosomes and therefore all the genetic information required to perform the functions of any cell. Cells do not differentiate because they receive different genes, or because certain genes are lost as cells divide. Rather, cell specialization is accomplished via *differential gene expression*. All cell types have the same genome but differ in which genes they express and to what degree. There are two important mechanisms of differential gene expression: cytoplasmic segregation and induction.

The foundation of cell specialization in an embryo is laid even before fertilization. Ova contain mRNA and proteins as *maternal determinants* that influence the course of development. One method of differential gene expression, *cytoplasmic segregation,* parcels out the maternal determinants as cleavage occurs. Different blastomeres receive different concentrations of maternal determinants, which lead to differential gene expression. Experimentally, the cytoplasm of a frog zygote is not uniform in content. After the first division of the frog zygote, only a daughter cell that receives a portion of the *gray crescent* develops into a complete embryo.

Induction is a more common mechanism of differential gene expression, in which one group of cells influences development by changing the behavior of an adjacent group of cells. Hans Spemann, who received the Nobel Prize in 1935 for his work on embryonic induction, found that particular chemical signals within the gray crescent activate the genes that regulate development.

Tissue types

The tissues of animals can be divided into four types: epithelial tissue, connective tissue, nervous tissue, and muscle tissue. Each tissue type contains specialized cells and extracellular matrix that the cells secrete.

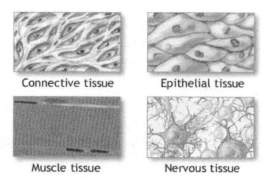

Connective tissue   Epithelial tissue

Muscle tissue   Nervous tissue

*Epithelial tissue*s may be multilayered, or one cell thick, and usually function in secretion, absorption or protection. Epithelial tissues are in the linings of glands, organs, body cavities, and blood vessels, and in the skin and mucous membranes. Various epithelia are originally derived from all three germ layers.

*Connective tissue*s often connect tissues and organs but can have structural and support functions. These are the most versatile and widespread of the body tissues. Cartilage, bone, blood and adipose tissue (fat) are special connective tissues. Proper connective tissues are divided into *loose* and *dense connective tissues*, which are divided into subtypes. These tissue types are originally derived from the *mesenchyme,* a portion of the mesoderm.

Nervous tissue includes the neurons and glial cells of the nervous system. Neurons (nerve cells) transmit electrical signals, while glial cells protect and support the neurons. All nervous tissue, both central and peripheral, arises from the ectoderm during neurulation.

Muscle tissue is a contractile tissue that can be divided into skeletal muscle, smooth muscle, and cardiac muscle. *Skeletal muscle* tissue makes up all of the muscles of the body under voluntary control and is attached to the skeletal system by tendons. *Smooth muscle* is under involuntary control, and is in the walls of hollow organs (e.g., the uterus, stomach and blood vessels). Cardiac muscle is under involuntary control but is only in the heart, where it creates the powerful contractions which move blood throughout the body. All three muscle types are originally derived from the mesoderm.

Embryonic and fetal development during gestation

Human gestation is 9 months, calculated by adding 280 days to the start of the last menstrual cycle. About 5% of infants arrive on their forecasted birth date, due to some complicating variables. Embryonic development occurs during months 1 and 2 of pregnancy and involves early cleavage of the zygote and initial organ development (organogenesis). Fetal development occurs during months 3 through 9 when organ systems grow, mature and increase in size and weight by a factor of nearly 600.

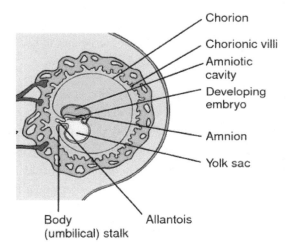

Chorion
Chorionic villi
Amniotic cavity
Developing embryo
Amnion
Yolk sac
Body (umbilical) stalk
Allantois

After fertilization, implantation of the blastocyst in the endometrium occurs about 1 week into pregnancy. Upon implantation, the trophoblast begins to secrete human chorionic gonadotropin (HCG) to stimulate the corpus luteum to maintain the endometrial lining. As the week progresses, the inner cell mass detaches from the trophoblast to become the *embryonic disc*, while the yolk sac forms below. The *yolk sac* is a membranous sac attached to the embryo. It is one of several *extraembryonic membranes* which lie outside of the embryo and protect and nourish the embryo and later the fetus.

The evolution of extraembryonic membranes in reptiles made development on land possible. An embryo that develops in water receives oxygen from the water and wastes float away. The surrounding water prevents desiccation and provides a protective cushion. However, for an embryo on land, these functions must be performed by extraembryonic membranes. These membranes are modified depending on whether the organism undergoes internal or external development.

In birds and reptiles, the yolk sac contains yolk which nourishes the developing embryo. In placental mammals, the umbilical cord and placenta are responsible for nourishing the embryo, so the "yolk sac" is, in fact, empty of yolk and rather functions as an early blood supply for the embryo before becoming part of the primitive gut later in gestation. Therefore, it is an *umbilical vesicle* to distinguish it from the yolk sacs of egg-laying animals.

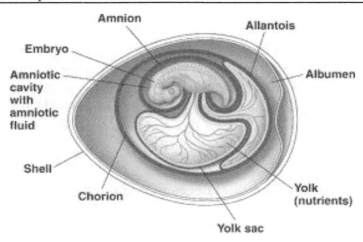

Egg-laying animals have large yolk sac with nutrients needed during embryogenesis

As the yolk sac forms, a second extraembryonic membrane of the *amnion* encloses the embryo. The amnion surrounds the amniotic cavity, which contains protective amniotic fluid that bathes the developing embryo. Amniotic fluid is a buffer to mechanical and chemical disturbances, and for thermal regulation. Some genetic disorders can be diagnosed by sampling this amniotic fluid in a prenatal test, like *amniocentesis*. Whether externally or internally, chordate animals develop in water, either in bodies of water or within the amniotic fluid.

The *chorion* is another extraembryonic membrane which, in birds and reptiles, lies next to the shell of an egg and carries on gas exchange. In placental mammals, the chorion implants into the endometrium and projects treelike extensions of *chorionic villi* into the maternal tissues. The maternal blood circulates these villi so that exchange of molecules between the fetal and maternal blood may take place. CO_2 and wastes move across from the fetus, and O_2 and nutrients flow from the maternal side. As the embryo develops, the interface of the chorionic villi and the uterine tissue becomes an organ the *placenta.* Placental formation begins at about week 1 after the embryo is fully implanted in the endometrium. The placenta is the life support structure of the developing embryo, providing oxygen and nutrients while removing wastes (e.g., CO_2). In humans, maternal blood directly contacts the chorion, but gas, nutrients and wastes across the chorion without the maternal and fetal blood mixing. Thus the placenta facilitates gas and nutrient exchange, but not fluid exchange.

The *umbilical cord* stretches between the placenta and the fetus and contains 2 fetal arteries and 1 fetal vein. The umbilical cord allows fetal blood to reach the placenta and exchange molecules with the maternal blood. Umbilical arteries transport CO_2 and other waste molecules to the placenta for disposal; the umbilical vein transports O_2 and nutrient molecules from the placenta to the rest of the fetal circulatory system. Although the fetal umbilical artery transports deoxygenated blood, it is an artery because it carries blood away from the fetus's heart. Likewise, the fetal umbilical vein carries oxygenated blood and is a vein because it brings blood to the fetus's heart.

The *allantois* is an extraembryonic membrane that evaginates from the archenteron of the gastrula. In birds and reptiles, the allantois initially stores waste products such as uric acid. It later fuses with the chorion. In mammals, the allantois initially transports waste products to the placenta but regresses in later stages of gestation. After birth, it may remain as a vestigial structure or disappear altogether. In birds and reptiles, the allantois and chorion form a double membrane that lies next to the shell and functions in gas exchange.

During week 3 of gestation, gastrulation occurs, followed by neurulation that initiates the development of the nervous system. The neural tube is visible as a thickening along the entire dorsal length of the embryo as the ectoderm develops into neural folds forming the neural tube. Towards the end of the first month, the heart starts to form, while designated cells begin to form the basic structure of the limbs, spine, nervous and circulatory systems. *Limb buds* appear for arms and legs. The head enlarges, the sense organs become more prominent, and the rudiments of the eyes, ears, and nose are evident.

During week 3, the development of the heart begins and continues into week 4. During this period, the right and left heart tubes fuse and the heart begins pumping blood, although the chambers are not fully formed. The veins enter this largely tubular heart posteriorly, and the arteries exit anteriorly. Later the heart twists so that all of the major vessels are located anteriorly.

During week 4 and 5, a bridge of mesoderm as the *body stalk* connects to the caudal (tail) end of the embryo via the chorionic villi. The head and tail then lift, and the body stalk moves anteriorly by constriction. Once this process is complete, the allantois, found in the body stalk, has extended into the blood vessels of the umbilical cord. Throughout, organs such as the heart, lungs, and liver are continuing to develop.

Through weeks 6 through 8, the developing embryo becomes more human-like in appearance. As the brain develops, the head achieves its normal relationship with the body as the neck region develops. At this point, the nervous system is developed well enough to permit reflex actions such as the startle response to touch. By week 8, the embryo is about 38 millimeters long and weighs no more than an aspirin tablet; however, all organs have been established.

During week 7, the fetus, regardless of sex, initially exhibits the same external and internal genital structures. Externally, the fetus exhibits a *genital tubercle,* a pair of *urogenital folds,* and a pair of *labioscrotal folds.* By week 12 of prenatal development, the genital tubercle differentiates into the glans (head) of the penis or clitoris, the urogenital folds enclose the urethra of a male or become the labia minora of a female, and the labioscrotal folds become the scrotum of a male or labia majora of a female. Now, the gender of the fetus can be identified anatomically.

Internally, the fetus has a pair of *gonadal ridges, mesonephric,* or *Wolffian ducts,* and *paramesonephric,* or *Müllerian ducts.* In an XY (male) fetus, the *SRY gene* encodes for the protein *testis-determining factor,* which initiates the development of male genitalia. The gonadal ridges become a pair of testes, which secrete testosterone and *Müllerian-inhibiting factor* (anti-Müllerian hormone).

By week 8, these hormones cause the Müllerian duct to degenerate at week 8 and induce the Wolffian duct to develop into the male reproductive tract. In a female fetus, where there is no Y chromosome, the gonads become ovaries. Around week 10, the absence of anti-Müllerian hormone, the Wolffian ducts degenerate, and the Müllerian ducts develop into the female reproductive tract.

By week 10, the placenta is fully formed and begun to produce progesterone and estrogen. Due to negative feedback control by the hypothalamus and anterior pituitary, no new ovarian follicles mature. Rather, ovarian follicles maintain the lining of the uterus and ensure that there is no menstruation during pregnancy. Facial characteristics of the fetus become recognizable at this time.

By week 14, all the major characteristics of the fetus have mostly developed. A fetus at first is only able to flex its limbs; later, it can move its limbs so vigorously that the mother can feel movements beginning in the fourth month. A fetus soon acquires hair, eyebrows, eyelashes, and nails. Fine, downy hair *lanugo* covers the limbs and trunk of the fetus but sheds shortly before or after birth. The skin grows so fast that it wrinkles. *Vernix caseosa* is a waxy substance that protects the skin from the watery amniotic fluid.

After 16 weeks, a fetal heartbeat can be heard by a stethoscope.

At 24 weeks, a fetus born prematurely may survive, however, the lungs are still immature and often cannot capture O_2 adequately.

From this time onwards, the fetus continues to grow in size rather than in complexity. At 38 weeks after fertilization, the fetus is mature enough to enter the world.

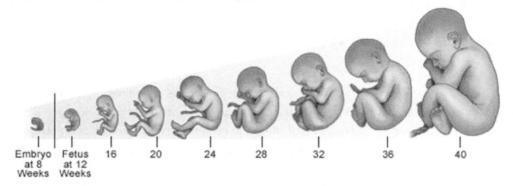

Fetal growth from week 8 to week 40. The gestational period in humans is 38 weeks; however, in obstetrics, the term is counted from the first day of the last menstruation (which is usually 2 weeks earlier), making the length of pregnancy to be 40 weeks.

Cell-to-cell communication in development

Development of the embryo is mediated by communication between cells. Induction is an important mechanism of differentiation that involves cell-to-cell signaling. The inducer is the cell that sends the signal for others to change, while the responder (or target) is the cell that receives the signal and changes accordingly. For example, *optic vesicles* are embryonic brain structures that induce the ectoderm to develop into the lenses of the eye.

Induction may result from juxtacrine mechanisms or paracrine mechanisms. *Juxtacrine* signaling involves communication between two cells that directly touch. *Paracrine* signaling occurs when an inducer cell releases a chemical signal which diffuses through space and is received by a target cell a short distance away.

In the early 20th century, amphibian embryologists Hans Spemann and Hilde Mangold experimented on the dorsal side of the embryo, where the notochord and the nervous system develop. They found that notochord tissue induces the formation of the nervous system, even when placed in a separate tissue environment such as beneath the belly ectoderm. Spemann showed that the dorsal lip of the blastopore, as the *primary organizer*, is necessary for development. The primary organizer directs the formation of germ layers during gastrulation. Those cells which invaginate first are closest to the primary organizer, so they become endoderm. The next cells at an intermediate distance from the primary organizer become mesoderm. Finally, the cells farthest away from the primary organizer become ectoderm.

Cell migration

For development, it is not enough for cells to differentiate; they must move. This movement of cells to predetermined areas in a controlled, organized manner as *cell migration*. Migrating cells have polarity and a distinguishable front and back (posterior). Without polarity, cells would be unable to maintain controlled movement; all sides of the cell would attempt to move at once in various directions. Researchers hypothesize that no matter the exact model behind cell migration, cytoskeletal filaments help establish and maintain a cell's polarity. The mechanisms of cell migration are not fully understood, but the foremost models are the cytoskeletal model and the membrane flow model.

The *cytoskeletal model theory* is based on the action of cytoskeletal elements. These elements interact to support and alter a cell's plasma membrane. Experimentation has shown that rapid actin polymerization of the cytoskeleton occurs at a migrating cell's front edge. Some researchers propose that the formation of actin filaments (microfilaments) at the front edge of a cell is the driving force behind cell movement. Furthermore, microtubules may act to contract the trailing edge of the cell. Thus, this model relies on a collaboration of the cytoskeletal filaments: microfilaments which push the front edge of the cell forward, and microtubules which retract the trailing edge.

The *membrane flow theory* is based on changes in the plasma membrane, rather than the cytoskeleton. It is known that cells undergo membrane recycling by continually returning membrane sections that are brought into the cell during endocytosis back to the plasma membrane. This process is the basis for the membrane flow theory, which posits that cell migration occurs by the addition of plasma membrane to the front of a cell. Under this hypothesis, the microfilaments at the front edge of the cell are stabilizing agents. The membrane flow model states that integrin proteins attached to the plasma membrane "walk" the cell along its migratory path. Integrins are continually recycled by endocytosis to the rear of the cell as fresh integrins are brought to the front of the cell by exocytosis.

The two models are not mutually exclusive, so a hybrid of the two may explain cell migration.

Stem cells and pluripotency

Stem cells in the early stages of an embryo are *totipotent* and capable of becoming any cell in the organism. A single totipotent stem cell can give rise to an entire organism, provided the organism display indeterminate cleavage. Only the zygote and subsequent cells of the morula are totipotent.

After just a few cell divisions, the embryonic stem cells become *pluripotent*, as each cannot individually divide into an entire organism but may differentiate into any embryonic cell. As the pluripotent embryonic cells continue to divide, they gradually become committed to a specific path. These cells, which have become committed to a single cell type, are *unipotent* and represent the majority of cells in an adult human.

However, adults do have stem cells. Their pluripotency is debated, and most adult stem cells are considered *multipotent*. Multipotent stem cells are between pluripotent and unipotent; they have a narrow range of possible cell types, but they still greatly benefit the organism for regeneration and repair of damaged tissue. Adult stem cells are in body tissues that must be frequently replaced (e.g., skin, as well as in places such as the bone marrow and the liver). Stem cells, both adult and embryonic, are the subject of intense research into their medical applications for tissue repair and treatment of degenerative diseases.

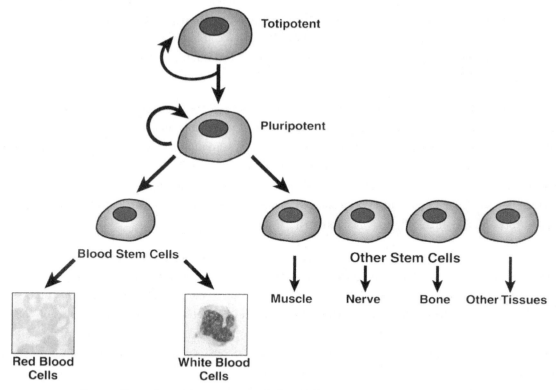

Stem cells undergo determination before differentiation where they exhibit
the morphology and biochemistry of the specialized cell

Gene regulation in development

Gene regulation of embryonic cells is the basis for cellular migration and differentiation during development. Knowledge of developmental gene regulation has been primarily based on research on *Caenorhabditis elegans* (roundworm) and *Drosophila melanogaster* (fruit fly), along with other animal models.

Maternal determinants are usually homogenously distributed throughout the ooblast of the ovum. However, *Drosophila melanogaster* exhibits cytoplasmic segregation of maternal determinants. The developing *Drosophila* ovum (oocyte) asymmetrically distributes *maternal effect genes.* These genes then produce mRNA and proteins (maternal determinants) in the mature ovum, which remain asymmetrically distributed. Maternal determinants directly influence the expression of certain zygotic genes. In this way, the genotype of the mother directly influences the phenotype of the zygote.

Gap genes are the first class of zygotic genes regulated by maternal effect genes. The proteins produced by gap genes regulate the transcription of *pair-rule genes*, which then influence the transcription of *segment polarity genes.*

The proteins produced by gap genes, pair-rule genes and segment polarity genes interact to influence *homeotic genes.* Together, gap genes, pair-rule genes, segment polarity genes, and homeotic genes partition the developing *Drosophila* embryo into body segments (i.e., pattern formation). Their function has been studied extensively by mutating these genes and observing the ensuing malfunctions in the body plan.

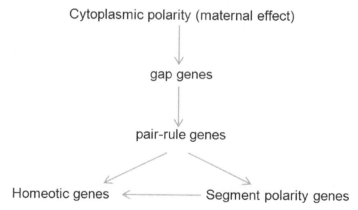

Gene regulation in development follows the proposed sequence of
gene expression and production of protein products

The initial gradients of maternal determinants are not the only method of gene regulation in a developing embryo. As in the cells of adult organisms, gene regulation is a highly complex process due to the multitude of ways in which genes may be regulated. At any step along the path from DNA to protein, and even after protein synthesis, gene expression can be regulated. At the highest level is transcriptional regulation, which includes histone modifications that make chromatin accessible (or inaccessible) to RNA polymerase, along with modulation of transcription factors that facilitate (inhibit) the binding of RNA polymerase during transcription.

Post-transcriptional regulation occurs after transcription and before translation, targeting the mRNA transcript as it moves from the nucleus to the cytoplasm. This may be accomplished by modulation of nuclear export, RNA splicing, RNA editing and modifications such as capping and polyadenylation, which protect the RNA from degradation while in the cytoplasm.

Once mRNA is processed (splicing of exons and removal of introns, the addition of a 5' G cap and 3' poly−A tail) and exported from the nucleus, the cell may continue to alter gene regulation at the translational level. This is accomplished by promoting (or inhibiting) ribosome recruitment to the mRNA.

Gene regulation does not cease once a protein is translated. Post-translational gene regulation acts upon proteins that have been synthesized. For example, proteins may be sequestered to certain parts of the cell or assembled with other proteins to form a working unit. Additionally, *zymogens* are enzymes that must be cleaved (or modified) to become active.

Post-translational modifications to histones, such as phosphorylation and acetylation, are responsible for the rearrangement of chromatin and is an active area of research.

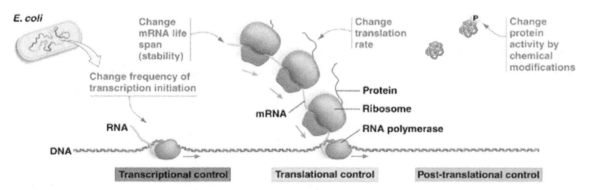

Regulation of gene expression: transcriptional control (DNA → RNA), post-transcriptional control (e.g., splicing), post-translational control (e.g., glycosylation). These examples of regulation are only a limited number of all mechanisms used by the cell during development.

Environment-gene interaction in development

Both genetic and environmental factors contribute to the complex etiology of congenital disabilities by disrupting the highly regulated embryonic developmental processes. The intrauterine environment of the developing embryo and fetus is easily altered by maternal factors such as health and disease status, lifestyle, medications, and maternal genotype, as well as exposure to environmental teratogens.

Teratogens are agents which cause congenital abnormalities. They may cause mild to severe congenital disabilities or terminate the pregnancy altogether. Radiation, harmful chemicals, certain drugs, and pathogens may be classified as teratogens. Congenital disabilities (1 in 33 babies in the US) continue to be the leading cause of infant death (about 5,500 per year) in the U.S., and therefore is a field of intense biomedical and clinical research. The field of teratology has focused on the causes and underlying mechanisms of congenital disabilities for decades, yet our understanding of these critical issues remains relatively vague.

Unfortunately, while the placenta is an effective mediator between the fetus and the mother, it is highly susceptible to penetration by chemicals. This is of particular concern during the embryonic period (fertilization through the 10th week of gestation or 8th week of embryonic age), when crucial structures are in the vulnerable first stages of development. Each organ has a sensitive period when toxic substances can alter normal development. For example, between 1958 and 1961 it was common for a pregnant woman in England and other countries to take the tranquilizer thalidomide between days 27 and 40 of pregnancy, which likely resulted with an infant born with deformed limbs, but after this period, the fetus is resistant to the effects of thalidomide and the child would be born healthy.

Studies conducted over the last few decades have suggested that the relationship between genes and environmental factors plays a major role in the propagation of these congenital disabilities. It is now widely believed that the study of gene-environment interactions leads to a better understanding of the biological mechanisms and pathological processes that contribute to the development of complex congenital disabilities.

Programmed cell death

Apoptosis is programmed cell death and is a part of normal cell development. Apoptosis is a central player in gametogenesis, embryogenesis, removal of damaged cells and aging.

During apoptosis, *caspases* (proteases) and nucleases digest the cell from within. These enzymes are inactive in a normal cell and are activated by a complex signal transduction pathway, which may be initiated by external commands or by internal recognition of cell malfunction, infection or DNA damage. The basis for apoptosis in embryogenesis and fetal development has been extensively studied in animal models. Fate maps of *C. elegans* show that 131 cells must undergo apoptosis for normal worm development. The cell-death protein in these cells is normally inhibited, but when the cells are signaled via induction, the cell-death cascade becomes active and produces proteases and nucleases to slice proteins and DNA.

Regardless of the reason for apoptosis, all apoptotic cells undergo characteristic changes: protrusion of the plasma membrane (blebbing), shrinkage, nuclear fragmentation, chromatin condensation, and chromosomal DNA fragmentation. An apoptotic cell attracts nearby cells, which consume the dying cell to avoid harmful cell contents from spilling out into the extracellular fluid and infecting other cells or causing inflammation.

During embryonic development, apoptosis removes harmful, abnormal or unneeded cells. Dysregulation of apoptosis during gestation can lead to severe congenital disabilities. Apoptosis is present from the beginning of development, as when cells in the blastula die off to form the gastrula. Morphogenesis is also accomplished through apoptosis; organs overproduce cells and then "prune" the excess cells to sculpt the intricate shape of the organ. This is most readily visible in the fingers and toes, which are webbed early in gestation but later lose their webbing and become recognizable digits. Another example of apoptosis-assisted morphogenesis occurs in the center of ducts and tubes to hollow them out.

Many fetal structures must be shed via apoptosis before birth. For example, female fetuses must degrade the Wolffian duct, while males must degrade the Mullerian duct.

Apoptosis continues its function after birth as a method of tissue remodeling and immune system maintenance in the adult organism. Cancer, autoimmune diseases and neurodegenerative diseases are all the result of dysregulated apoptosis. The absence of proper apoptosis allows dangerous cells to divide and invade other tissues, possibly leading to malignant tumors. However, excessive apoptosis leads to tissue degeneration. Both malfunctions are implicated in aging and disease.

The existence of regenerative capacity in various species

Regeneration is the reactivation of development to regrow a missing (or damaged) body structure. While regenerative abilities are limited in humans, many species of animals retain the capacity to regrow entire limbs after embryonic development.

For example, most echinoderms have robust regenerative powers. Some sea stars can regrow an entirely new organism from a single arm, while others can only replace lost limbs. Regrowth is not always guaranteed; it takes months or years, and sea stars are vulnerable to infection during the regrowth period. Separated limbs must survive off nutrient stores until they regenerate the rest of the sea star and begin to feed.

Newts and salamanders (particularly the axolotl) are noted for their regeneration. Newt and salamander regeneration proceeds in two stages. First, adult cells at the site of the limb's separation de-differentiate into *progenitor cells*. Second, the progenitor cells proliferate and differentiate until they replace the missing tissue.

Humans exhibit regeneration via progenitor cells, although regrowth of entire limbs is impossible. The most remarkable regeneration is in the liver. Through the use of several complicated signaling cascades, the liver manages to restore any lost mass and adjust its size to that of the organism, all while fulfilling its duties. Human embryos do have the ability to regenerate complete organs and limbs. Like many other vertebrates, however, this ability is lost during embryogenesis, leaving adult humans with a limited capacity to restore tissue damage.

Senescence and aging

Senescence is the gradual deterioration of function in an organism, eventually leading to death. All organisms, except for a few with remarkable properties of immortality, undergo senescence. One of the most puzzling questions humans currently face is *why* they age at all. Some have suggested that senescence results to avoid cancer, while others blame environmental factors such as radiation and oxidative agents, which cause DNA damage and cellular damage. The reasons for senescence are likely a combination of these and other factors.

At the cellular level, senescence refers to a cell which is no longer able to divide. Generally, cells are limited to 50 to 70 divisions, a threshold of the *Hayflick limit*. A senescent cell is not dead; rather, it actively continues its metabolic functions. The condition may be initiated by activation of oncogenes or by the cell's recognition of its DNA damage. *Oncogenes* signal abnormal cells to avoid senescence or undergo apoptosis to avoid the development of tumors.

Causes of DNA damage include radiation, oxidation (free radical damage) and shortening of telomeres. *Telomeres* are nonsensical DNA sequence repeats that are added to the 3' ends of DNA strands by the enzyme *telomerase*. Replication inevitably incurs shortening of the DNA, which eventually leads to DNA damage so severe that the cell is unable to continue dividing. Telomeres act as fodder for the inevitable losses from DNA replication to protect the encoding regions from destruction. Telomerase allows cells to proliferate for incredibly long periods of time because they are not affected by the loss of DNA incurred by replication. Most cells are subject to the Hayflick limit, but embryonic cells and certain adult cells contain high levels of telomerase so that they may divide continually.

The shortening of telomeres is associated with a wide variety of diseases involving premature aging, such as pulmonary fibrosis (scarring of the lungs). Shortened telomeres prevent the cells from properly dividing without losing genes, leading to cellular senescence. The lack of sufficient telomeres may cause chromosomes to fuse, which corrupts the genetic blueprint of a cell, making the chromosomes appear broken. The cell eventually recognizes what appears to be DNA damage, and thus enters apoptosis.

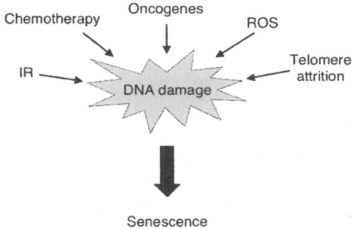

Environmental and biochemical factors contributing to senescence

Certain hydra, jellyfish, and flatworms are *biologically immortal*, so they exhibit negligible senescence and thus do not age. However, these organisms cannot truly escape death, because they are still subject to disease, injury, and predators. There is intense research into modeling therapies based on these organisms and their "immortal" cells, in the hope that aging and mortality can be slowed. For the present, however, they are inevitable parts of human development.

To access online SAT tests at a special pricing visit:
http://sat.sterling-prep.com/bookowner.htm

Notes

We want to hear from you

Your feedback is important to us because we strive to provide the highest quality prep materials. Email us if you have any questions, comments or suggestions, so we can incorporate your feedback into future editions.

Customer Satisfaction Guarantee

If you have any concerns about this book, including printing issues, contact us and we will resolve any issues to your satisfaction.

info@sterling-prep.com

We reply to all emails – please check your spam folder

Thank you for choosing our products to achieve your educational goals!

Notes

Please, leave your Customer Review on Amazon

Made in the USA
Middletown, DE
29 March 2019